Aeroecology

Phillip B. Chilson • Winifred F. Frick
Jeffrey F. Kelly • Felix Liechti
Editors

Aeroecology

 Springer

Editors
Phillip B. Chilson
School of Meteorology, Advanced
Radar Research Laboratory, and Center
for Autonomous Sensing and Sampling
Norman, Oklahoma
USA

Winifred F. Frick
Senior Director of Conservation Science
Bat Conservation International
Austin, Texas
USA

Department of Ecology
and Evolutionary Biology
University of California Ecology
and Evolutionary Biology
Santa Cruz, California
USA

Jeffrey F. Kelly
University of Oklahoma
Oklahoma Biological Survey
Norman, Oklahoma
USA

Felix Liechti
Swiss Ornithological Institute
Sempach, Switzerland

ISBN 978-3-030-09841-4   ISBN 978-3-319-68576-2   (eBook)
https://doi.org/10.1007/978-3-319-68576-2

Dedicated to Thomas H. Kunz

Preface

This book consists of a diverse collection of chapters that seeks to broaden our fundamental understanding of the ecological function and biological importance of the Earth's lower atmosphere, which provides a huge living space for billions of animals moving within and across continents. Their migration, dispersal, and foraging activities connect water and land habitats within and across continents. Drawing upon the wide-ranging experience of the authors, the book takes an inherently interdisciplinary approach that serves to introduce the reader to the topic of aeroecology, frame some of the basic biological questions that can be addressed within the context of aeroecology, and highlight several existing and emerging technologies that are being used to promote aeroecological studies. The book begins with several background chapters that provide introduction into such topics as atmospheric science, the concept of the habitat, animal physiology, and methods of navigation. It then continues with a broad discussion of observational methods available to and used by aeroecologists. Finally, several targeted examples of aeroecological studies are presented. Following the development of the chapters, the reader is provided with a unifying framework for investigating how the dynamic properties of meteorological conditions at local, regional, and global scales affect the organisms that depend on the air for foraging and movement. Material presented in the book should be of interest to anyone wishing to gain a comprehensive understanding of the aerosphere itself and the myriad airborne organisms that inhabit and depend upon this environment for their existence. The material should be accessible to a diverse set of readers at all stages of training and across a range of research expertise.

Norman, OK   Phillip B. Chilson
Austin, TX   Winifred F. Frick
Norman, OK   Jeffrey F. Kelly
Sempach, Switzerland   Felix Liechti

Contents

Part I   Background

1   **Aeroecology: An Integrative View of the Atmosphere**   3
Phillip B. Chilson, Winifred F. Frick, Jeffrey F. Kelly,
and Felix Liechti

2   **Atmospheric Environment Associated with Animal Flight**   13
J.K. Westbrook and R.S. Eyster

3   **Extending the Habitat Concept to the Airspace**   47
Robert H. Diehl, Anna C. Peterson, Rachel T. Bolus,
and Douglas H. Johnson

4   **Track Annotation: Determining the Environmental Context of
Movement Through the Air**   71
Renee Obringer, Gil Bohrer, Rolf Weinzierl, Somayeh Dodge,
Jill Deppe, Michael Ward, David Brandes, Roland Kays,
Andrea Flack, and Martin Wikelski

5   **Physiological Aeroecology: Anatomical and Physiological
Adaptations for Flight**   87
Susanne Jenni-Eiermann and Robert B. Srygley

6   **Properties of the Atmosphere in Assisting and Hindering
Animal Navigation**   119
Verner P. Bingman and Paul Moore

7   **Riders on the Wind: The Aeroecology of Insect Migrants**   145
Don R. Reynolds, Jason W. Chapman, and V. Alistair Drake

8   **Facing the Wind: The Aeroecology of Vertebrate Migrants**   179
Felix Liechti and Liam P. McGuire

Part II   Methods of Observation

9   **Aeroecological Observation Methods**   201
V. Alistair Drake and Bruno Bruderer

10 **Multi-Camera Videography Methods for Aeroecology** 239
    Margrit Betke, Tyson Hedrick, and Diane Theriault

11 **Using Agent-Based Models to Scale from Individuals to
    Populations** 259
    Eli S. Bridge, Jeremy D. Ross, Andrea J. Contina,
    and Jeffrey F. Kelly

12 **Radar Aeroecology** 277
    Phillip B. Chilson, Phillip M. Stepanian, and Jeffrey F. Kelly

13 **Inferring the State of the Aerosphere from Weather Radar** 311
    Eric Jacobsen and Valliappa Lakshmanan

Part III Aeroecological Applications

14 **Linking Animals Aloft with the Terrestrial Landscape** 347
    Jeffrey J. Buler, Wylie C. Barrow, Jr, Matthew E. Boone,
    Deanna K. Dawson, Robert H. Diehl, Frank R. Moore, Lori A. Randall,
    Timothy D. Schreckengost, and Jaclyn A. Smolinsky

15 **The Lofty Lives of Aerial Consumers: Linking Population Ecology
    and Aeroecology** 379
    Winifred F. Frick, Jennifer J. Krauel, Kyle R. Broadfoot,
    Jeffrey F. Kelly, and Phillip B. Chilson

16 **The Pulse of the Planet: Measuring and Interpreting Phenology
    of Avian Migration** 401
    Jeffrey F. Kelly, Kyle G. Horton, Phillip M. Stepanian,
    Kirsten de Beurs, Sandra Pletschet, Todd Fagin, Eli S. Bridge,
    and Phillip B. Chilson

17 **The Aerosphere as a Network Connector of Organisms and
    Their Diseases** 427
    Jeremy D. Ross, Eli S. Bridge, Diann J. Prosser, and John Y. Takekawa

18 **Sharing the Aerosphere: Conflicts and Potential Solutions** 465
    Judy Shamoun-Baranes, Hans van Gasteren, and Viola Ross-Smith

Editors and Contributors

About the Editors

Phillip B. Chilson is a professor in the School of Meteorology and Advanced Radar Research Center at the University of Oklahoma and director of the University's Center for Autonomous Sensing and Sampling. He received a BS and PhD in physics from Clemson University and an MS in physics from the University of Florida. Dr. Chilson has held positions at the Max-Planck Institut für Aeronomie in Germany, the Swedish Institute of Space Physics in Sweden, and the University of Colorado. His current research interests include investigations of the atmospheric boundary layer, aeroecology, the advancement of remote sensing technologies with a focus on radar, and the development of unmanned aircraft systems for atmospheric studies. Dr. Chilson has also worked in solid-state physics, ultralow-temperature physics, and aeronomy.

Winifred F. Frick is the senior director of Conservation Science at Bat Conservation International and adjunct faculty in ecology and evolutionary biology at the University of California Santa Cruz. She received a BA in environmental studies from UC Santa Cruz and a PhD at Oregon State University. Dr. Frick has studied bat ecology and conservation for over 15 years and has published over 40 research papers. Her research focuses on understanding how bat populations respond to natural and anthropogenic stressors and how science can be used to inform conservation decisions. She currently leads BCI's conservation science department and directs global bat conservation initiatives that target reducing threats and protecting globally endangered bats through research innovation and increasing capacity and local stakeholder involvement in conservation.

Jeffrey F. Kelly is the director of the Oklahoma Biological Survey and a professor of biology at the University of Oklahoma. He received a BS in wildlife management from the University of Maine, an MS in zoology from Oklahoma State University, and a PhD in biology from Colorado State University. Dr. Kelly's current research focuses on migratory macrosystems and interdisciplinary studies of human–wildlife conflicts in the lower atmosphere that emerge from rapid growth of telecommunication, energy, transportation, and commerce infrastructure in this habitat.

Felix Liechti is the director of the bird migration research department at the Swiss Ornithological Institute, a private science foundation. He received his MS in zoology at the University of Zurich and a PhD in biology from the University of Basel. Dr. Liechti's research focuses on quantifying the general patterns in bird migration as well as investigating the individual patterns and flight behavior of self-powered trans-Sahara migrants. Both topics aim on identifying the importance of environmental factors for the life history traits of bird migrants. This also includes contracted research on to the impact of wind turbines on bird movements.

Contributors

Wylie C. Barrow, Jr U.S. Geological Survey, Wetland and Aquatic Research Center, Lafayette, LA, USA

Margrit Betke Boston University, Boston, MA, USA

Verner P. Bingman Department of Psychology, Bowling Green State University, Bowling Green, OH, USA

J.P. Scott Center for Neuroscience Mind and Behavior, Bowling Green State University, Bowling Green, OH, USA

Gil Bohrer Department of Civil, Environmental and Geodetic Engineering, The Ohio State University, Columbus, OH, USA

Rachel T. Bolus Department of Biology, Southern Utah University, Cedar City, UT, USA

Matthew E. Boone Department of Entomology and Wildlife Ecology, University of Delaware, Newark, DE, USA

David Brandes Civil and Environmental Engineering, Lafayette College, Easton, PA, USA

Eli S. Bridge Oklahoma Biological Survey, University of Oklahoma, Norman, OK, USA

Kyle R. Broadfoot Oklahoma Biological Survey and Department of Biology, University of Oklahoma, Norman, OK, USA

Bruno Bruderer Swiss Ornithological Institute, Sempach, Switzerland

Jeffrey Buler Department of Entomology and Wildlife Ecology, University of Delaware, Newark, DE, USA

Jason W. Chapman Centre for Ecology and Conservation, University of Exeter, Penryn, Cornwall, UK

Phillip B. Chilson School of Meteorology, Advanced Radar Research Center, and Center for Autonomous Sensing and Sampling, University of Oklahoma, Norman, OK, USA

Andrea J. Contina Oklahoma Biological Survey, University of Oklahoma, Norman, OK, USA

Deanna K. Dawson U.S. Geological Survey, Patuxent Wildlife Research Center, Laurel, MD, USA

Kirsten de Beurs Department of Geography and Environmental Sustainability, University of Oklahoma, Norman, OK, USA

Jill Deppe Department of Biological Sciences, Eastern Illinois University, Charleston, IL, USA

Robert H. Diehl U.S. Geological Survey, Northern Rocky Mountain Science Center, Bozeman, MT, USA

Somayeh Dodge Department of Geography, Environment and Society, University of Minnesota, Twin Cities, Minneapolis, MN, USA

V. Alistair Drake School of Physical, Environmental and Mathematical Sciences, University of New South Wales at Canberra, Canberra, ACT, Australia

Institute for Applied Ecology, University of Canberra, Canberra, ACT, Australia

R. S. Eyster USDA, Agricultural Research Service, College Station, TX, USA

Todd Fagin Oklahoma Biological Survey, University of Oklahoma, Norman, OK, USA

Andrea Flack Max Planck Institute for Ornithology, Vogelwarte Radolfzell, Radolfzell, Germany

Winifred F. Frick Bat Conservation International, Austin, TX, USA

Department of Ecology and Evolutionary Biology, University of California, Santa Cruz, CA, USA

Tyson Hedrick University of North Carolina at Chapel Hill, Chapel Hill, NC, USA

Kyle G. Horton Oklahoma Biological Survey, Department of Biology, and Advanced Radar Research Center, University of Oklahoma, Norman, OK, USA

Eric Jacobsen University of Oklahoma School of Meteorology, Norman, OK, USA

Susanne Jenni-Eiermann Schweizerische Vogelwarte, Sempach, Switzerland

Douglas H. Johnson US Geological Survey, Northern Prairie Wildlife Research Center, Jamestown, ND, USA

Roland Kays NC State University, NC Museum of Natural Sciences, Raleigh, NC, USA

Jeffrey F. Kelly Oklahoma Biological Survey and Department of Biology, University of Oklahoma, Norman, OK, USA

Jennifer J. Krauel Department of Ecology and Evolutionary Biology, University of Tennessee, Knoxville, TN, USA

Valliappa Lakshmanan Google Inc., Seattle, WA, USA

Felix Liechti Swiss Ornithological Institute, Sempach, Switzerland

Liam P. McGuire Department of Biological Sciences, Texas Tech University, Lubbock, TX, USA

Frank R. Moore Department of Biological Sciences, University of Southern Mississippi, Hattiesburg, MS, USA

Paul Moore Department of Biological Sciences, Bowling Green State University, Bowling Green, OH, USA

J.P. Scott Center for Neuroscience Mind and Behavior, Bowling Green State University, Bowling Green, OH, USA

Renee Obringer Department of Civil, Environmental and Geodetic Engineering, The Ohio State University, Columbus, OH, USA

Anna C. Peterson Medicine Lake, MN, USA

Sandra Pletschet Oklahoma Biological Survey, University of Oklahoma, Norman, OK, USA

Diann Prosser USGS Patuxent Wildlife Research Center, Beltsville, MD, USA

Lori Randall U.S. Geological Survey, Wetland and Aquatic Research Center, Lafayette, LA, USA

Don R. Reynolds Natural Resources Institute, University of Greenwich, Kent, UK

Jeremy D. Ross Oklahoma Biological Survey, University of Oklahoma, Norman, OK, USA

Viola Ross-Smith British Trust for Ornithology, The Nunnery, Thetford, Norfolk, UK

Timothy D. Schreckengost Department of Entomology and Wildlife Ecology, University of Delaware, Newark, DE, USA

Judy Shamoun-Baranes Theoretical and Computational Ecology, Institute for Biodiversity and Ecosystem Dynamics, University of Amsterdam, Amsterdam, The Netherlands

Jaclyn A. Smolinsky Department of Entomology and Wildlife Ecology, University of Delaware, Newark, DE, USA

Robert B. Srygley U.S. Department of Agriculture, Northern Plains Research Lab, Agricultural Research Service, Sidney, MT, USA

Smithsonian Tropical Research Institute, Balboa, Republic of Panama

Phillip M. Stepanian School of Meteorology and Advanced Radar Research Center, University of Oklahoma, Norman, OK, USA

John Y. Takekawa Audubon California, Richardson Bay Audubon Center and Sanctuary, Tiburon, CA, USA

Diane Theriault Boston University, Boston, MA, USA

Hans van Gasteren Theoretical and Computational Ecology, Institute for Biodiversity and Ecosystem Dynamics, University of Amsterdam, Amsterdam, The Netherlands

Royal Netherlands Air Force, Breda, The Netherlands

Michael Ward Department of Natural Resources and Environmental Sciences, University of Illinois, Urbana, IL, USA

Rolf Weinzierl Max Planck Institute for Ornithology, Vogelwarte Radolfzell, Radolfzell, Germany

J. K. Westbrook USDA, Agricultural Research Service, College Station, TX, USA

Martin Wikelski Max Planck Institute for Ornithology, Vogelwarte Radolfzell, Radolfzell, Germany

Department of Biology, University of Konstanz, Konstanz, Germany

Part I

Background

Aeroecology: An Integrative View of the Atmosphere

1

Phillip B. Chilson, Winifred F. Frick, Jeffrey F. Kelly, and Felix Liechti

Abstract

The aerosphere is a fluid and rapidly evolving medium, which provides habitat for an abundance of organisms. The evolution of life on earth is replete with examples of key adaptations that enable all forms of life to use the atmosphere to move in ways that are simply impossible otherwise. The evolution of volancy is shaped by the habitat in which it occurs, the aerosphere. Understanding the atmosphere is central to understanding the movements and behavior of the animals that use it. Aeroecology represents a multidisciplinary approach to study animals in the aerosphere that taps into advanced observation, modeling, and visualization tools. This book on "Aeroecology" aims to provide the reader with a better understanding of the interdisciplinary field of aeroecology and its relevance by providing fundamental yet highly accessible material in several key areas. Moreover, the book includes several chapters that address a variety of observational methodologies actively used within aeroecology and how they are

P.B. Chilson (✉)
School of Meteorology, Advanced Radar Research Center, and Center for Autonomous Sensing and Sampling, University of Oklahoma, Norman, OK, USA
e-mail: chilson@ou.edu

W.F. Frick
Bat Conservation International, Austin, TX, USA

Department of Ecology and Evolutionary Biology, University of California, Santa Cruz, CA, USA
e-mail: wfrick@batcon.org

J.F. Kelly
Oklahoma Biological Survey and Department of Biology, University of Oklahoma, Norman, OK, USA
e-mail: jkelly@ou.edu

F. Liechti
Swiss Ornithological Institute, Sempach, Switzerland
e-mail: felix.liechti@vogelwarte.ch

© Springer International Publishing AG, part of Springer Nature 2017
P.B. Chilson et al. (eds.), *Aeroecology*,
https://doi.org/10.1007/978-3-319-68576-2_1

being applied to help us to better understand volant animals and their interaction with society. This overview chapter not only provides a broad introduction to the concept of aeroecology but also offers brief synopses of the other chapters in the book.

1 Introduction

In April of 2007, a group of about 50 biologists gathered in Sweetbriar, Virginia, USA, to discuss the future of research on animal migration. This meeting emphasized open discussions and there were very few research presentations. One of those presentations, however, was by Professor Thomas Kunz from Boston University. Kunz described his ideas for an emerging interdisciplinary field of science called Aeroecology (Kunz et al. 2008). For most of the biologists in attendance, this was the first time they had heard Kunz's new ideas and they generated a lot of discussion. Does it really make sense to think of ecological pattern and process in the air (aka, aerosphere) as separate phenomena from terrestrial ecology? Many, perhaps most animals (i.e., arthropods), have adaptations to allow them to travel through the air. Predator–prey interactions occur in the air routinely and it seems obvious that trade-offs among maneuverability and flight speed are central pieces of the life history of many organisms. The parallels with marine biology leap quickly to mind. Yet, which species of animals spend their entire lives in the air in the way that marine fishes live in the oceans? In particular what species reproduce in this environment? It seems certain that vertebrates do not. Does this difference from aquatic and terrestrial ecology invalidate aeroecology as a field of study?

In the decade since Tom Kunz introduced these ideas to biologists, these arguments have been repeated, expanded, and deepened. Rapid progress has been made in some areas. For example, there has been a wide recognition recently that the aerosphere is a critical habitat for much of Earth's biodiversity and that this habitat is under threat of being physically fragmented by human activities and altered fundamentally through climate change (Diehl 2013; Kelly and Horton 2016; Horton et al. 2016). On other topics, a diversity of opinions remain about what aeroecology is and what makes it a distinct field of research. Should there be a unified framework for aeroecology? Is there a need for a distinct set of fundamental theories of aeroecology? There is no agreement among practitioners on these topics. Indeed, a discussion among the authors of the chapters contained here led to a lively debate about whether we could adhere to a single definition of "aerosphere" across the volume. We could not! So, rather than try to define the field, we hope to present the diversity of research inspired by Kunz's initial ideas.

One of the promises of the Kunzian view of aeroecology was that it presented an opportunity to pursue a truly interdisciplinary approach. Biologists are, in general, poorly prepared to understand the theory of atmospheric science, the data available that describes its structure and functions, or the implications of these processes for ecological patterns and function, let alone its societal impacts. What will be evident to readers of this volume is that although biologists remain fascinated by life in the aerosphere, we have done a poor job of engaging the wider community in this endeavor. We think that the future of aeroecology depends on improving the buy-in that we can achieve in working with other disciplines. Biologists can document the changes in the aerosphere that cause harm to animals, but working to avoid and mitigate these effects goes well beyond the purview of biology. Widely communicating our passion for ecological pattern and process even when it occurs out of sight above our heads is our responsibility. Regardless of how we define aeroecology or delineate the aerosphere, we are united in our concern about the integrity of this vast and critical habitat and biodiversity that has evolved to rely on it for continued survival.

2 Book Synopsis

2.1 Fundamental: Chaps. 1–8

A primary aim in assembling this collection of diverse perspectives on aeroecology is to help those interested in exploring and perhaps pursuing research in this highly interdisciplinary field of study with a working, foundational knowledge in some of the core subject areas. Therefore, the book begins by providing introductory chapters on such topics as atmospheric science, the concept of the habitat, animal physiology, and methods of navigation. Obviously, the number of topics, which can be covered in a single volume, is limited. However, we hope that the selected material serves to convey the spirit of what aeroecology encompasses. A brief summary of these introductory chapters is provided below.

An inherent component of aeroecology is the actual atmosphere itself. The atmospheric environment is host to myriad species of airborne animals and has been a major driver in behavioral and evolutionary biology. In the chapter "Atmospheric Environment Associated with Animal Flight," the reader is provided with a comprehensive treatment of meteorology and atmospheric physics, especially as it relates to volant vertebrates and invertebrates. It provides an overview of small- and large-scale circulations, which impact meteorological and climatological patterns. These in turn impact migration, foraging, predator–prey interactions, and similar behavioral processes. Moreover, the chapter includes a discussion of the thermodynamic and kinematic processes that contribute to the structure and evolution of the planetary boundary layer and atmospheric turbulence, both of which are highly relevant for aeroecology.

Having established the fundamentals of atmospheric physics in the previous chapter, the book then goes on to describe in greater detail how volant animals use the aerosphere in "Extending the habitat concept to the airspace." The concept of habitat is inextricably related to the fields of ecology, animal behavior, and wildlife conservation and management. However, habitat is usually discussed in terms of terrestrial or aquatic environments. In this chapter, it is demonstrated that the atmosphere can be arguably considered as habitat since many essential functions of volant animals are conducted in the airspace. For this reason, the authors outline the importance of preserving the aerial habitat and safeguarding it against adverse anthropogenic effects.

The next theme covered in the volume addresses the complex topic of aerial movement ecology. Researchers have struggled for centuries to develop effective methods of tracking large-scale movements of volant animals in the airspace and to interpret the results within an analytic framework. It has been equally challenging to relate these movements to prevailing natural and anthropological factors. In the chapter "Track annotation—determining the environmental context of movement through the air," the reader is presented with an overview of measurement and modeling techniques that have been and are being employed to advance aerial movement ecology and how they are benefitting through the integrative approach offered through aeroecology. The concepts presented are motivated and illustrated by examples taken from studies involving avian species.

At some point in the study of aeroecology, one comes to and must address the basic question of what motivates animals to take flight and how have they needed to adapt to do so? Clearly locomotion in the airspace allows animals to avoid certain terrestrial barriers and to mitigate predation by land-based animals. However, powered flight is energetically expensive and animals must confront challenges such as wind and turbulence, and for long-distance movements, coping with high altitudes, cooler temperatures, and lower air densities become factors. The chapter "Anatomical and physiological adaptations for flight" begins by exploring ecological drivers that have motivated animals to fly, either actively or passively, and then goes on to outline various evolutionary adaptations, which facilitate flight. The chapter also covers the fundamental aerodynamics and fluid mechanics related to animal flight and how different species have evolved according to their respective biological drivers, e.g., the need for long distance migration, agility in flight, soaring, and so forth.

Animal migration is clearly a central and important aspect of aeroecology. Many species of birds, bats, and insects undertake incredible annual journeys as part of the reproductive cycle. The need to travel across vast distances while compensating for changing wind patterns and other factors has led to sophisticated navigational capabilities within migrants, which are actively studied by researchers. In "Properties of the atmosphere in assisting and hindering animal navigation," the authors examine in detail many of the factors used by animals to navigate on large and small scales during flight. These include light, odors, sound, and geomagnetic signatures, which are transmitted through the atmosphere and available to the migrants in transit. The atmospheric state can significantly impact how these signals are received and interpreted. Moreover, the authors address our current understanding of how animals assess wind patterns, which can be quite complex, and then apply navigational corrections.

Insects comprise the vast majority of organisms that utilize the aerosphere. Although capable of locomotion within the airspace, many insects are largely reliant on atmospheric currents for movement and as such are more closely coupled to the atmosphere than most other airborne animals. "Riders on the wind: the aeroecology of insect migrants" offers a broad overview of insect movements across a wide range of spatial and temporal scales with an emphasis on long-distance migration. In particular, the authors focus on insect migration in relation to the height and duration of flight, direction and speed of movement, seasonal and diel patterns, and responses to atmospheric conditions and phenomena. The chapter also discusses interactions between airborne vertebrate and invertebrates and differences in their respective strategies regarding flight.

Complementary to the previous chapter, "Facing the wind: the aeroecology of vertebrate migrants" places an emphasis on how vertebrates use the aerosphere when engaged in foraging, commuting, mating, and seasonal movements. This chapter incorporates several of the themes already developed in the preceding chapters to convey an integrative view of how birds and bats function in and interact with the aerosphere. For example, it synthesizes several aeroecological concepts such as flight dynamics, impacts of weather and climate on vertebrate flight behavior, and natural and anthropogenic drivers of behavioral ecology. Many pressing questions remain regarding volant vertebrates. The chapter closes by exploring how research on these questions is being advanced by the interdisciplinary approach available through aeroecology.

2.2 Observational Methods: Chaps. 9–13

The label aeroecology to describe the study of organisms in the airspace is relatively new; however, the observation and study of volant animals are not. Research into animal flight, patterns of migration, and how airborne organisms utilize and respond to the atmosphere goes back centuries. Along the way, researchers have developed a variety of tools and methods to investigate airborne animals and environment in which they move. In the following chapters, the book provides a broad discussion of observational methods available to and used by aeroecologists. Here again, the selection of observational techniques available within the book is meant to be representative rather than exhaustive. Nevertheless, we feel that these chapters provide the reader with a basic understanding of experimental aeroecology.

This section of the book begins with the aptly named chapter "Aeroecological observation methods." It presents a general overview of the core empirical tools and techniques that have been implemented to observe aerial organisms, in particular while in flight. The chapter begins with a discussion of "low-technology" methods such as visual observations, listening for auditory cues, capturing and potentially tagging animals, and such. To be fair, we should insert that visual and aural methods as well as animal tagging have undergone considerable technological advancement and become increasingly powerful. The chapter then goes on to discuss the use of radar to observe and track the movement of flying organisms in the aerosphere. Other remote sensing devices discussed in the chapter include observations using ground-based lidar and radio techniques from satellites. As discussed by the authors, there have also been considerable advancements in our ability to track animals in situ using electronic tags and geologgers.

Our capacity to observe and study aerial organisms in flight has been enhanced through the use of imaging and video analysis tools. As outlined in "Multi-camera videography methods for aeroecology," these techniques facilitate detailed and precision tracking of individuals over a limited volume of space and interval of time. Multi-camera videography provides unprecedented insights into the intricate patterns of animal flight; however, the method requires carefully planning the instrument deployment and meticulous calibration of the cameras. Once the images have been recorded, they must be post-processed using specially adapted algorithms to spatially isolate the observed organisms and render 3D flight tracks from the multi-view video sequences. The chapter closes with examples of how these techniques have been used to study the flight tracks of Brazilian free-tailed bats, cliff swallows, and chimney swifts near their respective roosts.

A somewhat different take on the topic of aeroecological observation is that of computer-based behavioral modeling. This allows researchers to pose and address fundamental questions on the processes that govern individual or aggregate behavior based on empirical observations. In "Using agent-based models to scale from individuals to populations," the authors contend that agent-based models can serve as a "unifying link between knowledge about individuals and observations of broad, population-level phenomena." Within the framework of agent-based modeling, the decisions and behavior of individuals are simulated in terms of external drivers, including the behavior of other agents. The chapter provides the reader with a foundational understanding of how agent-based models work and how this powerful resource is being used among aeroecologists to study migration, flocking, foraging, and similar activities.

Remote sensing has been a widely used method of observing organisms in flight, especially within the field of radar. Initially, the task of integrating radar into aeroecological research fell on a small subset of researchers; however, advances in hardware, computing, and networking technology has made this resource more accessible to a wider range of individuals. The chapter titled "Radar aeroecology" serves as an introduction for those wishing to explore both the qualitative and quantitative value of radar data for the study of volant animals. It begins with a basic treatment of how radar works and the concept of a radar cross section as applied to aerial organisms. It then outlines some of the basic scanning strategies used by radar. The chapter goes on to describe how operational weather radar data can be used by aeroecologists. It closes with examples of how weather radar is being used to study bat emergences and bird migration.

Following on the previous chapter "Inferring the state of the aerosphere from weather radar" continues the discussion on the benefits of utilizing weather radar beyond the purpose for which it was originally developed. Weather radars have been carefully designed and optimized for meteorological observations. As we have seen, however, they can also detect the presence of flying animals. As such, radars can be used to study both the animals and the environment through which they are moving. This chapter leads the reader through several techniques being applied to weather radar observations to merge data from networks of radars; remove contamination and effects from calibration error; extract and transform meteorological and biological data from the native spherical coordinate system to a Cartesian system; and relate the radar data to meaningful biological metrics. It becomes apparent that the weather radar could more aptly be called an ecological radar.

2.3 Aeroecological Applications: Chaps. 14–18

The remaining chapters of the book showcase examples of different research topics and how they are being advanced as part of aeroecology. These are meant to offer the reader a better sense of what aeroecology as a discipline entails. That is, they help to promote aeroecology as a unifying framework in which to understand aerial organisms within the context of the dynamic properties of meteorological conditions at local, regional, and global scales. As examples of how aeroecologists pursue particular research questions, these chapters are necessarily narrower in scope and delve deeper into particular methodologies than preceding chapters of the book.

Although aeroecology focuses primarily on the activity of organisms in the atmosphere, it must be recognized that most of these animals are coupled to terrestrial habitats. Consequently, most aerial organisms are strongly impacted by factors that alter or disrupt conditions on the surface. For example, long-distant migrants rely on terrestrial habitats as stopover sites during their treks. Changes in the landscape may affect the migrant's ability to find safe shelter and to feed. Moreover, migrants must confront and contend with ecological barriers and even weather conditions. These and similar factors are considered in detail within "Linking animals aloft with the terrestrial landscape." The chapter demonstrates in part the impact of natural and anthropogenic changes to the Earth's surface on the vitality of aerial organisms. It is argued by the authors that the linkage between terrestrial landscapes and volant animals should be considered during terrestrial habitat management and restoration projects.

The chapter "The lofty lives of aerial consumers: linking population ecology and aeroecology" provides the reader with an overview of population ecology within the context of aerial consumers. That is, the chapter explores how foraging strategies of aerial consumers are impacted by short- and long-term patterns in the aerosphere and these in turn affect population dynamics. Population dynamics reflect an organism's life history traits within the context of interactions with environmental drivers. The authors demonstrate how radar data can be harnessed to establish qualitative and quantitative metrics to evaluate weather and climate impacts on animal populations. This is done using empirical data, simulation data, along with conceptual and analytic models. Examples involving emergences of birds and bats are provided to illustrate these ideas.

Tools and resources currently available to researchers under the banner of aeroecology are allowing us to push the envelope on the types of questions we can ask: one of these relates to the drivers that affect the phenology of avian migration. Such questions often require observational and modeling capabilities that span multiple temporal and spatial scales. The authors of "The pulse of the planet: measuring and interpreting phenology of avian migration" are able to tap into weather surveillance radar and bird data to study migration phenology at local and regional scales. They explore the application of these resources both to a single-species and to widespread nocturnal songbird migration. The authors go on to examine NDVI (normalized difference vegetation index) records to test hypotheses related to how the migration patterns are impacted by climate variability.

The study of factors influencing the behavior and movement of aerial organisms goes well beyond helping to address fundamental aeroecological questions. Many animals capable of powered flight are known to carry and transmit pathogens across large distances. Therefore, a better understanding of the movement patterns of these animals is needed to track potential vectors of animal-borne diseases. In "The aerosphere as a network connector of organisms and their diseases" the reader is presented with a scenario involving the transmission of an avian influenza strain (HPAI) into North America from East Asia via waterfowl. Agent-based modeling is used to simulate the scenario. Such tools are needed to investigate the complex mass movement of birds, while taking weather and terrestrial factors into account. The authors conclude by examining how such an approach can be applied to other aeroecological questions related to the connectivity of flying animals.

The closing chapter of the book addresses a timely and complex issue: the need to address increasing use of the aerosphere by humans and how this could impact aerial organisms. Examples of existing conflicts between human populations and wildlife discussed in "Sharing the aerosphere: Conflicts and potential solutions" include collisions between commercial and general aviation aircraft with birds and collisions of birds and bats with human-made structures such as buildings, power lines, and wind turbines. Potential conflicts will be further exacerbated with the accelerated development of the unmanned aircraft system enterprise. All of these factors are being actively examined by researchers, industry, federal agencies, and policy makers in an attempt to find acceptable and affordable solutions. The chapter provides a comprehensive view into these challenges and how aeroecological studies are helping to provide much needed data to facilitate informed decisions.

3   Closing

As the diversity and depth of work covered in this volume demonstrates, researchers have long explored the fundamental concepts key to aeroecology, made tremendous and sustained progress on methodological tools, and applied these to the study and conservation of aerial organisms in various capacities. The umbrella of aeroecology brings together these varied perspectives to draw attention to the importance of the aerosphere and the animals that depend on it, incorporating both historic work and framing new directions for inquiry. The diversity of the expertise of the authors of this volume testifies to a core strength of Kunz's original vision of aeroecology: collaboration across disciplines will yield new insights and progress on both basic and applied aspects of the biology of aerial animals. Sadly, Professor Tom Kunz suffered a tragic accident in 2011 and has not been able to continue his pioneering work to bring together experts and inspire interdisciplinary collaborations to advance aeroecology. However, his legacy of inspiration and collaboration endures, indeed some of the co-editors of this volume were introduced by Kunz, and we dedicate this volume to him.

References

Diehl RH (2013) The airspace is habitat. Trends Ecol Evol 28:377–379

Horton KG, Van Dorren BM, Stepanian PM, Farnsworth A, Kelly JF (2016) Where in the air? Aerial habitat use of nocturnally migrating birds. Biol Lett 12:20160591

Kelly JF, Horton KG (2016) Toward a predictive macrosystems framework for migration ecology. Glob Ecol Biogeogr. https://doi.org/10.1111/geb.12473

Kunz TH, Gauthreaux SA Jr, Hristov NI, Horn JW, Jones G, Kalko EK, Larkin RP, McCracken GF, Swartz SM, Srygley RB, Dudley R, Westbrook JK, Wikelski M (2008) Aeroecology: probing and modeling the aerosphere. Integr Comp Biol 48:1–11

Atmospheric Environment Associated with Animal Flight

2

J.K. Westbrook and R.S. Eyster

Abstract

The atmospheric environment can assist or restrict flight of animals (insects, birds, and bats), influencing their ability to extend their population range and find new habitats for food, mating, and shelter (Pedgley, Windborne pests and diseases: meteorology of airborne organisms. Ellis Horwood, 1982; Drake and Gatehouse, Insect migration: tracking resources through space and time. Cambridge University Press, 1995; Isard and Gage, Flow of life in the atmosphere: an airscape approach to understanding invasive organisms. Michigan State University Press, 2001; Drake and Reynolds, Radar entomology: observing insect flight and migration. CABI, 2012). Flight initiation may be influenced by air temperature, wind speed, wind direction, barometric pressure change, or other atmospheric variables and may require substantial convective lift for small invertebrates such as wingless insects and spiders. Flight displacement distance and direction largely depends on prevailing wind speed and wind direction at respective flight altitudes, although flying animals can add to their displacement speed and can alter their displacement direction by heading at an angle relative to the wind direction (Wolf et al., Southwestern Entomol Suppl 18:45–61, 1995), including upwind flight. While flying, animals may engage in predatory behaviors and conversely may be preyed upon. Long-distance flight of flying animals may be abruptly terminated by mid-latitude fronts, gust fronts, and precipitation. Ultimately, movement of animal populations in the atmosphere ranges across spatial and temporal scales spanning the microscale to macroscale and is vulnerable to changes in short-term weather and long-term climate patterns.

J.K. Westbrook (✉) · R.S. Eyster
USDA, Agricultural Research Service, College Station, TX, USA
e-mail: john.westbrook@ars.usda.gov; ritchie.eyster@ars.usda.gov

© Springer International Publishing AG, part of Springer Nature 2017
P.B. Chilson et al. (eds.), *Aeroecology*,
https://doi.org/10.1007/978-3-319-68576-2_2

1   Scales of Atmospheric Structure and Motion

Animal flight is fundamentally a process that crosses vast spatial and temporal scales. This chapter examines the atmospheric properties, in time and space, that influence how diverse and numerous animal species engage in flight or wind-borne transport (Westbrook and Isard 1999). We discuss various scales of atmospheric structure and motion to describe atmospheric processes that lead to the distribution and displacement of flying animals. The definition of aeroecological boundary layers is based not only on the atmospheric environment but constrained to the spatial and temporal extent in which species of interest occupy that environment (Westbrook 2008). Aeroecological interest in particular species, or particular life activities, will dictate relevant atmospheric scales. We stress that flying animals do not readily fit into an aeroecologically scaled "box" but rather fit within an aeroecological continuum. Application of an aeroecological continuum is clearly needed as animals ascend into the atmosphere, and for directed flights that extend either continuously or discontinuously. We discuss atmospheric structure and motions that are relevant to all animal species that fly through the atmosphere within a range of spatial and temporal scales, by describing the atmospheric environment cascading from the macroscale to the mesoscale and microscale.

Several key terms must be defined to describe the atmospheric environment as it relates to animal flight. First, the scales of motion range from microscale (≤ 2 km), to mesoscale (2 to several hundred km), and to macroscale (≥ 1000 km). An individual animal could conceivably engage in flight that is influenced by the entire scale of atmospheric motions for example by initiating flight in response to turbulent gusts; flying continuously in a low-level jet (i.e., maximum in the vertical profile of wind speed) for several hundred kilometers; and continuing on successive flights over a distance of more than 1000 km (and affected by synoptic weather features such as frontal systems). All of these atmospheric motions occur in the troposphere, which is the part of the atmosphere where most weather occurs and extends from the surface to 10–20 km above ground level. The lowest part of the troposphere is the atmospheric boundary layer which is of variable depth, ranging from tens of meters to several km depending on the height of low-altitude temperature inversions (i.e., air temperature increasing with height) which limit the vertical extent of convection. Animals may fly within or above the atmospheric boundary layer, depending on the species, activity phase, and atmospheric conditions.

Historically, the science of understanding animal flight in the atmosphere has been termed aerobiology, or a niche within the broader realm of biometeorology. While biometeorology describes atmospheric effects on fauna and flora, the subject has largely focused on atmospheric impacts while the fauna and flora are on the ground. Aerobiology focuses on the presence of airborne biota, but has principally addressed inanimate biota, including pollen, spores, and bacteria, rather than flying animals. However, Kunz et al. (2008) either coined or made popular the field of aeroecology to emphasize the behavior and activity of flying animals, including aerial interactions between taxa and species.

Table 2.1 Typical reflection of sunlight by various surfaces

Surface	Albedo (%)
New snow	90
Old snow	50
Average cloud cover	50
Light sand	40
Light soil	25
Concrete	25
Green crops	20
Green forests	15
Dark soil	10
Asphalt	8
Water	8

2   Atmospheric Thermodynamics and Kinematics

Atmospheric thermodynamics describes the conversion of heat into work energy and consequently from potential energy to kinetic energy (expressed as wind) in the atmosphere. The sun is the heat source that modifies the atmospheric structure of temperature, humidity, and pressure and generates atmospheric motions and circulations. Seasonal changes in the amount of insolation at the top of the atmosphere are caused by the elliptical solar orbit (and changing distances between the earth and sun), and local variations are due to solar angle (associated with location latitude). Further reductions in insolation at the earth surface are due to clouds and atmospheric contaminants that reflect or reemit incident solar radiation primarily back to space. The heterogeneous earth surface comprises a wide array of soil types, vegetation types, and water bodies, each of which has a specific reflectance characteristic known as albedo (percentage of reflected incident solar radiation across all wavelengths of energy). The albedo of a particular unit of land surface will change when the surface is wet or covered with ice or snow; vegetation cover increases; vegetation grows or senesces; and water bodies freeze (Table 2.1). Differential absorption of insolation and subsequent heating of individual units of land surface leads to surface gradients of temperature. In turn, horizontal temperature gradients cause barometric pressure gradients that shape global, regional, and local atmospheric circulations.

3   Macroscale

The atmosphere drives many of the movements of biota around the world. These movements can be considered as local, regional, continental, and global. This section will cover the continental and global scale atmospheric structure and air flow. To better understand the role of the atmosphere in driving actions in the aerosphere, we will begin at the largest scales and drive down to the small scale.

3.1   Global Circulation

We begin by looking at the atmosphere on a global scale (Fig. 2.1). There are large-scale circulations such as the Hadley Cell in the subtropics, the Ferrel or Mid-latitude Cell, and the Polar Cell. All of these are driven by differences in the heat from incident solar radiation at the surface and aloft. At the Equator, the atmospheric circulation is marked by the Inter-Tropical Convergence Zone (ITCZ). This area of the global circulation is known as the Doldrums where there is very little wind. Here the warm and humid air rises resulting in storms which transport warm air to higher levels of the atmosphere. The flow then moves poleward as the air cools and begins to sink. This sinking air comes back to the earth surface in the regions of the world's deserts. Figure 2.2 illustrates a cross-sectional view of the circulation associated with each of the cells. In the Northern and Southern Hemispheres, the Ferrel Cell is the area in which surface pressure systems are active. The Ferrel Cell is the dividing circulation between the cold polar air and the

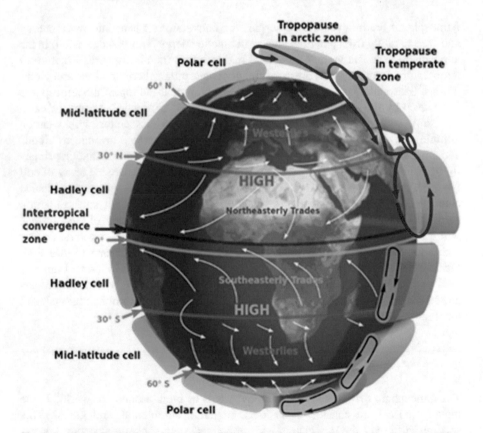

Fig. 2.1 Global Circulation with the three NS circulations: (1) Hadley, (2) Ferrel, and (3) Polar. The ITCZ, Easterly Trade Winds, Mid-Latitude Westerlies, and Polar Easterlies are indicated. http://commons.wikimedia.org/wiki/File:Earth_Global_Circulation_-_en.svg

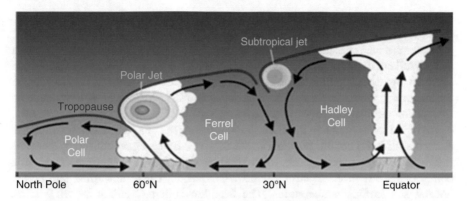

Fig. 2.2 Cross-sectional view of the Hadley, Ferrel, and Polar circulations with the tropopause, polar jet stream, and subtropical jet stream locations indicated. http://www.srh.noaa.gov/jetstream/global/jet.htm

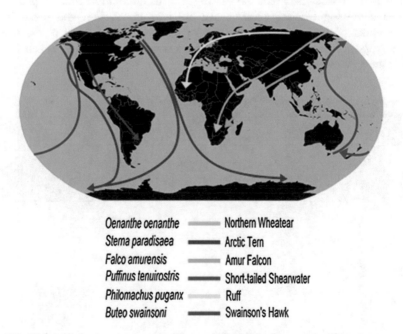

Fig. 2.3 Global bird migration routes (Shyamal, L. 2008. Own work: Migration routes of birds based on Newton, I. The Migration Ecology of Birds. Academic Press)

warm tropical air. The Ferrel Cell is bound at the tropopause by the subtropical and polar jet streams. Convection occurs at the equator and at the location of the polar jet. The subtropical jet is located over the sinking air at 30°N and 30°S. The jet stream locations vary season to season. At the poles, the air sinks resulting in cold high pressure. There are several bird species which migrate between hemispheres. Shyamal adapted a figure (Fig. 2.3) showing macroscale migration routes of birds.

The macroscale circulation cells may be a factor in these long-distance hemispheric migrations.

There are other circulations which transport energy from East to West. The first is the Walker Circulation which is associated with the El Niño/Southern Oscillation (ENSO) circulation in the Pacific Ocean. The Walker Circulation is East to West at the surface and West to East aloft. Figures 2.4 and 2.5 illustrate how the strength of the Walker circulation determines the strength of the Easterly Trade Winds and the depth of the thermocline in the Eastern Pacific near the coasts of Central and South America. El Niño conditions are a result of weak trade winds which suppress the thermocline and result in above normal sea surface temperatures. La Niña conditions are a result of strong trade winds which cause upwelling and below normal sea surface temperatures. The Madden–Julian Oscillation (MJO) is an eastward moving circulation or wave along the ITCZ (Madden and Julian 1971, 1972). The MJO is most commonly associated with increased convection in the Indian and Pacific Oceans. The MJO can also be seen in the Atlantic Ocean and over Africa (Madden and Julian 1971, 1972). This circulation when in phase with westward moving (easterly) waves can enhance convection and result in the development of strong tropical systems in the subtropical areas of the oceans. These tropical systems have been known to transport birds, and beneficial and pest insects (Westbrook 2008).

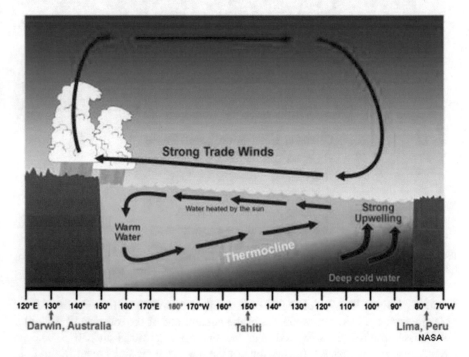

Fig. 2.4 Walker circulation associated with La Niña. http://oceanservice.noaa.gov/yos/resource/ JetStream/tropics/enso_patterns.htm

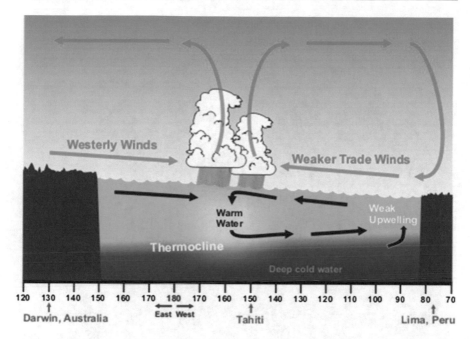

Fig. 2.5 Walker circulation associated with El Niño. http://oceanservice.noaa.gov/yos/resource/ JetStream/tropics/enso_patterns.htm

3.1.1   The Coriolis Effect

The general global flow comprises the Easterly Trade Winds in the subtropical latitudes, Westerlies in the mid-latitudes, and polar Easterlies in the Arctic and Antarctic regions. These wind patterns result from the rotation of the earth and the direction of air movement. Figure 2.6 illustrates the Coriolis Effect on meridional flow (poleward) and zonal flow (west to east). When air flows poleward, the apparent motion is eastward resulting from the Coriolis force. When the air flow is toward the equator, the apparent motion is westward. The Coriolis force is greatest at the poles and near zero at the equator. The global wind patterns transport warm and cold air across the globe. The formation and movement of large-scale atmospheric waves can be observed through cloud motions. These large-scale or Rossby waves are the general drivers of mid-latitude weather systems. In the tropics and subtropics, easterlies drive the movement of tropical systems such as hurricanes, typhoons, and cyclones over the world's oceans.

3.2   Tropical and Subtropical Systems

In the tropical and subtropical regions of the earth, summer and fall brings the development of tropical storms. These storms are called hurricanes, cyclones, or

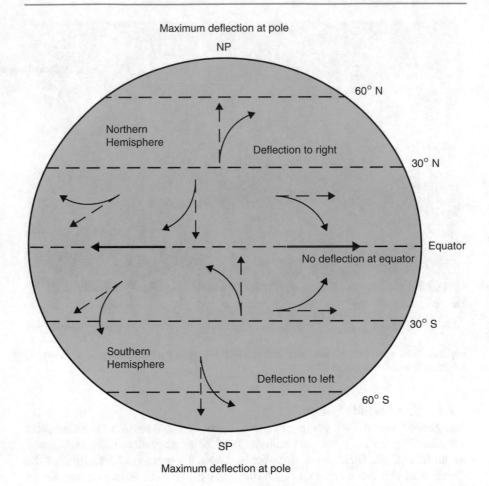

Fig. 2.6 Coriolis Effect on wind flows in both the Northern and Southern hemispheres. Airflow is deflected to the right in the northern hemisphere and to the left in the southern hemisphere. Image/photo courtesy of the National Snow and Ice Data Center, University of Colorado, Boulder

typhoons depending in which part of the world they develop. The time of year in which these storms occur may result in enhanced movement by the circulation in and around the storm. Figure 2.7 is a case study of a tropical storm which moved through South and West Texas over a 5-day period in August. The Southern Rolling Plains eradication zone (red counties on Fig. 2.7) did not have any boll weevils present for several years before this event. After Tropical Storm Erin moved through Texas, boll weevils were found across the southern part of the Southern Rolling Plains eradication zone (Kim et al. 2010). The middle of August is the time of cotton harvest in the South Texas/Winter Garden eradication zone (yellow counties on Fig. 2.6) as well as in the Southern Blacklands eradication zone (green counties on Fig. 2.6). Over a 3-day period (August 17–19), backward atmospheric trajectories calculated using the HYSPLIT model (Draxler and Hess 1998) identified probable boll weevil origins in the South Texas/Winter Garden

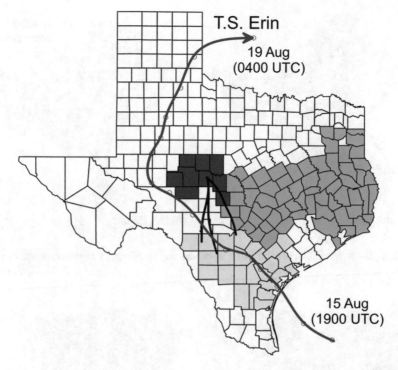

Fig. 2.7 Track of the center of circulation of Tropical Storm Erin (blue line) and backward atmospheric trajectories (black lines) from the location of a boll weevil reinfestation in the Southern Rolling Plains eradication zone (red counties). Boll weevil populations had not yet been eradicated in the Southern Blacklands eradication zone (green counties) and the South Texas/ Winter Garden eradication zone (yellow counties). Figure reprinted courtesy of the International Journal of Biometeorology

eradication zone. Analysis of DNA and pollen taken from boll weevils captured in the Southern Rolling Plains eradication zone strongly indicated that the boll weevils arrived from the Winter Garden district (Kim et al. 2010). This evidence suggests that tropical storms can move biota from tropical and subtropical regions of the world to temperate regions, which can result in infestations or reinfestations of pest insects.

3.3 Monsoons

In certain parts of the world such as India and southeast Asia, there is the seasonal monsoon flow. During the wet phase of the monsoon, air flow is from southeast though southwest directions (Pidwirny 2006; Dowling 2014). This air flow can be a mechanism to transport biota from tropical regions to agricultural regions northwest through northeast (Figs. 2.8 and 2.9). During the dry phase of the monsoon, air flow is from northeast through northwest (Figs. 2.8 and 2.9). This flow can act to help transport biota back to agricultural areas and warmer areas for overwintering. These

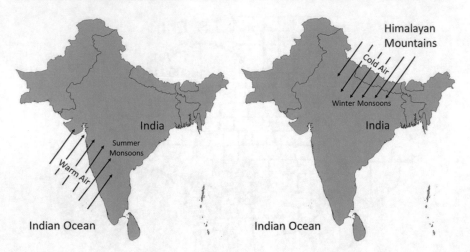

Fig. 2.8 Mean wind direction for India monsoons in the Northern Hemisphere summer and winter

Fig. 2.9 Monsoon wind flow across India and Southeast Asia in the Northern Hemisphere summer and winter

monsoonal flows occur in other regions as well and signal the start of growth of food sources. An example of a frequent summertime atmospheric pressure system in the USA is the Four Corners Low, which is centered over the four-state area (Utah, Arizona, Colorado, and New Mexico) in the southwestern USA (Fig. 2.10). The Four Corners Low creates an atmospheric pattern that entrains subtropical moisture and creates monsoon-type conditions in the USA desert southwest. There are other areas throughout the world that experience similar monsoonal flows. Most of these flows are around quasi-stationary subtropical high-pressure regions at and around 30°N and 30°S latitude (Figs. 2.11, 2.12, and 2.13). These areas include Australia, South America, and Africa. The monsoonal flow in these areas is driven by the location of the ITCZ. The Southwest USA monsoon is driven by the low surface pressure as a result of desert high heat. The ITCZ remains well south of the continental USA.

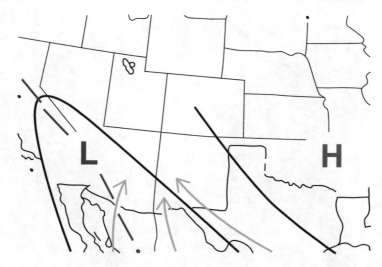

Fig. 2.10 Southwest USA summer monsoon air flow. http://www.srh.noaa.gov/elp/swww/v8n1/monsoonmapb.jpg

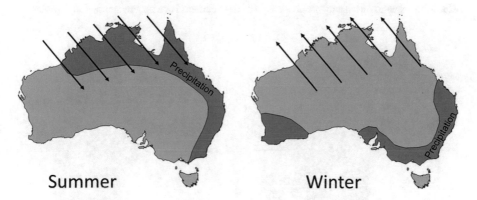

Fig. 2.11 Australian monsoon air flow in the Southern Hemisphere summer and winter

3.4   Quasi-stationary Mid-Latitude Pressure Systems

Regions of quasi-stationary or persistent atmospheric pressure systems are often named to identify atmospheric features that influence mesoscale and macroscale flow (Figs. 2.14 and 2.15). An example of a quasi-stationary atmospheric pressure system is the Bermuda High, which generally resides over Bermuda Island near the southeastern coast of the USA. This feature is the western part of the Azores High. The Bermuda High drives southerly flow over the southern USA, but often retrogresses and displaces the southerly flow westward. This extension of the Bermuda High guides tropical storms into and through the Gulf of Mexico and guides surface low-pressure systems north and away from the southeastern USA.

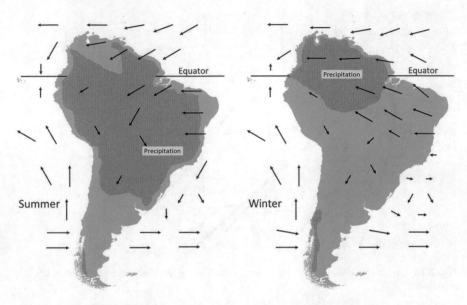

Fig. 2.12 South America monsoon air flow in the Southern Hemisphere summer and winter

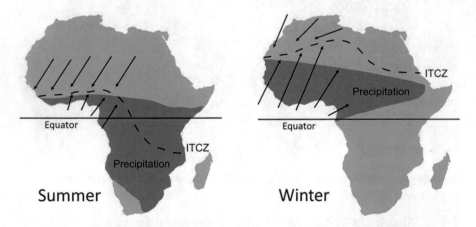

Fig. 2.13 Africa monsoon air flow in the Southern Hemisphere summer and winter

Each continent has air flows that are conducive to migration and dispersal of biota. Australia and parts of South America and southern Africa are the only large land areas in the Southern Hemisphere that support large-scale agricultural production. Figures 2.14 and 2.15 show summer and winter macroscale flows across South America, southern Africa, and Australia. High pressure is dominant on the western side of South America and southern Africa in January, resulting in deserts along the coastal areas. July has a similar pressure pattern with high pressure over the Indian Ocean moving west to Southeast Africa and high pressure developing over

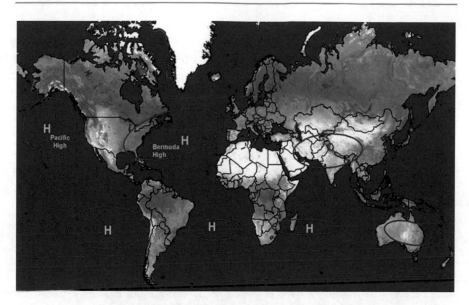

Fig. 2.14 Quasi-stationary atmospheric pressure systems during summer in the Northern Hemisphere. *Ellipses* indicate the extent of high-pressure (H) and low-pressure (L) systems. The Pacific High and Bermuda High are labeled. Basemap from ESRI Inc

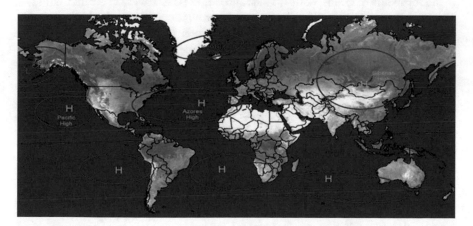

Fig. 2.15 Quasi-stationary atmospheric pressure systems during winter in the Northern Hemisphere. *Ellipses* indicate extent of high-pressure and low-pressure (L) systems. The Aleutian Low, Icelandic Low, Pacific High, and Azores High are labeled. Basemap from ESRI Inc

Australia (Fig. 2.14). The high-pressure area to the west of South America remains in the same general vicinity (Haltiner and Martin 1957). The locations of these high-pressure systems are quasi-stationary throughout the year. Similar atmospheric flows occur in the Southern Hemisphere as in the Northern Hemisphere and result in similar seasonal migration and dispersal of biota.

3.4.1   Flow to and from Desert Regions

There are other large-scale circulations that are able to move biota from tropical and subtropical latitudes to temperate areas. Some of these flows are around the world deserts. Deserts are associated with strong high pressure aloft and strong low-pressure systems at the surface due to extreme heat. The resulting air flow moves biota from the periphery of the surface low pressure toward the north on the east side and toward the south on the west side. These flows have been associated with massive movements of biota from Africa to the Middle East and Europe. Figure 2.16 illustrates the major wind flows around the Mediterranean Sea for summer and winter. Strong winds can blow in from the Sahara Desert as well as from Europe toward northern Africa. These air flow patterns can be conducive to long-range dispersal of biota.

3.5   Mid-Latitude Circulation

The evolution of Rossby waves is governed by differential heating between the poles and equator (Fig. 2.17). The flow around the Polar Low begins as zonal flow or winds from west to east. As differential heating occurs, energy balance requires the movement of warm air toward cold air and cold air toward warm air. This results in the formation of atmospheric troughs and ridges (i.e., waves) in the mid-latitudes. The number of these atmospheric waves around the hemisphere governs the stability of the global circulation. The mean winter wavelength is 7025 km, while the mean wavelength in summer is 4967 km. If the wavelength is shorter than the

Fig. 2.16 Descriptive names for hot, dry winds (*red*) and cold winds (*blue*) in the Mediterranean region during the Northern Hemisphere summer and winter, respectively. Basemap From ESRI, Inc

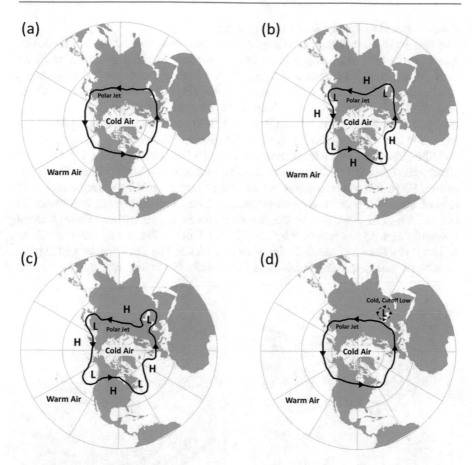

Fig. 2.17 Evolution of Rossby waves in the polar jet stream around the Northern Hemisphere polar vortex: (**a**) circulation relatively undisturbed; (**b**) incipient waves develop; (**c**) wave amplitude increases; and (**d**) circulation relatively undisturbed with new cutoff low-pressure cell

mean, the waves move eastward. If the wavelength is longer than the mean the waves move westward (Wallace and Hobbs 1977; (Haltiner and Martin 1957; Haltiner and Williams 1980). Superimposed on the Rossby waves are short waves. Short waves move rapidly downwind in the jet stream flow (Haltiner and Martin 1957). When the short wave and long wave are in phase, the overall wave becomes very high amplitude resulting in cold air moving far toward the equator and warm air moving far toward the poles. It is important to observe the hemispheric wave pattern to determine the potential for atmospheric wave amplification and subsequent changes in surface pressure and wind patterns that are conducive to migration of biota.

3.5.1 Mid-Latitude Blocking Patterns

The mid-latitude wave train can develop into what are known as blocking patterns. There are several upper air wind flows known as blocking patterns. Blocking patterns are caused by large and strong high pressure at 500 mb level (see Sect. 3.5.2). Some of these are the Cut-off Low (Fig. 2.18a), the Omega Block (Fig. 2.18b), the Rex Block (Fig. 2.18c), the Ring-of-Fire Block (Fig. 2.18d), and the Split-Flow Block (Fig. 2.18e) (Haby 2015). These blocking patterns can persist for 2–3 weeks and can develop in either hemisphere. Omega Blocks, Rex Blocks, and Split-Flow Blocks are most common in the late winter and spring. Cut-Off Low Blocks can occur at any time of year but are least common in the winter. The Ring-of-Fire Block is a common summer blocking feature, usually appearing on the west side of quasi-stationary high-pressure ridges. This blocking pattern is common in the southeastern USA during the summer and is an expansion of the Azores/Bermuda high to the west (Haby 2015). The Omega Block can break down to form either a Cut-Off Low, Split-Flow, or Rex Block. The Cut-Off Low and Omega Block can lead to sustained periods of flow conducive to animal migrations in the

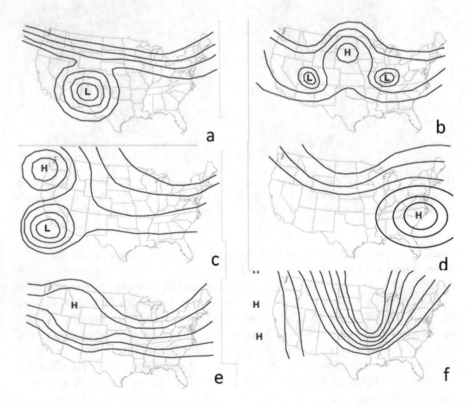

Fig. 2.18 Illustrations of atmospheric blocking patterns: (**a**) Cut-Off Low, (**b**) Omega Block, (**c**) Rex Block, (**d**) Ring-of-fire Block, (**e**) Split-Flow Block, and (**f**) El Niño blocking high pressure. Basemap from ESRI, Inc. Illustrations based on Haby (2015)

atmosphere. Surface low pressure develops underneath the upper low resulting in surface flow from the south. If the block sets up over the Four Corners area of the USA, southerly flow will be over the Central and Southern Plains. Many birds and insects overwinter in southern and coastal areas of Texas, Louisiana, Mississippi, Alabama, Florida, and Mexico. Southerly winds bring ideal conditions for movement of biota to the north (Muller 1977). One example is the yearly migration of the monarch butterfly. Figure 2.19 illustrates the known flyways of the monarch butterfly from overwintering sites in Mexico, California, and Florida. The monarch butterfly flies north throughout the spring and summer, reaching its breeding grounds in four to five generations (MonarchWatch.org). In the fall, this final generation migrates back to overwintering sites. Another common blocking pattern is associated with El Niño conditions in the eastern Pacific Ocean. Figure 2.18f illustrates the strong ridge of high pressure that forms across the western USA and adjacent waters of the Pacific Ocean in association with El Niño conditions. These conditions can exist for several months and result in severe drought conditions in the western USA.

3.5.2   Structure of Troughs and Ridges

A trough of low pressure is visualized by the height of a constant pressure surface (Fig. 2.20). At the center of the trough, the height of the pressure surface is lower than at the outer extent of the trough. Conversely, the height of the pressure surface at the center of a ridge is higher than at the outer extent of the ridge. When observing the vertical structure of the atmosphere from the earth's surface to the top of the troposphere (Tropopause), a trough tilts to the north and west (Haltiner and Martin 1957). This allows for divergence aloft associated with the jet stream to be over a surface low-pressure system. Short wave troughs are often indicated by jet

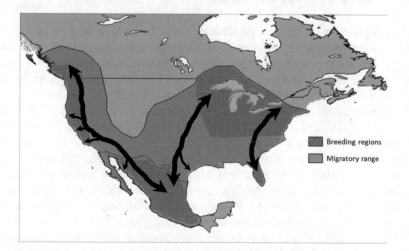

Fig. 2.19 Monarch butterfly migration routes in North America

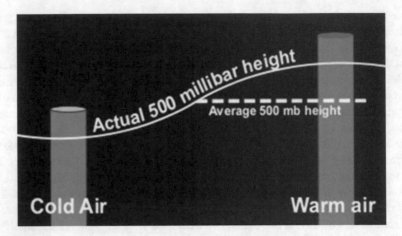

Fig. 2.20 Cross-sectional view of 500-mb height from cold air trough to warm air ridge. http://www.srh.noaa.gov/jetstream/global/pressure.htm

streaks. Jet streaks (areas of maximum wind) within the jet stream can occur on the upwind side of a trough or at the base or downwind side of a trough. Divergence in the exit region of a jet streak is situated directly over the surface convergence and low-pressure system. Surface lows under an upwind short wave are usually weak because there is little warm air available to promote surface convergence. Jet streaks that occur at the base or upwind side of a trough bring all the ingredients together to form strong surface low-pressure systems.

3.6   Surface Pressure Systems

Surface high-pressure systems are located along the leading edge of atmospheric ridges and low-pressure systems are located along the leading edge of atmospheric troughs. When a strong high-pressure system follows a strong low-pressure system, cold air is forced equatorward and warm air poleward. This helps to intensify atmospheric trough and ridge patterns at the 500 mb pressure surface.

The leading edge of cold air is indicated by a cold front and the leading edge of the warm air is indicated by a warm front. Figure 2.21 illustrates the fronts and air masses associated with a strong low-pressure system in the south-central USA. Figure 2.22 is a 3D diagram of the air flow around a low-pressure system. This is a common pattern during late winter to early summer. Frontal systems, as shown on weather maps, are significant catalysts behind the movement of flying animals in the atmosphere. During spring and summer, winds following warm fronts move poleward and are capable of transporting both flying biota such as birds and insects as well as flightless biota. Temperatures reach levels at which flying animals can initiate and sustain flight. For some common moth species, temperatures of 10 °C (50 °F) are the threshold for flight. At the onset of fall and winter in either

Fig. 2.21 Fronts and air masses associated with a strong low-pressure system in the south-central USA

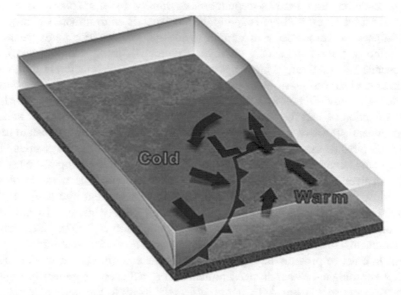

Fig. 2.22 Three-dimensional diagram of the air flow around a low-pressure system. http://www. srh.noaa.gov/srh/jetstream/synoptic/cyclone.htm

hemisphere, flying biota can move toward the equator. Several bird species are known to fly between the Northern and Southern Hemispheres for mating and to return as cold weather develops (Newton 2008). Most long-distance movement occurs in the spring as conditions at the earth surface begin to warm and allow for survival at points further poleward. It is during this time that surface low-pressure systems intensify, resulting in strong jet streams aloft and in the atmospheric boundary layer. The presence of low-level jet streams helps to maximize the potential migration displacements of all flying animals poleward where food sources are growing.

4   Mesoscale

The mesoscale encompasses the atmospheric depth and atmospheric processes that prominently impact migratory flights of animals.

4.1   Measurement of Thermodynamic and Wind Fields

We describe atmospheric measurements in the mesoscale section because of predominant flight activity within this atmospheric scale, and the same measurements can also be usefully applied to microscale and macroscale motions. Significant advances in atmospheric profiling instrumentation, widespread deployment of instrumentation, and increased public accessibility to atmospheric data have revolutionized the capability to explore the atmospheric environment and processes that impact the aeroecology of animal species of interest. For many years, the theodolite (a three-dimensional surveying telescope), pilot balloon (pibal), and radiosonde (alternatively named rawinsonde when wind velocity is derived by tracking the balloon) represented the principal equipment used to measure vertical profiles of barometric pressure, air temperature, relative humidity, wind speed, and wind direction. When the theodolite is used to track a pibal without an attached radiosonde, a profile of wind velocity can be computed based on a constant ascent rate for a pibal of known buoyancy and size. Detailed atmospheric profiles of the atmospheric boundary layer can be acquired by increasing the sampling rate of the radio receiver monitoring data transmitted by radiosondes. If a rawinsonde is tracked by radar, a vertical profile of the atmosphere is measured to the highest mandatory pressure altitude of 100 hPa (100 mb) (approximately 16 km MSL) (National Oceanic and Atmospheric Administration 1997). Rawinsonde measurements for operational monitoring networks require maximum vertical extent in order to provide atmospheric data for general circulation models. However, aeroecological research measurements using radiosondes generally limit the vertical extent to <3 km AGL but significantly increase the vertical resolution. Further, operational rawinsonde measurements are typically scheduled every 12 h and may not reveal significant evolution of the atmospheric boundary layer such as during the crepuscular periods of dusk and dawn when bats, noctuid insect species, and other animals are flying. The incorporation of global positioning satellite (GPS) into radiosonde systems obviates the need for theodolite or radar tracking of pibals with attached radiosondes.

In lieu of using rawinsondes to measure vertical profiles of wind velocity, remote sensing systems are currently available. For example, wind profilers and Doppler weather radars remotely detect wind velocities and create vertical profiles of wind velocity but use different detection and analysis techniques. Radar wind profilers transmit and receive pulsed radio waves in three directions: vertically and at 75° above ground along the east–west axis and along the north–south axis from which to compute three-dimensional wind velocities (Mitchell et al. 1995). Doppler weather radars transmit pulsed (and polarized) radio waves at a prescribed set of elevation angles and rotate antennas 360° to obtain radial components (i.e., along the radar beam) of wind velocity (Klazura and Imy 1993). Horizontal wind velocity values can be estimated as the mean of the azimuthal set of radial components of wind velocity. For both wind profilers and Doppler weather radars, the presence of flying animals confounds the measurement of wind velocity, but these remote sensing systems fortuitously measure the combined wind velocity and animal flight velocity (i.e., resultant ground velocity vector).

Kites and tethered blimps have been used as "sky hooks" to position radiosondes at selected heights and to collect series of measurements within a vertical profile. Further, tethered instrumentation platforms may carry aeroecological monitoring equipment such as cameras, ultrasonic radio detectors, and insect collection nets (McCracken et al. 2008; Krauel et al. 2015). A fixed frame of reference of tethered systems supports collection of local time series of data, but the relatively high speed (i.e., resultant ground speed) of flying animals can impose challenges for effective biological measurements of these animals.

Superpressure mylar balloons (e.g., tetrahedral-shaped balloons or tetroons) are neutrally buoyant at a selected altitude (or more accurately constant-density surface) based on the displacement volume and buoyant lift at the surface. Tetroons maintain altitude along an equilibrium atmospheric density surface. Because tetroons drift with the wind, they provide a moving frame of reference of the wind field that approximates the flight trajectory of migrating animals (Westbrook et al. 1995), and which results in minimum speed of flying animals relative to the tetroon. Similar to the leveraged benefit of tethered platforms, tetroons can carry aeroecological monitoring equipment such as cameras and ultrasonic radio detectors. When atmospheric profiling instrumentation is unavailable, publically-available atmospheric data sets (Table 2.2) can be accessed to reveal details about the dynamic atmospheric environment that are relevant to animal flight.

4.2 Modeling of the Atmosphere and Animal Flight Displacements

Simulation models (Table 2.3) and forecasts of atmospheric structure and motions have improved significantly with increased computing power. However, inadequate parameterization and modeling of atmospheric processes continue to introduce errors in estimated profiles of temperature, humidity, and wind velocity values, notably in stable atmospheric boundary layers (Holtslag et al. 2013).

Modeling of airborne displacements of flying animals has typically been simulated by trajectory or dispersion analysis. Trajectories describe a locus of points along the path of movement, with displacement from point to point based

Table 2.2 Atmospheric data

Name	Description	Scales	URL
National Climate Data Center (USA)	Land-based, marine, model, radar, weather balloon, and satellite datasets	Microscale, mesoscale, macroscale	http://www.ncdc.noaa.gov/data-access
Weather Underground	Land-based, marine, model, radar, weather balloon, and satellite datasets	Microscale, mesoscale, macroscale	http://www.wunderground.com
Unisys Weather data	Land-based, marine, model, radar, weather balloon, and satellite datasets		http://Unisys.weather.com
Bureau of Meteorology	Land-based, marine, model, radar, weather balloon, and satellite datasets		http://www.bom.gov.au/australia/charts/
National Center for Environmental Prediction	Model output		http://www.ncep.noaa.gov
European Center for Data	Model output		www.ecud.eu/download/ensembles/ensemble.php/dailydata/index.php
European Center for Medium Range Weather Forecasting	Model output		www.ecmwf.int/en/research/climate-reanalysis/browse-reanalysis-datasets
Accuweather	Land-based, marine, model, radar, weather balloon, and satellite datasets		http://www.accuweather.com
Weather online	Model output		http://www.weatheronline.co.uk/cgi-bin/expertcharts

Table 2.3 Atmospheric models

Name	Description	Scales	URL
HySPLIT	Atmospheric trajectory and dispersion model	Mesoscale to macroscale	http://ready.arl.noaa.gov/hysplit.php
ECMWF	European Global Spectral Model	Mesoscale to macroscale	http://www.weatheronline.co.uk/cgi-bin/expertcharts
EURO4	Global Weather Model from UKMET Office, North Atlantic European Model	Mesoscale to macroscale	http://www.weatheronline.co.uk/cgi-bin/expertcharts
EURO4h	High Resolution version of EURO4 Model	Mesoscale to macroscale	http://www.weatheronline.co.uk/cgi-bin/expertcharts
UKMET	United Kingdom Meteorology Office Unified Prediction Model	Mesoscale to macroscale	http://www.weatheronline.co.uk/cgi-bin/expertcharts
GME	German Weather Service (Global Weather Forecast Model)	Mesoscale to macroscale	http://www.weatheronline.co.uk/cgi-bin/expertcharts
RHMC	Global Weather Forecast Model from the Hydrometeorological Center of Russia	Mesoscale to macroscale	http://www.weatheronline.co.uk/cgi-bin/expertcharts
Unified	YRNO-MET Norway Global Model	Mesoscale to macroscale	http://www.weatheronline.co.uk/cgi-bin/expertcharts
HIRLAM	YRNO-MET Norway Global.Model—Highres	Mesoscale to macroscale	http://www.weatheronline.co.uk/cgi-bin/expertcharts
GSM	Japanese Global Spectral Model	Mesoscale to macroscale	http://www.jma.go.jp/jma/en/Activities/nwp.html
MSM	Japanese Mesoscale Model	Mesoscale	http://www.jma.go.jp/jma/en/Activities/nwp.html
LFM	Japanese Local Forecast Model	Mesoscale	http://www.jma.go.jp/jma/en/Activities/nwp.html

(continued)

Table 2.3 (continued)

Name	Description	Scales	URL
GEM	Global Environmental Multiscale model	Mesoscale to macroscale	http://weather.gc.ca/model_forecast/global_e.html
GRAPeS	Mainland China—Global and Regional Assimilation and Prediction System (in development)	Mesoscale to macroscale	
ACCESS-G	Australian Global Forecast Model	Mesoscale to macroscale	http://www.bom.co.au
ACCESS-R	Australian Regional Forecast Model	Mesoscale to macroscale	http://www.bom.co.au
GFS	NCEP Global Forecast System	Mesoscale to macroscale	http://www.ncep.noaa.gov
GFS .25°	NCEP Global Forecast System .25° resolution	Mesoscale to macroscale	http://www.ncep.noaa.gov
NAM	NCEP North America Mesoscale Model	Mesoscale to macroscale	http://www.ncep.noaa.gov
COAMPS[R]	Naval Research Laboratory Coupled Ocean/Atmosphere Mesoscale Prediction System	Mesoscale to macroscale	http://www.weatheronline.co.uk/cgi-bin/expertcharts
CFS	NCEP Climate Forecast System	Macroscale	http://www.ncep.noaa.gov
NGM	NCEP Limited Area Nested Grid Model	Mesoscale to macroscale	http://www.ncep.noaa.gov
RUC	NCEP Rapid Update Cycle model	Mesoscale to macroscale	http://www.ncep.noaa.gov

Fig. 2.23 Vector addition of wind velocity and flight velocity

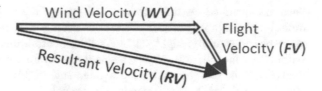

on the speed and direction of movement. Horizontal velocity is a vector that is comprised of magnitude (speed) and direction (relative to north). The horizontal velocity vector can be decomposed into two orthogonal unit vectors, each multiplied by the wind speed component in the respective orthogonal directions (west-to-east and south-to-north). The vector sum of wind velocity (*WV*) and flight velocity (*FV*) is termed the resultant (or ground-relative) velocity (*RV*) (Fig. 2.23), which is calculated as the sum of the corresponding zonal (westerly) and meridional (southerly) wind velocity components. Trajectories are generally modeled at a specific altitude (Chapman et al. 2010; Westbrook et al. 1997) or atmospheric level (e.g., isobaric or isentropic). Dispersion analysis is similar to trajectory analysis but adds the feature of three-dimensional spread about the trajectory centerline (Westbrook et al. 2011). While trajectory models can estimate the mean path and arrival time of an air parcel or group of migrating animals, dispersion models can also estimate the three-dimensional concentration of migrating animals. Both trajectory and dispersion models generally are used to estimate continuous animal migration. However, trajectory and dispersion models can also estimate discontinuous (e.g., layover stops for flying animals to obtain food, water, and shelter from predators) animal flight. Sufficiently long flight times may lead to trajectories and dispersion patterns that bridge scales of motion.

4.3   Atmospheric Boundary Layer Temperature Lapse Rates

The vertical profile of air temperature evolves through the diurnal and nocturnal periods. After sunrise, solar insolation heats the surface which erodes the surface stable layer. During the day, the surface layer heats and mixes through an increasing vertical depth. In the summer, atmospheric mixing can increase the depth of the atmospheric boundary layer to an altitude exceeding 1000 m above ground level (AGL). However, the presence of subsiding (warming) air within a high-pressure cell suppresses vertical mixing, and creates a subsidence inversion that caps the maximum depth of the daytime atmospheric boundary layer. The temperature profile of the atmosphere is described by the lapse rate, which is defined as the rate of decrease of temperature with height. The standard lapse rate of the troposphere is 6.5 °C km^{-1} (American Meteorological Society 2015). We introduce the concept of an adiabatic process in which an air parcel expands (cools) or compresses (warms) as it rises or descends, respectively, with no exchange of heat with the surrounding atmosphere (American Meteorological Society 2015). The dry adiabatic lapse rate (9.8 °C km^{-1}) applies to the rate of temperature change within unsaturated air parcels that are ascending or descending. After an air parcel becomes saturated, vertical motions will change the air parcel temperature at the moist adiabatic lapse rate (5.5 °C km^{-1} on average). There are four broad classes of atmospheric stability: highly stable, stable, neutral, and unstable. Various layers within the atmosphere are termed stable if temperature decreases with height at less than the bulk environmental lapse rate, and highly stable if the temperature increases with height (i.e., inversion). Although quite rare, neutral stability occurs when the lapse rate equals the adiabatic lapse rate. Any vertical force on air parcels within a neutral layer will cause air parcels to rise (e.g., from mechanical lifting of air flowing over mountains) or descend (e.g., from downdrafts) without restraint. Unstable atmospheric layers occur when the lapse rate exceeds the adiabatic lapse rate, leading to thermal convection (Haltiner and Martin 1957). Unstable atmospheric layers tend to be transient because vertical convection is induced, which leads to atmospheric mixing and erosion of the unstable lapse rate.

A description of the vertical profile of air temperature and absolute humidity is critical to understanding the current state of the atmosphere. Such a description is presented by the skew-T log-p diagram (Fig. 2.24), which plots the dry-bulb air

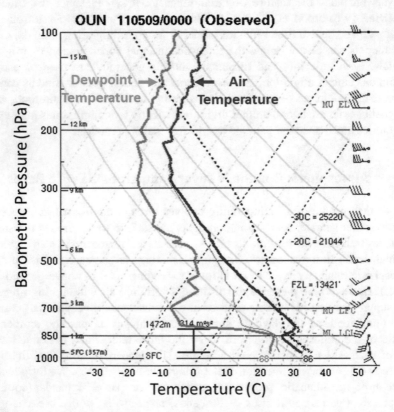

Fig. 2.24 Skew-T diagram depicting vertical profiles of air temperature, dew point temperature, and wind velocity

temperature and dew point temperature relative to a logarithmic scale of barometric pressure. The skew-T log-p diagram effectively reveals the relative stability or instability of atmospheric layers and identifies the altitude at which convection will result in condensation and cloud formation. Further, wind velocity vectors are plotted along a vertical margin of the skew-T log-p diagram to indicate vertical profiles of wind speed, wind direction, and wind shear. In meteorological convention, wind velocity vectors are identified as arrows pointing in the direction of flow, and for which speed is denoted by the sum value of barbs (i.e., a half-barb = 5 knots and a full barb = 10 knots) and pennants (i.e., 50 knots) on the arrow shaft. However, the skew-T log-p diagram is generally based on national networks of rawinsonde observations that measure atmospheric conditions up to and above 100 hPa (100 mb), and additional measurements (e.g., by low-level rawinsondes and wind profilers) are needed to detail the vertical profile of barometric pressure, air temperature, absolute humidity, and wind velocity within the atmospheric boundary layer.

4.4   Atmospheric Boundary Layer Wind Velocity Profiles

Horizontal air temperature gradients create gradients of barometric pressure and air density which, in turn, lead to atmospheric motions. Above a smooth surface, atmospheric motions may be laminar and follow the contour of the surface. However, the earth surface is rough, especially in forest, agricultural, and urban environments. Frictional forces caused by atmospheric motions over rough surfaces will reduce near-surface wind speed and generate turbulence. Frictional effects diminish at increasing altitude and consequently affect both wind speed and wind direction. Wind direction tends to rotate within the vertical profile owing to the reduced frictional effects. This feature of the vertical profile of wind direction is known as the Ekman layer (spiral layer). The rotation of wind direction within the atmospheric boundary layer does not exceed 45° (Hess 1959). The Ekman spiral is described by wind direction that veers clockwise as altitude increases in the Northern Hemisphere and counterclockwise in the Southern Hemisphere. The rotation of wind velocity within the atmospheric boundary layer can be displayed as a hodograph by mapping wind velocity vectors from a common origin (Fig. 2.25). The absence of frictional force at the top of the Ekman spiral results in the attainment of a wind velocity vector (known as the geostrophic wind vector) for which the Coriolis acceleration balances the horizontal pressure force, and the wind flows along the contour lines of an isobaric surface at a speed that is directly proportional to the contour gradient (American Meteorological Society 2015).

In addition to thermally induced changes in the depth and wind velocity profile in the atmospheric boundary layer, overlaying of air parcels from diverse sources (e.g., maritime or continental) can create or enhance the strength of temperature inversions and vertical shear (i.e., speed and/or direction) of the wind velocity profile.

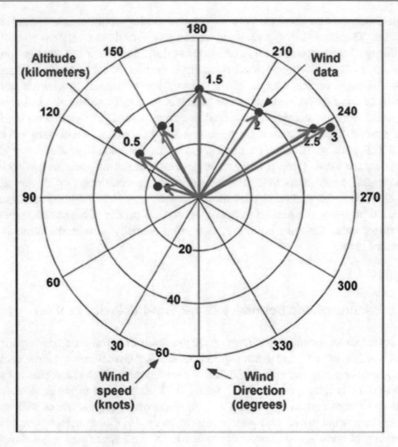

Fig. 2.25 Hodograph illustrating wind shear within the atmospheric boundary layer. http://www.crh.noaa.gov/Image/lmk/Hodographs_Wind-Shear.pdf

4.5   Episodic Atmospheric Boundary Layer Wind Features

Macroscale flow above the atmospheric boundary layer is not affected by surface friction. After sunset, a low-altitude temperature inversion often develops between approximately 200 and 600 m AGL that reduces vertical mixing. Immediately above the low-altitude temperature inversion, the wind speed accelerates as frictional effects are suppressed. The wind maximum in the atmospheric boundary layer is alternatively referred to as the low-level wind jet, as the low-level jet (LLJ), and frequently as the low-level nocturnal jet (Bonner 1968). Low-level jets frequently occur at many geographic locations and are named to identify these specific circulations. Migration of many animal species has been associated with the Great Plains LLJ in the central USA, which often extends for 3000 km from southern Texas to northern Minnesota. The Caribbean LLJ (Muñoz et al. 2008), South American LLJ (Vera et al. 2006), Somali LLJ (Ardanuy 1979), and many other LLJ are distributed globally (Stensrud 1996).

Cold air descending from the top of thunderstorms is known as thunderstorm outflow, which can create gust fronts that propagate away from the thunderstorm. Gust fronts are often identified by a narrow band of clouds that can be detected by satellite and radar. Atmospheric convergence along the gust front can concentrate flying animals as the gust front moves for tens or hundreds of kilometers. Flying animals that are caught in the gust front circulation may be unable to escape the gust front before it dissipates.

A phenomenon known as a dry line generates effects similar to those of thunderstorm outflow (Geerts 2008). The dry line boundary separates moist and dry air masses In the USA, the dry line typically orients from south to north during the spring and early summer, where it separates moist air from the Gulf of Mexico (to the east) and dry desert air from the southwestern states (to the west). The dry line typically moves eastward during the afternoon and returns westward at night. A typical dry line passage results in a sharp drop in humidity, clearing skies, and a wind shift from southerly or southeasterly to westerly or southwesterly. Rising temperatures also may follow, especially if the dry line passes during the daytime. These changes occur in reverse order when the dry line retreats westward. Severe and sometimes tornadic thunderstorms often develop along a dry line or in the moist air just to the east of it, especially when it begins moving eastward.

4.6   Geographically-Fixed Boundary Layer Flows

Regions with high contrast in surface heat capacity and topographic features develop distinct atmospheric circulations that occur repeatedly and are often defined as a quasi-stationary phenomenon. Contrasts in heat capacity of land and water surfaces often lead to the development of land-sea breeze or land-lake breeze circulations. Diurnal insolation and heating of land generates thermal convection which lowers the surface barometric pressure. Cool air from above the water surface displaces toward the surface low pressure over land. The rising warm air over land spreads laterally and descends over water to close the sea-breeze circulation. Differential nocturnal cooling reverses the circulation and forces a surface land breeze with return flow aloft.

Mountainous regions profoundly influence atmospheric circulations within the atmospheric boundary layer and mixing with the macroscale atmospheric circulation aloft. Greater cooling of mountain slopes can generate substantial horizontal gradients of air temperature and horizontal barometric pressure gradients that lead to a mountain breeze of cool downslope (katabatic) wind (Fig. 2.26). For example, as surface air along the mountain slopes cools by nocturnal radiative heat transfer or by contact with snow and ice, air becomes more dense and descends within a shallow layer along the mountain slope. Conversely, (anabatic) air flow rises as a valley breeze along mountain slopes when heated by insolation or contact with a relatively warm surface (Fig. 2.26). The warm surface air along the slope ascends as thermals, causing a horizontal pressure gradient from the valley toward the mountain slopes. The pressure gradient forces air up the mountainous slope. The speed of

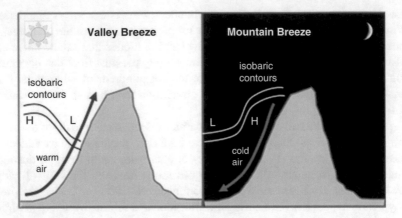

Fig. 2.26 Vertical cross-section of valley- and mountain-breeze circulations

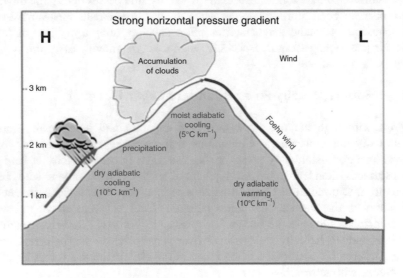

Fig. 2.27 Diagram of foehn wind process

anabatic wind is often less (3–5 m s^{-1}) than that of katabatic wind and extends farther above the slope due to convective mixing. When a mountain valley is capped by a stable atmospheric layer, a mountain-valley air circulation rotor may develop in which cross-valley wind at the surface reverses direction aloft similar to that of land-sea breeze circulations. However, if the upper-level atmospheric circulation aligns with the longitudinal axis of the mountain valley and the atmospheric stability within the valley decreases, the mountain-valley rotor will dissipate and surface wind will flow along the longitudinal axis of the valley.

Foehn winds are warm, dry downslope winds that develop when a strong macroscale gradient of barometric pressure forces air flow over a mountain range. Examples of foehn wind (Fig. 2.27) are those in the Alps Mountains of central

Europe, as well as the Santa Ana wind in the Sierra Nevada mountains of southern California (USA), and the chinook wind in the Rocky Mountains of the USA and Canada. Foehn winds can accelerate to high speeds and rapidly warm the air and can rapidly sublimate snow, evaporate water, and dessicate plant material.

5   Microscale

The microscale atmospheric environment is typically confined to the surface boundary layer of the atmosphere where surface friction and thermal effects dominate, but microscale atmospheric features are also important between distinct atmospheric layers such as at the top of the convective atmospheric boundary layer where layers of small insects may be concentrated (Eaton et al. 1995). Microscale measurements are commonly acquired by automated meteorological stations on the ground or instrumented towers, by instrumented neutrally buoyant balloons, or by instrumented aircraft, all of which can record air temperature, absolute humidity, barometric pressure, and three-dimensional wind velocity at a fast sampling rate (<1 s^{-1}).

Heat transfer between the earth surface and the surface boundary layer occurs via radiative and conductive processes. During daytime, solar insolation heats the surface according to the solar angle, cloud cover, surface albedo, and surface heat capacity. During nighttime, the cloud cover and surface heat capacity modify the rate of surface cooling. These factors lead to nonuniform heating and cooling of the surface.

Heterogeneous thermal distributions of the surface create air density gradients that induce vertical and horizontal atmospheric motions. The surface lapse rate varies widely during the day and night. A strong (unstable) temperature lapse rate often develops during the day, and strong temperature inversions develop under clear skies at night. During daytime, land surfaces warm quickly causing rising motions (thermals), but water surfaces warm much slower. Conversely, land surfaces cool quickly causing descending motions, and water surfaces remain at nearly constant temperature. Another aspect of the transient nature of the heterogeneous heating and cooling is the development of variable wind speed (i.e., gusts) and direction.

Thermal properties of the surface and thermodynamic processes within the mixed layer cause microscale motions that lead to atmospheric mixing. The mixing generates turbulence that cascades from (i.e., erodes) larger eddies to smaller eddies. At the surface, kinetic energy is dissipated through friction. Similar exchange of turbulent kinetic energy occurs at the interface of distinct atmospheric layers. Thermals create convective air columns that may displace downwind and generally persist and regenerate for periods on the order of minutes, providing lift for soaring flight of animals including Monarch butterflies (Gibo and Pallett 1979).

Disclaimer Mention of trade names or commercial products in this article is solely for the purposes of providing specific information and does not imply recommendation or endorsement by the U.S. Department of Agriculture.

References

American Meteorological Society (2015) Adiabatic; geostrophic wind; standard atmosphere. Glossary of Meteorology. Available online at http://glossary.ametsoc.org/wiki/"standard atmosphere"

Ardanuy P (1979) On the observed diurnal oscillation of the Somali Jet. Mon Weather Rev 107: 1694–1700

Bonner WD (1968) Climatology of the low level jet. Mon Weather Rev 96:833–850

Chapman JW, Nesbit RL, Burgin LE, Reynolds DR, Smith AD, Middleton DR, Hill JK (2010) Flight orientation behaviors promote optimal migration trajectories in high-flying insects. Science 327:682–685

Dowling M (2014) Monsoon at www.mrdowling.com. Updated 30 Dec 2014. Web. Downloaded 25 Feb 2015. http://www.mrdowling.com/612-monsoon.html

Drake VA, Gatehouse AG (1995) Insect migration: tracking resources through space and time. Cambridge University Press, Cambridge, 478 pp

Drake VA, Reynolds DR (2012) Radar entomology: observing insect flight and migration. CABI, Oxfordshire, 489 pp

Draxler RR, Hess GD (1998) An overview of the HYSPLIT_4 modeling system of trajectories, dispersion, and deposition. Aust Meteorol Mag 47:295–308

Eaton FD, McLaughlin SA, Hines JR (1995) A new frequency-modulated continuous wave radar for studying planetary boundary layer morphology. Radio Sci 30:75–88

Geerts B (2008) Dryline characteristics near Lubbock, Texas, based on radar and West Texas Mesonet data for May 2005 and May 2006. Weather Forecast 23:392–406

Gibo DL, Pallett MJ (1979) Soaring flight of monarch butterflies, *Danaus plexippus* (Lepidoptera: Danaidae) during the late summer migration in southern Ontario. Can J Zool 57:1393–1401

Haby J (2015) Atmospheric blocking. http://www.Theweatherprediction.com/blocking/: Blocking Pattern Images

Haltiner GJ, Martin FL (1957) Dynamical and physical meteorology. McGraw-Hill Book, New York, 470 pp

Haltiner GJ, Williams RT (1980) Numerical prediction and dynamic meteorology. Wiley, Hoboken, 477 pp

Hess SL (1959) Introduction to theoretical meteorology. Holt, Rinehart and Winston, New York, 362 pp

Holtslag AAM, Svensson G, Baas P, Basu S, Beare B, Beljaars ACM, Bosveld FC, Cuxart J, Lindvall J, Steeneveld GJ, Tjernström VDWBJH (2013) Stable atmospheric boundary layers and diurnal cycles: challenges for weather and climate models. Bull Am Meteorol Soc 94: 1691–1706

Isard SA, Gage SH (2001) Flow of life in the atmosphere: an airscape approach to understanding invasive organisms. Michigan State University Press, East Lansing, 240 pp

Kim KS, Jones GD, Westbrook JK, Sappington TW (2010) Multidisciplinary fingerprints: forensic reconstruction of an insect reinvasion. J R Soc Interface 7:677–686

Klazura GE, Imy DA (1993) A description of the initial set of analysis products available from the NEXRAD WSR-88D system. Bull Am Meteorol Soc 74:1293–1311

Krauel JJ, Westbrook JK, McCracken GF (2015) Weather-driven dynamics in a dual-migrant system: moths and bats. J Anim Ecol 84:604–614

Kunz TH, Gauthreaux SA Jr, Hristov NI, Horn JW, Jones G, Kalko EKV, Larkin RP, McCracken GF, Swartz SM, Srygley RB, Dudley R, Westbrook JK, Wikelski M (2008) Aeroecology: probing and modeling the aerosphere. Integr Comp Biol 48:1–11

Madden R, Julian P (1971) Detection of a 40-50 day oscillation in the zonal wind in the Tropical Pacific. J Atmos Sci 28:702–708

Madden R, Julian P (1972) Description of global-scale circulation cells in the tropics with a 40-50 day period. J Atmos Sci 29:1109–1123

McCracken GF, Gillam EH, Westbrook JK, Lee Y-F, Jensen ML, Balsley BB (2008) Brazilian free-tailed bats (*Tadarida brasiliensis*: Molossidae, Chiroptera) at high altitude: links to migratory insect populations. Integr Comp Biol 48:107–118

Mitchell MJ, Arritt RW, Labas K (1995) A climatology of the warm season great plains low-level jet using wind profiler observations. Weather Forecast 10:576–591

Muller RA (1977) Synoptic climatology for environmental baseline analysis: New Orleans. J Appl Meteorol 16:20–33

Muñoz E, Busalacchi AJ, Nigam S, Ruiz-Barradas A (2008) Winter and summer structure of the Caribbean low-level jet. J Clim 21:1260–1276

National Oceanic and Atmospheric Administration (1997) Rawinsonde and Pibal observations. Federal Coordinator for Meteorological Services and Supporting Research, Federal Meteorological Handbook No. 3, FCM-H3-1997, Washington, DC, 191 pp

Pedgley D (1982) Windborne pests and diseases: meteorology of airborne organisms. Ellis Horwood, Chichester, 250 pp

Pidwirny M (2006) Local and regional wind systems. Fundamentals of physical geography, 2nd edn. Date Viewed 3/11/2015. Monsoon Southeast Asia. http://www.physicalgeography.net/fundamentals/7o.html

Stensrud (1996) Importance of low-level jets to climate: a review. J Clim 9:1698–1711

Vera C, Baez J, Douglas M, Emmanuel CB, Marengo J, Meitin J, Nicolini M, Nogues-Paegle J, Paegle J, Penalba O, Salio P, Saulo C, Silva Dias MA, Silva Dias P, Zipser E (2006) The South American low-level jet experiment. Bull Am Meteorol Soc 87:63–77

Wallace JM, Hobbs PV (1977) Atmospheric science: an introductory survey. Academic Press, New York, 467 pp

Westbrook JK, Eyster RS, Wolf WW, Lingren PD, Raulston JR (1995) Migration pathways of corn earworm (Lepidoptera: Noctuidae) indicated by tetroon trajectories. Agric For Meteorol 73:67–87

Westbrook JK, Wolf WW, Lingren PD, Raulston JR, López JD Jr, Matis JH, Eyster RS, Esquivel JF, Schleider PG (1997) Early-season migratory flights of corn earworm (Lepidoptera: Noctuidae). Environ Entomol 6:12–20

Westbrook JK, Isard SA (1999) Atmospheric scales of biotic dispersal. Agric Forest Meteorol 97:263–274

Westbrook JK (2008) Noctuid migration in Texas within the nocturnal aeroecological boundary layer. Integr Comp Biol 48:99–106

Wolf WW, Westbrook JK, Raulston JR, Pair SD, Lingren PD (1995) Radar observations of orientation of noctuids migrating from corn fields in the Lower Rio Grande Valley. Southwestern Entomol Suppl 18:45–61

Extending the Habitat Concept to the Airspace

3

Robert H. Diehl, Anna C. Peterson, Rachel T. Bolus, and Douglas H. Johnson

Abstract

Habitat is one of the most familiar and fundamental concepts in the fields of ecology, animal behavior, and wildlife conservation and management. Humans interact with habitats through their senses and experiences and education to such a degree that their perceptions of habitat have become second nature. For this reason, it may be difficult at first to accept the airspace as habitat, an area that is invisible, untouchable, highly dynamic, and its occupants difficult to see. Nonetheless, the habitat concept, by definition and in practice, applies readily to the airspace. Some ecological and behavioral processes including habitat selection, foraging, and reproduction are operational in the airspace, while others, particularly those mediated by resource limitation such as territoriality, are likely uncommon if present at all. The behaviors of flying animals increasingly expose them to anthropogenic hazards as development of the airspace accelerates. This exacerbates the need to identify approaches for managing these human–wildlife conflicts in aerial habitats, especially where human safety or at-risk populations are concerned. The habitat concept has proven useful in shaping environmental law and policy to help

R.H. Diehl (✉)
US Geological Survey, Northern Rocky Mountain Science Center, Bozeman, MT, USA
e-mail: rhdiehl@usgs.gov

A.C. Peterson
Department of Fisheries, Wildlife and Conservation Biology, University of Minnesota, Saint Paul, MN, US
e-mail: Pete1112@umn.edu

R.T. Bolus
Department of Biology, Southern Utah University, Cedar City, UT, USA
e-mail: rachelbolus@suu.edu

D.H. Johnson
US Geological Survey, Northern Prairie Wildlife Research Center, Jamestown, ND, USA
e-mail: douglas_h_johnson@usgs.gov

© Springer International Publishing AG, part of Springer Nature 2017
P.B. Chilson et al. (eds.), *Aeroecology*,
https://doi.org/10.1007/978-3-319-68576-2_3

mitigate these conflicts. It remains to be seen whether current law can bend to include a more expansive concept of habitat that includes the airspace.

1   What Is Habitat?

Consider an organism moving through a vast and apparently featureless yet dynamic fluid medium. Its behavioral and physical phenotype is designed for life in this medium, and its survival contingent upon its ability to seek out favorable currents in which to move about, find food, sleep, and avoid predators. This might describe the rudimentary natural history of countless pelagic fish species and we would not hesitate to call the open water they occupy a habitat. This might also describe the natural history of numerous bird species and the airspaces they occupy, yet most would claim a bird in flight is not in habitat. The notion of habitat is firmly linked to terrestrial or aquatic places, yet nothing about the ecology of habitats excludes the airspace which shares many ecological parallels with its terrestrial and aquatic counterparts.

Ecologists have struggled to define habitat. Linnæus (1754), his mind on plants, informally referred to "the native places" (where one finds a given species of plant) as "station," his nascent term for habitat. This concept was developed with principles that apply to sessile plants and which have continued to influence scientific understanding of habitat; however, highly mobile organisms defy our attempts to discretely categorize their habitats by plant communities (e.g., forest, grassland) and/or geological formations (e.g., river, alpine). Linnæus' general perception of habitat remains widely accepted today, even as a persistent dissatisfaction surrounds a precise definition. Odum's (1959) definition of habitat as "the place an organism lives, or where one would go to find it" is perhaps the most familiar to ecologists. A few years later, perhaps dissatisfied with this and other definitions, Elton (1966) stated that, "definition of habitat, or rather lack of it, is one of the chief blind spots in zoology," and efforts to define it continue. Many of the more generalizable definitions of habitat offer some reference to occupancy or resources that promote occupancy (e.g., Hall et al. 1997; Dennis et al. 2003, 2006; Morrison et al. 2006). Of these, Morrison et al.'s is among the more precise and generalizable, calling habitat an "area with a combination of resources and environmental conditions that promotes occupancy by individuals of a given species that allows individuals to survive and reproduce." More context specific definitions usually involve additional criteria and represent special cases and are therefore less broadly applicable to the life history of organisms as a whole (e.g., by requiring the presence of vegetation). The airspace satisfies criteria set forth in these and other broadly defined conceptions of habitat in that it contains resources that promote occupancy. These include food, mates, opportunities to avoid predators, and a low friction, relatively uncluttered medium enabling efficient travel. The idea that the airspace should be considered habitat (Diehl 2013) is gaining acceptance and is a fundamental tenet of the emerging field of aeroecology (Kunz et al. 2008).

It was unclear how long airborne resources could sustain life during flight, particularly of a vertebrate, until Alpine and Common Swifts (*Tachymarptis melba* and *Apus apus*) were documented remaining aloft for several months without landing (Liechti et al. 2013; Hedenström et al. 2016). The airspace alone provided all the resources required to sustain the swift's biology, including food and sleep. Much as a terrestrial amphibian uses the water only to breed, these swifts appear to need land only to support their nests and eggs. (The word "amphibious" means "leading a double life" in Greek, and had early taxonomists not applied this word exclusively to animals that exist between the land and water, we might also label swifts as "amphibious.") In addition, recent research shows that the microscopic biological flotsam of the air is composed not only of inert cells and bits of feathers but also of life: pollen, bacteria, fungi, and protists (Deprés et al. 2012). The airspace is teeming with microbial communities that spend generations aloft before settling back to the surface after metabolizing, undergoing selection, and reproducing (Womack et al. 2010). Such phenomena challenge biologists to accept that the airspace is habitat. The alternative—that such organisms are sustained through large proportions of their lives outside habitat—suggests that the current habitat model be rejected altogether, although it is unclear what might replace it (Dennis et al. 2003). A revised understanding of habitat that includes the airspace offers a new context to consider the habitat concept. Relative to its terrestrial and aquatic counterparts, the airspace is highly dynamic yet presumably lacking in much of the structure and organizational complexity that comes with higher plant and animal abundances, species diversity, and physical substrates typically associated with habitat.

Despite a strong ecological basis that the airspace should be considered habitat, the notion remains counterintuitive and references to the airspace as habitat are vague where they exist at all (e.g., Cadenasso et al. 2003). Through formal education and personal experience, we learn about habitat strictly in terms of land and water. Indeed, among all vertebrates and macroinvertebrates, humans and a relative handful of other species are actually unusual for their inability to fly. Our perceptions shape how we understand and apply the habitat concept in ways that may have consequences for wildlife conservation and management and the education of future generations.

Early hunter-gatherer societies likely cultivated a deep understanding of the natural history of plants and animals, so the awareness of habitat (in the sense that certain resources can be found in certain places at certain times) predates scientific recognition and definition by millions of years. The knowledge of when and where to gain access to resources was critical to survival throughout evolutionary history and therefore linked to resources that were accessible (that is, by land and water, but seldom through the air). As such, perceptions of habitat may be deeply ingrained, possibly even codified in our DNA, and indeed, a genetic basis for habitat preference is well documented among some species (Jaenike and Holt 1991).

Evolution shaped our senses for survival on land and water to a degree that almost certainly influences our perceptions of what constitutes habitat, so considering the airspace as a habitat poses several challenges. The airspace is largely invisible as are many of the features that give it structure (e.g., wind, temperature, pressure, geomagnetism). Even those features which are detectable to humans through nonvisual channels (e.g., wind) likely do not carry the same meaning as they do with flying animals. We lack certain sensory receptive organs (e.g., cilia) and the neurological infrastructure that interprets signals from those organs. Moreover, certain characteristics of aerial habitat seem to defy common sense. That a habitat could be invisible or is capable of being present one moment and gone the next is far outside most personal and professional experiences. For example, the convective thermals used by raptors represent aerial habitats that are both invisible and highly ephemeral. Despite these limitations, extensions to our senses offered by access to the airspace though manned and unmanned aircraft and advances in remote sensing technology are beginning to reveal the intricacies of an ecological airspace (Helms et al. 2016; Bridge et al. 2011; Robinson et al. 2009).

2   The Nature of Aerial Habitat

The term "aerial habitat" has been used more in reference to plants than flying animals, specifically in relation to the phyllosphere where the ecology of epiphytes and other arboreal organisms plays out, usually high in the trees (e.g., Lindow and Brandl 2003). What constitutes aerial habitat in the context of aeroecology remains a somewhat open question. In atmospheric terms, aerial habitat includes those parts of the troposphere where aerobiological activity occurs, within the planetary boundary layer and lower parts of the free atmosphere (Davy et al. 2017). Typically, most activity occurs within the first several hundred meters above ground level. In biological terms, aerial habitat occurs when and where an organism's presence indicates use of an airborne resource (e.g., for food or as a movement corridor).

Many properties of the airspace give rise to variability in the structure of aerial habitat. These include wind speed and direction and their diverse manifestations (e.g., convection, shear, turbulence); various forms of moisture including humidity and cloud; precipitation; temperature; pressure and air density (see Chap. 2); magnetic and gravitational field strengths; odor; light conditions (e.g., polarized light, also Van Doren et al. 2017); and natural and anthropogenic gases, chemicals, and particulate matter. Other properties that may not be considered structural in relation to the airspace itself but are nonetheless influential include aerial predators and prey, topography, and anthropogenic structure (Lambertucci et al. 2015; Shipley et al. 2017). These characteristics influence the likelihood of occupancy for a given species within the occupiable airspace, which has vague upper and lower boundaries.

Animals occupying aerial habitats respond primarily to conditions aloft and are often behaviorally decoupled from land and water surfaces. However, at low altitudes, aerial habitat segues into terrestrial or aquatic habitat and animal behavior aloft is increasingly influenced by surface features. A large literature exists concerning the ecology of habitat interfaces, usually between patches of vegetation (Cadenasso et al. 2003). Where obvious transitions between the primary physical habitat types (land–water, land–air, air–water) have received some attention, the ecology of land–water interfaces dominates. Land–air and air–water interfaces are considered hard boundaries with comparatively abrupt transitions. Research on these boundaries focuses primarily on the physics and chemistry of gas exchange and microbial communities (Belnap et al. 2003) with little attention given to higher trophic levels. In these areas, the airspace may be a functional extension of surface terrestrial or aquatic habitat rather than aerial habitat *per se*, and time spent aloft by vertebrates and macroscopic invertebrates can be brief and usually occurs irrespective of conditions in the interstitial airspaces (e.g., nectar feeding bats, terrestrial territoriality in birds).

As altitude increases, conditions supporting buoyancy and powered flight diminish and the upper limit for a particular species is determined by the extent to which its behavior, anatomy, and physiology have adapted to decreasing air density, temperature, oxygen, and increasing radiation. Where these upper boundary conditions lie for most species is unknown owing to the difficulty of tracking small organisms. Winds are capable of lifting and dispersing invertebrates, seeds, and numerous microorganisms including pollen, spores, and fungi to altitudes exceeding 50 km (e.g., Drake and Reynolds 2012; Westbrook and Isard 1999; Womack et al. 2010). Among high-flying vertebrates, the bar-headed goose (*Anser indicus*) may be the most well-studied and most extreme. These geese routinely reach altitudes exceeding 6000 m when transiting the Himalayan Mountains of the Tibetan Plateau (Hawkes et al. 2011, 2013), although these geese and other species have been documented at higher altitudes, including a Ruppell's Vulture (*Gyps rueppellii*) which was struck by an aircraft at >11,000 m (Laybourne 1974). At 6000 m, air density and the partial pressure of oxygen is approximately half that at sea level under standard atmospheric conditions. Therefore, oxygen availability to power aerobic flight is considerably reduced just as the demands of flight increase as a result of reduced lift from low air density. Despite these constraints, bar-headed geese are not known to benefit from supportive winds in their trans-Himalayan migrations, instead relying on physiological adaptation to altitude and favoring calm atmospheric conditions (Hawkes et al. 2011).

3 Ecology of the Airspace

Terrestrial and aquatic habitats are home to a long list of ecological and behavioral processes including predation, competition, reproduction, and territoriality. The extent to which many of these processes are operational in the airspace is poorly known and, even if present, may be uncommon, ephemeral, or difficult to detect and observe. Certain behaviors are unambiguously at work including habitat selection, reproductive behavior, and predator–prey interactions. Others, especially those that partition resources such as exploitative competition and territoriality, even when operational, may be too short-lived or ambiguous to be useful.

Persistent resource defense territoriality generally occurs only when resources are defendable, limited, and predictable. In most cases, the energetic cost of occupying the airspace may be high relative to that in terrestrial and aquatic habitats as the airspace has no substrate and offers little unassisted buoyant support (at least for most vertebrates). Flying animals must power movement using their own energy reserves or by harvesting energy conveniently available from the wind (Sachs 2016; Pennycuick 1998), either of which limits the defendability of airborne resources. Also, most properties of the airspace that might serve as airborne resources (e.g., favorable wind, airborne invertebrates as food) may be both unpredictable and effectively unlimited at any given place and time. As a result, resource defense territoriality of airspaces is likely either uncommon or assumes an alternative form than that which has been previously recognized. For example, instead of defending a particular static area bounded in three dimensions, perhaps territoriality manifests as competition in space relative to other individuals (e.g., location in a flock or breeding swarm), which may be more spatially dynamic than the traditional view of a territory. Flying animals, particularly some invertebrates, do use the airspace to defend underlying terrestrial or aquatic habitats and their associated resources. We regard this use of airspace as an extension of surface territoriality as opposed to strict defense of an aerial resource.

Like other difficult-to-study habitats, the airspace offers a relatively unexplored context for examining the generality of ecological theory. If a given ecological process is present because the airspace and organism meet known criteria required for that process to be operational, then the durability of that concept is reinforced. Alternatively, an ecological process that is present despite the absence of expected criteria may call the generality of theory into question. Similar reasoning applies for ecological processes not present in the airspace. We discuss the occurrence of some of these ecological processes to emphasize the point that the airspace functions as habitat and that the ecology of the airspace in many ways differs little from the ecology of the land and the water.

3.1   Habitat Quality, Use, and Selection

Aerial habitats, like their terrestrial and aquatic counterparts, vary in quality in ways that influence survival and reproductive success (Jones 2001). Organisms react to this variation, although reactions differ among species. Microorganisms may be limited in their ability to "select" habitat, but distinct microbial communities exist in the airspace. Community composition is influenced by adjacent source terrestrial/aquatic habitats (Bowers et al. 2011; Brodie et al. 2007), weather (Harrison et al. 2005), and microhabitat differences in the airspace (Womack et al. 2010). Winds aerosolize microorganisms upward from the land (especially from plants; Vaïtilingom et al. 2012) or water and then can horizontally transport the larger microbes and pollen, sometimes for long distances (e.g., from the Sahara to the Alps; Meola et al. 2015), mixing communities. Once they have been passively dispersed, species differ in their ability to survive extreme temperature shifts, radiation, osmotic shocks, and desiccation (Joly et al. 2015), conditions that vary spatially and temporally (Womack et al. 2010). Eventually microorganisms return to earth due to wind drift, sedimentation, or precipitation (Deprés et al. 2012). Adaptations to the harsh life of the atmosphere (e.g., pigments to reduce the stress from radiation, cold shock proteins; Joly et al. 2015) have likely contributed to the successful inoculation of life in some of its most extreme aquatic and terrestrial habitats (e.g., polar ice sheets; Gonzalez-Toril et al. 2009).

In contrast to microorganisms' passive habitat use, flying animals often select habitats because they provide resources in support of some critical behavior such as foraging, reproduction, or migration. For example, migrating birds choose routes and weather conditions (e.g., Fig. 3.1) that allow them to manage the often conflicting goals of minimizing travel time, energy expenditure, and risk (Alerstam and Lindström 1990); these birds often respond to wind support more than temperature and humidity when selecting the routes and altitudes for their movements (Liechti et al. 2000). For instance, to assist their long-distance migrations over the ocean, bar-tailed godwits time departures from Alaska to coincide with weather likely to produce tailwinds (Gill et al. 2009). In contrast, some migratory insects choose their altitudes primarily based on atmospheric turbulence rather than tail wind assistance (Aralimarad et al. 2011). Diurnal soaring birds, such as the golden eagle, restrict their flight to a narrow band of low-altitude airspace as wind speeds increase (Lanzone et al. 2012). These and other forms of aerial habitat selection appear to be non-density dependent, since the resources being selected (e.g., favorable winds) are not thought to be limiting so as to create conditions for competitive exclusion.

A feature of habitat selection that distinguishes it from mere habitat use is that it results from the relationship between the costs and benefits of occupancy. For instance, during nocturnal migration, birds appear to select the lowest altitude where they first encounter supportive wind conditions even if "better" winds occur at higher altitudes (Mateos-Rodríguez and Liechti 2012). Presumably birds are unwilling to cross a cost barrier to seek out potentially more advantageous flying conditions suggesting that frequent altitude changes in search of optimal flying conditions may be costly (but see Bowlin et al. 2015).

3.2 Reproduction

The airspace provides many resources supporting reproduction, ranging from the concrete (space for asexual division, gamete dispersal, or copulation; protection from predators) to the abstract (a clear channel for communicating with potential mates). Many plants are pollinated through the action of the wind either directly or indirectly by affecting the flight paths of their pollinators. Bacteria have shorter reproductive cycles than their typical residency times in the air column and may have multiple generations between their ascents and descents (Burrows et al. 2009; Sattler et al. 2001).

Most aerial reproductive behavior in flying animals occurs among invertebrates. Insect mating swarms can take many forms. Under seasonally appropriate weather conditions, Diptera form highly localized mating swarms and may maintain

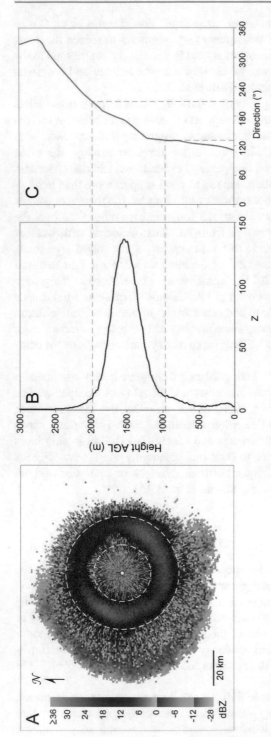

Fig. 3.1 (**a**) Weather radar reflectivity at 3.51° antenna elevation from Brownsville, TX (KBRO) showing layered bird migration on 12 May 2006, 2330 CST. A white dot shows the location of the radar. (**b**) The vertical profile of reflectivity computed from the radar reflectivity volume scan on this day, and the (**c**) associated wind profile showing the direction from which the wind was blowing that day at 1900 CST. The dashed circles in (**a**) represent 16 and 32 km range rings and correspond to the ~1000 and ~2000 m height AGL dashed lines, respectively, in (**b**) and (**c**)

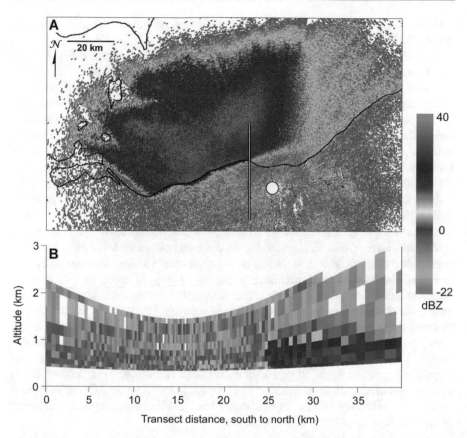

Fig. 3.2 Emergence of *Hexagenia* mayflies from western Lake Erie on 18 June 2015, 2204 CST as captured by Cleveland, Ohio weather radar, KCLE (white dot). Radar reflectivities (dBZ) show varying mayfly densities aloft. The line segment in the upper image (**a**) west of the radar identifies the south-to-north cross section through the radar's volume scan shown below (**b**)

consistent locations with respect to conspicuous "swarm markers" in the landscape (e.g., a geographic feature or patch of sunlight; Sullivan 1981). Alternatively, mating swarms may occupy enormous areas (Dublon and Sumpter 2014). The mayfly swarming event depicted in Fig. 3.2 occupies more than 3000 km^2 of Lake Erie's surface, at least 12% of its total area. Emergence generally occurs when both water temperatures exceed 20 °C (Corkum et al. 2006) and winds are moving toward the shore, as molting occurs on land after emergence; therefore, aerial habitat selection affects reproductive timing. Aerial reproductive swarming of this magnitude also serves as predator defense, as the massive pulse of food for local predators overwhelms the predator community to the point of satiation (Sweeney and Vannote 1982; but see Allan and Flecker 1989). Aerial copulation is common among invertebrates (e.g., Scott 1972; Allan and Flecker 1989), but appears to be rare among vertebrates. However, there is some evidence that red bats

(*Lasiurus borealis*) and hoary bats (*Lasiurus cinereus*) copulate in flight (Cryan and Brown 2007).

The airspace is an important area for mating displays across taxa. Aerial lekking has been described for some insects (e.g., Lepidoptera, Andersson et al. 1998). Eusocial insects, particularly the Hymenoptera and Blattodea (e.g., termites), often make nuptial flights, although many of these occur close to the ground, where the role of the airspace may be limited (e.g., Dial and Vaughan 1987). Some bats may respond to turbulence when choosing mating display locations; this hypothesis has been suggested as an explanation for why bats congregate in the lee of wind turbines, a behavior which unfortunately often results in mortality (Cryan et al. 2014). Many birds have song flights and elaborate aerial displays, including the roding display of the Eurasian woodcock (*Scolopax rusticola*; Hoodless et al. 2008), the dive and shuttle of Anna's hummingbird (*Calypte anna*; Stiles 1982), and the sky dancing of the Montagu's harrier (*Circus pygargus*; Arroyo et al. 2013).

In addition to visual displays, many acoustic and olfactory breeding signals are transmitted through the air, including an assortment of vocalizations, stridulations, airborne pheromones, and pollinator attractors. Many of these signals evolved characteristics that allow them to travel efficiently through the air to receivers, taking forms that propagate with minimal degradation (Morton 1975). The signals themselves compete for airspace, or at least clear transmission through it, by varying in ways that minimize overlap within a community (Nelson and Marler 1990). This variation can either be an adaptive characteristic of a species- or population-specific signal (e.g., a distinct floral scent; Wright and Schiestl 2009) or the result of an individual modulating its signal in real time in response to current acoustic or olfactory "clutter" (e.g., modulating a bird song frequency to avoid overlap with noise; Goodwin and Podos 2013).

3.3 Foraging, Competition, and Resource Partitioning

The airspace supports considerable foraging habitat for many aerial specialists and obligate insectivores (e.g., swallows, flycatchers, insectivorous bats). Other flying animals depend upon predictable seasonal resource pulses (Yang et al. 2008). Usually these occur as the invertebrate mating swarms described above, but some aerial predators have adapted to exploit seasonal songbird migrations as a food source. The giant noctule bat (*Nyctalus lasiopterus*) will opportunistically switch diets to prey almost exclusively upon migrating songbirds (Popa-Lisseanu et al. 2007; Fukui et al. 2013). Eleonora's falcon (*Falco eleonorae*) also takes migrating songbirds in flight and appears to synchronize its reproduction with the seasonal availability of this food source (Wink and Ristow 2000). Similarly, peregrine falcons (*Falco peregrinus*) and red-tailed hawks (*Buteo jamaicensis*) hunt outside the large breeding colonies of insectivorous Brazilian free-tailed bats (*Tadarida brasiliensis*; Lee and Kuo 2001), which in turn prey on migrating noctuid moths. This sets up an aerial tritrophic interaction (raptors > bats > noctuid moths) that occurs each summer and is mediated by abiotic conditions in the airspace that determine the favorability of conditions for moth migration (McCracken et al. 2008).

Intra- and interspecific competition over aerial food resources also occurs in the airspace, although the difficulty in defending airborne resources together with limited niche overlap among competitors reduces the likelihood that such competition will arise (Mills 1986). Bats and common nighthawks (*Chordeiles minor*) have been observed in competitive interaction over insects at a human light source (Shields and Bildstein 1979). Altitudinal resource partitioning is hypothesized to explain differences in niche breadth among some avian aerial insectivores, wherein physical adaptations have allowed certain swift species to forage at higher altitudes and take advantage of larger and higher flying insects, thus reducing competition with swifts at lower altitudes (Collins 2015). Several species of desert bats differed in the locations and timing of foraging bouts over water sources in the Central Negev Highlands of Israel, suggestive of aerial resource partitioning due to competition for a shared insect resource (Razgour et al. 2011).

4   Anthropogenic Impacts

Even as human encroachment into the airspace advances, it is difficult to know whether certain taxa experience non-compensatory levels of mortality (Drewitt and Langston 2006; Kunz et al. 2007) which complicates how development of the airspace can be managed to reduce impacts to flying animal populations. In this context, recognizing the airspace as habitat may have meaningful consequences for species conservation and management. The habitat concept is necessarily central to international species-based environmental law (e.g., US Endangered Species Act and National Environmental Policy Act, EU's Habitats Directive and Birds Directive) and influences how biologists that work for regulatory agencies interpret the law in practice. That many parts of the airspace fall within the habitat concept could change the scope of application of these laws to include vast new areas. However, regulatory references to the airspace as habitat are scant; we could find only one and it argued against recognizing the airspace as habitat. In the context of a review of New Zealand's Resource Management Act which recognizes protection of significant habitats, Wallace (2007) specifically argues against characterizing flight corridors or migration routes as habitat because those areas are "disconnected from breeding and feeding spaces." Aside from being imprecise (many organisms do breed and feed in the airspace, even if the waterbirds considered by this Act do not), the point is also irrelevant since breeding and foraging are not required characteristics of habitat and are not the only habitat-related behaviors critical to survival (e.g., hibernating, roosting; Cryan and Veilleux 2007, Mehlman et al. 2005). Even if waterbirds were the only taxon of concern for a regulatory agency, they repeatedly travel similar routes between feeding and roosting locations (Everaert and Stienen 2007; Cox and Afton 1996), so any anthropogenic disruption in their path could have negative consequences on their populations by increasing mortality and decreasing reproductive success. Recognizing the airspace as habitat is critical for species conservation and management that consider entire life histories (Davy et al. 2017).

4.1   Threats to Aerial Organisms

The actions of society through development, commerce, transportation, policy, and resource extraction increasingly result in conflicts with flying animals that have impacts ranging from the safety of airline travel (Dolbeer and Wright 2008) to the price of corn (Boyles et al. 2011). The increasing construction of tall structures may result in long-term disruptions of aerial habitats, including wind turbines (Arnett et al. 2010); power lines (Martin 2011; Bevanger 1998); buildings, especially those using glass construction (Loss et al. 2014; Klem 1990); and some communications towers (Gehring et al. 2011; Gehring et al. 2009). One of the newest anthropogenic entrants to the airspace, unmanned aircraft systems (UAS), pose possible threats in that they may interfere with natural animal behavior and in other ways disrupt aerial ecosystems (e.g., Lambertucci et al. 2015). Current US federal guidelines allow UAS operations in all classes of airspace except those in the vicinity of airports (Class B; Federal Aviation Administration 2015a), but because proposed guidelines require operators to keep UAS within visual line-of-sight (Federal Aviation Administration 2015b), UAS use generally occurs relatively close to the ground within the region of most flying animal activity.

Anthropogenic development of the airspace creates novel challenges for flying animals, many of which are not adapted to flying in cluttered environments. Put another way, "perceptually [birds] have no prior for human artifacts such a buildings, power wires or wind turbines" (Martin 2011). In addition, not only may organisms lack the sensory capabilities to avoid development, but their senses may trick them into responding inappropriately to it. Ecological traps occur when rapid changes in the environment produce misleading cues about the quality of habitats. Based on these cues, organisms may select lower quality habitat which leads to lower survival or reproductive success, even though higher quality habitat may be available. Numerous examples of ecological traps have been documented for terrestrial habitats, and anthropogenic factors are usually responsible (Robertson and Hutto 2006; Schlaepfer et al. 2002). The airspace is no exception. Migrating birds drawn into circling illuminated communications towers, particularly those with guy wires, in low cloud conditions die in large numbers (Gehring et al. 2009; Larkin and Frase 1988). Under these conditions, airspace in the vicinity of lighted towers might be considered low-quality aerial habitat. Nonetheless, birds in these circumstances remain in flight, presumably because the hazy glow of a lighted tower in cloud is in some way attractive. Similarly, bats that die in the vicinity of wind turbines may select this aerial habitat, perhaps as a mating location or because turbines concentrate insects as food (Cryan et al. 2014).

Anthropogenic structures and machines in the airspace are not the only threat to aerial organisms. Chemicals and particulate matter as air pollutants also degrade aerial habitats, and the impacts of air pollution on animals in flight are poorly understood. Is the health of flying animals disproportionately affected owing to their relatively high metabolic rates during flight? Does air pollution interfere with olfactory or visual cues necessary for foraging, reproduction, and orientation? Might pollution that impacts aerial invertebrates have a trophic effect on their aerial predators? Does extreme air pollution impede visibility enough to increase risk of collisions with structure? Does pollution interfere with the composition and metabolic processes of aerial microbial communities in ways that affect climate and human health?

Anthropogenic structures that intrude upon aerial habitats can be designed, refitted, or adapted to limit their impacts on flying animals. Communications towers can be designed with lighting and structural support that reduces collision risk (Gehring et al. 2009). Curtailment of wind turbines has been shown to reduce bat mortality (Arnett et al. 2010), and current research is exploring ultraviolet (UV) illumination of turbines as a way of dissuading bats from approaching altogether (Gorresen et al. 2015). Combination camera/acoustic detection and deterrence systems are being explored as a means of reducing eagle mortality at wind facilities (Sinclair and DeGeorge 2016). A variety of markers are now available which increase the visibility of powerlines to birds in flight (e.g., Sporer et al. 2013). UV reflecting patterns on glass, which are visible to birds but not humans, may be effective at deterring birds from colliding with glass structure (Klem 2009).

Where technological deterrents are impractical or otherwise ineffective, the airspaces themselves may require protection. Airspace reserves may be needed to support the conservation of species as human development of aerial habitats continues and populations face compounding pressures across all phases of their life cycle (e.g., terrestrial habitat loss, mismatched phenology through climate change, predation and competition from nonnative species). Such reserves might focus on regulating development of the airspace near traditional movement corridors or in other areas where high densities of aerial organisms commonly occur or where species of concern are prone to fly (Diehl 2013; Lambertucci et al. 2015). Aerial reserves may be appropriate in areas where human–wildlife conflicts become a human safety concern or development presents a population-level risk to flying organisms or the ecological services they provide.

The preservation of aerial habitats may be quite different than traditional approaches for terrestrial and aquatic habitats. Most terrestrial and aquatic habitat preservation focuses on particular areas of concern and is intended to be continuous and long term, as disturbance can have lasting effects. In contrast, many human–wildlife conflicts are temporally dependent. Where animal use of aerial habitat is predictable, preservation may involve the timing of human activities rather than permanent restrictions on a particular area. For instance, proper lighting and curtailment of wind turbines when flying animals are most active can reduce or eliminate systematic impacts (Arnett et al. 2010; Duerr et al. 2012). Turbine curtailment could conceivably occur over even short timescales such as when fog or low cloud interferes with animals' ability to avoid structures (Kirsch et al. 2015). Many North American cities participate in seasonal lights-out programs, turning off the lights of large city buildings during peak bird migration times. These examples illustrate how anthropogenic hazards to flying animals can be reduced by timing the use of airspaces to complement animal phenology.

Brazilian free-tailed bats offer an interesting conservation case study. Brazilian free-tailed bats are migratory and roost during summer in caves across the southern tier of the continental USA, with exceptionally large concentrations in central Texas (Wilkins 1989). Bracken, Frio, and other caves in this region host summer maternity colonies containing millions of these bats; the colony in Bracken Cave is one of the largest single bat colonies in the world (McCracken 2003). In the evenings, they depart the caves *en masse* to forage on noctuid moths and other insects near the planetary boundary layer (Westbrook 2008; Fig. 3.3).

Conservation organizations acquired land to protect terrestrial habitat in the immediate vicinity of the cave entrance. Protecting these areas from human development indirectly offers overlying airspaces some associated protection by excluding potentially hazardous anthropogenic structures (e.g., wind turbines, communications towers) from the immediate area. Protection of Bracken Cave and its airspace preserves the phenomenon of mass exodus (*sensu* Brower and Malcolm 1991), the biological integrity of its populations (e.g., Allee effects, Langwig et al. 2015), the ecological services these bats provide to agriculture (Boyles et al. 2011; Davidai et al. 2015), and the educational opportunities afforded by these conspicuous movements (Pennisi et al. 2004). Even though Bracken and other caves in the region are occupied only half of the year (March through October), protecting the integrity of the cave itself from any form of development or even routine visitation is a year-round necessity as the cave represents a nonrenewable resource for bats. The immediate overlying airspace is also critical habitat as millions of bats transit this space when moving to and from foraging habitat. In contrast to the protection of the cave, aerial habitat in the vicinity of the cave could be considered renewable and its preservation seasonal.

The reserve concept is often oriented toward nonhuman species protection, but understanding and perhaps regulating airspace use with respect to flying animals has financial and human safety implications as well. From 1990 to 2007, bird strikes by civilian aircraft accounted for the overwhelming majority of monetary losses associated with wildlife strikes (~$125 to ~$625 million annually in the USA) and resulted in 10 fatalities and 164 injuries (Dolbeer and Wright 2008). After US Airways Flight 1549 struck a flock of Canada Geese (*Branta canadensis*) and crash landed into the Hudson River, there were calls for improved monitoring of bird movements (Marra et al. 2009).

Bird-avoidance models exist that combine both historic and current data to identify times and places of increased risk of bird strikes for low-flying aircraft (Shamoun-Baranes et al. 2008). These warning systems forecast hazards using near real-time feedback from biological data collected from weather radar. However, classification algorithms designed to identify and extract biological data from weather radar data streams are rudimentary at best. Currently, no published algorithms exist that can differentiate among types of biological echoes (i.e., taxa) on weather radars, although some effort has been made to distinguish biological from nonbiological radar echoes (Park et al. 2009; Bachmann and Zrnić 2007).

Fig. 3.3 Series of weather radar sweeps showing the evening exodus of Brazilian free-tailed bats from Frio Cave (yellow dot) near Concan, Texas, during the evening of 22 July 2014. The white dot shows the location of the weather radar (KDFX) at Laughlin AFB, Texas. Colors show differential phase, a radar polarimetry measure used here because it best distinguishes emerging bats (browns, yellows, and dark greens) from other radar bioscatterers (light green). The first four sweeps are separated by 10 min; the fifth sweep occurred 30 min after the fourth and shows the movement more developed and bats emerging from other roosts in the vicinity

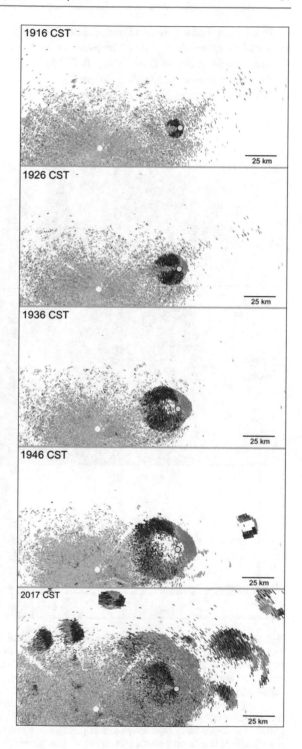

Identifying radar echoes to taxon is critical to the operational usefulness of any near real-time warning system, since large birds pose the greatest hazard to manned aircraft (Dolbeer and Wright 2008). Reliable information on the passage of large birds would enable responders to manage aircraft altitudes and approach and departure times and directions as needed. Figure 3.4 shows a typical autumn

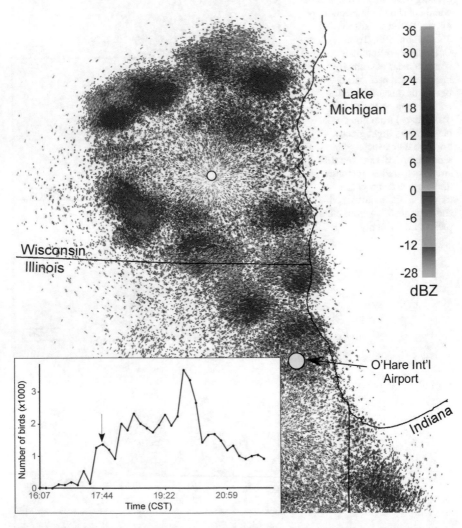

Fig. 3.4 Weather radars capture a moment during the onset of waterfowl migration toward the SSE across southeastern Wisconsin and northeastern Illinois on 22 Nov 2010, at ~17:45 CST. A 10 km diameter airspace centered on O'Hare International Airport is represented by a yellow circle. The inset shows changes in the estimated number of birds within this circle as the migration begins and progresses. (The abundance calculation assumes mallard-sized birds for simplicity and because good estimates of radar cross section are available for Mallards (*Anas platyrhynchos*) to convert radar reflectivity to bird numbers (O'Neal et al. 2010; Chilson et al. 2012b); the likely presence of larger birds such as geese and perhaps some cranes reduces these estimates of bird abundances.). The arrow indicates when in the time series this radar capture occurred

waterfowl migration toward the SSE with ducks, Canada Geese, and some Sandhill Cranes (*Grus canadensis*) departing habitats throughout southeastern Wisconsin and northeastern Illinois. Waves of waterfowl passed through the airspace of O'Hare International Airport, near Chicago, Illinois (Fig. 3.4, time series). In this image, the northern-most band of waterfowl radar echoes running west of the "Lake Michigan" label passed through O'Hare airspace at ~19:50 and was responsible for the large pulse of waterfowl seen in the time series. These migratory pushes out of southeastern Wisconsin occur routinely each autumn, so during this time O'Hare becomes both heavily used commercial airspace and seasonally predictable waterfowl aerial habitat.

5   Conclusions

Definitions matter. How habitat is defined influences law, policy, and the actions of resource managers in interpreting policy. Frequent efforts to redefine habitat in ways that capture perceived shortcomings testify to the continuing importance of the habitat concept. The airspace easily satisfies widely recognized definitions of habitat and shares many ecological properties of other more familiar habitat types. Considering the airspace as habitat also encourages new perspectives on the habitat concept. For example, airspace lacks many of the familiar visible cues we associate with habitat such as vegetation, substrate, or topographic relief. The energetic costs of occupying aerial habitats are often high, and critical aerial resources can be simultaneously abundant, ephemeral, and unpredictable with respect to any one location. Identifying properties shared across all major habitat types—terrestrial, aquatic, and aerial—clarifies our understanding of habitat by focusing our attention on its most fundamental and essential elements.

Considering the airspace as habitat also reminds us that time as much as place is a critical component of habitat, since many of the airspace's resources for flying animals depend upon the weather and are therefore highly dynamic. Yet, many widely recognized definitions of habitat explicitly refer only to place and tend to reinforce a rather static notion of habitat (but see Dingle 2014; Southwood 1977). This conforms with human perception of habitats as generally stable through time, at least in the manner we are accustomed to instinctively experiencing them, even if academically, we might recognize that the suitability of nearly any habitat for a given species varies through time, often seasonally if not through ecological succession or some other biotic or abiotic process. In short, that habitat is "where one would go to find it" (Odum 1959) is no more important than *when* one would go to find it.

The habitat concept has proven to be influential in shaping and informing environmental law and natural resource management in many countries. It is unknown whether aerial habitat represents an expansion of the habitat concept that is applicable under existing law, for example, if the need should arise to create airspace reserves or otherwise regulate the airspace to ensure human safety or protect animal populations. Accelerating human occupancy and development of aerial habitats has created a greater need for regulation of the airspace, and clarification of the habitat concept is critical for developing or interpreting laws and regulatory practices that protect species when and where they are present.

We have much to learn about the ecology of the airspace. The dynamism of aerial habitat leads to dynamism of aeroecological processes generally, making them challenging to study, even with the aid of contemporary technology (Chilson et al. 2012a; Bridge et al. 2011). Recognizing that the airspace is habitat allows the known ecology of terrestrial and aquatic habitats to serve as models for predicting the conditions required for a wide range of ecological and behavioral processes to be operational in the airspace.

Acknowledgements We thank S. Soileau, H. Hoffmann, and the editors for helpful input on this chapter. The USGS Northern Rocky Mountain Science Center Ecolunch group provided valuable feedback as the concept of aerial habitat was developed. Any use of trade, product, or firm names is for descriptive purposes only and does not imply endorsement by the U.S. Government.

References

Alerstam T, Lindström A (1990) Optimal bird migration: the relative importance of time, energy, and safety. In: Gwinner E (ed) Bird migration. Springer, Berlin, pp 331–351

Allan JD, Flecker AS (1989) The mating biology of a mass-swarming mayfly. Anim Behav 37: 361–371

Andersson S, Rydell J, Swensson GE (1998) Light, predation and the lekking behavior of the ghost swift *Hepialus humuli* (L.) (Lepidoptera: Hepialidae). Proc R Soc Lond B 264:1345–1351

Aralimarad P, Reynolds AM, Lim KS, Reynolds DR, Chapman JW (2011) Flight altitude selection increases orientation performance in high-flying nocturnal insect migrants. Anim Behav 82: 1221–1225

Arnett EB, Huso MM, Schirmacher MR, Hayes JP (2010) Altering turbine speed reduces bat mortality at wind-energy facilities. Front Ecol Environ 9:209–214

Arroyo B, Mougeot F, Bretagnolle V (2013) Characteristics and sexual functions of sky-dancing displays in a semi-colonial raptor, the Montagu's Harrier (*Circus Pygargus*). J Raptor Res 47: 185–196

Bachmann S, Zrnić DS (2007) Spectral density of polarimetric variables separating biological scatterers in the VAD display. J Atmos Ocean Technol 24:1186–1198

Belnap J, Hawkes CV, Firestone MK (2003) Boundaries in miniature: two examples from soil. Bioscience 53:739–749

Bevanger K (1998) Biological and conservation aspects of bird mortality caused by electricity power lines: a review. Biol Conserv 86:67–76

Bowers RM, McLetchie S, Knight R, Fierer N (2011) Spatial variability in airborne bacterial communities across land-use types and their relationship to the bacterial communities of potential source environments. ISME J 5:601–612

Bowlin MS, Enstrom DA, Murphy BJ, Plaza E, Jurich P, Cochran J (2015) Unexplained altitude changes in a migrating thrush: Long-flight altitude data from radio-telemetry. Auk 132: 808–816

Boyles JG, Cryan PM, McCracken GF, Kunz TH (2011) Economic importance of bats in agriculture. Science 332:41–42

Bridge ES, Thorup K, Bowlin MS, Chilson PB, Diehl RH, Fleron RW, Hartl P, Kays R, Kelly JF, Robinson WD, Wikelski M (2011) Technology on the move: recent and forthcoming innovations for tracking migratory birds. Bioscience 61:689–698

Brodie EL, DeSantis TZ, Parker JPM, Zubietta IX, Piceno YM, Andersen GL (2007) Urban aerosols harbor diverse and dynamic bacterial populations. Proc Natl Acad Sci 104:299–304

Brower LP, Malcolm SB (1991) Animal migrations: endangered phenomena. Am Zool 31:265–276

Burrows SM, Elbert W, Lawrence MG, Pöschl U (2009) Bacteria in the global atmosphere–Part 1: review and synthesis of literature data for different ecosystems. Atmos Chem Phys 9:9263–9280

Cadenasso ML, Pickett STA, Weathers KC, Bell SS, Benning TL, Carreiro MM, Dawson TE (2003) An interdisciplinary and synthetic approach to ecological boundaries. Bioscience 53:717–722

Chilson PB, Frick WF, Kelly JF, Howard KW, Larkin RP, Diehl RH, Westbrook JK, Kelly TA, Kunz TH (2012a) Partly cloudy with a chance of migration: weather, radars, and aeroecology. Bull Am Meteorol Soc 93:669–686

Chilson PB, Frick WF, Stepanian PM, Shipley JR, Kunz TH, Kelly JK (2012b) Estimating animal densities in the aerosphere using weather radar: To Z or not to Z? Ecosphere 3:72

Collins CT (2015) Food habits and resource partitioning in a guild of Neotropical swifts. Wilson J Ornithol 127:239–248

Corkum LD, Ciborowski JJH, Dolan DM (2006) Timing of *Hexagenia* (Ephemeridae: Ephemeroptera) mayfly swarms. Can J Zool 84:1616–1622

Cox RR, Afton AD (1996) Evening flights of female northern pintails from a major roost site. Condor 89:810–819

Cryan PM, Brown AC (2007) Migration of bats past a remote island offers clues toward the problem of bat fatalities at wind turbines. Biol Conserv 139:1–11

Cryan PM, Veilleux JP (2007) Migration and use of autumn, winter, and spring roosts by tree bats. In: Lacki MJ, Hayes HJP, Kurta A (eds) Bats in forests: conservation and management. The Johns Hopkins University Press, Baltimore, pp 153–175

Cryan PM, Gorrensen PM, Hein CD, Schirmacher MR, Diehl RH, Huso MM, Hayman DTS, Fricker PD, Bonaccorso FJ, Johnson DH, Heist K, Dalton DC (2014) Behavior of bats at wind turbines. PNAS 111:15126–15131

Davidai N, Westbrook JK, Lessard J, Hallam TG, McCracken GF (2015) The importance of natural habitats to Brazilian free-tailed bats in intensive agricultural landscapes in the Winter Garden region of Texas, United States. Biol Conserv 190:107–114

Davy CM, Ford AT, Fraser KC (2017) Aeroconservation for the fragmented skies. Conserv Lett. https://doi.org/10.1111/conl.12347

Dennis RLH, Shreeve TG, Van Dyck H (2003) Towards a functional resource-based concept of habitat: a butterfly biology viewpoint. Oikos 102:417–426

Dennis RLH, Shreeve TG, van Dyck H (2006) Habitats and resources: the need for a resource-based definition to conserve butterflies. Biodivers Conserv 15:1943–1966

Deprés VR, Huffman JA, Burrows SM, Hoose C, Safatov AS, Buryak G, Fröhlich-Nowoisky J, Elbert W, Andreae MO, Pöschl U, Jaenicke R (2012) Primary biological aerosol particles in the atmosphere: a review. Tellus B 64. https://doi.org/10.3402/tellusb.v64i0.15598

Dial KP, Vaughan TA (1987) Opportunistic predation on alate termites in Kenya. Biotropica 19:185–187

Diehl RH (2013) The airspace is habitat. Trends Ecol Evol 28:377–379

Dingle H (2014) Migration: the biology of life on the move, 2nd edn. Oxford University Press, Oxford

Dolbeer RA, Wright S (2008) Wildlife strikes to civil aircraft in the United States 1990–2007. Bird Strike Committee Proceedings

Drake VA, Reynolds DR (2012) Radar entomology: observing insect flight and migration. CABI, Wallingford

Drewitt AL, Langston RHW (2006) Assessing the impacts of wind farms on birds. Ibis 148:29–42

Dublon IAN, Sumpter DJT (2014) Flying insect swarms. Curr Biol 24:R828–R830

Duerr AE, Miller TA, Lanzone M, Brandes D, Cooper J, O'Malley K, Maisonneuve C, Tremblay J, Katzner T (2012) Testing an emerging paradigm in migration ecology shows surprising differences in efficiency between flight modes. PLoS One 7:e35548. https://doi.org/10.1371/journal.pone.0035548

Elton CS (1966) The pattern of animal communities. Chapman and Hall, London

Everaert J, Stienen EWM (2007) Impact of wind turbines on birds in Zeebrugge (Belgium): significant effect on breeding tern colony due to collisions. Biodivers Conserv 16:3345–3359

Federal Aviation Administration (2015a) Notice JO 7210.882. Unmanned aircraft operations in the national airspace system (NAS)

Federal Aviation Administration (2015b) 14 CFR Chapter I. Clarification of the applicability of aircraft registration requirements for unmanned aircraft systems (UAS) and request for information regarding electronic registration for UAS

Fukui D, Dewa H, Katsuta S, Sato A (2013) Bird predation by the birdlike noctule in Japan. J Mammal 94:657–661

Gehring J, Kerlinger P, Manville AM (2009) Communication towers, lights, and birds: successful methods of reducing the frequency of avian collisions. Ecol Appl 19:505–514

Gehring JL, Kerlinger P, Manville AM (2011) The role of tower height and guy wires on avian collisions with communication towers. J Wildl Manage 75:848–855

Gill RE Jr, Tibbitts TL, Douglas DC, Handel CM, Mulcahy DM, Gottschalck JC, Warnock N, McCaffery BJ, Battley PF, Piersma T (2009) Extreme endurance flights by landbirds crossing the Pacific Ocean: ecological corridor rather than barrier? Proc R Soc Lond B 276:447–457

Gonzalez-Toril E, Amils R, Delmas RJ, Petit J-R, Komarek J, Elster J (2009) Bacterial diversity of autotrophic enriched cultures from remote, glacial Antarctic, Alpine and Andean aerosol, snow and soil samples. Biogeosciences 6:33–44

Goodwin SE, Podos J (2013) Shift of song frequencies in response to masking tones. Anim Behav 85:435–440

Gorresen PM, Cryan PM, Dalton DC, Wolf S, Johnson JA, Todd CM, Bonaccorso FJ (2015) Dim ultraviolet light as a means of deterring activity by the Hawaiian hoary bat *Lasiurus cinereus semotus*. Endanger Species Res 28:249–257

Hall LS, Krausman PR, Morrison ML (1997) The habitat concept and a plea for a standard terminology. Wildl Soc Bull 25:173–182

Harrison RM, Jones AM, Biggins PD, Pomeroy N, Cox CS, Kidd SP, Hobman JL, Brown NL, Beswick A (2005) Climate factors influencing bacterial count in background air samples. Int J Biometeorol 49:167–178

Hawkes LA, Balachandran S, Batbayar N, Butler PJ, Frappell PB, Milsom WK, Tseveenmyadag N, Newman SH, Scott GR, Sathiyaselvam P, Takekawa JY, Wikelski M, Bishop CM (2011) The trans-Himalayan flights of bar-headed geese (Anser indicus). Proc Natl Acad Sci 108:9516–9519

Hawkes LA, Balachandran S, Batbayar N, Butler PJ, Chua B, Douglas DC, Frappell PB, Hou Y, Milsom WK, Newman SH, Prosser DJ, Sathiyaselvam P, Scott GR, Takekawa JY, Natsagdorj T, Wikelski M, Witt MJ, Yan B, Bishop CM (2013) The paradox of extreme high-altitude migration in bar-headed geese Anser indicus. Proc R Soc Lond B: Biol Sci 280. https://doi.org/10.1098/rspb.2012.2114

Hedenström A, Norevik G, Warfvinge K, Anderson A, Bäckman J, Åkesson S (2016) Annual 10-month aerial life phase in the common swift Apus apus. Curr Biol 26. https://doi.org/10.1016/j.cub.2016.1009.1014

Helms JA, Godfrey AP, Ames T, Bridge ES (2016) Predator foraging altitudes reveal the structure of aerial insect communities. Sci Rep 6:28670. https://doi.org/10.1038/srep28670

Hoodless AN, Inglis JG, Doucet J-P, Aebischer NJ (2008) Vocal individuality in the roding calls of woodcock Scolopax rusticola and their use to validate a survey method. Ibis 150:80–89

Jaenike J, Holt RD (1991) Genetic variation for habitat preference: evidence and explanations. Am Nat 137:S67–S90

Joly M, Amato P, Sancelme M, Vinatier V, Abrantes M, Deguilluame L, Delort A-M (2015) Survival of microbial isolates from clouds towards simulated atmospheric stress factors. Atmos Environ 117:92–98

Jones J (2001) Habitat selection studies in avian ecology: a critical review. Auk 118:557–562

Kirsch EM, Wellik MJ, Suarez M, Diehl RH, Lutes J, Woyczik W, Krapfl J, Sojda R (2015) Observation of sandhill cranes' (*Grus canadensis*) flight behavior in heavy fog. Wilson J Ornithol 127:281–288

Klem D Jr (1990) Collisions between birds and windows: mortality and prevention. J Field Ornithol 61:120–130

Klem D Jr (2009) Avian mortality at windows: the second largest human source of bird mortality on Earth. Tundra to tropics: connecting birds, habitats and people. In: Proceedings of the 4th international partners in flight conference: Tundra to Tropics. USDA, Forest Service, McAllen, Texas, pp 244–251

Kunz TH, Arnett EB, Erickson WP, Hoar AR, Johnson GD, Larkin RP, Strickland MD, Thresher RW, Tuttle MD (2007) Ecological impacts of wind energy development on bats: questions, research, needs, and hypotheses. Front Ecol Environ 5:315–324

Kunz TH, Gauthreaux SA Jr, Hristov NI, Horn JW, Jones G, Kalko EK, Larkin RP, McCracken GF, Swartz SM, Srygley RB, Dudley R, Westbrook JK, Wikelski M (2008) Aeroecology: probing and modeling the aerosphere. Integr Comp Biol 48:1–11

Lambertucci SA, Shepard ELC, Wilson RP (2015) Human-wildlife conflicts in a crowded airspace. Science 348:502–504

Langwig KE, Frick WF, Reynolds R, Parise KL, Drees KP, Hoyt JR, Cheng TL, Kunz TH, Foster JT, Kilpatrick AM (2015) Host and pathogen ecology drive the seasonal dynamics of a fungal disease, white-nose syndrome. Proc R Soc Lond B: Biol Sci 282. https://doi.org/10.1098/rspb.2014.2335

Lanzone M, Miller TA, Turk P, Brandes D, Halverson C, Maisonneuve C, Tremblay J, Cooper J, O'Malley K, Brooks RP, Katzner T (2012) Flight responses by a migratory soaring raptor to changing meteorological conditions. Biol Lett 8:710–713

Larkin RP, Frase BA (1988) Circular paths of birds flying near a broadcasting tower in cloud. J Comp Psychol 102:90–93

Laybourne RC (1974) Collision between a vulture and an aircraft at an altitude of 37,000 feet. Wilson Bull 86:461–462

Lee Y, Kuo Y (2001) Predation on Mexican free-tailed bats by peregrine falcons and red-tailed hawks. J Raptor Res 35:115–123

Liechti F, Klaasen M, Bruderer B (2000) Predicting migratory flight altitudes by physiological migration models. Auk 117:205–214

Liechti F, Witvliet W, Weber R, Bachler E (2013) First evidence of a 200-day non-stop flight in a bird. Nat Commun 4. https://doi.org/10.1038/ncomms3554

Lindow S, Brandl M (2003) Microbiology of the phyllosphere. Appl Environ Microbiol 69:1875–1883

Linnæus C (1754) Genera plantarum, 5th edn

Loss SR, Will T, Loss SS, Marra PP (2014) Bird–building collisions in the United States: estimates of annual mortality and species vulnerability. Condor: Ornithol Appl 116:8–23

Marra PP, Dove CJ, Dolbeer RA, Dahlan N, Heacker M, Whatton J, Diggs N, France C, Henkes G (2009) Migratory Canada geese cause crash of US Airways Flight 1549. Front Ecol Environ 7:297–301

Martin GR (2011) Understanding bird collisions with man-made objects: a sensory ecology approach. Ibis 153:239–254

Mateos-Rodríguez M, Liechti F (2012) How do diurnal long-distance migrants select flight altitude in relation to wind? Behav Ecol 23:403–409

McCracken GF (2003) Estimates of population sizes in summer colonies of Brazilian free-tailed bats (*Tadarida brasiliensis*). Monitoring trends in bat populations of the US and territories: problems and prospects. In: O'Shea TJ, Bogan MA (eds) US geological survey, biological resources discipline, information and technology report, USGS/BRD/ITR-2003-003, 21–30

McCracken GF, Gillam EH, Westbrook JK, Lee Y, Jensen ML, Balsley BB (2008) Brazilian free-tailed bats (Tadarida brasiliensis: Molossidae, Chiroptera) at high altitude: links to migratory insect populations. Integr Comp Biol 48:107–118

Mehlman DW, Mabey SE, Ewert DN, Duncan C, Abel B, Cimprich DA, Sutter RD, Woodrey MS (2005) Conserving stopover sites for forest-dwelling migratory landbirds. Auk 122:1281–1290

Meola M, Lazzaro A, Zeyer J (2015) Bacterial composition and survival on Sahara dust particles transported to the European Alps. Front Microbiol 6:1454

Mills AM (1986) The influence of moonlight on the behavior of goatsuckers (Caprimulgidae). Auk 103:370–378

Morrison ML, Marcot BG, Mannan RW (2006) Wildlife-habitat relationships: concepts and applications, 3rd edn. Island Press, Washington, DC

Morton ES (1975) Ecological sources of selection on avian sounds. Am Nat 109:17–34

Nelson DA, Marler P (1990) The perception of birdsong and an ecological concept of signal space. In: Stebbins WC, Berkley MA (eds) Comparative perception, Complex signals, vol 2. Wiley, New York, pp 443–478

O'Neal BJ, Stafford JD, Larkin RP (2010) Waterfowl on weather radar: applying ground-truth to classify and quantify bird movements. J Field Ornithol 81:71–82

Odum EP (1959) Fundamentals of ecology, 2nd edn. W. B. Saunders, Philadelphia

Park H, Ryzhkov AV, Zrnic DS, Kim K-E (2009) The hydrometeor classification algorithm for the polarimetric WSR-88D: description and application to an MCS. Weather Forecast 24:730–748

Pennisi LA, Holland SM, Stein TV (2004) Achieving bat conservation through tourism. J Ecotour 3:195–207

Pennycuick CJ (1998) Field observations of thermals and thermal streets, and the theory of cross-country soaring flight. J Avian Biol 29:33–43

Popa-Lisseanu AG, Delgado-Huertas A, Forero MG, Rodríguez A, Arlettaz R, Ibáñez C (2007) Bats' conquest of a formidable foraging niche: the myriads of nocturnally migrating songbirds. PLoS One 2. https://doi.org/10.1371/journal.pone.0000205

Razgour O, Korine C, Saltz D (2011) Does interspecific competition drive patterns of habitat use in desert bat communities? Oecologia 167:493–502

Robertson BA, Hutto RL (2006) A framework for understanding ecological traps and an evaluation of the existing evidence. Ecology 87:1075–1085

Robinson WDBM, Bisson IA, Shamoun-Baranes J, Thorup K, Diehl RH, Kunz TH, Mabey SE, Winkler DW (2009) Integrating concepts and technologies at the frontiers of bird migration. Front Ecol Environ 8:354–361

Sachs G (2016) In-flight measurement of upwind dynamic soaring in albatrosses. Prog Oceanogr. https://doi.org/10.1016/j.pocean.2016.01.003

Sattler B, Puxbaum H, Psenner R (2001) Bacterial growth in supercooled cloud droplets. Geophys Res Lett 28:239–242

Schlaepfer MA, Runge MC, Sherman PW (2002) Ecological and evolutionary traps. Trends Ecol Evol 17:474–480

Scott JA (1972) Mating of butterflies. J Res Lepid 11:99–127

Shamoun-Baranes J, Bouten W, Buurma L, Defusco RP, Dekker A, Sierdsema H, Sluiter F, van Belle J, van Gasteren H, van Loon E (2008) Avian information systems: developing web-based bird avoidance models. Ecol Soc 13:38

Shields WM, Bildstein KL (1979) Birds versus bats: behavioral interactions at a localized food source. Ecology 60:468–474

Shipley JR, Kelly JF, Frick WF (2017) Toward integrating citizen science and radar data for migrant bird conservation. Remote Sens Ecol Conserv. https://doi.org/10.1002/rse2.62

Sinclair K, DeGeorge E (2016) Wind energy industry eagle detection and deterrents: research gaps and solutions workshop summary report. National Renewable Energy Laboratory, Golden, CO, Technical report NREL/TP-5000-65735

Southwood TRE (1977) Habitat, the templet for Ecological strategies? J Anim Ecol 46:337–365

Sporer MK, Dwyer JF, Gerber BD, Harness RE, Pandey AK (2013) Marking power lines to reduce avian collision near the Audubon National Wildlife Refuge, North Dakota. Wildl Soc Bull. https://doi.org/10.1002/wsb.329

Stiles FG (1982) Aggressive and courtship displays of the male Anna's hummingbird. Condor 84: 208–225

Sullivan R (1981) Insect swarming and mating. Fla Entomol 64:44–65

Sweeney BW, Vannote RL (1982) Population synchrony in mayflies: a predation satiation hypothesis. Evolution 36:810–821

Vaïtilingom M, Attard E, Gaiani N, Sancelme M, Deguillaume L, Flossmann AI, Amato P, Delort AM (2012) Long-term features of cloud microbiology at the puy de Dôme (France). Atmos Environ 56:88–100

Van Doren BM, Horton KG, Dokter AM, Klinck H, Elbin SB, Farnsworth A (2017) High-intensity urban light installation dramatically alters nocturnal bird migration. PNAS 114:11175–11180

Wallace P (2007) The nature of habitat. NZ J Environ Law 11:211–240

Westbrook JK (2008) Noctuid migration in Texas within the nocturnal aeroecological boundary layer. Integr Comp Biol 48:99–106

Westbrook JK, Isard SA (1999) Atmospheric scales of biotic dispersal. Agric For Meteorol 97: 263–274

Wilkins KT (1989) *Tadarida brasiliensis*. In: Best TL, Anderson S (eds) Mammalian species, no 331. American Society of Mammalogists, pp 1–10

Wink M, Ristow D (2000) Biology and molecular genetics of Eleonora's falcon (*Falco eleonorae*), a colonial raptor of Mediterranean Islands. In: Chancellor, Meyburg (eds) Raptors at risk. Hancock House, Blaine, pp 653–668

Womack AM, Bohannan BJM, Green JL (2010) Biodiversity and biogeography of the atmosphere. Philos Trans R Soc Lond B 365:3645–3653

Wright GA, Schiestl FP (2009) The evolution of floral scent: the influence of olfactory learning by insect pollinators on the honest signaling of floral rewards. Funct Ecol 23:841–851

Yang LH, Bastow JL, Spence KO, Wright AN (2008) What can we learn from resource pulses. Ecology 89:621–634

Track Annotation: Determining the Environmental Context of Movement Through the Air

4

Renee Obringer, Gil Bohrer, Rolf Weinzierl, Somayeh Dodge, Jill Deppe, Michael Ward, David Brandes, Roland Kays, Andrea Flack, and Martin Wikelski

Abstract

Volant organisms are adapted to atmospheric patterns and processes. Understanding the lives of animals that inhabit this aerial environment requires a detailed investigation of both the animal's behavior and its environmental context—i.e., the environment that it encounters at a range of spatial and temporal scales. For aerofauna, it has been relatively difficult to observe the environment they encounter while they move. Large international efforts using

R. Obringer (✉) · G. Bohrer
Department of Civil, Environmental and Geodetic Engineering, The Ohio State University, Columbus, OH, USA
e-mail: obringer.21@osu.edu

R. Weinzierl · A. Flack
Max Planck Institute for Ornithology, Vogelwarte Radolfzell, Radolfzell, Germany

S. Dodge
Department of Geography, Environment and Society, University of Minnesota, Twin Cities, Minneapolis, MN, USA

J. Deppe
Department of Biological Sciences, Eastern Illinois University, Charleston, IL, USA

M. Ward
Department of Natural Resources and Environmental Sciences, University of Illinois, Urbana, IL, USA

D. Brandes
Civil and Environmental Engineering, Lafayette College, Easton, PA, USA

R. Kays
NC State University, NC Museum of Natural Sciences, Raleigh, NC, USA

M. Wikelski
Max Planck Institute for Ornithology, Vogelwarte Radolfzell, Radolfzell, Germany

Department of Biology, University of Konstanz, Konstanz, Germany

© Springer International Publishing AG, part of Springer Nature 2017
P.B. Chilson et al. (eds.), *Aeroecology*,
https://doi.org/10.1007/978-3-319-68576-2_4

satellite and weather model reanalysis now provide some of the environmental data on atmospheric environments throughout the globe. Track annotation—the approach of merging the environmental data with the movement track measured via telemetry—can be conducted automatically using online tools such as Movebank-Env-DATA or RNCEP. New parameterization approaches can use environmentally annotated tracks to approximate specific atmospheric conditions, such as uplift and tail wind, which are not typically observed at the exact locations of the movement, but are critical to movement. Reducing the complexity of movement to single-dimensional characteristic (such as flight speed, elevation, etc.) and defining the temporal scope of the movement phenomenon in the focus of the analysis (seasonal, daily, minutely, etc.) makes it possible to construct empirical models that explain the movement characteristic as driven by the environmental conditions during flight, despite the highly dynamic, complex, and scale-dependent structures of both the flight path and atmospheric variables. This chapter will provide several examples for such empirical movement models from different species of birds and using several resources for atmospheric data.

1   Introduction

Aerial movement ecology is challenged by the ability of researchers to access environmental data at the appropriate scale, especially as many animals traverse remote environments that are difficult to measure (Bowlin et al. 2010). An additional challenge rises due to the high spatiotemporal variability of the atmospheric environment, coupled with the complex role that atmospheric conditions play in the movement. Furthermore, with any predictive movement model, it is difficult to determine which of the interdependent environmental variables is most probably affecting the migration pattern (Dodge et al. 2014).

As GPS and other tracking technologies have developed, collecting data about the migration route has gotten easier. Remote sensing and model reanalyses, which combine remote sensing and ground-based observations and modeling to generate gridded data products and are a common tool for weather forecast, now provide a wealth of information about environmental conditions worldwide. Capitalizing on advances in the collection of both animal tracking and environmental data, the track-annotation approach (Mandel et al. 2011) provides a surrogate for missing observations of environmental conditions en route by linking state-of-the-art environmental data with observed tracks of precise animal locations. An annotated track includes the original location coordinates (in space and time) of the observed movement and additional environmental variable values from external observations. These variables must be interpolated in space and time to the observed movement coordinates. Emerging tools for track annotation, such as Movebank's Environmental-Data Automated Track Annotation (Env-DATA) (Dodge et al. 2013) and RNCEP (Kemp et al. 2012), provide the means to automatically annotate large movement datasets with a large number of environmental variables from different datasets and data sources. Typical datasets that provide observations of variables that may affect aerial movement include wind speed and direction and other weather variables from global weather reanalyses, [such as the European Centre for Medium-Range Weather Forecasts (ECMWF) ERA-Interim reanalysis], vegetation greenness and other land surface properties from the Moderate Resolution Imaging Spectro-radiometer Satellites (MODIS), and precipitation from the Tropical Rainfall Measuring Mission (TRMM) or the new global precipitation mission (GPM).

Provided this wealth of environmental data, analyzing the underlying biological and environmental effects that drive the movement as it is expressed in the observed movement tracks is getting easier. Approaches such as home-range analyses (Dodge et al. 2014; Bohrer et al. 2014), habitat selection (Cimino et al. 2013; Phillips and Dudík 2008), preferential use analysis (Bohrer et al. 2012), and individual-based models (McLaren et al. 2012; Bartlam-Brooks et al. 2013) can be conducted given movement tracks annotated with environmental variables.

Here, we provide two different test cases to illustrate the possibilities of track annotation driven analysis of aerial migration movement. We analyze flight data from two bird species—Swainson's thrushes (*Catharus ustulatus*) and white storks (*Ciconia ciconia*). We particularly include a very coarse resolution dataset from a small migrant species (thrush) and a very high resolution dataset from a large migrant (stork) to contrast the challenges, limitations, and opportunities associated with each movement data type. For the both species, we hypothesize that the flight speed, which we consider one of the key characteristics of the movement, is dependent on environmental conditions. When analyzing migration data, it is important to consider the movement as the observable result of both internal and external factors (Nathan et al. 2008). Under this approach, environmental variables will directly interact with the internal capacity to move by governing the energetic cost of moving. For both species, certain environmental conditions will require larger energy expenditure to keep up a certain speed. Specifically, we demonstrate how to justify selection of a small number of environmental variables out of many available variables that could hypothetically affect the ground speed of flying bids.

2 Methods

2.1 Environmental Drivers of Swainson's Thrush Flight Speed Crossing the Gulf of Mexico

The study species—Swainson's thrush (*Catharus ustulatus*)—is a small Nearctic-Neotropical songbird species that breeds in Canada and the northern United States and migrates to Central and South America in the winter. There are two major populations of Swainson's thrushes: the coastal population and the continental population. The coastal population migrates down the pacific coast to Mexico or Costa Rica and the continental population migrates along an eastern route to

Fig. 4.1 Swainson's thrush
with radio transmitter
attached to back Photo credit:
William Cochran and Bill

Panama and South America. The migration of the continental population often involves a long, nonstop flight across the Gulf of Mexico (Cochran and Wikelski 2005). This trans-Gulf migration typically starts at dawn and continues throughout the night and next day for an average of about 20 h. The nature of this migration makes tracking and observing these birds difficult. The birds generally weigh between 23 and 45 g, and therefore any tracking device attached to a bird must be lighter than 2 g in order not to affect the bird's flight. GPS transmitters are larger than this size limit. This poses a strong limitation of the technology used for tracking these and similarly small sized birds. Radio telemetry provides the opportunity to locate the birds using a very small radio transmitter (Fig. 4.1). The observation, however, is limited to the locations of the telemetry antennas, and typically these types of observations provide very sparse information of the movement track (Cochran and Wikelski 2005; Bowlin and Wikelski 2008).

We studied the Swainson's thrush population that migrates from the Bon-Secour National Wildlife Refuge on the Fort Morgan Peninsula in Alabama in the USA (30°13'N, 88°00''W) to the northern coast of the Yucatan Peninsula in Mexico (21°31'N, 87°40''W) in the fall. The birds were captured using mist nets while flying through the wildlife refuge in Alabama. After capture, the birds were weighed, measured, and fitted with radio transmitters. These transmitters are glued to the back of the birds. The glue wears off in several days. This method allows the data to be collected automatically without having to recapture the birds in Mexico to recover the transmitter.

The study area included ten radio-telemetry towers: three in the Fort Morgan Peninsula in Alabama and seven in the Yucatan Peninsula (Fig. 4.2). Each tower in

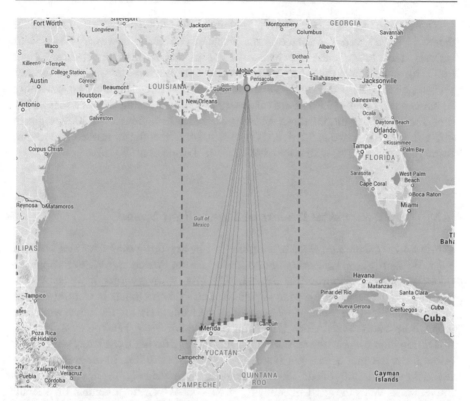

Fig. 4.2 Assumed tracks of Swainson's Thrushes recorded in the Yucatan Peninsula from Movebank.org. Dashed box illustrates the area that was annotated with gridded environmental information

Alabama had an Automated Receiving Unit (ARU, Sparrow systems, Inc.) attached to six antennas (Kays et al. 2011). The towers in Alabama recorded when the birds departed on their flight across the Gulf of Mexico. The towers in the Yucatan each consisted of an ARU with two antennas and recorded when the birds arrived on the continent after the flight across the Gulf of Mexico. The Yucatan towers were located in Sisal, Chicxulub, Dzilam de Bravo, Rio Lagartos Field Station, El Cuyo Field Station, Holbox, and Isla Contoy, creating a "fence" of receivers.

For each bird that was detected on at least one receiver in Alabama and one in Yucatan, we determined the crossing time by calculating the time difference between the last detection in Alabama and first detection in Yucatan. In cases where the bird was observed by more than one tower in Yucatan, we assigned an average between-towers location to the arrival detection location. We defined the migration-track bearing over the entire trans-Gulf migration section (following a similar approach to Thorup et al. 2007), as the exact path was unknown, assuming a straight flight path between the departure and arrival locations (Fig. 4.2). The mean flight speed could then be calculated as the distance between the two detection locations divided by the crossing time.

There were 145 birds tagged from September 26, 2009 to October 21, 2014. Of these, 42 (29%) were redetected in the Yucatan Peninsula; however, 19 (45% of arrivals) took longer than expected to arrive at the Yucatan via a direct flight (<35 h), suggesting that they took circuitous routes around the Gulf or left Ft. Morgan and stopped over elsewhere before making direct flights. The remaining 23 successful arrivals (55% of arrivals) were recorded crossing within a reasonable time for a direct flight. Of these 23 birds, six were documented in 2009, eight in 2010, four in 2012, two in 2013, and three in 2014. Direct flights ranged in duration from 14.85 to 34.55 h, with a mean of 21.77 h (Deppe et al. 2015).

2.2   Environmental Drivers of Stork Flight Speed

White storks (*Ciconia ciconia*) breed from Europe to Northwest Africa and Western Asia. Western European birds migrate southwest via France and Western Spain to Gibraltar, where they cross the Strait of Gibraltar to overwinter in West Africa. Storks from the eastern European population leave their colonies in Central Europe along a south-eastern route. They fly through the Middle East to their overwintering areas in East and South Africa (Hagemeijer and Blair 1997; Watson et al. 1978).

During migration, white storks congregate in large flocks, a tendency which may allow faster thermal detection (Shamoun-Baranes et al. 2003). In order to stay in a flock, however, individuals need to respond appropriately to their neighbor's movements. With the goal of studying group behavior, the juveniles of a small stork colony in the South of Germany (47° 45′ 10.8″N, 8° 56′ 2.4″E) which consists of 22 nests was followed during migration. All birds belong to the western subpopulation that migrates through Western Europe to the northwestern parts of Africa.

Storks are large enough to carry a transmitting GPS with a relatively large battery that can obtain and report location and instantaneous speed and compass heading measurements at high frequency. In total, 61 juveniles, the colony's entire offspring, were equipped with high-resolution, solar GSM-GPS-ACC loggers (E-Obs GmbH; Munich, Germany) one week prior to fledging. The transmitters (weight 43 g) were attached using a Teflon-nylon harness (weight ~12 g). The total weight of transmitter and harness was 65 g, corresponding to approximately 2% of the mean body mass of white storks. The GPS is set to provide every 15 min high frequency (1 s) observations for 5 min.

We used one 5-min set of high-frequency data from 10 tagged birds, flying in the same area (within several km) during noontime of August 14, 2014. We calculated the flight speed using the incremental observations between two consecutive GPS locations. We did not calculate wind support as the wind dataset is much coarser in space and time than the secondly movements of the storks and the turbulent eddies that affect their direction. We averaged flight speed and all other annotated environmental variables over all the observations within the 4-min period, per bird.

2.3 Environmental Annotation: The Movebank Env-DATA System

Movebank is an online database of animal tracking data, which provides users with a place to store, manage, analyze, and share animal movement data (Kranstauber et al. 2011). The Env-DATA system is an extension of Movebank that automatically annotates tracks with environmental data (Dodge et al. 2013) (see full list of annotation variables the system enables in: https://www.movebank.org/node/7471).

There are two types of annotation interfaces available through Env-DATA: (1) a set trajectory, which uses the Env-DATA online graphic user interface (GUI) and the movement track stored in Movebank to annotate the observed movement coordinates and (2) a gridded geographical area that corresponds to a given movement track, but includes a gridded set of locations around the observed track and is annotated using an R program function. The thrush data only contained two locations for each bird, the last radio-telemetry identification upon departure and the first detection upon arrival. The actual path, between these two points, was unknown. We, therefore, used the gridded area annotation method to get an estimate of the mean environmental conditions the birds would experience during the flight in the approximate area of the migration (Fig. 4.2). The storks, on the other hand, had exact locations of each bird at 1-Hz resolution during the section of flight that we analyzed. We, therefore, used the trajectory annotation method for this dataset.

Env-DATA provides several interpolation methods. The bi-linear approach fits best with regularly gridded data of continuous variables and was used for all variables in this study. For variables that are provided on a 3-D mesh that includes a vertical coordinate (e.g., wind speed from reanalysis), bi-linear interpolation include eight points at the two heights above and below the reported elevation of the movement observation. In both bi-linear and weighted distance interpolations, data are first interpolated in space and then two spatially interpolated values from the last available environmental data before and the first after the movement observation time are further interpolated in time to match the spatiotemporal coordinate of the movement point (Dodge et al. 2013).

2.4 Annotation Variables

Remote-Sensing data and data from global weather reanalyses were used to generate information about specific environmental conditions during the flight. Weather reanalysis datasets are the products of atmospheric models that are forced with a large number of satellite, meteorological station, and weather balloon observations. For annotating the gridded area used for the thrush study, we used data from the NCEP North American Regional Reanalysis [NARR, at a spatial resolution of ~32 × 32 km, and at 3-hourly time resolution (Mesinger et al. 2006)): the U (East-West) and V (North-South) components of wind speed at 10 m above sea level, air temperature and humidity at 2 m above sea level, and the height of the planetary boundary layer (HPBL). Additionally, we used the cumulative daily precipitation from TRMM [with a spatial resolution of 0.25° and a temporal resolution of 3-hourly (Kummerow et al. 1998)]. Though the birds fly high above the sea surface, humidity and temperature tend to be well correlated within the atmospheric boundary layer. However, and particularly at nighttime, narrow wind jets can develop above the thin nightly boundary layer. Nonetheless, because the flight height of the thrushes was unknown, we were limited to choosing an arbitrary elevation. We chose the surface values of the atmospheric variables, as representatives to the values throughout the atmospheric column, rather than the values at any particular height. Unlike Deppe et al. (2015), which provide a more conclusive analysis to an expanded version of the same dataset, we did not include any internal variables, such as sex, age, and fat reserves in the analysis of the effects of environmental variables we conduct here.

Variables were interpolated to a gridded array of location, covering the region of interest (dashed box in Fig. 4.2). The coordinates of four corners of the annotated grid were: 91°W 31°N; 91°W 20°N; 85°W 20°N; and 85°W 31°N, with time varying between birds. We gridded that region at a resolution of 25 × 25 km. This resolution was chosen somewhat arbitrarily, to provide fine and uniform coverage across all the source datasets (each with a different grid and resolution). We define the flight time as the period between the departure time and arrival time. Depending on the dataset, data was available at temporal sampling intervals of 6, 4, or 3 h. We used all the data observation times that were available during the flight time. For each bird, data for all the locations within the region were averaged in each data time step, and then the regional mean was averaged in time over the Gulf-crossing event period per bird. The storks' tracks were annotated using the set trajectory approach, with U and V components of wind speed from ECMWF at the observed height of their flight and with thermal uplift, derived from ECMWF variables.

Wind-speed components data were further processed into three wind characteristics: wind speed, direction, and support defined as:

$$w_s = \sqrt{U^2 + V^2} \tag{4.1}$$

$$w_d = \tan^{-1}(U, V) \tag{4.2}$$

$$w_p = w_s \cos(w_d - h_d) \tag{4.3}$$

where w_s is wind speed in m/s, w_d is wind direction in degrees, w_p is wind support in m/s, and h_d is the movement bearing. Wind support is defined as the component of the wind speed at the direction of movement.

2.5   Hierarchal Modeling

We tested the pairwise correlation between each environmental variable and the flight speed. We only used environmental variables for which we had a hypothesis of how they may affect flight speed. A list of the variables we tested appears in Tables 4.1 and 4.2. We assume that wind support (as defined above, Eq. 4.3) assists the bird to fly faster with lower energy expenditure, uplift assists the bird in saving energy while maintaining or gaining elevation, and temperature and humidity affect the physiological state of the birds at flight. Additionally, some variables can provide indirect indications for conditions that are supportive of fast flight; for example, the high planetary boundary layers (PBLs) are typical of conditions with stronger uplift than during times of low PBLs, and a certain combination of temperature and humidity over the Gulf of Mexico is associated with the movement of weather fronts that provide preferable wind direction (as found for our study area, Deppe et al. 2015). We then discarded the variables that did not have a significant pairwise correlation with flight speed. We used all the variables that significantly correlated with the total Gulf-crossing flight time to construct an empirical model using a step-wise hierarchal approach. We built a set of

Table 4.1 Summary of correlation data (coefficient of determination, R^2 and significance, P) between listed variables and the flight speed of the thrushes

Model	R^2	AICc	P
HPBL	**0.4676**	**104.07**	**<0.001**
Wind speed	**0.4574**	**104.05**	**<0.001**
Temperature	0.3526	107.40	0.0028
Humidity	0.2227	107.44	0.0230
Precipitation			NS
Wind support			NS

NS marks nonsignificant results ($P > 0.05$). The Akaike information criterion, AICc, for the whole model indicates the justification for adding each incremental variable. Variables were added in the order of their pairwise R^2 if they had a significant correlation with flight speed. Variables that had a significant and justified effect on the overall model and were included in the final model are marked in bold type

Table 4.2 Summary of correlation data (R^2 and significance, P) between listed variables and the flight speed of the storks

Model	Coefficient	R^2	AICc	P
Thermal uplift	**−1534**	**0.90**	**−7.50**	**<0.001**
HPBL	**4.38**	**0.88**	**−11.11**	**<0.001**
Orographic uplift		0.84	−10.01	<0.001
Topographic elevation		0.63	−9.58	0.006
Wind speed	**1.53**	**0.44**	**−13.13**	**0.036**

Variables that had a significant and justified effect on the overall model (according to the decrease in whole-model AICc compared to the simpler model) were included in the final model and are marked in bold type. The final model coefficient for each variable is also listed

increasingly complex models using the aforementioned variables. For each new model, a comparison of the goodness-of-fit, significance, and the information criterion were used to determine if the last variable that was incrementally added provided a justified improvement to the model. A typical concern with environmental variables is cross-correlation. Most variables have similar diurnal, seasonal, and spatial variation that creates a strong cross-correlation between them. We use Akaike's information criterion to determine the most parsimonious model. For example, in a case where two environmental variables are strongly correlated, including both correlated variables, will create a less parsimonious model compared to an alternative one that only includes one of these correlated variables, because both variables include the same information. We used the Akaike information criterion ($AICc$) to reconcile the goodness of the fit with the number of parameters used in the model to determine the most justified model (Akaike 1974). Regressions and stepwise model statistics were calculated with the JMP.11 software (SAS Corporation, Cary, NC).

3 Results

3.1 Swainson's Thrushes

For the thrushes crossing the Gulf of Mexico, the height of the PBL (NARR variable—HPBL) was the variable most strongly correlated with the flight speed of the birds (Table 4.1). As the height of the boundary layer increases the birds tend to fly faster, as shown in Fig. 4.3b. In addition to the boundary layer height, the wind speed was also significantly correlated with flight speed (Table 4.1). Finally, both humidity and temperature were also correlated with flight speed (Table 4.1). In both cases, the flight speed decreases as the temperature and humidity increase (Fig. 4.3). Precipitation and wind support were not significantly correlated with mean flight speed (Table 4.1). The final model included HPBL and wind speed and did not include temperature, humidity, wind support, or precipitation.

3.2 White Storks

For storks flying over central Europe in their southern migration, we found that thermal uplift and wind speed were the major drivers of flight speed. While orographic uplift, boundary layer height, and ground elevation also had significant pairwise correlation with flight speed, they did not add significant information to the model, probably because they are strongly cross-correlated with wind speed and thermal uplift.

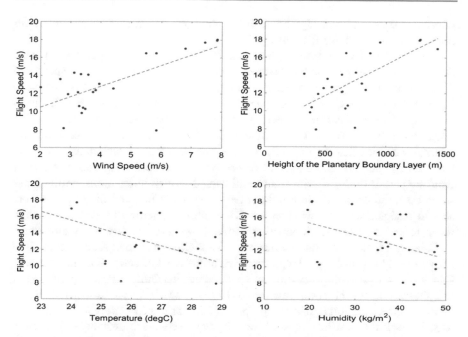

Fig. 4.3 Correlations between thrush flight speed and (**a**) wind speed, (**b**) height of the PBL, (**c**) temperature, and (**d**) Humidity

4   Discussion

The final model for the thrushes included boundary layer height and wind speed. As shown in Fig. 4.3b, the flight speed increases with increasing boundary layer height. The planctary boundary layer is the atmospheric layer that is directly influenced by the planetary surface. Over the ocean, the height of this layer is mostly affected by the ocean-surface temperature (von Engeln and Teixeira 2013). A warmer ocean relative to the atmosphere creates strong sensible heat flux that drives thermal uplift and increases the height of the boundary layer. Conditions such as these would be beneficial to the thrushes as they fly across the Gulf of Mexico. Although the thrushes aren't soaring birds, like the storks, the thermals could still help them conserve energy by gliding through the buoyant eddies instead of flapping their wings. This may be particularly true over the ocean at nighttime where thermals in general are particularly weak. A slightly thermally conductive boundary layer may make a big difference than a shallower stable boundary layer that suppresses thermal uplift. The algorithm used in Movebank to determine thermal uplift is only applied on land. The NARR-based PBL height is, therefore, the only environ- mental variable available over the Gulf of Mexico that can indirectly infer the availability of thermal uplift. PBL height may also be indicative of flight height, as often, and particularly at nighttime, strong wind jets which may be preferable for flight develop just above the PBL.

We found that wind speed reduces the Gulf-crossing time regardless of wind direction. Surprisingly, the wind support, which incorporates both wind speed and direction and the migration heading, did not have a significant effect on the mean flight speed over the Gulf. This could potentially be explained by the fact that most birds timed their departure to periods when the winds were generally oriented towards the bearing of the Gulf crossing. Deppe et al. (2015), who accounted for internal variables that significantly affected the departure decision, and included more individuals and more species of passerines, did find a significant effect of wind support. It is also possible that small wind direction fluctuations, during times when birds did cross the Gulf, did not have a strong effect and the thrushes were able to navigate with the wind while still, eventually, getting to their destination, regardless of the exact wind direction within the general south-western direction.

Finally, it is possible that the mean wind direction over the entire Gulf at 10 m above ground was not a good representative of the wind direction the birds actually flew with, at a particular elevation and path across the Gulf. A strong limitation of this analysis for the thrushes, which stems from the availability of thrush movement data, is the assumption that the flight path of the thrushes is straight. The method used to track the thrushes allowed for only two recorded timestamps—departure from Alabama and arrival in Mexico, everything in between these points is unknown. A 3D movement model (Rachel T. Bolus, personal communication) for the thrushes indicates that selection of flight elevation may provide additional wind support, compared to the conditions near sea the surface, that is used by the birds but cannot be estimated by the 10-m wind data. The small size of these birds does not permit GPS sensors, and information on the track is not available beyond the departure and arrival times. The area that we annotated was relatively small and the time span matched each bird's flight time, and therefore the averaged values of that sub-domain provided a reasonable, though limited, representation of the conditions met by the bird. In future studies, individual-based movement models (as done for soaring birds by van Loon et al. 2011) could provide predictions for hypothetical routs for the observed Gulf-crossing events, and these could be used to improve the time and space resolution of the environmental data used.

The final model for the storks included thermal uplift, boundary layer height, and wind speed. Two of these variables are closely related, as the thermal uplift directly affects the boundary layer height. The storks are soaring birds, and it was expected that the thermal uplift would be a large factor that affected flight speed (Kemp et al. 2010; Shamoun-Baranes et al. 2003; van Loon et al. 2011; Flack et al. 2016). However, it is interesting to note that the stronger thermals actually led to a slower flight speed (negative coefficient, Table 4.2). Over long time periods, storks will typically spend less time in updrafts when thermal uplift is strong and produces strong thermals, than when thermal uplift is weak, because they gain height faster. However, the very short time period sampled here represents a typical period of rising within a single thermal. Therefore, for this limited dataset and high temporal resolution but very short period, we hypothesize to see the opposite relationship.

Over just a few minutes (i.e., during a single thermal and at a very small domain), individuals in locations with stronger uplift were probably engaged in thermal soaring within that thermal. At the same time, individuals over areas with lower thermal uplift were probably gliding in a straight line on their way from one thermal, which they used a short time before the observation period, to the next one they will use after the observation period. Because we used the horizontal distance covered by the birds over the 5-min period to calculate the flight speed, birds that were riding thermals mostly moved in circles and their overall speed would not be as large as the birds that were gliding in a straight line. As mentioned above, higher boundary layer is associated with stronger uplift. However, over land the boundary layer height is also associated with topographic elevation. This explains the fact that after including HPBL in the model, the topographic elevation did not contribute new information to the model and was not included in the final model, despite having a very high pairwise correlation with flight speed.

Contrary to the thrushes study, there was an abundance of location observation data for the storks—but it was all during a very short period (4 min) and over a small domain. The limiting factor here was the available resolution of the environmental data. The dataset we used is at roughly 75 × 75 km. Therefore, it included very small variation over the area where the 10 storks' flights took place.

5   Conclusions

Track-annotation tools provide information that helps us understand movement ecology more fully and particularly help quantify the interactions between internal and external states of migrating birds and other animals. This type of analysis can identify the environmental drivers of migration paths, including variables that were impossible to observe directly at the time and place the bird flew. These data have numerous applications, each working to increase the general understanding of aerial migrants and the factors that affect them.

It is important to point out that track annotation only provides information about the location and time of the observed track. As such, it is strongly dependent on the resolution of the movement locations and the resolution of the environmental data sources. As the two test cases analyzed here show, our conclusions may have been different if the thrush movement data was at higher spatiotemporal resolution and included elevation information and more observed locations during the Gulf crossing. Similarly the stork flight analysis could have been improved with wind and uplift data at much higher resolution than possibly available from global reanalyses models. New tag technology and approaches to estimate wind speed, direction, and uplift from the high resolution GPS measurements provided by the bird-borne tags (Treep et al. 2016) may offer a future solution to this need. The limited scope of track-annotation analysis precludes it from being directly used for analysis of habitat selection, flight strategy, and migration departure decisions. These analyses are intrinsically dependent on comparing the conditions along the observed track locations, with conditions when the birds were and were not present. In such cases, the statistics and distribution of selected environmental conditions in a prescribed reference area or a set virtual null observations or simulated tracks can be compared with the distributions of the same environmental variables that were annotated to the tracks to draw conclusions about the location, flight, or habitat preferences of migrating birds (e.g., Bohrer et al. 2012).

An increased understanding of the interactions between the environment and aerial migratory movement will be increasingly important as our planet's climate continues to change. A change in climate will likely affect weather patterns in the areas that these birds migrate.

Acknowledgements The authors would like to acknowledge advice and comments from Rachel Bolus. The studies described here were supported by the National Science Foundation (IOS Award #1146832, 1147096, 1145952, and 1147022) and NASA (grant #NNX11AP61G). Additional support for the thrush study was provided by the National Geographic Society Committee on Research and Exploration (Award # 8971-11), Eastern Illinois University (Research and Creative Activity Awards to J.L.D.), University of Illinois Urbana-Champaign, the Max-Planck Institute for Ornithology, and The University of Southern Mississippi. The data on the white storks was collected with the help of Wolfgang Fiedler.

References

Akaike H (1974) A new look at the statistical model identification. IEEE Trans Autom Control 19 (6):716–723. https://doi.org/10.1109/tac.1974.1100705

Bartlam-Brooks HLA, Beck PSA, Bohrer G, Harris S (2013) When should you head for greener pastures? Using satellite images to predict a zebra migration and reveal its cues. J Geophys Res 118:1427–1437. https://doi.org/10.1002/jgrg.20096

Bohrer G, Brandes D, Mandel JT, Bildstein KL, Miller TA, Lanzone M, Katzner T, Maisonneuve C, Trembley JA (2012) Estimating updraft velocity components over large spatial scales: contrasting migration strategies of golden eagles and turkey vultures. Ecol Lett 15:96–103. https://doi.org/10.1111/j.1461-0248.2011.01713.x

Bohrer G, Beck PSA, Ngene SM, Skidmore AK, Douglas-Hamilton I (2014) Elephant movement closely tracks precipitation-driven vegetation dynamics in a Kenyan forest-savanna landscape. Mov Ecol 2:2. https://doi.org/10.1186/2051-3933-2-2

Bowlin MS, Wikelski M (2008) Pointed wings, low wingloading and calm air reduce migratory flight costs in songbirds. Plos One 3(5):e2154. https://doi.org/10.1371/journal.pone.0002154

Bowlin MS, Bisson IA, Shamoun-Baranes J, Reichard JD, Sapir N, Marra PP, Kunz TH, Wilcove DS, Hedenström A, Guglielmo CG, Åkesson S, Ramenofsky M, Wikelski M (2010) Grand challenges in migration biology. Integr Comp Biol 50(3):261–279. https://doi.org/10.1093/icb/icq013

Cimino MA, Fraser WR, Irwin AJ, Oliver MJ (2013) Satellite data identify decadal trends in the quality of *Pygoscelis* penguin chick-rearing habitat. Glob Change Biol 19(1):136–148. https://doi.org/10.1111/gcb.12016

Cochran WW, Wikelski M (2005) Individual migratory tactics of New World Catharus thrushes: current knowledge and future tracking options from space. In: Greenberg R, Marra PP (eds) Birds of two worlds: the ecology and evolution of migration. Johns Hopkins University Press, Baltimore, pp 274–289

Deppe JL, Ward MP, Bolus RT, Diehl RH, Celis-Murillo A, Zenzal TJ, Moore FR, Benson TJ, Smolinsky JA, Schofield LN, Enstrom DA, Paxton EH, Bohrer G, Beveroth TA, Raim A, Obringer RL, Delaney D, Cochran WW (2015) Fat, weather, and date affect migratory songbirds' departure decisions, routes, and time it takes to cross the Gulf of Mexico. Proc Natl Acad Sci 112(46):E6331–E6338. https://doi.org/10.1073/pnas.1503381112

Dodge S, Bohrer G, Weinzierl R, Davidson SC, Kays R, Douglas D, Cruz S, Han J, Brandes D, Wikelski M (2013) The environmental-data automated track annotation (Env-DATA) system: linking animal tracks with environmental data. Mov Ecol 1:3. https://doi.org/10.1186/2051-3933-1-3

Dodge S, Bohrer G, Bildstein K, Davidson SC, Weinzierl R, Bechard MJ, Barber D, Kays R, Han J, Wikelski M (2014) Environmental drivers of variability in the movement ecology of turkey vultures (Cathartes aura) in North and South America. Philos Trans R Soc B 369 (1643):20130195. https://doi.org/10.1098/rstb.2013.0195

Flack A, Fiedler W, Blas J, Pokrovsky I, Kaatz M, Mitropolsky M, Aghababyan K, Fakriadis I, Makrigianni E, Jerzak L, Azafzaf H, Feltrup-Azafzaf C, Rotics S, Mokotjomela TM, Nathan R, Wikelski M (2016) Costs of migratory decisions: a comparison across eight white stork populations. Sci Adv 2(1):e1500931. https://doi.org/10.1126/sciadv.1500931

Hagemeijer WJM, Blair MJ (1997) The EBCC atlas of European breeding birds: their distribution and abundance. Poyser, London

Kays R, Tilak S, Crofoot M, Fountain T, Obando D, Ortega A, Kuemmeth F, Mandel J, Swenson G, Lambert T, Hirsch B, Wikelski M (2011) Tracking animal location and activity with an automated radio telemetry system in a tropical rainforest. Comput J 54(12):1931–1948. https://doi.org/10.1093/comjnl/bxr072

Kemp MU, Shamoun-Baranes J, van Gasteren H, Bouten W, van Loon EE (2010) Can wind help explain seasonal differences in avian migration speed? J Avian Biol 41(6):672 677. https://doi.org/10.1111/j.1600-048X.2010.05053.x

Kemp MU, van Loon EE, Shamoun-Baranes J, Bouten W (2012) RNCEP: global weather and climate data at your fingertips. Methods Ecol Evol 3:65–70

Kranstauber B, Cameron A, Weinzierl R, Fountain T, Tilak S, Wikelski M, Kays R (2011) The Movebank data model for animal tracking. Environ Model Softw 26(6):834–835. https://doi.org/10.1016/j.envsoft.2010.12.005

Kummerow C, Barnes W, Kozu T, Shiue J, Simpson J (1998) The tropical rainfall measuring mission (TRMM) sensor package. J Atmos Ocean Technol 15(3):809–817. https://doi.org/10.1175/1520-0426(1998)015<0809:TTRMMT>2.0.CO;2

Mandel JT, Bohrer G, Winkler DW, Barber DR, Houston CS, Bildstein KL (2011) Migration path annotation: cross-continental study of migration-flight response to environmental conditions. Ecol Appl 21(6):2258–2268

McLaren JD, Shamoun-Baranes J, Bouten W (2012) Wind selectivity and partial compensation for wind drift among nocturnally migrating passerines. Behav Ecol 23(5):1089–1101

Mesinger F, DiMego G, Kalnay E, Mitchell K, Shafran PC, Ebisuzaki W, Jovic D, Woollen J, Rogers E, Berbery EH, Ek MB, Fan Y, Grumbine R, Higgins W, Li H, Lin Y, Manikin G, Parrish D, Shi W (2006) North American regional reanalysis. Bull Am Meteorol Soc 87 (3):343–360

Nathan R, Getz WM, Revilla E, Holyoak M, Kadmon R, Saltz D, Smouse PE (2008) A movement ecology paradigm for unifying organismal movement research. Proc Natl Acad Sci USA 105 (49):19052–19059. https://doi.org/10.1073/pnas.0800375105

Phillips SJ, Dudík M (2008) Modeling of species distributions with Maxent: new extensions and a comprehensive evaluation. Ecography 31(2):161–175. https://doi.org/10.1111/j.0906-7590.2008.5203.x

Shamoun-Baranes J, Leshem Y, Yom-Tov Y, Liechti O (2003) Differential use of thermal convection by soaring birds over central Israel. Condor 105(2):208–218

Thorup K, Bisson IA, Bowlin MS, Holland RA, Wingfield JC, Ramenofsky M, Wikelski M (2007) Evidence for a navigational map stretching across the continental US in a migratory songbird. Proc Natl Acad Sci USA 104(46):18115–18119. https://doi.org/10.1073/pnas.0704734104

Treep H, Bohrer G, Shamoun-Baranes J, Duriez O, Frasson RPdM, Bouten W (2016) Using high resolution GPS tracking data of bird flight for meteorological observations. Bull Am Meteorol Soc (in press). doi:https://doi.org/10.1175/BAMS-D-14-00234.1

van Loon EE, Shamoun-Baranes J, Bouten W, Davis SL (2011) Understanding soaring bird
    migration through interactions and decisions at the individual level. J Theor Biol 270
    (1):112–126. https://doi.org/10.1016/j.jtbi.2010.10.038
von Engeln A, Teixeira J (2013) A planetary boundary layer height climatology derived from
    ECMWF reanalysis data. J Clim 26(17):6575–6590. https://doi.org/10.1175/JCLI-D-12-
    00385.1
Watson GE, Cramp S, Simmons KEL (1978) Handbook of the birds of Europe, the Middle East,
    and North Africa: the birds of the Western Palearctic, Ostrich to ducks, vol 1. Oxford
    University Press, Oxford

Physiological Aeroecology: Anatomical and Physiological Adaptations for Flight

5

Susanne Jenni-Eiermann and Robert B. Srygley

Abstract

Flight has evolved independently in birds, bats, and insects and was present in the Mesozoic pterosaurians that have disappeared. Of the roughly one million living animal species, more than three-quarters are flying insects. Flying is an extremely successful way of locomotion. At first glance, this seems surprising because leaving the ground and moving in the air is energetically expensive. We will therefore start with the question: why do some animals spend a substantial proportion of their life in the air? To generate lift, a few key features are required, and yet, animals show incredible diversity in their flight mechanics. We will review constraints imposed by body size including anatomical adaptations of the skeleton, muscles, and organs necessary to stay airborne with a special focus on the wings. Ecology of the aerial organism, such as diet or migration, has diversified flight styles and the physiological adaptations required to optimize performance. For example, animals are exposed to low temperatures and low oxygen pressure at high altitude, whereas overheating can pose a problem at low altitudes. Moreover, aerial prey can be particularly apparent to aerial predators resulting in selection on flight speed and maneuverability of predators and prey. Flight is energetically costly, much more costly than walking, with the majority of the cost dictated by body mass. Hence, adding weight load to fuel flight also adds to the cost of flight. We review energy supply

S. Jenni-Eiermann (✉)
Schweizerische Vogelwarte, Seerose 1, 6204 Sempach, Switzerland
e-mail: Susi.jenni@vogelwarte.ch

R.B. Srygley
U.S. Department of Agriculture, Northern Plains Research Lab, Agricultural Research Service, Sidney, MT, USA

Smithsonian Tropical Research Institute, Balboa, Republic of Panama

© Springer International Publishing AG, part of Springer Nature 2017
P.B. Chilson et al. (eds.), *Aeroecology*,
https://doi.org/10.1007/978-3-319-68576-2_5

for flight, and special adaptations for long-term flights. Aeroecology has resulted in extraordinary visual and aural sensory systems of predators, which in coordination with the locomotor system are under strong selection to detect and intercept prey in flight.

Box with Definitions

Aspect ratio of a wing is the ratio of its length to its breadth. A high aspect ratio indicates long, narrow wings (swift, albatross), whereas a low aspect ratio indicates short, stubby wings.

Drag is a slowing force caused by viscosity.

Gliding is powered by gravity. Gliding occurs with no active thrust; a glider must always descend relative to the air it moves through.

Gravity of Earth The relationship between gravitational acceleration and the downward weight force experienced by objects on earth is given by the equation force = mass × acceleration. Hence the more a body weighs, the more is the gravity.

Maneuverability is the ability to turn in a small radius.

Reynolds Number gives the ratio of pressure (inertial) and viscous contributions to drag, formally defined as (density * speed * characteristic length)/dynamic viscosity. The Reynolds number is important aerodynamically, because at low Reynolds number (small size and/or low speeds), wings tend to have lower lift-to-drag ratios.

Thrust is the forward force produced by the wings. During flight at uniform speed, air meets the bird at the flying speed. To keep flying at the same speed, the bird must generate an amount of thrust equal to the total drag on body and wing in horizontal direction. Therefore, the wings accelerate the air under their control backward to give it a higher speed behind the bird.

Vertical Lift is the upward force produced by wings that counteracts gravity. Because flapping is mostly not horizontal, the upward force is only a fraction of the lift produced by flapping wings.

Wing load is body weight divided by wing area.

1   Introduction

This chapter provides a very basic overview of anatomical and physiological adaptions to a life in the air. It is not within the scope of this chapter (and book) to deal with these subjects in depth. For further studies, the reader is referred to the textbooks of aerodynamics of flight and morphology (e.g., Dudley 2000; Alexander 2002; Videler 2005; Pennycuick 2008). There are representatives in nearly all vertebrate classes, such as squirrels, marsupials, lizards, snakes, big-footed frog, and flying fish, which have developed membranes enabling them to glide or to do controlled descents. In addition, several invertebrates, such as flying squid, canopy ants (Yanoviak et al. 2005), and bristletails (Yanoviak et al. 2009), are capable of controlled descents, whereas some caterpillars and spiders disperse long distances by ballooning (Taylor and Reling 1986). However, in this chapter, we will treat only animals able to use powered flight and spending a substantial proportion of their life cycle airborne.

1.1 Why Are Animals in the Air?

Why do so many animals move into the air? Judging by numbers, flying is extremely successful. Of the roughly one million living animal species, more than three-quarters are flying insects (Schmidt-Nielsen 1997). To leave the ground which supports their weight, they need force to move in the air and to overcome gravity. The advantages of being in the aerosphere are the wide space, which—outside forests—contains fewer barriers than the ground. Therefore, fast locomotion without a need to maneuver around obstacles is possible. This is of importance when escaping from predators, especially from predators unable to fly. It also allows the rapid location of potential mates, such as the male moth seeking the pheromone-emitting female or the male firefly seeking the luminescent female. However, animals in the air or on the ground are often easier to detect by airborne predators in hunting flights. Signals sent in the open space of the aerosphere, such as the aerial displays of the winnowing snipe or the color pattern on a butterfly's wings, transmit long distances to both conspecifics and potential predators.

The aerosphere is not only used for short flights; many species spend days or even months exclusively airborne. They take advantage of thermic conditions and forage while riding on the wind, such as the petrels. Others migrate each year to their wintering grounds and back, thereby covering thousands of kilometers over partially inhospitable grounds, such as oceans or deserts. Covering such distances within a relatively short time would not be possible on the ground. Flying animals typically move 10–20 times faster than a similar sized animal moves on the ground (Alexander 2002). Also crossings of oceans would be impossible for terrestrial animals.

Finally, some species rely on flight for most of their life. Alpine swifts for instance use the air as their main habitat and spend 200 days continuously in the air (Liechti et al. 2013). They are able to hunt, sleep, and copulate in flight and return to the ground just for breeding. Apparently they do not experience fatigue, sleep deprivation (see Schwilch et al. 2002), muscle damage (only little; Guglielmo et al. 2001), suppression of immunity (none found; Hasselquist et al. 2007), excessive stress hormone levels (none found in healthy birds; Falsone et al. 2009; Jenni-Eiermann et al. 2009), or elevated oxidative stress (Costantini et al. 2007; Jenni-Eiermann et al. 2014).

2   Anatomical Adaptations: What Is Necessary to Fly?

The ability to fly requires some general adaptations of the skeleton, since flight is dominated by severe aerodynamic and physiological constraints common to all flyers. For a better understanding of the functional morphology, knowledge of aerodynamics is required. Here, only some very basic rules of aerodynamics will be provided. A flying animal has to generate lift, a force in the upward direction, to offset gravity, a force in downward direction (Fig. 5.1). To move forward through the air, the flyer has to produce a horizontal force, the thrust, thereby overcoming a slowing force, called drag. The direction of the lift is perpendicular to the movement of air over the wing. Newton's laws predict that on average over time the total vertical component of force must equal weight and the sum of the drag forces must be balanced by the generated thrust (Norberg 1990; Videler 2005). Hence, in order to fly, it is crucial that the vertical forces weight and lift and the horizontal forces drag and thrust are balanced.

The following paragraphs will show various morphological adaptations of the skeleton, the forelimbs, and the body shape which different vertebrate classes and families have evolved.

2.1   Body Shape

As described above, an efficient flyer should produce as little drag as possible to keep the costs for thrust low. The more streamlined a body is the lower is the drag. And indeed, birds and bats have—in spite of a large number of species—maintained to a great deal structural similarity. Birds have streamlined bodies, their forelimbs evolved to wings, and only their hindlimbs are used for walking or perching. The more time a species spends in the air, the more streamlined is its body shape. Bats are not as aerodynamically shaped as birds, but they all have wings and flight muscles (Alexander 2002).

Fig. 5.1 For an animal, such as this pigeon, flying forward at a steady speed without changes in altitude, vertical lift, and thrust counter the pigeon's weight and drag on the body and wings

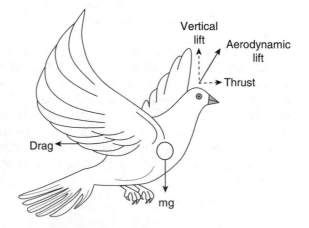

As with birds and bats, insect wings attach to the thorax and the thoracic musculature powers the flapping wings. However, insects evolved their wings independently of their legs, and this may have allowed greater variation of body and wing forms than is observed in bats and birds. The bodies of some insects are fusiform, in part to minimize drag but also to maximize load carried near to the wings. For example, hawkmoths are streamlined to such an extent that they appear like hummingbirds when hovering at flowers. By positioning more mass near to the center of gravity, the rotational moment of inertia is decreased allowing the insect to pitch forward more readily to accelerate its speed or yaw more readily in maneuvers. This is evident in the short, squat body shape of butterflies that fly fast and maneuver to escape predators (Chai and Srygley 1990; Srygley and Chai 1990a; Srygley and Dudley 1993; Srygley and Kingsolver 2000; Kingsolver and Srygley 2000). However, most insects are so small that they operate in a Reynolds number where viscosity dominates over inertial contributions to drag and large variation in body form has little aerodynamic effect.

2.2   Body Mass

Flight costs strongly depend on body mass, in bats as well as in birds (Speakman and Thomas 2003). In geometrically similar birds (isometric scaling), theory predicts that the power required for flapping flight increases with about the 7/6 power of body mass, while available power increases only with the 5/6 power of body mass, because larger birds beat their wings at a lower frequency than smaller ones (Pennycuick 2008). This would result in a theoretical limit of about 16 kg body mass for a flapping flying bird. Indeed, body mass of the heaviest birds known for prolonged powered flight is up to 16 kg (great bustard *Otis tarda*), 20 kg (kori bustard *Ardeotis kori*), and 22.4 kg (mute swan *Cygnus olor*), but this does not mean that heavier birds (and pterosaurs) cannot fly. They might reduce the power needed for flight by soaring in upwinds (in thermals or behind waves). However, it seems that large flapping flyers are not geometrically similar to small flapping flyers and the power for flapping flight actually scales to the power of 0.67 with body mass (Videler 2005).

Therefore, reduction of body mass by developing pneumatic bones is an efficient anatomical adaptation to save flight costs, a strategy found in birds only. Pneumatic bones are filled with air spaces instead of bone and marrow and therefore lighter in weight than similar sized bones of other vertebrates. In skull bones, air spaces arise from nasal passageways, whereas those in the vertebrae, sternum, ribs, pelvis, humerus, and femur are connected to either lung sacs or the lungs directly.

Adding weight load to fuel flight also adds to the cost of flight. When flight fuel is at a premium, as often occurs in long-distance migrants, then natural selection operates to conserve energy (Alerstam et al. 2003). Power saved by minimizing weight (low wing loadings) is compromised by the needs to have mass allocated to flight muscles and fuel to power the long-distance flights. In the bar-tailed godwit *Limosa lapponica* and the red knot *Calidris canutus islandica,* birds with exceptionally long, nonstop flights, another adaptation to reduce body mass was found. The gut is reduced shortly before departure, apparently to reduce the mass that is not required to sustain flight (Piersma and Gill 1998; Piersma et al. 1999; Landys-Ciannelli et al. 2003). For insects, the oogenesis flight syndrome (Johnson 1969) characterizes many migrants that first migrate and then reproduce. By delaying reproduction, mass that is not allocated to muscle or fuel to power flight is minimized.

Limitations on load might also constrain the diets of flying birds and bats. For example, power required for flight might constrain the evolution of folivory in birds and bats not only because of the added weight of a bolus of leaves but also because leaf digestion typically requires an evolutionary extension of the gut which adds bulk (Dudley and Vermeij 1992). Only a few birds, such as the hoatzin, specialize on leaves and these only fly short distances if at all. However, bats from at least 16 species extract and swallow the juices from leaves and then expel the residual fibers to reduce weight (Kunz and Ingalls 1994).

2.3   Wings

2.3.1   Properties of Wings

All animals that fly under power do so by flapping. The key innovations required for powered flight in animals are therefore adaptations that permit and improve the flapping movements. The wings generate lift by the Bernoulli principle: typically air passes over the dorsal surface faster than it moves under the wing, which generates a low pressure above the wing to offset gravity and generate thrust. This mechanism is known as the bound vortex because the vortex encircles the wing's upper and lower surfaces (Fig. 5.2). Although airplanes also utilize a bound vortex to generate lift, flapping wings are different in that they must generate a new vortex with each downstroke and sometimes also again during the upstroke of the wing (compare Fig. 5.3a, b). Hence, the constraint of stopping and starting the wing

Fig. 5.2 (a) More rapid flow over the wing than under the wing generates lift; the resultant of this differential flow is a bound vortex that encircles the wing. (b) Due to the conservation of momentum, a starting vortex equal and opposite to the bound vortex also develops

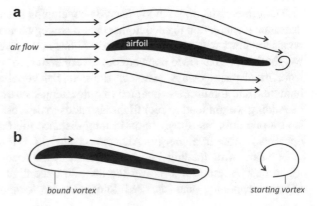

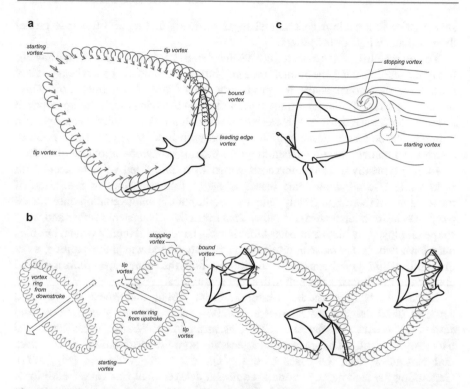

Fig. 5.3 (a) In gliding flight, a swift has a bound vortex encircling the wing along its chord, tip vortices formed from leakage over the wing tips, and a starting vortex left far behind from where the glide first began. Together these make an elongated vortex ring just like a fixed wing aircraft. The swift also has a leading edge vortex (on the upper surface of each wing but shown on only one) that generates additional lift. The swept back wings of the swallow are believed to stabilize the leading-edge vortex at even low angles of attack, as also seen in delta winged aircraft. (b) In flapping flight, sequential views of a bat are drawn with the first at the beginning of the downstroke, the second after it has completed the downstroke and is beginning the upstroke, and the third near the top of the upstroke. Unlike a gliding animal, the starting vortex, tip vortices, and bound vortex are generated with each downstroke of the wing and often during the upstroke as well, as shown here. From the beginning to the end of the downstroke and from the beginning to near the top of the upstroke, the bat is shown generating tip vortices that stretch between the sequential positions of the wing tips. The starting vortex and the bound vortex complete the donut-like ring through which most of the force generated by the circulating air is thrust. At the end of each half stroke, the bat sloughs off the bound vortex as the stopping vortex, one of which is to its left forming a part of the vortex ring generated from the sweep of the wings during the upstroke. Complete rings from the previous downstroke and upstroke form the vortex wake to the left of the bat. The arrow drawn through the vortex ring is a rough estimate of the direction of the force generated by the halfstroke. The downstroke generates most of the vertical lift and some of the thrust. (c) When wings of many butterflies and some other insects clap together at the top of the upstroke, they force a pocket of air out from between them and generate a vortex ring that rotates inward and backward, mainly generating thrust. These tracings of smoke trails from Srygley and Thomas (2002) show a cross section through the ring that is on a plane parallel to the page

stroke twice in a cycle makes animal flight unsteady and more difficult to model than airplane wings or propellers.

To generate lift, wings have to fulfill some general properties. The wings have to be rigid enough to carry the animal's weight, but they should not be too heavy. They must have the proper shape for producing lift to keep the animal aloft while minimizing drag. The animal must be able to flap the wings with the appropriate motions to produce the thrust that drives it through air. And finally, the shape of a wing should be aerodynamic and be powered with little energy, but also good to control and maneuver and—depending on foraging behavior—allow rapid flights.

Insects typically have membranous wings with fluid-filled veins to keep them rigid while also allowing some insect wings to fold when not in use. Lack of musculature in the wings greatly reduces weight but the wing venation only allows passive conformational changes, rather than the active changes in camber and wing shape employed by birds and bats. Most insects have four flapping wings, but flies only have two. Insect musculature can act directly on the wings for flapping, as in dragonflies and grasshoppers, or indirectly flap the wings by conformational changes in the thorax, as in butterflies, bees, and flies.

Some insects (hawkmoths: Ellington et al. 1996; butterflies: Srygley and Thomas 2002; dragonflies: Thomas et al. 2004; bees: Bomphrey et al. 2009) and some birds (swifts: Videler et al. 2004; hummingbirds: Warrick et al. 2005) and bats (Mujires et al. 2008) are known to generate extra lift with leading-edge vortices that corkscrew from the bases of the wings out to the wing tips (Fig. 5.3a). Generating the leading edge vortex requires a delayed stall that causes additional drag. So the leading edge vortex is utilized mostly during take-offs, slow flights, or maneuvers and less in fast, steady flight when sufficient air is moving across the wing to generate the lift required. However, the swept-back wings of swifts may allow maintenance of the leading-edge vortex at low angles of attack so that it can even be used at fast flight speeds (Fig. 5.3a, Videler et al. 2004).

There are other mechanisms that insects use to generate lift more readily. Some butterflies (Ellington 1984) and flies (Lehmann 2008) clap their wings together at the top of the upstroke (Fig. 5.3c), which causes air to begin to move across each wing pair's dorsal surfaces, and then they fling their wing apart so that a bound vortex equal and opposite to one another in circulation is generated on each of the two wing pairs (the clap and fling, Ellington 1984). Flies (Dickinson et al. 1999; Birch and Dickinson 2003) and butterflies (Srygley and Thomas 2002) are able to capture energy from their wake, which depending on the passing of the wing across the wake can generate lift in nearly any direction and potentially result in rapid maneuvers. The independent movement of the dragonfly's four wings allows control over the fore- and hindwing stroke cycles so that each wing can interact with the wake of its pair (Lehmann 2008). Wakes of insects persist only briefly for a short distance behind the wings. However, large or fast flying birds generate wakes that extend several wing spans and persist for tens of wingbeats (Swartz et al. 2008). These wakes are exploited to enhance lift of neighboring birds in the characteristic V-formation of migrating flocks (e.g., ibises, Portugal et al. 2014) but not in the less structured flocks of pigeons (Usherwood et al. 2011).

2.3.2   How Does Ecology Affect Flight Type and Wing Morphology?

The ecology of the aerial organism, such as foraging or migration, has diversified flight styles and the anatomical and physiological adaptations required to optimize flight performance. A given species might use different flight types under different circumstances. However, one flight type mostly dominates depending on the habitat where the animal lives (Alexander 2002) and its ecology. Aeroecology uniquely provides an environment where strong selective forces on traits that pertain to flight have been identified and the evolution of power and performance in flying organisms can be investigated. This approach proves useful because it provides empirical data to test biomechanical theory. In addition, theory may be used to generate models of optimality and ideally map traits on an adaptive landscape to identify trade-offs and other evolutionary constraints on flight.

As discussed above, flight is energetically costly with the majority of the cost dictated by body mass. Therefore, body mass is a factor affecting flight style. Large birds expend energy at rates close to the limits of their abilities while sustaining flight in still air. It is therefore not surprising that most large flying birds spend more time on low-cost gliding flight than in expensive flapping flight. Smaller birds in comparison have a surplus of power that can be used to improve acceleration, load carrying ability, or maneuverability. Another way to save energy by flight style is the intermittent mode of flight as finches or skylarks use. They alternate between flapping and folding the wings. Whenever the drag due to the air friction on the outstretched wings exceeds the drag due to lift, they can save energy when the wings are folded a part of the time. This can be true only for birds that fly at speeds well in excess of their minimum-drag speed and therefore it can only be true for small birds with plenty of power to spare (Bicudo et al. 2010).

Predator–prey interactions provide a context where maneuverability is under strong selection. Bats are unable to detect echoes from small insects until 1–5 m distance, which makes maneuvering to capture extremely important (Jones and Rydell 2003). Both a small turning radius and the ability to roll into a bank quickly are aspects of maneuverability (Norberg and Rayner 1987). However, in the capture of mobile prey, linear acceleration is another important feature of capture success.

For birds and bats, large, long wings reduce the turning radius in banked turns, whereas rolling into the turn is improved by positioning more wing mass near to the axis of rotation. In swifts and swallows, linear acceleration is favored by high wing loading and low aspect ratio, which puts these features in direct conflict with those that minimize turning radius (Warrick 1998). For bats, maneuvering has probably been a strong selective force to limit body mass (Jones and Rydell 2003). Maneuverability is inversely related to mass and wing span of leaf-nosed bats, whereas it improves in those with wings that permit deeper wing cambering (Stockwell 2001). Maneuverability is also achieved by low wing loading and positioning more wing mass near to the wing base, lowering the rotational moment of inertia and improving acceleration in high speed turns. Bats that fly more slowly may maneuver well at low speeds if wing area is positioned far from the wing base so that strong asymmetrical forces can be generated in turns (Swartz et al. 2003).

Fast and highly maneuverable flight that characterizes many insect species is one means of escape for those species that are highly sought by predators. Perhaps less well known is the slow, steady flight of many distasteful species, which may be a cue for predators that the insect is not worth pursuing (Chai and Srygley 1990; Srygley and Dudley 1993; Srygley and Kingsolver 1998). Allocation of more mass to the thoracic musculature and less to the abdomen is a common feature of butterflies that are palatable to birds, whereas those that are distasteful have smaller muscles and consequentially are able to allocate more mass to reproduction (Srygley and Chai 1990a; Marden and Chai 1991). Selection by aerial predators also constrains both the palatable species' spatial niches and their thermal physiology so that they operate at higher body temperatures relative to those of distasteful butterflies (Srygley and Chai 1990b). Moths also come in two flavors—palatable and distasteful—with some of the latter possessing visual signals to birds and audible warning signals to bats (Ratcliffe and Nydam 2008). Hence, we would predict that palatable moths that need to take evasive maneuvers and distasteful ones also have distinct flight styles and morphologies related to maneuverability.

3   Constraints of Altitude

In a stationary atmosphere from sea level to 11 km, mean air temperature drops monotonically (6.5 °C per km above ground level) with altitude at a given time and location, air pressure and the partial pressure of oxygen decline exponentially, and air density also declines (Fig. 5.4). More typically, the aerosphere is not stationary and is dotted with warm air rising in thermals, downdrafts of cooler air, and inversion layers where temperature increases with altitude. Birds and insects use currents generated by spatial differences in air temperature and the occurrence of thermals to reduce costs of flight (Westbrook 2008). However, the lower temperatures and limited O_2 aloft might also constrain flight height. For example at 5500 m above sea level, air pressure and hence O_2 partial pressure is only 50% of that at sea level. Hummingbirds are unable to hover when the oxygen concentration is reduced to 11–14%, corresponding to the O_2 partial pressure at approximately 5500 m (Chai and Dudley 1996; Altshuler and Dudley 2003). Other birds spend long periods at these elevations and hence with about half the O_2 partial pressure available to sea level residents. They are highly tolerant to hypoxia to levels that are deleterious to mammals. The reduction of O_2 partial pressure is to some extent counteracted by an increase in the gaseous diffusion coefficient which varies in inverse proportion to total pressure. Nevertheless, the question is: how do birds and insects manage to meet high aerobic demands at high altitudes?

How to avoid hypoxia Birds have evolved numerous traits for enhanced O_2 uptake and delivery from the pulmonary system to the circulating blood and the muscles fibers. Here we will pick out only some traits. For a review, please see, e.g., Scott et al. (2015).

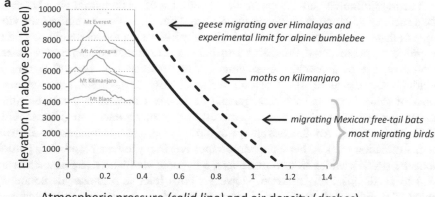

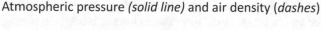

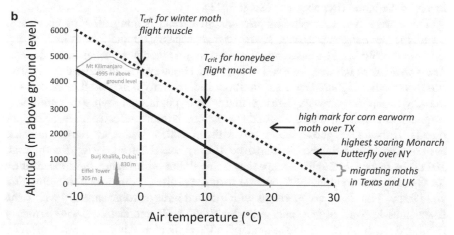

Fig. 5.4 (a) The decline in air pressure (atm) and air density (kg/m³) with elevation. Air density is also a function of temperature and relative humidity, and so to illustrate the change with elevation, we have assumed a temperature of 30 °C at sea level and 20% relative humidity. Profiles show elevations of select mountains around the world. Representative elevations for bar-headed geese migrating over the Himalayas and Mexican free-tailed bats migrating in Colorado are indicated with arrows, and the range at which most birds migrate is indicated with brackets. The maximum elevation that moths are active on Mt. Kilimanjaro is indicated with an arrow as is the experimental limit for bumblebee flight. (b) The decline in ambient temperature (°C) with altitude in a stationary atmosphere, assuming a temperature of 30 and 20 °C at ground level. Representative altitudes for ground-based radar-detected moth migrations are indicated with brackets. The highest records for a *Heliothis zea* moth (Westbrook 2008) and a *Danaus plexippus* butterfly (Gibo 1981) are indicated with arrows. Profiles show heights of human-made constructions and also the height of Mt. Kilimanjaro, a free-standing volcano that rises from its base at 900–5895 m. As a rough indication of the maximum altitude at which the winter moth and honeybee can fly above ground, we indicate the lower critical temperature (T_{crit}) for flight muscles of winter moths and honeybees to contract (Esch 1988). However, T_{crit} is based on muscle temperatures and not air temperatures, and so it is possible that these insects could continue to fly at altitudes with air temperatures lower than these lower critical temperatures (T_{crit}), as long as their muscle temperature remained above T_{crit} and the muscles were generating sufficient power for flight

The most important advantage of birds over most other terrestrial vertebrates is the structure and function of their respiratory system. Bird lungs communicate with air sacs that extend throughout the body, thereby increasing the total volume of the respiratory system about three times to that of a mammal of comparable size. Moreover, the parabronchi (finest branches of the bronchial system) permit a continuous, passage of air. The unidirectional flow through the gas-exchange units of avian lungs and the arrangement of airway and vessels create a cross-current gas exchanger attaining a superior efficiency for gas exchange (Scheid et al. 1972; Maina 2006). Another advantage of birds—in contrast to mammals—is their high ventilation rates in hypoxia. More rapid breathing lowers blood CO_2 partial pressure (PCO_2), which in turn restrains the hypoxic ventilatory response and can lead to an alkalosis of the blood. However, birds tolerate a greater depletion of blood CO_2, either with the capacity to restore blood pH rapidly and offset changes in blood PCO_2 or with a brain vasculature that is insensitive to low carbon dioxide levels in the blood (hypocapnia) (Scott 2011).

These were two examples of physiological traits helping all bird species to overcome low oxygen pressure. There exist also special adaptations in birds flying at extreme altitudes. The best studied species in this respect is the bar-headed goose *Anser indicus,* which was observed flying in the Himalayas at an altitude of 9000 m. For this or other highland species, additional adaptations were found in all steps of the O_2 transport pathway. During flight in hypoxia, the hypoxic ventilatory response is enhanced and the breathing pattern is more effective due to deeper breaths (bar-headed goose; Scott and Milsom 2007). Lungs are larger and the surface area for diffusion is increased (Carey and Morton 1976; Scott et al. 2011). In several high-altitude species, hemoglobin with higher baseline affinity for O_2 (McCracken et al. 2009) and an increase of the O_2 flux to the mitochondria were found. Finally, in comparisons with lowland birds, aerobic capacity within the flight muscle was higher and mitochondrial ATP production more strongly regulated, such as in the movement of ATP equivalents between sites of ATP supply and demand (reviewed in Scott 2011).

Bats do not have these adaptations; nevertheless, they fly with about the same oxygen consumption as birds (eight- to tenfold increase of resting metabolic rate). However, they do not reach altitudes as some bird species do and there are also no bat species known flying for months without resting.

Insect flight muscle is one of the most active muscle types and has a large density of mitochondria providing this exceptional aerobic capacity (Crabtree and Newsholme 1975). Gas exchange occurs in a branching system of tracheae that originate at the body surface in spiracle openings and interact directly with the mitochondria via the tracheoles. Changes in thoracic volume associated with flapping compress air sacs and trachea to actively ventilate the muscles. In some insects such as bees and wasps, abdominal pumping is used to augment ventilation independent of flight muscle activity (Harrison and Roberts 2000). At rest, American locusts *Schistocerca americana* double abdominal pumping frequency and increase tidal volume to tolerate O_2 partial pressure as low as 1.8 kPa (Greenlee and Harrison 2004). Dragonfly flight was not affected by oxygen levels as low as 10%, but few insects initiated flight in mixtures of 5% O_2 (Harrison and Lighton 1998).

Adaptations to low air density Flying in less dense air requires expansion of wing area (lower wing loading) or changes in wing kinematics to generate sufficient lift. In air made less dense while O_2 partial pressure was held constant (normoxic), hummingbirds increased stroke amplitude but showed only a slight increase in wing beat frequency when hovering (Altshuler and Dudley 2003). Wing beat frequency and airspeed of nocturnal migrating birds increased with altitude (Schmaljohann and Liechti 2009). Honeybees compensated for low-density, normoxic air by increasing stroke amplitude while holding wing beat frequency constant (Altshuler et al. 2005). By increasing stroke amplitude, high-altitude bumblebees were capable of hovering in air with density and O_2 partial pressure corresponding to an altitude of 9000 m (Dillon and Dudley 2014).

Adaptations to extreme temperatures In high altitudes, low oxygen pressures and low temperatures might constrain flight ability. In contrast, at low altitudes high temperatures and low relative humidity might limit endurance flight. Therefore, birds migrating long distances prefer to fly at night. However, temperatures of more than 30 °C and relative humidities as low as 30% can also occur at night. For example, these stressful conditions were measured between 0.5 and 1.5 km above sea level over the Western Sahara during the autumn migration of 64% of the nocturnal songbirds at that location (Schmaljohann et al. 2009). In wind tunnel experiments, an ambient temperature of 25 °C already seems to constrain flight. Birds either refuse to fly (e.g., Ward et al. 1999; Engel et al. 2006) or were dehydrated after a 2 h flight (Biesel and Nachtigall 1987). Migrants crossing deserts or oceans have no possibility to drink and rely on the production of metabolic water. Water gain produced by fat and protein oxidation should compensate for water loss via the lungs (see also Sect. 4). To avoid water loss, birds could choose altitudes with low temperatures. However, this might not always be the better solution. In the Western Sahara, the most favorable winds for autumn migration were found at low altitudes where temperatures were high, while unfavorable winds prevailed at high altitudes. According to Schmaljohann et al. (2009), wind direction was the key factor determining flight altitude. How the migrants managed not to become dehydrated is still under debate. One explanation might be that wind tunnel experiments are more stressful than free flights and overestimate water loss.

The foraging activities of bats are predominantly (but not exclusively) restricted to twilight and night. One hypothesis to explain the nocturnal activity of bats is the "hyperthermia hypothesis" presuming that solar radiation might drive body temperature to critical levels (e.g., Speakman et al. 1994). In contrast to birds, bat wing membranes are not insulated by feathers, have a low degree of reflectance, and indeed absorb up to 90% of short-wave solar radiation (Thomson and Speakman 1999). The core body temperature actually increases by 2 °C when flying at daylight (Voigt and Lewanzik 2011). Hence, solar radiation might be one factor driving bats into the nocturnal niche.

Larger insects generate substantial metabolic heat in flight and can maintain body temperatures elevated above ambient temperature, whereas the body temperature of small insects varies with that of the surroundings. Nocturnal insects are often layered at altitudes with the warmest airstreams (Reynolds et al. 2005; Westbrook 2008). Aphids fly in warmer updrafts sufficiently to remain neutrally buoyant and free-fall in cooler downdrafts (Reynolds and Reynolds 2009). Even larger insects in updrafts will reach an altitude at which convective cooling is too great to maintain an elevated body temperature, and they will cease flapping and free-fall.

Note that reduced atmospheric pressure and cold temperature may interact to further reduce flight performance. For example, the decline in flight performance of *Drosophila melanogaster* in atmospheric pressures that were similar to high-altitude conditions was more pronounced at colder temperatures (Dillon and Frazier 2006). Developmental temperature may also broaden the temperatures and hence the altitudes at which insects may be active. For example, tephritid flies that develop at lower temperatures have greater flight performance across a wider range of temperatures than those developing at high temperatures (Esterhuizen et al. 2014).

Effects of winds Wind speeds increase exponentially with height above ground up to several 100 m. Because most insects are small, wind speeds typically exceed maximum flight speed of the majority of insects, which have little chance to move long distances above the vegetation other than with the winds. The selection of favorable winds for take-off indicates that aphids and other small insects are not simply dispersing passively.

In contrast, larger insects, birds, and bats fly faster and can direct their movement independently of winds to a greater height above the ground (i.e., their flight boundary layer is broader; Srygley and Dudley 2008). However, they all require a ground reference to know their wind displacement. Hence, activity during the daytime may be necessary for drift compensation. In addition, an animal's visual system may limit the altitude at which a ground reference may be resolved. For example, the limited ability of the insects' compound eyes to resolve landmarks from a distance limits the height at which they are able to compensate for drift more so than the visual acuity of birds.

As flight height above ground increases, an animal's ability to compensate for wind drift decreases. However, insects, birds, and bats may all fly above their respective flight boundary layer to take advantage of favorable winds. For example, noctuid moths, principally *Spodoptera frugiperda* and *Helicoverpa zea* migrating seasonally in Texas, are most abundant at 400–600 m above ground level where their migration is assisted by a low-level wind jet and where they also attract foraging bats (Westbrook 2008; McCracken et al. 2008). Even above their flight boundary layer, large nocturnal moths, such as *Autographa gamma*, compensate in part for being blown off course by the wind (Chapman et al. 2008). However, they too might use a ground reference in the contemporary English landscape where artificial lights at ground level could serve as beacons for partial course correction.

4   Energy Requirements of Flight and Flight Physiology

Flight is a very costly form of locomotion, because it is performed at a very high metabolic rate. In contrast to walking, the relationship between speed and energy expenditure during flapping flight is U-shaped. A walking animal can slow down to reduce energy expenditure. A flying bird, however, cannot escape a certain level of power output, determined by the lowest point of the U-shaped power curve. Even at this minimum, the metabolic rate of flying birds is about twice the maximum rate of exercising small mammals and is among the highest in all vertebrates (Butler et al. 1998). Oxygen consumption rates of flying insects are 3–30 times greater than those of terrestrial ones of the same body mass at the same body temperature (Harrison and Roberts 2000).

4.1   Powering Flight

Powering short burst flight The amount of energy and the type of fuel used for flight depend on the flight duration but also on the flight style. For short burst flights, typically hunting or escape flights, birds have to rely on energy that is available instantaneously. An immediate source of energy is adenosine triphosphate (ATP). ATP is the only substance the muscle proteins can use directly. However, only small amounts are present in the muscle (Schmidt-Nielsen 1997). Another source of immediate energy is the organic compound creatine phosphate. It is present in larger amounts than ATP, but it might last only for some seconds of flight (Schmidt-Nielsen 1997). For short flights, glycogen, stored in skeletal muscles, will be mobilized. It contributes substantially to energy expenditure at the onset of flight before lipids are fully available (Rothe et al. 1987; Schwilch et al. 1996; Gannes et al. 2001; Jenni-Eiermann et al. 2002). However, the amount of glycogen stored is very small (e.g., Marsh 1983) and can quantitatively not serve as a fuel complementing lipids during endurance flight (see below). In the absence of sufficient oxygen, the muscles can depend on anaerobic glycolysis and oxidize glycogen to lactic acid. The quick muscle contractions are made possible by the fast glycolytic fibers. However, anaerobic glycolysis cannot be continued for long because lactic acid accumulates and impedes muscular activity (Schmidt-Nielsen 1997).

Most insects have flight muscle that is directly stimulated to contract by nerve impulse (called non-fibrillar or synchronous). Flies, bees, wasps, and some beetles possess fibrillar (or asynchronous) flight muscle that contracts faster than the nervous stimulation with nerve impulses only required to initiate and maintain muscle contraction. Contraction rates and wing beat frequencies are higher in insects with asynchronous flight muscles, and yet this enhanced performance comes with little cost, because mass-specific metabolic rates are similar to those with synchronous flight muscles (Harrison and Roberts 2000).

Many insects fuel flight with carbohydrates, which are stored as glycogen in the fat body and other tissues and converted to trehalose to pass into the hemolymph.

Circulating trehalose is then converted to glucose in the muscles. Relative to using lipids, fueling of flight with carbohydrates requires more frequent stops to ingest sugars. Assuring rapid uptake of glucose from the gut, the low concentration of blood glucose is maintained by absorption of glucose into the fat body and conversion to trehalose or glycogen for storage (Steele 1985).

Powering endurance flight Migratory endurance flight of birds is performed at a high metabolic rate and can be maintained for hours and up to days in certain species. Moreover, most bird species (except aerial feeders) do not feed or drink during endurance flight and, thus, have to rely exclusively on body stores of energy and water. The power needed to carry additional weight in flight is much higher than when walking or swimming (Schmidt-Nielsen 1984). Hence, it is of paramount importance that the energy density of stored fuel is high. Lipids are the optimal fuel for flight because the energy density of stored lipids is more than seven times higher than that of glycogen and protein. In terms of high energy phosphate (e.g., ATP), fat from adipose tissue yields eight times more chemical energy than wet protein, and 8.2–10.3 times more than glycogen (Jenni and Jenni-Eiermann 1998). This is chiefly because fat stored in adipose tissue contains only about 5% water (Piersma and Lindström 1997), compared with 70% or more for muscle tissue or stored glycogen. Another advantage of adipose tissue is its comparatively low maintenance costs (e.g., Scott and Evans 1992).

However, there are a number of shortcomings of lipids as a fuel during flight. Lipids are not readily available at the onset of flight but need some time to be mobilized from adipose tissue. With their low aqueous solubility, free fatty acids must be transported by soluble protein carriers at every step from adipose tissue to the mitochondria of the flight muscles (Fig. 5.5a). Lipid transport is likely to limit fatty acid oxidation in birds, as it does in mammals (Weber 1992; McWilliams et al. 2004). Finally, fatty acids cannot cover all of the needs (e.g., glucose for the central nervous system) and therefore a minimum amount of requirements needs to be met by glucose (either derived from glycogen stores or via gluconeogenesis from amino acids or glycerol) and protein. Also intermediates of the citric acid cycle, needed to oxidize fatty acids, are constantly drained away and need to be replaced from carbohydrates or certain amino acids (anaplerotic flux; Dohm 1986; Sahlin et al. 1990). Therefore, the energy supplied by lipids needs to be complemented by an inevitable minimum amount of protein or carbohydrates.

For insects, lipids are stored as triglycerides in the fat body. Both lipids and carbohydrates are used to fuel the initial flight of the migratory locust *Locusta migratoria*. However, during the first 30 min, hemolymph trehalose levels decrease by 50%, while blood lipid titers increase to its highest levels (Steele 1985). Triglycerides are converted to diglycerides for release from the fat body into the hemolymph where adipolipoproteins reversibly attach and transport each diglyceride molecule from fat body to muscle (Fig. 5.5b). After steady-state flight is reached for locusts, trehalose provides 25% and lipids 75% of the fuel.

A few insects fuel flight with the amino acid proline. Some beetles, such as *Leptinotarsa* and *Pachnoda*, use a combination of carbohydrates and proline for

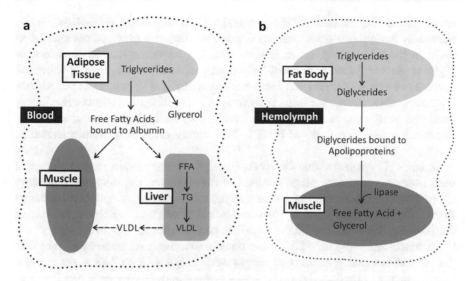

Fig. 5.5 The pathways for conversion of stored lipids (triglycerides, TG) to forms that can be transported to the flight muscles are shown for (**a**) birds and (**b**) insects. Due to constraints on the density of free fatty acids (FFA) bound to albumin carried by the blood and the high metabolic demands of migratory flights, very-low-density lipoproteins (VLDL) manufactured in the liver are also used by small passerine birds to transport lipids to the flight muscles

flight, whereas the tsetse fly *Glossina* oxidizes proline almost exclusively (Goldsworthy and Joyce 2001). Proline oxidation in the flight muscles yields alanine, which circulates to the fat body where stored lipids are utilized to recycle alanine back to proline. Proline is highly soluble in hemolymph allowing it to fuel flight without the transport mechanism required for lipids. Hence, proline can serve to fuel both trivial and prolonged flights. In addition, proline yields about the same energy as lipids and several times more than glucose (Goldsworthy and Joyce 2001).

Proteins have no special storage form. Apart from a small free amino acid pool, all proteins serve specific functions in the body. Therefore, an appreciable amount of protein catabolism inevitably results in functional loss. In birds, it was shown that proteins are catabolized as a complement to lipids for various reasons (Jenni and Jenni-Eiermann 1998): to supplement endogenous protein turnover and repair (there is moderate flight-induced muscle damage; Guglielmo et al. 2001), to provide gluconeogenic precursors, to refill citric acid cycle intermediates (anaplerotic flux) for oxidation of fatty acids, and to provide water (see below). Finally by catabolizing flight muscles during flight, the mass of the flight muscles is constantly adapted to the decreasing body mass and energy needs as fuel stores are depleted (Pennycuick 1998).

The optimal composition of fuel types for migratory endurance flight appears therefore to be a maximum of lipids and a minimum of protein; glycogen does not play a critical role. The question is how well can birds, during their strenuous endurance flight, maximize lipid use and minimize protein catabolism. While mammals during strenuous endurance exercise run on carbohydrates and fat as major fuels, with negligible amounts of protein (Roberts et al. 1996), endurance flight of birds is fueled with about 95% energy derived from lipids and about 5% derived from protein (Jenni and Jenni-Eiermann 1998). This pattern is particularly interesting since endurance flight is performed at 60–85% of maximum respiratory rate, and even more when birds are loaded with fuel (Guglielmo et al. 2002). Therefore, endurance flight of birds is fueled very differently from mammalian running.

In birds, it is expected that the optimum ratio of fat to protein stored before, and used during, migratory flight primarily depends on the migration strategy (in particular on the length of nonstop flights) and the risk of dehydration during flight. The higher the need to carry large fuel stores and the lower the risk of dehydration, the better should the organism be tuned for high lipid utilization.

In conclusion, migrating birds are unique in that they fuel endurance flight with the highest possible percentage of energy derived from fat oxidation. As a consequence, their metabolism shows remarkable adaptations to push lipid mobilization, fatty acid transport, and oxidation during flight to a maximum.

4.2 Adequate Energy Supply During Flight

During endurance flapping flight, the metabolism is greatly increased to meet the high energetic costs of flight, which are 8–18 times higher than basal metabolic rate for flapping flyers (lower in aerial feeders and seabirds; Videler 2005). On the other hand, a small reduction of flight costs may be obtained by hypothermia (Battley et al. 2001; Butler and Woakes 2001).

In contrast to mammals, which derive most of lipids and glycogen from stores within muscle cells (Weber et al. 1996a, b), birds store negligible amounts of fuel within the working muscles. Therefore, bird flight is fueled via the circulatory system from stores outside the flight muscles. However, at the onset of flight, when extramuscular lipids and protein are not yet available, intramuscular and hepatic carbohydrates are mainly used, while fatty acids from adipose tissues reach their steady-state contribution after about 1–2 h of flight and proteins after 4–5 h (Rothe et al. 1987; Schwilch et al. 1996; Jenni-Eiermann et al. 2002).

Fatty acid supply may be constrained by the enzymatic system of mobilization, proteins transporting fatty acids in the bloodstream and cytoplasm, and translocation across cell membranes (Weber 1992). Many aspects of fuel supply to the working muscles have not yet been studied in birds. However, several avian adaptations are known. In general, the supply of fatty acids is not constrained by adipocyte mobilizations or by mitochondrial oxidation capacity, but rather by perfusion limitations in adipose tissue, circulatory transport, and sarcolemmal uptake (Vock et al. 1996; McWilliams et al. 2004). Although these constrain lipid use in mammals, birds seem to have evolved special mechanisms to overcome these limitations.

Triglycerides stored in adipocytes need to be hydrolyzed into free fatty acids and glycerol, in order to be released into the blood (Fig. 5.5a). The rate of lipolysis in adipose tissue apparently does not limit the supply of fatty acids, because a large fraction of the fatty acids does not even leave the adipocyte and is re-esterified (Wolfe et al. 1990). Circulatory transport capacity seems to be elevated in birds during endurance flight by several means. Whether an increase in heart size (as shown in several species; Piersma et al. 1996, 1999; Guglielmo and Williams 2003), which may increase cardiac output, facilitates fatty acid transport remains to be shown. The flight muscles of long-distance migrants have a higher capillary density than those of partial migrants and sedentary species and are thus well prepared to supply oxygen and fuel at maximum rates (Lundgren and Kiessling 1988; Maillet and Weber 2007).

Free fatty acids are insoluble in the blood and are thus transported bound to albumin. It is unlikely that albumin concentration can be increased, because of limitations set by blood viscosity and plasma osmotic pressure. Indeed, plasma albumin decreased during flight in pigeons (George and John 1993).

In small passerines, we were surprised to observe elevated concentrations of triglycerides bound in very-low-density lipoproteins VLDL (Jenni-Eiermann and Jenni 1992). We suggested that apart from flight muscles, fatty acids would also be taken up by the liver with its high lipid processing capacity. This would enable plasma albumin to transport more free fatty acids per unit time. Fatty acids taken up by the liver would be re-esterified and released into the plasma in VLDL, a metabolic pathway known from other physiological contexts. This conversion would allow fluxes of large amounts of fatty acids to the flight muscles without a large burden on the plasma and would circumvent the limitations set by albumin to fatty acid transport. However, elevated triglycerides and VLDL have not been observed in other species sampled under experimental flight conditions (Schwilch et al. 1996; Jenni-Eiermann et al. 2002; Pierce et al. 2005). However, flux rather than plasma concentrations should be measured and preferably in birds under natural conditions.

The supply of fatty acids may also be constrained by other steps in the transport system, e.g., into the cell and across the cell membranes (Weber 1992). In birds, high concentrations of heart-type fatty acid binding protein on the membranes of the muscles (FAT/CD36 and FABPpm) and in the cytosol (H-FABP) were found and upregulated during migratory seasons (Guglielmo et al. 1998, 2002; Pelsers et al. 1999; McFarlan et al. 2009). In the flight muscle of a typical migrant (*Calidris mauri*), the H-FABP concentration was about tenfold greater than in mammalian muscles and 70% more abundant during the migratory season (Guglielmo et al. 1998, 2002; Guglielmo 2010).

During flight, fatty acids are oxidized by the upregulated enzyme system (CAT, β-oxidation enzymes such as hydroxyacyl-CoA-dehydrogenase HOAD, citrate synthase, cytochrome oxidase; McWilliams et al. 2004; Maillet and Weber 2007). Fatty acid chain length, degree of unsaturation, and placement of double bonds can affect the rate of mobilization of fatty acids from adipose tissue, utilization by muscles, and possibly performance (Price 2010).

4.3   Water Supply During Endurance Flight

Water may limit endurance flight (Carmi et al. 1992; Klaassen 1996). Birds during experimental flight have been estimated to be sensitive to dehydration in ambient temperatures above about 20 °C (Giladi and Pinshow 1999; Engel et al. 2006). However, birds flying readily over the Sahara under quite unexpected ambient conditions (mean 30 °C, 27% relative humidity) show that current estimates must be refined (Schmaljohann et al. 2008). Birds incurring dehydration during flight may alleviate or compensate water loss by shifting the composition of fuel types from lipids to more protein which produces 5.9 times more water than an isocaloric amount of lipids (Jenni and Jenni-Eiermann 1998). Recently, an increase in protein catabolism has indeed been demonstrated in water-restricted house sparrows (Gerson and Guglielmo 2010). Migrants caught in the Sahara with emaciated flight muscles and ample lipid stores also hint on such a mechanism.

4.4   Hormonal Regulation of Flight Metabolism

Coordinating mechanisms must exist to achieve an adaptive balance between lipid and protein catabolism throughout avian flight. Corticosterone may be involved in the regulation of metabolism during flight and the composition of fuel types used (Jenni et al. 2000). Elevated levels of corticosterone promote gluconeogenesis from amino acids and, thus, increase breakdown of muscle protein (Gwinner et al. 1992; Kettelhut et al. 1988). Studies in free-living small passerines caught during migratory flight (Falsone et al. 2009), in homing pigeons after a 180 km flight (Haase et al. 1986), and in shorebirds just after landing from a long spring migratory flight (Landys-Cianelli et al. 2002; Reneerkens et al. 2002), showed slightly increased, intermediate corticosterone levels in comparison with resting birds. These data support the hypothesis of an upregulation of corticosterone as a response to increased energy demands. However, in laboratory studies, results were equivocal showing either slightly increased corticosterone levels (ducks and pigeons exercising in a treadmill; Harvey and Phillips 1982; Rees and Harvey 1987) or unchanged levels (pigeons electrically stimulated for 2 h, John and George 1973; red knots flying in a wind tunnel for 2 and 10 h, respectively, Jenni-Eiermann et al. 2009). Whether the latter results might be explained by the experimental situation remains to be shown. In any case, corticosterone increases to variable degrees during flight, but only rises to the very high concentrations typical of an acute stressor when fat deposits are nearly depleted and birds in endurance flight enter phase III of fasting (Gwinner et al. 1992; Jenni et al. 2000). In this case, corticosterone seems to trigger protein catabolism as suggested by the emaciated breast muscles and the high uric acid levels in those individuals (Gwinner et al. 1992; Jenni et al. 2000). The role of other hormones in the regulation of flight metabolism has been investigated in free-flying homing pigeons but not in migrant birds. After a flight of 48 km or about 1.5 h, significant increases of glucagon, adrenaline, noradrenaline, and growth hormone and either significantly reduced or unchanged concentrations of T3 and T4 were found (George et al. 1989; John et al. 1988; Viswanathan et al. 1987). George et al. (1989) concluded that increased sympathetic activity increases glucagon, a lipid-mobilizing hormone. The inconsistent results for T3 and T4 might be explained by differing training protocols. Only the untrained pigeons showed a decrease for thyroid hormone, most probably as a consequence of the activation of the HPA axis, which in turn reduces peripheral T3 formation and suppresses thyroidal T4 secretion (see George et al. 1989 and citations therein).

For insects, octopamine is a hormone that, among other responses to stress, increases trehalose concentration in the hemolymph. Octopamine operates to fuel flight initially with carbohydrates while the flight activity stimulates release of adipose kinetic hormone (AKH) from the corpus cardiacum (Goldsworthy and Joyce 2001). AKH acts directly on the fat body to increase hemolymph lipid concentration, and within 30 min of flight initiation, fatty acids are the main fuel for flight. Octopamine and AKH also act directly on flight muscle to enhance metabolism (Steele 1985; Orchard et al. 1993). Although there can be more than one form of AKH in an insect, e.g., the migratory locust has three, the molecules seem to have the same functions and differ only in their half-lives (Goldsworthy and Joyce 2001).

5 Adaptions to Hunt Out of the Air: Visual Acuity of Birds and Insects

Hunting out of the air requires extremely good vision. Birds are known to have the best vision among vertebrates, but, depending on the foraging ecology, special features of the eye are required. Species such as raptors for instance, hunting from a considerable height, must be able to detect and focus rapidly moving prey at changing distances. Not surprisingly the wedge-tailed eagle *Aquila audax* and the American kestrel *Falco sparverius* exceed the visual acuity of humans more than twice. Species such as swifts, hunting against the bright sky, need a fine-tuned color vision to detect small insects, and nocturnal birds such as owls have to be able to fly and hunt in very low light conditions. Accordingly, a variety of adaptations can be observed in the avian eye.

The first, very obvious, difference between species is the placement of the eye. In most birds, the eyes are situated laterally, enabling only monocular vision but a wide field of view, important for the detection of predators. In contrast, birds of prey typically have their eyes situated more forward and therewith a high overlap of the right and left visual fields enabling binocular vision and depth perception (Fig. 5.6). This is crucial to determine distances and indispensable to successfully capture prey (Evans and Heiser 2004; Martin 2012). Accommodation of the eye is essential to bring objects at varying distances into clear focus. The mechanism of accommodation in the avian eye is mediated by ciliary muscles which move the ciliary processes attached directly at the lens. They exert pressure on the unusually soft and—compared with other vertebrates—pliable lens. By constriction of the iris

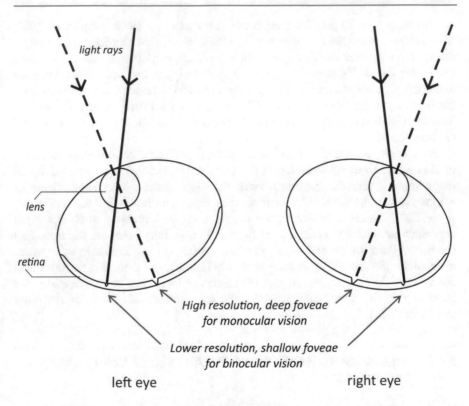

Fig. 5.6 Bird eyes are unique in having two foveae, which are shown in this cross-sectional representation of the eyes (adapted from Tucker 2000). The cornea and lens are combined into a single circular idealized lens. Although somewhat controversial, the deeper, centrally located fovea centralis is believed to have higher resolution. For example, a sparrowhawk is thought to have sufficient resolution to see a fly on a branch at 250 m, but its central placement in the retina and the orientation of the eyes on the bird's head generally makes one deep fovea unable to interact with the visual system from the bird's other eye. The forward eyes of owls are an exception. In contrast, the shallow fovea temporalis is positioned laterally such that the foveae from the two eyes interact in binocular vision, improving depth perception during target approach

sphincter, the lens can even be pushed through the pupil. In some bird species, the cornea is also implicated in accommodation, a unique feature of the avian eye. Forming the lens and the cornea to varying shapes together with the varying aperture of the pupil achieves the best possible vision (Martin 1985).

The composition of the retina also varies between species of different foraging ecology. The retina consists of light-sensitive cells, rods, and cones. The closer the packing of the bright-light-sensitive cone cells, the more acuity (resolving power) the animal has (Fig. 5.7a). Raptors for instance have regions on their retina with as much as one million cone cells per square millimeter, enabling their high resolution. Cones enable color vision and are found in high numbers in diurnal birds. The cones of birds are more complex than those of humans. They have four to five distinctive photoreceptor pigments with peak sensitivities in the red, green, blue,

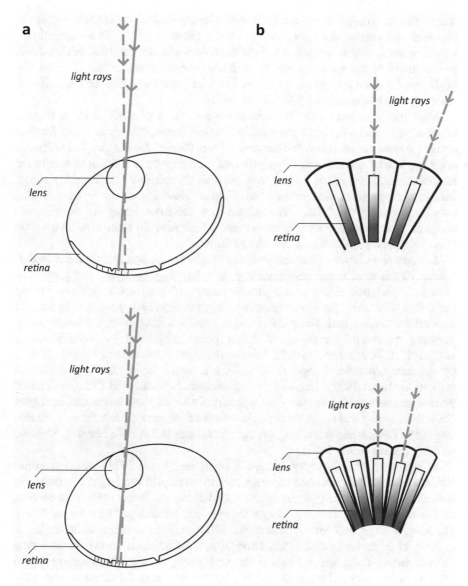

Fig. 5.7 (a) Birds and (b) insects can both increase spatial resolution of their vision by increasing the density of visual sensors. For a cross section of a bird's eye, we show a doubling of retinal cells decreases the angle between the two light rays passing through the lens that are incident on adjacent cells. For a cross section through an insect compound eye, a doubling of ommatidia achieves the same result

and ultraviolet spectrum. Moreover, they contain highly pigmented oil droplets that function as filters, altering and focusing color sensitivity. They may also contain very high concentrations of carotenoids that are thought to act as selective cutoff filters, absorbing short-wave light (Martin 2012; Evans and Heiser 2004). Swallows and swifts have only few red and orange droplets, probably to avoid darkening the sky and to enhance the detection of insects (Martin 1985). The highly light-sensitive rods, which do not discriminate colors but do function in dim light, predominate in the retina of nocturnal birds. Nevertheless, acuity of nocturnal birds can be poor and species such as the barn owl *Tyto alba* rely heavily on auditory cues to detect prey (Güntürkün 2000).

Most birds, as mammals, have in the central part of their retina a circular depression of variable depth and size, the central fovea, where the cones (rods in owls) are most concentrated and the neural layer thinned for the sharpest vision. In addition, hawks and other fast flying diurnal predators have a second fovea in the temporal area of the retina. Hence, they see well on both sides while seeing ahead, because images from each side of their field of view fall on the *fovea centralis*, providing monocular vision, while images from the front, being the area they are about to enter, fall on the fovea temporalis, which provide acute binocular vision (Fig. 5.6, Tucker 2000; Evans and Heiser 2004).

The compound eyes of insects have poor spatial resolution, and yet many insects pursue targets with remarkable accuracy. In many flies, such as some hoverflies, blowflies, and houseflies, males pursue conspecifics in territorial and mating behaviors. The male eyes have regions with high visual acuity improved by smaller ommatidial angles and larger facets with corresponding retinal adaptations to improve speed and sensitivity of the response (reviewed by Nordström and O'Carroll 2009; Nordström 2012). The threshold for detection of a target is limited by the inter-ommatidial angle (Fig. 5.7b), but larger facets also improve image resolution (Land 1997). Because this acute zone is located in the dorso-frontal portion of the visual field, the flies typically fixate on their target and then give chase from below and behind to copy the moves of the leader. Male fly eyes are also larger and have acute zones with greater spatial resolution than females, allowing stealthy pursuit (Land 1997).

In contrast, other hoverfly species (Collett and Land 1978) and dragonflies (Olberg et al. 2000) calculate the trajectory to anticipate the target's position for contact. They also mainly pursue targets from behind and below to increase contrast of the target against the sky. Dragonfly eyes are unusually large for an insect, allowing a large number of ommatidia with small inter-ommatidial angles in regions of high acuity (0.24–1.8, Land 1997). In addition to the dorso-frontal zone of visual acuity where facets are larger (Olberg et al. 2007), a second frontal zone of high acuity in both sexes might play a role in target fixation nearer to the horizon (Land and Eckert 1985).

6 Concluding Remarks and Future Directions

This chapter gives background information of how animals adapted anatomically and physiologically to the use of the aerosphere as their main habitat. Only some very conspicuous adaptations were picked out with a special focus on the wings, because they are essential to perform powered flight, and the eyes, because they enable foraging in the air. As examples for physiological adaptations we focused on the extraordinary abilities of birds powering endurance flight, unsteady aerodynamics, limitations on weight loading, adaptations for maneuvering, and flying at high altitudes. Of course, there are numerous other specializations such as the hearing of the owls, the echolocation of bats, or orientation and navigation. For detailed studies, the reader is encouraged to deepen his knowledge in other chapters of this book or in specific textbooks.

There are still many gaps in our knowledge of flight physiology or migration behavior, mainly because it is so difficult for a terrestrial creature to investigate and follow a flying animal. Wind tunnels offer the possibility to investigate aerodynamic and physiological aspects of flight under controlled experimental conditions. However, not all species can be trained to fly in a wind tunnel and contradicting results suggest that the conditions in wind tunnels do not always mirror the complex conditions of natural surroundings (e.g., Jenni-Eiermann et al. 2002; Schmaljohann et al. 2008). While migrating naturally across lakes and seas, day-flying moths, butterflies, and dragonflies have been tracked and measured from small boats to address questions concerning where, how, and why they migrate (e.g., Srygley 2001; Srygley et al. 2010). Ever more sophisticated and pervasive ground-based Doppler weather radar have been employed to track migrating bats, birds, and insects (Gauthreaux and Belser 1998; Chilson et al. 2012). Geo-locators (Liechti et al. 2013) or implantable instruments (Bishop et al. 2015) are becoming smaller, lighter, and more refined and open up possibilities to record migration routes or physiological and biomechanical data during flight. Tracking migratory flights of passerines from the International Space Station may also soon be realized (Bridge et al. 2011). These technical developments will help us to investigate new questions and solve the many unknowns of physiological aeroecology.

References

Alerstam T, Hedenström A, Akesson S (2003) Long-distance migration: evolution and determinants. Oikos 103:247–260

Alexander DE (2002) Nature's flyers. John Hopkins University Press, Baltimore

Altshuler DL, Dudley R (2003) Kinematics of hovering hummingbird flight along simulated and natural elevational gradients. J Exp Biol 206:3139–3147

Altshuler DL, Dickson WB, Vance JT et al (2005) Short-amplitude high-frequency wing strokes determine the aerodynamics of honeybee flight. Proc Nati Acad Sci 102:18213–18218

Battley PF, Dekinga A, Dietz MW et al (2001) Basal metabolic rate declines during long-distance migratory flight in great knots. Condor 103:838–845

Bicudo JEPW, Buttemer WA, Chappell MA et al (2010) Ecological and environmental physiology of birds. Oxford University Press, New York

Biesel W, Nachtigall W (1987) Pigeon flight in a wind tunnel. IV Thermoregulation and water homeostasis. J Comp Physiol B 157:117–128

Birch JM, Dickinson MH (2003) The influence of wing-wake interactions on production of aerodynamic forces in flight. J Exp Biol 206:2257–2272

Bishop CM, Spivey RJ, Hawkes LA et al (2015) Th roller coaster flight strategy of bar-headed geese conserves energy during Himalayan migrations. Science 347:250–254

Bomphrey RJ, Taylor GK, Thomas ALR (2009) Smoke visualization of free-flying bumblebees indicates independent leading-edge vortices on each wing pair. Exp Fluids 46:811–821

Bridge ES, Thorup K, Bowlin MS, Chilson PB, Diehl RH, Fleron RW, Hartl P, Kays R, Kelly JF, Robinson WD, Wikelski M (2011) Technology on the move: recent and forthcoming innovations for tracking migratory birds. Bioscience 61:689–698

Butler PJ, Woakes AJ (2001) Seasonal hypothermia in a large migrant bird: saving energy for fat deposition? J Exp Biol 204:1361–1367

Butler PJ, Woakes AJ, Bishop CM (1998) Behaviour and physiology of Svalbard barnacle geese Branta leucopsis during their autumn migration. J Avian Biol 29:536–545

Carey C, Morton ML (1976) Aspects of circulatory physiology of montane and lowland birds. Comp Biochem Physiol 54A:61–74

Carmi N, Pinshow B, Porter WP et al (1992) Water and energy limitations on flight duration in small migrating birds. Auk 109(2):268–276

Chai P, Dudley R (1996) Limits to flight energetics of hummingbirds hovering in hypodense and hypoxic gas mixtures. J Exp Biol 199:2285–2295

Chai P, Srygley RB (1990) Predation and the flight, morphology, and temperature of Neotropical rainforest butterflies. Am Nat 135:48–765

Chapman JW, Reynolds DR, Mouritsen H et al (2008) Wind selection and drift compensation optimize migratory pathways in a high-flying moth. Curr Biol 18:514–518

Chilson PB, Frick WF, Kelly JF, Howard KW, Larkin RP, Diehl RH, Westbrook JK, Kelly TA, Kunz TH (2012) Partly cloudy with a chance of migration: weather, radars, and aeroecology. Bull Am Meteorol Soc 93:669–686

Collett TS, Land MF (1978) How hoverflies compute interception courses. J Comp Physiol 125:191–204

Costantini D, Cardinale M, Carere C (2007) Oxidative damage and anti-oxidant capacity in two migratory bird species at a stop-over site. Comp Biochem Physiol C 144:363–371

Crabtree B, Newsholme EA (1975) Comparative aspects of fuel utilization and metabolism by muscle. In: Usherwood PNR (ed) Insect muscle. Academic, New York, pp 405–491

Dickinson MH, Lehmann FO, Sane SP (1999) Wing rotation and the aerodynamic basis of insect flight. Science 284:1954–1956

Dillon ME, Dudley R (2014) Surpassing Mt. Everest: extreme flight performance of alpine bumble-bees. Biol Lett 10:20130922

Dillon ME, Frazier MR (2006) Drosophila melanogaster locomotion in cold thin air. J Exp Biol 209:364–371

Dohm GL (1986) Protein as a fuel for endurance exercise. Exerc Sport Sci Rev 14:143–173

Dudley R (ed) (2000) The biomechanics of insect flight: form, function, evolution. Princeton University Press, Princeton

Dudley R, Vermeij GJ (1992) Do the power requirements of flapping flight constrain folivory in flying animals? Funct Ecol 6:101–104

Ellington CP (1984) Aerodynamics of flapping flight. Am Zool 24:95–105

Ellington CP, Van den Berg C, Willmott AP et al (1996) Leading-edge vortices in insect flight. Nature 384:626–630

Engel S, Biebach H, Visser GH (2006) Water and heat balance during flight in the rose-colored starling (Sturnus roseus). Physiol Biochem Zool 79:763–774

Esch H (1988) The effects of temperature on flight muscle potentials in honeybees and cuculiinid winter moths. J Exp Biol 135:109–117

Esterhuizen N, Clusella-Trullas S, van Daalen CE et al (2014) Effects of within-generation thermal history on the flight performance of Ceratitis capitata: colder is better. J Exp Biol 217:3545–3556

Evans HE, Heiser JB (2004) What's inside: anatomy and physiology. In: Podulka S, Rohrbaugh RW Jr, Bonney R (eds) Handbook of bird biology, 2nd edn. Cornell Lab of Ornithology, Ithaca (chapter 4)

Falsone K, Jenni-Eiermann S, Jenni L (2009) Corticosterone in migrating songbirds during endurance flight. Horm Behav 56:548–556

Gannes LZ, Hatch KA, Pinshow B (2001) How does time since feeding affect the fuels pigeons use during flight? Physiol Biochem Zool 74:1–10

Gauthreaux SA Jr, Belser CG (1998) Display of bird movements on the WSR-88D: patterns and quantification. Weather Forecast 13:453–464

George JC, John TM (1993) Flight effects on certain blood parameters in homing pigeons (Columba livia). Comp Biochem Physiol 106A:707–712

George JC, John TM, Mitchell MA (1989) Flight effects on plasma levels of lipid, glucagon and thyroid hormones in homing pigeons. Horm Metab Res 21:542–545

Gerson AR, Guglielmo CG (2010) House sparrows (Passer domesticus) increase protein catabolism in response to water restriction. Am J Regul Integr Comp Physiol 300:R925–R930

Gibo DL (1981) Altitudes attained by migrating monarch butterflies, Danaus p. plexippus (Lepidoptera: Danaidae), as reported by glider pilots. Can J Zool 59:571–572

Giladi I, Pinshow B (1999) Evaporative and excretory water loss during free flight in pigeons. J Comp Physiol B 169:311–318

Goldsworthy GJ, Joyce M (2001) Physiology and endocrine control of flight. In: Woiwod IP, Reynolds DR, Thomas CD (eds) Insect movement: mechanisms and consequences. CABI, Wallingford, pp 65–86

Greenlee KJ, Harrison JF (2004) Development of respiratory function in the American locust Schistocerca Americana. I. Across-instar effects. J Exp Biol 207:497–508

Guglielmo CG (2010) Move that fatty acid: fuel selection and transport in migratory birds and bats. Integr Comp Biol 50:336–345

Guglielmo CG, Williams TD (2003) Phenotypic flexibility of body composition in relation to migratory stage, age, and sex in the Western sandpiper (Calidris mauri). Physiol Biochem Zool 76:84–98

Guglielmo CG, Haunerland NH, Williams TD (1998) Fatty acid binding protein, a major protein in the flight muscle of the Western Sandpiper. Comp Biochem Physiol 119B:549–555

Guglielmo CG, Piersma T, Williams TD (2001) A sport-physiological perspective on bird migration: evidence for flight-induced muscle damage. J Exp Biol 204:2683–2690

Guglielmo CG, Haunerland NH, Hochachka PW, Williams TD (2002) Seasonal dynamics of flight muscle fatty acid binding protein and catabolic enzymes in a long-distance migrant shorebird. Am J Physiol Regul Intgr Comp Physiol 282:R1405–R1413

Güntürkün O (2000) Sensory physiology: vision. In: Whittow GC (ed) Sturkie's avian biology, 5th edn. Academic, San Diego, pp 1–14

Gwinner E, Zeman M, Schwabl-Benzinger I et al (1992) Corticosterone levels of passerine birds during migratory flight. Naturwissenschaften 79:276–278

Haase E, Rees A, Harvey S (1986) Flight stimulates adrenocortical activity in pigeons (Columba livia). Gen Comp Endocrinol 61:424–427

Harrison JF, Lighton JR (1998) Oxygen-sensitive flight metabolism in the dragonfly Erythemis simplicicollis. J Exp Biol 201:1739–1744

Harrison JF, Roberts SP (2000) Flight respiration and energetics. Annu Rev Physiol 62:179–205

Harvey S, Phillips JG (1982) Adrenocortical responses of ducks to treadmill exercise. J Endocrinol 94:141–146

Hasselquist D, Lindström Å, Jenni-Eiermann S et al (2007) Long flights do not influence immune responses of a long-distance migrant bird: a wind-tunnel experiment. J Exp Biol 210:1123–1131

Jenni L, Jenni-Eiermann S (1998) Fuel supply and metabolic constraints in migrating birds. J Avian Biol 29:521–528

Jenni L, Jenni-Eiermann S, Spina F et al (2000) Regulation of protein breakdown and adrenocortical response to stress in birds during migratory flight. Am J Physiol Reg Integr Comp Physiol 278:R1182–R1189

Jenni-Eiermann S, Jenni L (1992) High plasma triglyceride levels in small birds during migratory flight: a new pathway for fuel supply during endurance locomotion at very high mass-specific metabolic rates. Physiol Zool 65:112–123

Jenni-Eiermann S, Jenni L, Kvist A et al (2002) Fuel use and metabolic response to endurance exercise: a wind tunnel study of a long-distance migrating shorebird. J Exp Biol 205:2453–2460

Jenni-Eiermann S, Hasselquist D, Lindström Å et al (2009) Are birds stressed during long-term flights? A wind tunnel study on circulating corticosterone in the red knot. Gen Comp Endocrinol 164:101–106

Jenni-Eiermann S, Jenni L, Smith S et al (2014) Oxidative stress in endurance flight: an unconsidered factor in bird migration. PLoS One. https://doi.org/10.1371/journal.pone.0097650

John TM, George JC (1973) Effect of prolonged exercise on levels of plasma glucose, free fatty acids and corticosterone and muscle free fatty acids in the pigeon. Arch Int Physiol Bioch 81:421–425

John TM, Viswanathan M, George JC et al (1988) Flight effects on plasma levels of free fatty acids, growth hormone and thyroid hormones in homing pigeons. Horm Metab Res 20:271–271

Johnson CG (1969) Migration and dispersal of insects by flight. Methuen, London

Jones G, Rydell J (2003) Attack and defense: Interactions between echolocating bats and their insect prey. In: Kunz TH, Fenton MB (eds) Bat ecology. University of Chicago, Chicago, pp 301–345

Kettelhut IC, Wing SS, Goldberg AL (1988) Endocrine regulation of protein breakdown in skeletal muscle. Diabetes Metab Rev 4:751–772

Kingsolver JG, Srygley RB (2000) Experimental analyses of body size, flight and survival in pierid butterflies. Evol Ecol Res 2:593–612

Klaassen M (1996) Metabolic constraints on long-distance migration in birds. J Exp Biol 199:57–64

Kunz TH, Ingalls KA (1994) Folivory in bats: an adaptation derived from frugivory. Funct Ecol 8:665–668

Land MF (1997) Visual acuity in insects. Annu Rev Entomol 42:147–177

Land MF, Eckert H (1985) Maps of the acute zones of fly eyes. J Comp Physiol A 156:525–538

Landys-Cianelli MM, Ramenofsky M, Piersma T et al (2002) Baseline and stress-induced plasma corticosterone during long-distance migration in the bar-tailed godwit, Limosa lapponica. Physiol Biochem Zool 75:101–110

Landys-Ciannelli MM, Piersma T, Jukema J (2003) Strategic size changes of internal organs and muscle tissue in the bar-tailed godwit during fat storage on a spring stopover site. Funct Ecol 17:151–179

Lehmann FO (2008) When wings touch wakes: understanding locomotor force control by wake-wing interference in insect wings. J Exp Biol 211:224–233

Liechti F, Witvliet W, Weber R et al (2013) First evidence of a 200-day non-stop flight in a bird. Nat Commun 4:2554. https://doi.org/10.1038/ncomms3554

Lundgren BO, Kiessling KH (1988) Comparative aspects of fibre types, areas, and capillary supply in the pectoralis muscle. J Comp Physiol B 158:165–173

Maillet D, Weber JM (2007) Relationship between n-3 PUFA content and energy metabolism in the flight muscles of a migrant shorebird: evidence for natural doping. J Exp Biol 210:413–423

Maina JN (2006) Development, structure, and function of a novel respiratory organ, the lung-air sac system of birds: to go where no other vertebrate has gone. Biol Rev 81:545–579

Marden JH, Chai P (1991) Aerial predation and butterfly design: how palatability, mimicry and flight constrain mass allocation. Am Nat 138:15–36

Marsh R (1983) Adaptations of the Gray Catbird (Dumetella carolinensis) to long distance migration: energy stores and substrate concentrations in plasma. Auk 100:70–179

Martin GR (1985) Eye. In: King AS, McLelland J (eds) Form and function in birds, vol 3. Academic, London, pp 311–374

Martin GR (2012) Through birds' eyes: insights into avian sensory ecology. J Ornithol 153 (Suppl 1):S23–S48

McCracken GF, Gillam EH, Westbrook JK et al (2008) Brazilian free-tailed bats (Tadarida brasiliensis: Molossidae, Chiroptera) at high altitude: links to migratory insect populations. Integr Comp Biol 48:107–118

McCracken KG, Barger CP, Bulgarella M et al (2009) Parallel evolution in the major haemoglobin genes of eight species of Andean waterfowl. Mol Ecol 18:3992–4005

McFarlan JT, Bonen A, Guglielmo CG (2009) Seasonal up-regulation of protein mediated fatty acid transport in flight muscles of migratory white-throated sparrows (Zonotrichia albicollis). J Exp Biol 212:2934–2940

McWilliams SE, Guglielmo CG, Pierce B et al (2004) Flying, fasting, and feeding in birds during migration: a nutritional and physiological ecology perspective. J Avian Biol 35:377–393

Mujires F, Johansson LC, Barfield R et al (2008) Leading-edge vortex improves lift in slow-flying bats. Science 319:1250–1253

Norberg UM (ed) (1990) Vertebrate flight. Springer, Berlin

Norberg UM, Rayner JMV (1987) Ecological morphology and flight in bats (Mammalia: Chiroptera): wing adaptations, flight performance, foraging strategy and echolocation. Philos Trans R Soc Lond B 316:335–427

Nordström K (2012) Neural specializations for small target detection in insects. Curr Opin Neurobiol 22:272–278

Nordstrom K, O'Carroll DC (2009) Feature detection and the hypercomplex property in insects. Trends Neurosci 32:383–391

Olberg RM, Worthington AH, Venator KR (2000) Prey pursuit and interception in dragonflies. J Comp Physiol A 186:155–162

Olberg RM, Seaman RC, Coats MI et al (2007) Eye movements and target fixation during dragonfly prey-interception flights. J Comp Physiol A 193:685–693

Orchard I, Ramirez J-M, Lange AB (1993) A multifunctional role for octopamine in locust flight. Annu Rev Entomol 38:227–249

Pelsers MMAL, Butler PJ, Bishop CM et al (1999) Fatty acid binding protein in heart and skeletal muscles of the migratory barnacle goose throughout development. Am J Phys 276:R637–R643

Pennycuick CJ (1998) Computer simulation of fat and muscle burn in long-distance bird migration. J Theor Biol 191:47–61

Pennycuick CJ (ed) (2008) Modelling the flying bird. Elsevier, Amsterdam

Pierce B, McWilliams SR, O'Connor TP et al (2005) Effect of dietary fatty acid composition on depot fat and exercise performance in a migrating songbird, the red-eyed vireo. J Exp Biol 208:1277–1285

Piersma T, Gill RE Jr (1998) Guts don't fly: small digestive organs in obese Bar-tailed Godwits. Auk 115:196–203

Piersma T, Lindström Å (1997) Rapid reversible changes in organ size as a component of adaptive behaviour. Trends Ecol Evol 12:134–138

Piersma T, Everaarts JM, Jukema J (1996) Build-up of red blood cells in refuelling Bar-tailed Godwits in relation to individual migratory quality. Condor 98:363–370

Piersma T, Gudmundsson GA, Lilliendahl K (1999) Rapid changes in the size of different functional organ and muscle groups during refueling in a long-distance migrating shorebird. Physiol Biochem Zool 72:405–415

Portugal SJ, Hubel TY, Fritz J et al (2014) Upwash exploitation and downwash avoidance by flap phasing in ibis formation flight. Nature 505:399–402

Price ER (2010) Dietary lipid composition and avian migratory flight performance: development of a theoretical framework for avian fat storage. Comp Biochem Physiol A 157:297–309

Ratcliffe JM, Nydam ML (2008) Multimodal warning signals for a multiple predator world. Nature 455:96–99

Rees A, Harvey S (1987) Adrenocortical responses of pigeons (Columba livia) to treadwheel exercise. Gen Comp Endocrinol 65:117–120

Reneerkens J, Morrison RIG, Ramenofsky M et al (2002) Baseline and stress-induced levels of corticosterone during different life cycle substages in a shorebird on the high arctic breeding grounds. Physiol Biochem Zool 75:200–208

Reynolds AM, Reynolds DR (2009) Aphid aerial density profiles are consistent with turbulent advection amplifying flight behaviours: abandoning the epithet 'passive'. Proc R Soc B 276:137–143

Reynolds DR, Chapman JW, Edwards AS (2005) Radar studies of the vertical distributions of insects migrating over southern Britain: the influence of temperature inversions on nocturnal layer concentrations. Bull Entomol Res 95:259–274

Roberts TJ, Weber JM, Hoppeler H et al (1996) Design of the oxygen and substrate pathways II. Defining the upper limits of carbohydrates and fat oxidation. J Exp Biol 199:1651–1658

Rothe HJ, Biesel W, Nachtigall W (1987) Pigeon flight in a wind tunnel. II Gas exchange and power requirements. J Com Physio B 157:99–10

Sahlin K, Katz A, Broberg S (1990) Tricarboxylic acid cycle intermediates in human muscle during prolonged exercise. Am J Phys 259:C834–C841

Scheid P, Slama H, Piiper J (1972) Mechanisms of the unidirectional flow in parabronchi of avian lungs: measurements in duck lung preparations. Respir Physiol 14:83–95

Schmaljohann H, Liechti F (2009) Adjustments of wingbeat frequency and air speed to air density in free-flying migratory birds. J Exp Biol 212:3633–3642

Schmaljohann H, Bruderer B, Liechti F (2008) Sustained bird flights occur at temperatures far beyond expected limits. Anim Behav 76:1133–1138

Schmaljohann H, Liechti F, Bruderer B (2009) Trans-Sahara migrants select flight altitudes to minimize energy costs rather than water loss. Behav Ecol Sociobiol 63:1609–1619

Schmidt-Nielsen K (ed) (1984) Scaling: why is animal size so important? Cambridge University Press, Cambridge

Schmidt-Nielsen K (ed) (1997) Animal physiology, 5th edn. Cambridge, University Press

Schwilch R, Jenni L, Jenni-Eiermann S (1996) Metabolic responses of homing pigeons to flight and subsequent recovery. J Comp Physiol B 166:77–87

Schwilch R, Piersma T, Holmgren NMA et al (2002) Do migratory birds need a nap after a long non-stop flight? Ardea 90(1):149–154

Scott GR (2011) Elevated performance: the unique physiology of birds that fly high altitudes. J Exp Biol 214:2455–2462

Scott I, Evans PR (1992) The metabolic output of avian (Sturnus vulgaris, Calidris alpina) adipose tissue liver and skeletal muscle: implications for BMR/body mass relationships. Comp Biochem Physiol 103A:329–332

Scott GR, Milsom WK (2007) Control of breathing and adaption to high altitude in the bar-headed goose. Am J Phys Regul Integr Comp Phys 293:R379–R391

Scott GR, Schulte PM, Egginton S et al (2011) Molecular evolution of cytochrome c oxidase underlies high-altitude adaptation in the bar-headed goose. Mol Biol Evol 28:351–363

Scott GR, Hawkes LA, Frappell PB (2015) How bar-headed geese fly over the Himalayas. Physiology 30:107–115

Speakman JR, Thomas DW (2003) Physiological ecology and energetics of bats. In: Kunz TH, Fenton MB (eds) Bat ecology. University of Chicago, Chicago, pp 430–490

Speakman JR, Hays GC, Webb PI (1994) Is hyperthermia a constraint on the diurnal activity of bats? J Theor Biol 171:325–341

Srygley RB (2001) Compensation for fluctuations in crosswind drift without stationary landmarks in butterflies migrating over seas. Anim Behav 61:191–203

Srygley RB, Chai P (1990a) Flight morphology of Neotropical butterflies: palatability and distribution of mass to the thorax and abdomen. Oecologia 84:491–499

Srygley RB, Chai P (1990b) Predation and the elevation of thoracic temperature in brightly colored Neotropical butterflies. Am Nat 135:766–787

Srygley RB, Dudley R (1993) Correlations of the position of center of body mass with butterfly escape tactics. J Exp Biol 174:155–166

Srygley RB, Dudley R (2008) Optimal strategies for insects migrating in the flight boundary layer: mechanisms and consequences. Integr Comp Biol 48:119–133

Srygley RB, Kingsolver JG (1998) Red-wing blackbird reproductive behaviour and the palatability, flight performance, and morphology of temperate pierid butterflies (Colias, Pieris, and Pontia). Biol J Linn Soc 64:41–55

Srygley RB, Kingsolver JG (2000) Effects of weight loading on flight performance and survival of palatable Neotropical Anartia fatima butterflies. Biol J Linn Soc 70:707–725

Srygley RB, Thomas ALR (2002) Unconventional lift-generating mechanisms in free-flying butterflies. Nature 420:660–664

Srygley RB, Dudley R, Oliveira EG et al (2010) El Niño and dry season rainfall influence hostplant phenology and an annual butterfly migration from Neotropical wet to dry forests. Glob Chang Biol 16:936–945

Steele JE (1985) Control of metabolic processes. In: Kerkut GA, Gilbert LI (eds) Endocrinology II, Comprehensive insect physiology, biochemistry, and pharmacology, vol 8. Pergamon Press, Oxford, pp 99–145

Stockwell EF (2001) Morphology and flight manoeuvrability in New World leaf-nosed bats (Chiroptera: Phyllostomidae). J Zool 254:505–514

Swartz SM, Freeman PW, Stockwell EF (2003) Ecomorphology of bats: comparative and experimental approaches relating structural design to ecology. In: Kunz TH, Fenton MB (eds) Bat ecology. University of Chicago, Chicago, pp 257–300

Swartz SM, Breuer KS, Willis DJ (2008) Aeromechanics in aeroecology: flight biology in the aerosphere. Integr Comp Biol 48:85–89

Taylor RAJ, Reling D (1986) Density/height profile and long-range dispersal of first-instar gypsy moth (Lepidoptera: Lymantriidae). Environ Entomol 15:431–435

Thomas ALR, Taylor GK, Srygley RB et al (2004) Dragonfly flight: free-flight and tethered flow visualizations reveal a diverse array of unsteady lift generating mechanisms, controlled primarily via angle of attack. J Exp Biol 207:4299–4323

Thomson SC, Speakman JR (1999) Absorption of visible spectrum radiation by the wing membranes of living pteropodid bats. J Comp Physiol B 169:187–194

Tucker VA (2000) The deep fovea, sideways vision and spiral flight paths in raptors. J Exp Biol 203:3745–3754

Usherwood JR, Stavrou M, Lowe JC et al (2011) Flying in a flock comes at a cost in pigeons. Nature 474:494–497

Videler JJ (ed) (2005) Avian flight. Oxford University Press, Oxford

Videler JJ, Stamhuis EJ, Povel GDE (2004) Leading-edge vortex lifts swifts. Science 306:1960–1962

Viswanathan M, JohnTM GJC et al (1987) Flight effects on plasma glucose, lactate, catecholamines and corticosterone in homing pigeons. Horm Metab Res 19:400–402

Vock R, Weibel ER, Hoppeler H et al (1996) Design of the oxygen and substrate pathways. V. Structural basis of vascular substrate supply to muscle cells. J Exp Biol 199:1675–1688

Voigt CC, Lewanzik D (2011) Trapped in the darkness of the night: thermal and energetic constraints of daylight flight in bats. Proc R Soc B 278:2311–2317

Ward S, Rayner JMV, Möller U et al (1999) Heat transfer from starlings Sturnus vulgaris during flight. J Exp Biol 202:1589–1602

Warrick DR (1998) The turning- and linear-maneuvering performance of birds: the cost of efficiency for coursing insectivores. Can J Zool 76:1063–1079

Warrick DR, Tobalske BW, Powers DR (2005) Aerodynamics of the hovering hummingbird. Nature 435:1094–1097

Weber JM (1992) Pathways for oxidative fuel provision to working muscles: ecological consequences of maximal supply limitations. Experientia 48:557–564

Weber JM, Roberts TJ, Vock R et al (1996a) Design of the oxygen and substrate pathways. III Partitioning energy provision from carbohydrates. J Exp Biol 199:1659–1666

Weber JM, Brichon G, Zwingelstein G et al (1996b) Design of the oxygen and substrate pathways. IV. Partitioning energy provision from fatty acids. J Exp Biol 199:1667–1674

Westbrook JK (2008) Noctuid migration in Texas within the nocturnal aeroecological boundary layer. Integr Comp Biol 48:99–106

Wolfe RR, Klein S, Carraro F et al (1990) Role of triglyceride-fatty acid cycle in controlling fat metabolism in humans during and after exercise. Am J Phys 258:E382–E389

Yanoviak SP, Dudley R, Kaspari M (2005) Directed aerial descent in canopy ants. Nature 433:624–626

Yanoviak SP, Kaspari M, Dudley R (2009) Gliding hexapods and the origins of insect aerial behaviour. Biol Lett 5:510–512

Properties of the Atmosphere in Assisting and Hindering Animal Navigation

6

Verner P. Bingman and Paul Moore

Abstract

Airborne birds, bats, and insects carry out some of the animal world's most spectacular migrations, migrations often supported by complex behavioral-navigational mechanisms. As aerial navigators, these animals must confront potentially disastrous wind conditions that can carry them far from their migratory route. But the same aerial migrators, as well as aerial residents, can also extract spatial information from the atmosphere to support navigation or just locating a goal across a range of spatial scales. The goal of our chapter is to first highlight some of the physical challenges associated with extracting navigational, principally olfactory, information from the air. We then go on to look at the complex role wind plays as a factor influencing navigation in the air and the documented occurrence of both short- and long-distance navigation and goal localization reliant on atmospheric chemicals/odors. The new field of aeroecology offers an exciting opportunity to revisit some of the classic questions examining the relationship among airborne migrations, wind, wind drift, and the need to carry out corrective reorientations, as well as opening up new investigations into the relationship between the properties of atmospheric stimuli and their potential use in supporting navigation.

V.P. Bingman (✉)
Department of Psychology, Bowling Green State University, Bowling Green, OH, USA

J.P. Scott Center for Neuroscience Mind and Behavior, Bowling Green State University, Bowling Green, OH, USA
e-mail: vbingma@bgsu.edu

P. Moore
Department of Biological Sciences, Bowling Green State University, Bowling Green, OH, USA

J.P. Scott Center for Neuroscience Mind and Behavior, Bowling Green State University, Bowling Green, OH, USA

© Springer International Publishing AG, part of Springer Nature 2017
P.B. Chilson et al. (eds.), *Aeroecology*,
https://doi.org/10.1007/978-3-319-68576-2_6

1   Introduction

The birds, bats, and insects that spend much of their time in the air carry out some of the animal world's most spectacular migrations. These migrations are often associated with a remarkable accuracy in navigational ability, a navigational ability that at least in experienced migratory birds enables them to reorient their flights when displaced from a familiar migratory route by unfavorable winds. It is perhaps because of the potential of drift by wind that aerial navigators evolved their sophisticated navigational mechanisms (Bingman and Cheng 2005). The same aerial navigators also extract information from the atmosphere to find goal locations, such as food sources, over distances that are magnitudes shorter than those experienced during migration. But what physical properties of the atmosphere may assist or hinder navigation? Sources of navigational information such as light, sound, and geomagnetism are transmitted through the air, but are generally not considered properties of the atmosphere. By contrast, chemical odors present in the air are considered properties of the atmosphere and are exploited to support both short- and, perhaps surprisingly, long-distance navigation. The goal of our chapter is to first highlight some of the physical challenges associated with extracting navigational, principally olfactory, information from the air. We then go on to look at the complex role wind plays as a factor influencing navigation in the air and the documented occurrence of both short- and long-distance navigation reliant on atmospheric chemicals/odors.

2   Navigational Information in the Aerial Habitat

2.1   The Importance of Scales

To navigate within an aerial habitat requires the extraction of relevant information from environmental stimuli (Shöne 1984; Dusenbery 1992). The underlying physical processes that structure the spatial and temporal distribution of information in air are fairly well understood (Vickers 2000). Within this environment, the movement and interaction of air with physical features such as forests, fields, or land masses, e.g., geological formations, determines the types of wind flow and chemical distribution from which animals may extract relevant information (Vogel 1996). The key element to deriving some semblance of a theory of information for an aerial habitat needs to rely first and foremost on understanding the relevance of scale.

The two scales that need to be considered when applying physical principles to animal navigation are spatial and temporal. When considering the important spatial and temporal scales under consideration, we need to keep in mind that the typical domains of processes acting at each scale are orders of magnitude. This means that spatial scales of centimeters to tens of centimeters will have similar aerodynamics, and those of kilometers and hundreds of kilometers will be different than those forces acting at the centimeter scale. For example, the types of wind movement and physical phenomena experienced by an insect locating a suitable mate by following a pheromone plume are significantly different than the physical phenomena experienced by homing pigeons (*Columba livia*) navigating to a home loft over hundreds of kilometers or monarch butterflies (*Danaus plexippus*) migrating thousands of kilometers to Mexico (Vickers 2000; Moore and Crimaldi 2004). In an identical fashion to this spatial scale analysis, the temporal detection, sampling, extraction of information, and the rate of movement of an animal through a habitat are order of magnitude phenomena based on the behavior and animal under analysis. To return to the example above, an insect navigating locally extracts temporal information from wind and odor stimuli at the rate of tenths of seconds to seconds, whereas migrating insects and birds operate on much longer time scales (Rumbo and Kaissling 1989; Masson and Mustaparta 1990; McKeegan 2002). Thus, scaling factors determine the nature of the important physical phenomena that need to be considered to answer the behavioral questions of navigation discussed below.

2.1.1 Spatial Scales

With respect to spatial scale, there are two intertwined and equally important size/distance issues. First, one spatial aspect refers to the size range over which an animal samples the environment (Matsumoto and Hildebrand 1981; Murlis et al. 1992). For olfactory signals, this is the space over which the sensory system integrates disparate spatial locations into a single spatial sample. Local differences in odor concentrations in space are ignored because they are processed as a single sample. Conversely, odor patches distributed by air that are larger than what can be combined as a single sample are processed as two distinct odor samples. Most animals have paired olfactory receptors (naris for vertebrates and antennae for insects), but most evidence shows that vertebrates do not distinguish between each naris whereas insect do discriminate across different antennae (Craven et al. 2007, 2010). For smaller animals, such as insects, these spatial scales (in relation to single olfactory samples) are on the order of centimeters to tens of centimeters (Schneider 1964). For larger organisms (bats and birds), inter-naris distance is of a similar magnitude, from around centimeters to tens of centimeters (Allison 1953). The implication of this first spatial scale for navigational purposes is the extraction of navigational information applicable to the distance to be travelled. For some navigational challenges, animals extracting directional information from signals require a comparison of signal intensity using two samples acquired in different spatial regions. The comparison of signal intensity differences on the order of centimeters is only relevant for organisms navigating at smaller spatial scales, i.e., odor plume tracking in insects (Vickers 2000). By contrast, animals tracking odor plumes over larger distances are more likely extracting relevant directional information by comparing samples taken over significantly larger distances than the inter-naris distance and thus comparing samples taken at different time points as they move through the sensory landscape (Moore and Crimaldi 2004).

For wind (or mechanical) cues, the same concept applies. The detection of air flow across different mechanosensory receptors can be integrated and therefore are considered a single sample in space (Newland et al. 2000). Unlike olfactory receptors, which tend to be centralized, mechanoreceptors are often distributed across the body making an estimate of local spatial scales difficult (Anderson 1985; Gnatzy and Tautz 1980; Kumagai et al. 1998). As a conservative estimate, differential air flow across the body can be detected. Thus, animals have the ability to process spatial information that is on the order of a body length, i.e., centimeters for insects and tens of centimeters to a meter or more for birds. The body surface sets the physical parameters of a single information sample and is the smallest spatial scale that needs to be considered for defining the mechanosensory active space.

The second spatial scale is the movement pattern of a flying organism. Aeroecology takes place within a habitat that ranges wildly in size. At one end of the spatial scales is an insect's foraging or mating searches that can occur over hundreds of meters (For a review, see Vickers 2000) and within visual contact with the ground. As with the previous scale discussion, spatial scales should be considered order of magnitude phenomena. Therefore, all searches that occur on a scale of hundreds of meters co-occur with an experience of sensory stimuli that are dispersed by similar physical processes, which are discussed below. Bat return flights to caves and some foraging/homing runs of birds that occur over the span of kilometers to tens of kilometers (bats, Henry et al. 2002; pigeons, Wallraff 2005) can be considered medium-scale spatial movements. Even at such horizontal distances, most of the ecological interactions relevant for navigation occur within visual contact of the ground. The final and largest spatial scale that needs to be considered is the long-distance (or extent) migration and homing of birds, bats, and insects that occurs from hundreds to thousands of kilometers and can take place at altitudes exceeding thousands of meters (Wallraff 2005). The types of stimuli and the physical processes that distribute stimuli at these larger surface and altitudinal extents are significantly different than the processes that occur closer to the ground. The processes involved with the physical dispersion of chemical signals will be covered below.

The implications of spatial grain and extent for "aerial navigation" can be considered in the context of the concept of sensory landscapes (developed more below). Different physical and chemical processes that structure sensory information in the air operate at specific size/distance scales. For example, molecular diffusion is a small-scale phenomenon, laminar air flow is a mesoscale phenomenon, and turbulence is a process that occurs at large spatial ranges. Insects navigating during foraging runs are going to be influenced more heavily by turbulence and laminar flow at small spatial extents, which structure the distribution of stimuli (Dudley 2002; Combes and Dudley 2009). By contrast, animals navigating at distances of kilometers are moving through a sensory landscape structured in part by large-scale turbulence (Nisbet 1955; Greenewalt 1975).

2.1.2 Temporal Scales

The temporal scale, as opposed to the spatial scale, is slightly more complex and less well understood. There are two important time scales to consider: the sampling and integration time of the olfactory or visual system and the olfactory/visual memory of the organism (Uchida et al. 2006; Campbell and Borden 2006).

(Although we do not consider visual cues, unlike chemical-based odors, to be a property of the atmosphere, we consider them here to help illustrate the properties of temporal scale.) The integration time of a sensory system refers to the period of time where the chemical fluctuations are combined into a single event by the nervous system, or when visual images are combined into a single visual perception. For vision, this sampling rate can be considered a flicker-fusion rate or the rate at which individual visual images are fused into moving images. Fortunately, flicker-fusion rate for a wide range of animals is known and varies in proportion to relative body speed. For example, relatively slow flying organisms have much slower flicker-fusion rates than faster flying organisms. Flicker-fusion rates are around 100 Hz for pigeons and around 200 Hz for bees (Campbell and Borden 2006; Hershberger and Jordan 1998). The time scale over which olfactory inputs are integrated into a single perception is understudied, but initial findings show that flicker-fusion rates for olfaction are an order of magnitude slower than for vision. The best studies have been performed in moths, whose odor flicker-fusion rate is approximately 10 Hz (Rumbo and Kaissling 1989). These sampling rates determine the maximum temporal resolution (or grain) of changing information that these organisms are capable of tracking in their environment.

The other temporal scale that needs to be determined is the sensory memory of the organism. The use of the term memory here actually refers to the comparison time of the sensory system. Essentially, this is a measure of the time difference needed by an organism to be capable of comparing two different sensory samples (Gomez and Atema 1996). In olfactory terms, searching or attempting to locate a source of a sensory stimulus, for example, in the context of tracking an odor plume, requires an animal to compare both the chemical composition and the concentration of odor samples taken at two different time points (Vickers 2000). In visual terms, the same idea is applicable. As an organism compares two visual images for differences, how current does the older scene need to be in order to effect decisions on differences in images (Vanrullen and Thorpe 2001). In essence, the organism is attempting to determine if the current sensory sample is different from one sampled at an earlier time, and based on this difference, what sort of behavior needs to be performed. The critical temporal scale in this comparison is how long the organism can "remember" the previous sensory sample in order to compare. Again, direct research in this area is rare, and the scaling factors are order of magnitude estimates. Most estimates in the field put the longest comparison time on the order of minutes to hours (Vanrullen and Thorpe 2001).

In summary, the spatial scales over which sensory information can be acquired to support navigation to a goal can range from a spatial grain of centimeters (distance between olfactory appendages) to meters (small extent searches and flights) to kilometers (large extent homing/migration). The temporal scales that need to be considered are on the order of seconds (flicker fusion and integration of stimuli) to minutes (comparison between two sensory samples). Both scales determine how animals extract information from their environment and are influenced by how that information is structured across the landscape they are traversing. Some Antarctic seabirds, for example, use variation in odor patches over large spatial extents generated by krill to navigate to feeding grounds in an otherwise featureless environment (Nevitt et al. 1995). The large distances necessarily require these birds to compare odor concentrations and patterns over very large distances and over relatively long time frames. The ability to store in memory the concentration of an odor sampled some time and distance ago, and to compare that previous sample with one at a bird's current location, would also dictate the type of neuroarchitecture necessary to process the information to support navigation (Nevitt et al. 1995). By contrast, male hawkmoths (*Manduca sexta*) travelling upwind to a female releasing pheromone are operating under much faster temporal and smaller spatial grain (Vickers 2000).

2.2   Physical Constraints on Odor Transmission in Air

Atmospheric odors/chemicals are a rich source of exploitable navigational information (see below). Because of the spatial and temporal scales described above, a set of parameters can be used to quantify the transmission and movement of chemical signals in aerial habitats. These physical parameters of dispersion allow one to look past the superficial characteristics of habitats and understand chemical signal dispersion as a more abstract physical construct, which allows one to develop a broader theory of a chemical landscape. A much more detailed development of plume parameters can be found elsewhere (Moore and Crimaldi 2004). But briefly, the transport of chemicals in the aerial environment consists of two distinct physical processes: advection and dispersion. Advection is defined as the macroscopic, bulk-chemical transport by air flow. At the spatial and temporal scales outlined above, advection is a dominant process in the movement of chemical signals even in extremely low wind conditions, and at the spatial scales under consideration, turbulence is the single most important physical force structuring the dispersion of sensory information.

Turbulence is the fluctuation in velocity around a mean air movement vector. For animals navigating in an aerial habitat, the important factor is the fluctuation of velocity (turbulence) rather than the mean air speed. This statement is, of course, dependent upon the organism under study, thus, consideration of spatial and temporal scales is critical. An insect attempting to locate a home is buffeted by the turbulence operating at the "active" spatial scale of meters. Conversely, in navigation across large spatial extents, for example, an albatross tracking an odor plume to a food source at distances of kilometers, animals would be constrained more by the bulk or mean flow of air rather than turbulent fluctuations.

Because the diffusion of chemicals operates at smaller spatial and slower temporal grains than the distances/times over which navigational behavior typically occurs, it becomes reasonable to focus solely on stirring as the major process that determines the dispersion of chemical signals as a source of aerial information. To further demonstrate this concept, the relative scales (both grain and extent) of flow and diffusion need to be compared. The process of stirring in air is governed by the nature and degree of the turbulence within the air flow. Stirring is the mechanical process (due to chaotic flow patterns) of intermingling of flow patches with differing concentrations. In the interaction between diffusion and stirring, stirring causes odor patches to combine and redistribute the odors whereas diffusion acts to eradicate differences in chemical concentrations along spatial gradients that are created by stirring. In other words, diffusion acts to homogenize the chemical signal whereas stirring acts to disperse the chemical signal. As such, diffusive processes set a size limit on the smallest possible spatial gradient in concentration differences and are crucial for navigation when gradients are followed or need to be learned (see below). This size limit is important in relation to the various spatial scales discussed above. The "smallest spatial scale" (or spatial grain) means that an organism sampling at a size/distance smaller than this will detect no concentration differences. Thus, for navigation, organisms would need to sample odor patches that are farther apart than the smallest detectable concentration difference. The spatial scales of the smallest velocity and chemical structures in any aerial odor plume are controlled by the two diffusivities: ν and D_m (the kinematic viscosity of the fluid and the molecular diffusion of the compound). The smallest spatial scale of chemical concentration differences is determined by the smallest velocity structures within a flow. The Kolmogorov microscale is a quantification of the smallest velocity structures within a flow. The Kolmogorov scale is

$$\eta = \left(\frac{\nu^3}{\varepsilon}\right)^{1/4}$$

where ε is the rate at which turbulent kinetic energy is dissipated into heat and ν is the kinematic viscosity of the medium. As energy within the air flow increases by increases in the flow velocity (and its variance), increases in length, or decreases in the kinematic viscosity, the flow becomes more turbulent. As the flow becomes more turbulent, the dissipation ε increases and the smallest velocity structures in the flow, η, become smaller. Thus, in more turbulent air flow, eddies of air flow are smaller and the small-scale concentration differences are smaller also.

In a similar derivation to the Kolmogorov scale, the Batchelor microscale is the scale of the smallest chemical heterogeneity within a flow. The Batchelor microscale introduces chemical diffusion into the calculation and is a measure of the smallest detectable differences within a flow. The Batchelor microscale is

$$\eta_B = \left(\frac{\nu D_m^2}{\varepsilon}\right)^{1/4}$$

For a typical terrestrial air flow patterns at the smallest scale of insect local navigation, the Kolmogorov microscale would be on the order of microns, with the Batchelor microscale being approximately 30 times smaller. These are the scales that are critical to small-scale flight searches, but in reality, can serve a similar purpose for large-scale odor tracking, again for example in seabirds. Given the much larger spatial extent of long-distance foraging, the mean concentrations of odorants and mean advection of the bulk flow will be far more important than the small-scale differences described above. Still, the derivations are critical because they reveal the physical landscape structure of sensory stimuli that are present for each organism at each scale.

These two physical parameters describe the grain at which information is resolvable by navigating organisms. Thus, the Kolmogorov scale equations can be used to determine the grain of velocity gradients that exist within air flow. Organisms sensitive to velocity gradients, either in the context of navigation or for flight itself (such as gliding), are navigating through a sensory landscape of air patterns. These patterns, at the smaller end, can be quantified and illustrated using the Kolmogorov scales. In a similar fashion, those organism using spatial or temporal chemical gradients for navigational purposes are moving through a landscape that can be quantified using the Batchelor scale equations. Taken together, these two equations can be used to quantify the sensory landscape and the information contained within that landscape that allows spatial navigation to occur.

2.2.1 Boundary Layer

The final aspect of the physical derivations of chemical dispersion and the movement of air flow is termed the boundary layer. Odor dispersion and air movement is influenced by the interaction between air flow and a solid surface. With respect to the terrestrial environment of concern for aeroecology, boundary layers are generated as air flows over woods, grass, and all types of terrain (Garratt 1994). When air flows over a solid structure such as the ground, the interface between a solid structure and the air flow is called a "boundary layer." At the solid surface (such as the ground), air flow is nonexistent, but as distance away from the structure increases, air velocity also increases. Thus, the boundary layer is associated with a velocity gradient, and because signals are dispersed primarily by air movement, the boundary layer has significant impacts on the spatial and temporal structure of chemical signals and the spatial and temporal nature of turbulent eddies. Chemicals within a near-surface region, whether a wall, floor, or ground, exhibit a streaky, persistent characteristic (Crimaldi et al. 2002) and air flow within this region often has dampened eddies. Further from the surface, turbulent energy increases and eddies become larger and larger. As eddy size increases, stirring intensifies and chemicals are dispersed more rapidly. Within most terrestrial environments, the atmospheric boundary layer is approximately 1 km in height from the ground. As such, the importance of boundary layer dynamics for navigation would depend on the altitude where the navigational information is acquired. Given that most if not all olfactory-based navigational behavior will take place within the boundary layer (see below), boundary layer air and chemical dynamics are of particular relevance. However, birds flying above the boundary layer, if they are indeed actively processing navigational information at such altitudes, would be confronted with very different air and chemical dynamics.

2.3   Sensory Landscapes at Varying Scale

All of the physical processes outlined above cause chemical sources to be dispersed in a three-dimensional distribution that is dynamic in both space and time, and along with this, develop different turbulent air structures based on the scale under consideration. When considered from both the space and time scales outlined above, the distribution of air currents and odor plumes creates a sensory landscape where the height of the landscape (valleys and peaks) can be considered different concentrations or air velocities, whose spatial distribution of valleys and peaks are dynamically changing over time and space (Fig. 6.1). The exact profile of these changes would depend upon the scale at which an organism samples and moves through the landscape.

This sensory landscape can be compared to physical landscapes for organisms that vary in size. The movement of an ant through a forest floor is dictated by different physical constraints and sensory information than a deer walking through that same habitat. Their sensory systems are adapted to extract the spatial and temporal information relevant for their size and rate of movement (Wehner 1987). In a similar fashion, animals at different ends of the aeroecology spectrum (i.e., short vs long distance or low altitude vs high altitude navigators) may both be moving through/over the same physical habitat (e.g., meadow), but the sensory landscape that each organism perceives and responds to is significantly different. A bee foraging in an open meadow is not hundreds of meters off the surface of the ground and thus is subjected to the turbulent air flow moving through the atmospheric boundary layer, which is heavily influenced by the plants and flowers of the habitat. The odor plume that may guide search behavior is significantly heterogeneous in space and time because of the odor(s) sampled by its antennae is structured by this small-scale turbulence. Conversely, a bird flying over the meadow at an altitude of several hundred meters is unaware of the odor plumes guiding the bee's navigation and the turbulence that is buffeting the bee's flight (Nisbet 1955; Combes and Dudley 2009). The highly heterogeneous odor plume of the bee is replaced by larger scale odor gradients or patches that are relatively homogeneous over the space normally engaged by short-distance navigating bees. Being several hundred meters off of the ground, the bird is subjected to significantly larger wind velocities that are not dampened by the boundary layer physics near the plants below. Understanding the importance of patterns and processes operating at different spatial and temporal scales offers the researcher insight into the structure of the sensory landscape within which navigation takes place.

3   Long-distance Navigation

As described above, atmospheric properties can provide information, operative over relatively short distances, which can influence or guide goal navigation. For example, small-scale navigation can be enabled by animals detecting the predictable spatial relationship between chemicals contained in the atmosphere and a goal

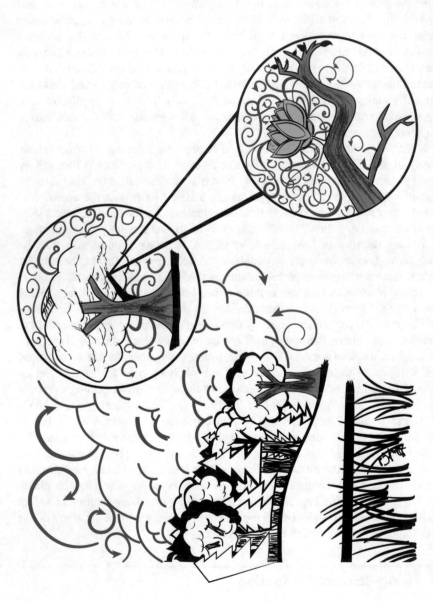

Fig. 6.1 Illustration showing the scales at which wind and chemical signals mix to create information. At the largest spatial scale (forest, left hand side) wind eddies (blue arrows) are large and comingle with plumes from many individual, local sources (purple). Thus, the finer-scale, filamentous nature of turbulent

location. However, given the lack of detailed quantification of many sensory signals at larger spatial scales, understanding the relationship between long-distance navigation and odors is considerably more challenging. It is, therefore, reasonable to ask to what extent chemicals contained in the atmosphere can be exploited as a source of navigational information to support longer distance navigation. What other properties of the atmosphere could support or influence navigation? We will explore these questions in the context of bird and insect migration (or homing in the case of pigeons), but first we need to clarify the often confusing concept of navigation as it applies to long-distance movements.

At its simplest, navigation can be described as any goal-directed behavior. As such, a simple magnetotactic response in a bacteria (Blakemore 1975), which would succeed in broadly bringing a bacteria to its target habitat, could be considered semantically equivalent to something vastly more complex and space-specific, like a "cognitive map" in a mammal or bird (Gagliardo et al. 2009; O'Keefe and Nadel 1978). Although somewhat obsolete, for long-distance movements, we still find useful the conceptual scheme of Griffin (1952, but see Able 2001 for a more current terminology of navigational mechanisms) for understanding long-distance navigation guided or influenced by atmospheric properties. Griffin (1952) described three types of navigational mechanisms. His Type 1 navigation was based on the use of familiar landmarks, the spatially richest manifestation of which could be the use of a hippocampal-dependent cognitive map (Gagliardo et al. 2009; O'Keefe and Nadel 1978). Type I navigation is generally thought to be guided by visual cues, but is not necessarily limited to them (see below). Type II navigation reflected an ability to maintain a compass bearing and there are aspects of Type II navigation that are relevant to questions of aeroecology. Type II navigation does not have the characteristics of a map; the maintenance of a compass bearing does not provide an animal with any information on position relative to some goal. The best known compass mechanisms are based on the sun (Schmidt-Koenig 1958), stars (Emlen 1967), and earth's magnetic field (Wiltschko and Wiltschko 1972), but at least for migratory birds and insects, atmospheric winds can have a profound effect on how animals orient their flights, sometimes even serving as an orientation cue themselves. Type III navigation is characterized by the ability to orient in the direction of a goal without reliance on familiar landmarks and is sometimes referred to as "true navigation." The navigational map of homing pigeons (Wallraff 2005; Wiltschko and Wiltschko 2003), which can be used to compute the direction home from even

←――――――――――――――――――――――――――――――――――――

Fig. 6.1 (continued) plumes is lost at larger spatial scales. At intermediate spatial scales (single trees, top bubble), the largest wind eddies are smaller and the individual nature of filaments of odor are present as possible sources of information. At the smallest spatial scale (single flower, bottom right bubble), the smallest wind eddies are of the same spatial magnitude as the filaments of odors, and thus, move the filaments as a whole creating turbulent odor plumes that are highly heterogeneous with respect to concentration differences. The atmospheric interactions occur simultaneously at the different spatial scales and operate at different temporal scales. Illustration by Alex Neal

distant, unfamiliar locations, would be the prototypical example of Type III navigation. Here, again, the atmosphere is relevant as there is an accumulating body of evidence demonstrating that, like the small-scale navigation best exemplified by insects, atmospheric odors can similarly support the landscape-scale, Type III navigation of homing pigeons and other birds (Gagliardo 2013; Wallraff 2005). It is, therefore, winds and odors that are properties of the atmosphere that are of most interest for the aeroecology of long-distance navigation.

3.1   The Multifarious Influence of Wind

For aerial navigators, winds can both interfere and assist the navigational mechanisms that guide migration and long-distance foraging. Flying in opposing headwinds would be associated with a tremendous energetic cost in birds and would be almost impossible for slow flying insects. Perhaps navigationally more damaging, flying in crosswinds could result in drift from a preferred route. By contrast, flying with tailwinds would result in energetic savings and a more rapid progression toward a goal destination. In fact, for some migratory birds, the winds appear to serve directly as an orientation cue, supporting what was described above as Type II navigation. As such, aerial navigation is inextricably bound to the challenge of wind as an agent of good or bad.

3.1.1   Insects

For migrating insects, whose air speeds are typically less than wind speeds and much of their migratory progress is a result of being "carried" by the wind, successful migration very much depends on selecting favorable winds that align closely with their preferred migratory directions. Indeed, a growing body of literature suggests that insects are surprisingly skilled at choosing winds favorable for migration and show a capacity to compensate for wind drift that is remarkably similar to that seen in migratory birds (Alerstam et al. 2011; Chapman et al. 2010, 2015).

The best known and perhaps most spectacular insect migration is that of the monarch butterfly (*Danaus plexippus*). During their autumn migration (not a true migration because the animals that migrate south will not return), individual monarchs leave their natal sites in temperate North America and travel to a surprisingly restricted overwintering site in central Mexico (Guerra and Reppert 2015). Although monarchs are likely not capable of the Type III, true navigation described above (Mouritsen et al. 2013), they possess a robust sun compass and a capacity to orient by the Earth's magnetic field, enabling a Type II navigational ability (Guerra and Reppert 2015; Mouritsen and Frost 2002). Monarchs exploit these compass mechanisms to orient their migration, but how do these compass mechanisms interact with either beneficial or interfering winds?

Similar to migratory birds, monarchs on their autumn migrations will preferentially fly with tailwinds with respect to their preferred migratory direction (Brower 1996). Behavior reflecting selection of favorable winds has also been reported in the painted lady butterfly (*Vanessa cardui*; Stefanescu et al. 2007; see also

Chapman et al. 2010). When flying with favorable tailwinds, monarchs will soar with updrafts, reaching often considerable heights where stronger winds facilitate often impressive ground speeds (Brower 1996; Gibo and Pallett 1979). By contrast, when confronted with unfavorable winds, monarchs will often suspend migration, but if they do fly, they will do so close to the ground where the wind will be weaker and rely on powered flight (Brower 1996). As observed in other Lepidopteran species (Srygley 2001), in navigationally confounding crosswinds monarchs likely rely on stationary landmarks to actively correct for drift, with the capacity for compensation diminished in areas where landmarks are not present (e.g., over a sea). Behavioral compensation for wind drift has also been described in a species of migratory dragonfly (Srygley 2003). Although nonmigratory, bumblebees are also capable of compensating for wind drift, again presumably relying on landmarks to assess displacement from the preferred direction or heading (Riley et al. 1999).

Insects that migrate during the day orient by their sun (or magnetic) compass and are able to use the sun compass to select winds that closely align with their preferred migratory direction. They can also use landmarks to alter headings as compensation for drift. By contrast, the wind-related challenges confronting the navigation of nocturnal migratory insects, given the naturally diminished availability of visual cues, appear prodigious. Despite such anticipated difficulties, the nocturnal migratory moth, *Autographa gamma*, is remarkably adept at behaviorally responding to winds to facilitate its migratory advance (Chapman et al. 2008), a capacity observed in other nocturnal, migratory insects (Aralimarad et al. 2011; Chapman et al. 2010). *Autographa gamma* selectively migrates on nights with favorable tailwinds, often flying at high altitudes where the winds would be most favorable. Even more surprising, when winds are not aligned with their preferred migratory direction, they alter their headings to actively compensate for wind drift. This capacity of compensation implies that individual moths are able to compare their direction of movement over the ground with their preferred migratory direction as determined by some compass mechanism. Given that this species migrates at night, Chapman et al. (2008) speculated that the compass reference might be derived from the Earth's magnetic field.

Smaller insects are more susceptible to the vagaries of wind, but even here behavioral adaptations can enable animals to most profitably use winds to assist migratory progress. Notable examples are aphids, which can engage in a range of active behavior to control their position in an air column (Reynolds and Reynolds 2009; Reynolds et al. 2009).

The absence of a positional or map sense (Type III navigation) during long distance, migration ostensibly makes insects less interesting than birds in the context of aerial navigation. However, migratory insects can move across vast distances and show an impressive degree of goal directedness. They do so relying on compass mechanisms to determine migratory direction only and limited power to generate air speed. As such, winds are of paramount importance not only energetically but also in how they can promote or interfere with navigation. The brief summary above describes an impressive array of behavioral adaptations insects can apply to exploit tailwinds, which would promote progress in the preferred, compass-defined migratory direction, and compensate for drift by crosswinds.

3.1.2 Birds

In contrast to insects, migrant birds fly at speeds that routinely exceed wind speed, and in the case of shorebirds and waterfowl, can considerably exceed wind speed. Nonetheless, ecologically challenging migratory routes, for example, crossing of the Gulf of Mexico or the remarkable flight of the blackpoll warbler (*Setophaga striata*) from the Maritime Provinces and New England to the islands of the Caribbean (Nisbet 1970), necessarily require that birds while migrating are, at least periodically, aided by favorable tailwinds. Indeed, as a consequence of the considerable energetic cost of migration, natural selection has promoted an adaptive sensitivity to weather, and winds in particular, that enables birds to determine when and how to engage in energetically propitious migrations (Alerstam and Lindström 1990; Erni et al. 2005; Liechti and Bruderer 1998). As such, birds routinely and selectively migrate with favorable winds. However, like migratory insects, crosswinds can displace birds from their migratory routes, requiring some kind of compensatory behavior to correct for such displacements. But perhaps most curious, it appears that under some conditions birds purposely orient downwind; in other words, the winds can be used to determine their migratory orientation.

Weather is known to be critical in determining if a bird will migrate on a given day or not. Captive birds will alter their migratory activity levels with changes in weather parameters such as barometric pressure and temperature (Hein et al. 2011; Metclafe et al. 2013; Walter and Bingman 1984), which are often predictive of changes in wind direction. In the field, birds can use information gathered on the ground to influence the decision to migrate or not (Åkesson et al. 2002; Eikenaar and Schmaljohann 2015). However, the decision to migrate or not does not speak directly to how winds can influence active navigation.

Migration in cross winds can result in navigational error as a bird's heading and direction of movement with respect to the ground deviate (see Richardson 1990). It should not be surprising, therefore, that migratory birds adopt behavioral strategies to compensate for any errors. Because of their powered flight, which can produce air speeds that typically exceed wind speeds, it is much easier for birds to maintain a preferred migratory direction in crosswinds compared to slow flying insects, which are more dependent on flying with favorable tail winds to advance their migrations (Alerstam 2011). For example, like insects, birds that migrate during the day can use visual landmarks to detect and eventually compensate for displacement by winds. However, birds that migrate at night or over water would seem to have little opportunity to detect and correct for displacement by crosswinds. However, at least under some conditions, nocturnal migrants can use conspicuous environmental features to compensate for wind drift (Bingman et al. 1982), and birds migrating over water may partially compensate by using wave patterns as a reference (Alerstam 1976). Migrant birds can also adjust the altitude of migration to take advantage of more favorable winds in certain atmospheric layers (Bruderer 1997; Dokter et al. 2013). Rather than compensate during an entire day or night migratory flight, it could be energetically more advantageous for migrants to allow themselves to be drifted by the wind and then carry out a compensatory reorientation later (Alerstam 1979). Consistent with such a strategy are observations of nocturnal migrants flying early in the morning in a direction that would correct for drift experienced the night before (Bingman 1980; Gauthreaux 1978; Van Doren et al. 2014). Although poorly studied compared to birds, at least one species of bat, the fruit bat (*Eidolon helvum*), has been shown to compensate for wind drift as individuals move between roosting and feeding sites separated by more than 40 km (Sapir et al. 2014).

The strong dependence of raptors on thermal soaring and inter-thermal gliding during migration renders them particularly susceptible to winds in shaping how they move with respect to the ground (see Kerlinger 1989; Kerlinger et al. 1985). It is not surprising, therefore, that raptors will preferentially fly with tailwinds, and overall, their often elliptical migratory paths are thought to reflect a compromise between the navigational consequences of wind drift and the energetic cost of compensation. Albatrosses engaged in dynamic soaring during foraging or return to a nest site may similarly negotiate with the wind to most efficiently reach their goal (Sachs et al. 2013).

Migratory birds time their migration in relation to weather predictive of wind direction and perhaps by even detecting winds near the ground directly (although it may be noteworthy that autumn migrants in central Europe frequently are forced to migrate against opposing winds; Erni et al. 2005). Migrants must also behaviorally compensate for crosswinds as a potential source of navigational error. However, these wind elements do not reveal any role for wind as a source of directional information that directly guides migration along the lines of Griffin's Type II navigation described above. However, evidence that wind could serve as an orientation cue has come from migrant birds in the southeastern United States (Able 1974; Gauthreaux and Able 1970). Here, migrating songbirds detected by radar invariably fly downwind even when the wind is blowing toward the opposite direction the birds need to fly to reach their migratory destination. By contrast, higher altitude, faster flying shorebirds and waterfowl consistently fly in the expected migratory direction independent of wind direction. The behavior of migrants in the northeastern United States is more complicated (Able 1982). On nights when celestial directional information (i.e., information from the setting sun or stars) is available, songbirds fly in a seasonally appropriate direction independent of wind direction. However, on overcast nights, migrants fly downwind even when that means flying in a seasonally inappropriate direction. The use of wind as an orientation cue under overcast is further supported by the finding that migratory white-throated sparrows (*Zonotrichia albicollis*), when experimentally released aloft with their vision obscured, also fly downwind (Able et al. 1982).

At least under some conditions, migrant birds can use winds as an orientation cue determining the direction of their migratory flight. In some sense then, wind can be discussed in the same conversation as the non-atmospheric, better known orientation cues of the setting sun, stars, and the Earth's magnetic field (see Able 1982; Wiltschko and Wiltschko 2003). It is not unexpected, therefore, that migrants are sensitive to weather predictive of wind direction; under conditions when orientation would be determined by wind direction, it would clearly be adaptive to fly on nights when wind direction would point in the seasonally appropriate, migratory direction or to choose to fly at altitudes where wind direction would correspond most closely to the preferred migratory direction. It probably has not been lost on the reader that the literature on wind as a candidate orientation cue, to the best of our knowledge, now dates back 30 years. Given the extraordinary advances in animal tracking technologies (Guilford et al. 2011) and global meteorological surveillance, it seems that now is a good time to revisit the question of how extensive wind may be used as an orientation cue.

3.2   Bird Navigation by Atmospheric Odors

3.2.1   Tracking Odor Plumes

Chemical cues or odors are certainly the most salient source of potential navigational information that is part of the atmosphere. As we saw in some examples above, local goal navigation in insects is often guided by predictable variation in the spatial distribution of odors between an animal and its goal. But as one of nature's foremost navigators, birds are historically considered unlikely "olfactory navigators." This prejudice primarily comes from the generally small olfactory bulbs of birds (but see Bang and Cobb 1968), which would appear insufficiently substantial to carry out the necessary odor discriminations to support navigation. However, recent evidence of a rich repertoire of olfactory receptor genes (Steiger et al. 2008) and an extensive neuroanatomical network of olfactory processing regions in the avian brain (Atoji and Wild 2014) suggest that birds in general may have far richer olfactory sensitivity than traditionally thought.

Over relatively short and long distances, numerous species of birds, particularly procelleriiform seabirds, have been shown to track odor plumes while navigating, as broadly defined above, to goal locations (DeBose and Nevitt 2008). Leach's petrel (*Oceanodromo leucorhoa*) rely on odors to locally steer a course to their nest once on land after returning from an overwater foraging trip (Grubb 1974). Over considerably longer distances, wandering albatrosses (*Diomedea exulans*) in search of patchily distributed prey in the open ocean engage in "olfactory search behavior" to pick up odor plumes that can be tracked to the food source (Nevitt et al. 2008).

3.2.2   True Navigation by Olfaction

Although impressive, the capacity of some species of birds to track odor plumes to a goal destination can be considered a variant of the Type I navigation described above. Griffin (1952) did not make a distinction between landmark navigation with little cognitive demand (e.g., attraction to a visual/landmark beacon neighboring a goal location) compared to something like a "cognitive map," which would be cognitively more challenging. In our view, the tracking of an odor plume would be comparable to reaching a goal location by approaching a visual-landmark beacon. By contrast, it is the capacity of birds, most notably homing pigeons, to use the distribution of atmospheric odors to support Type III, true navigation that has had the greatest impact on discussion of the relationship among odors, atmospheric conditions and navigation (Gagliardo 2013; Wallraff 2004, 2005; Wallraff and Andreae 2000).

The mechanisms by which birds routinely navigate to distant and remote goal locations from places they have never been before, following both natural and experimental displacements, remains one of nature's still unresolved mysteries. It is generally agreed that Type III, true navigation requires both a "map," or positional sense, and a "compass," or directional sense (Wallraff 2005; Wiltschko and Wiltscko 2003). While compass mechanisms are now well understood (see references above), even someone only remotely acquainted with animal navigation research is aware of the continued debate on the sensory basis of the map sense. Despite the considerable attention given to it (Wiltschko and Wiltschko 2003), there is still no compelling evidence that birds possess a map based on any geomagnetic parameter (but see Kishkinev et al. 2013), and this is particularly true of homing pigeons (Bingman and Cheng 2005; Gagliardo 2013; Wallraff 2005). By contrast, the evidence for an olfactory map based on the spatially heterogeneous distribution of atmospheric odors is, in our view, persuasive (Gagliardo 2013; Wallraff 2004, 2005).

Pigeons that are prevented from sensing odors by a variety of means, including olfactory nerve cuts and lesions of the piriform (olfactory) cortex, invariably fail to successfully navigate home (see Gagliardo 2013; Wallraff 2004, 2005 for reviews). Immediate early gene activation in olfactory processing areas of the brain during navigation also point to atmospheric odors as a crucial source of navigational information (Patzke et al. 2010). Perhaps more compelling, young pigeons can construct a meaningful navigational map based on the experimental manipulation of natural and experimental odors (Ioalé et al. 1978, 1990; Papi et al. 1974). And as adults, they can be fooled to orient in a "false direction" when released from one location but with the olfactory-navigational input originating from another (false) location (Benvenuti and Wallraff 1985). In other words, if a pigeon's navigational input originates from Location A, but they are actually released from Location B, it is more likely to fly off in a direction that would be homeward oriented with respect to Location A.

In the context of aeroecology, the important questions are related to the nature of the chemicals that are sensed and their distribution in the atmosphere both over the earth's surface and with respect to altitude. Unfortunately, much has yet to be learned. What is clear is that impeding determination of the directional origin (Wallraff 1966) or manipulating the direction of origin (Ioalé et al. 1978, 1990; Papi et al. 1974) of atmospheric surface odors can be sufficient to disrupt or alter, respectively, map learning. There are two frequently discussed models of the surface distribution of atmospheric odors that could enable the learning of a map (Gagliardo 2013), but they should not be viewed as mutually exclusive. The generally favored model is that of a gradient map whereby pigeons, and perhaps other birds, learn the directional axis along which the strength or concentration of an odor or odor profile increases or decreases. Note, although the essential navigational property of an odor would be its concentration gradient in the atmosphere, the gradient would *not* be followed to return home as was the case with the nest and food finding behavior described for the seabirds above. Rather, the gradient would allow the comparison between a pigeon's current location and its home and thus enable a computation of direction to its loft. The journey home would then be guided by one of the compass mechanisms (Gagliardo 2013; Wallraff 2005). The alternative "mosaic" model assumes that different directions or locations in space are characterized by different olfactory "signatures," and that pigeons learn a navigational map based on the discriminative odor signatures carried by winds that originate from different directions with respect to the home loft. Therefore, a pigeon could learn a mosaic map never leaving its home aviary, as long as the aviary is open to ambient air, simply by associating different odor signatures with the direction of the winds that bring them.

Wallraff and colleagues (Wallraff 2013; Wallraff and Andreae 2000) have made a first-order attempt at identifying how odors are structured in the surface atmosphere and whether or not that structure would be stable enough to allow for an olfactory map. In a breakthrough study, Wallraff and Andreae (2000) took samples from the atmosphere about 4 m above the ground from 96 locations across a large portion of Germany. They found that although winds could displace the gradients, across a number of anthropogenic hydrocarbons, the gradient relationship among the different chemicals was stable enough to model the navigational performance of homing pigeons. In other words, the *relative* concentrations of a *group* of atmospheric chemicals was sufficient to approximate locations in space at a spatial resolution compatible with the observed resolution of the homing pigeon navigational map (Fig. 6.2). For example, and drawing from Fig. 6.2, a pigeon sensing a relatively high concentration of butane and a low concentration of i-Hexane would indicate a displacement to the northwest, and the larger the relative concentration difference the farther away the pigeon would be.

It should be emphasized that the authors did not claim the anthropogenic hydrocarbons measured were the actual odors used in constructing a map. Indeed, naturally emitted, volatile hydrocarbons are considered a more "natural" source of atmospheric spatial information. An open question remains with respect to the altitudinal distribution of chemicals/odors that can support navigation. Are there atmospheric chemicals spatially distributed at 100 m above the surface whose structure homing pigeons could exploit for navigation? If yes, how stable would that structure be when compared to odors closer to the surface? More dramatic, could migratory birds flying at 500–2000 m above the surface take navigational advantage of odors at those atmospheric layers (Bingman and Cheng 2005)?

Is an olfactory map limited to homing pigeons or is it more diffusely found among birds? Here seabirds are again of interest. As noted above, albatrosses can use a kind of Type I navigation to track odors to concentrations of food. Because such food resources will move with the ocean currents and at some point become depleted, it is clear that an enduring map of such spatial and temporal unstable food locations would be of little survival value; it would be better to track the moment and then remember a location on a "mental map" of where things were last week. But what about breeding seabirds that leave the nest site to forage for food sometimes more than 1000 km away over the ocean, who then need to efficiently navigate back to the nest? Navigating to a spatially stable breeding island or nest

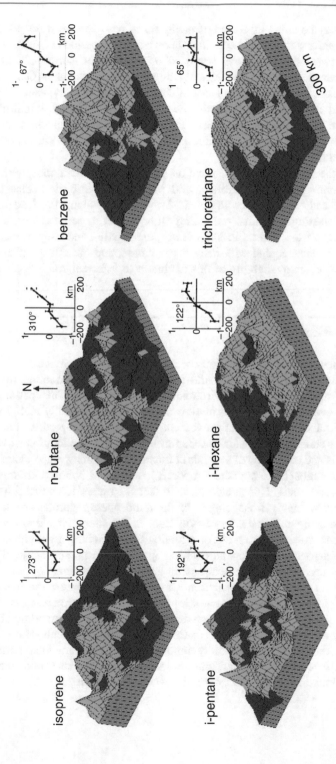

Fig. 6.2 Relief maps showing standardized ratios of six hydrocarbons, relative to the sum of all six, in the ground-level atmosphere near Würzburg, Germany. Regions in which the relative portion of a given compound was above its overall mean are drawn green, those with ratios below average red. Small diagrams show means at different distance points along the computed gradient, whose direction of increasing relative concentration is indicated (y-axis displays the Z difference from overall mean in units of standard deviation). (After Wallraff and Andreae 2000)

site would certainly be aided by a kind of navigational map described for homing pigeons, and indeed, recent experiments have shown that navigation to a nest site in the Cory's shearwaters (*Calonectris borealis*) of the eastern North Atlantic is dependent on atmospheric odors when visual landmarks are unavailable (Gagliardo et al. 2013). Tracked shearwaters, displaced 800 m from their breeding island, were notably impaired navigating back to the nest following olfactory impairment (magnetically impaired shearwaters were unimpaired). And even short-range, nest-homing in this species is disrupted at night when animals are rendered anosmic (Dell'Ariccia and Bonadonna 2013).

In summary, the importance of chemical odors dispersed in the atmosphere for navigation in homing pigeons, seabirds, and probably other bird species (see Gagliardo 2013) can no longer be doubted. As such, and taking the lead from Wallraff, the interesting questions remaining with respect to aeroecology is the identity of the odors used, and more importantly, their spatial and temporal structure both along the surface and with respect to altitude, and the effects of atmospheric dynamics on the availability of this navigational information.

4   Conclusion

Efficient movement among goal locations necessarily requires a navigational system that takes advantage of spatially and temporally stable information that is heterogeneously distributed in the environment (see Able 1980). The spatial and temporal scales over which any information needs to be sufficiently stable will depend on the distance and speed of movement. As such, the properties of aerial navigation will be considerably different for an insect tracking an odor plume over a distance of meters and a duration of seconds compared to a homing pigeon locating its position using atmospheric odors to support a journey of a hundred kilometers that will take several hours. Nonetheless, it is in fact chemicals/odors that provide the most exploitable "navigational input" of the atmosphere regardless of scale. Embedded in the context of "olfactory navigation" is the unpredictable effects of wind, which can serve as either an aid or a hindrance to reaching a goal location. The new field of aeroecology offers an exciting opportunity to revisit some of the classic questions investigating the relationship among wind, orientation, and the capacity to carry out corrective reorientation. Of perhaps more interest, the question of olfactory navigation is at the forefront of long-distance, navigation research, and it is clear that progress here will very much depend on a better understanding of the physical identity and relationship of navigationally relevant chemicals/odors with the atmosphere. And although generally ignored in our chapter, what remains practically unexplored are the underlying neural mechanisms that support aerial navigation. We look forward to the future discoveries that await us.

References

Able KP (1974) Environmental influences on the orientation of free-flying nocturnal bird migrants. Anim Behav 22(1):224–238

Able KP (1980) Mechanisms of orientation, navigation and homing. In: Gauthreaux S (ed) Animal migration, orientation and navigation. Academic, New York, pp 283–373

Able KP (1982) Field studies of avian nocturnal migratory orientation I. Interaction of sun, wind and stars as directional cues. Anim Behav 30(3):761–767

Able KP (2001) The concepts and terminology of bird navigation. J Avian Biol 32(2):174–183

Able KP, Bingman VP, Kerlinger P, Gergits W (1982) Field studies of avian nocturnal migratory orientation II. Experimental manipulation of orientation in white-throated sparrows (Zonotrichia albicollis) released aloft. Anim Behav 30(3):768–773

Åkesson S, Walinder G, Karlsson L, Ehnbom S (2002) Nocturnal migratory flight initiation in reed warblers Acrocephalus scirpaceus: effect of wind on orientation and timing of migration. J Avian Biol 33(4):349–357

Alerstam T (1976) Do birds use waves for orientation when migrating across the sea? Nature 259:205–207

Alerstam T (1979) Optimal use of wind by migrating birds: combined drift and overcompensation. J Theor Biol 79(3):341–353

Alerstam T (2011) Optimal bird migration revisited. J Ornithol 152(1):5–23

Alerstam T, Lindström Å (1990) Optimal bird migration: the relative importance of time, energy, and safety. In: Gwinner E (ed) Bird migration. Springer, Berlin, pp 331–351

Alerstam T, Chapman JW, Bäckman J, Smith AD, Karlsson H, Nilsson C, Reynolds DR, Klaassen RHG, Hill JK (2011) Convergent patterns of long-distance nocturnal migration in noctuid moths and passerine birds. Proc R Soc B Biol Sci 282(1804):rspb20110058

Allison AC (1953) The morphology of the olfactory system in the vertebrates. Biol Rev 28(2):195–244

Anderson H (1985) The distribution of mechanosensory hair afferents within the locust central nervous system. Brain Res 333(1):97–102

Aralimarad P, Reynolds AM, Lim KS, Reynolds DR, Chapman JW (2011) Flight altitude selection increases orientation performance in high-flying nocturnal insect migrants. Anim Behav 82(6):1221–1225

Atoji Y, Wild JM (2014) Efferent and afferent connections of the olfactory bulb and prepiriform cortex in the pigeon (Columba livia). J Comp Neurol 522(8):1728–1752

Bang BG, Cobb S (1968) The size of the olfactory bulb in 108 species of birds. Auk 85(1):55–61

Benvenuti S, Wallraff HG (1985) Pigeon navigation: site simulation by means of atmospheric odours. J Comp Physiol A 156(6):737–746

Bingman VP (1980) Inland morning flight behavior of nocturnal passerine migrants in eastern New York. Auk 97(3):465–472

Bingman VP, Cheng K (2005) Mechanisms of animal global navigation: comparative perspectives and enduring challenges. Ethol Ecol Evol 17(4):295–318

Bingman VP, Able KP, Kerlinger P (1982) Wind drift, compensation, and the use of landmarks by nocturnal bird migrants. Anim Behav 30(1):49–53

Blakemore R (1975) Magnetotactic bacteria. Science 190(4212):377–379

Brower L (1996) Monarch butterfly orientation: missing pieces of a magnificent puzzle. J Exp Biol 199(1):93–103

Bruderer B (1997) The study of bird migration by radar part 2: major achievements. Naturwissenschaften 84(2):45–54

Campbell SA, Borden JH (2006) Close-range in-flight integration of olfactory and visual information by a host-seeking bark beetle. Entomol Exp Appl 120(2):91–98

Chapman JW, Reynolds DR, Mouritsen H, Hill JK, Riley JR, Sivell D, Smith AD, Woiwod IP (2008) Wind selection and drift compensation optimize migratory pathways in a high-flying moth. Curr Biol 18(7):514–518

Chapman JW, Nesbit RL, Burgin LE, Reynolds DR, Smith AD, Middleton DR, Hill JK (2010) Flight orientation behaviors promote optimal migration trajectories in high-flying insects. Science 327(5966):682–685

Chapman JW, Reynolds DR, Wilson K (2015) Long-range seasonal migration in insects: mechanisms, evolutionary drivers and ecological consequences. Ecol Lett 18(3):287–302

Combes SA, Dudley R (2009) Turbulence-driven instabilities limit insect flight performance. Proc Natl Acad Sci 106(22):9105–9108

Craven BA, Neuberger T, Paterson EG, Webb AG, Josephson EM, Morrison EE, Settles GS (2007) Reconstruction and morphometric analysis of the nasal airway of the dog (Canis familiaris) and implications regarding olfactory airflow. Anat Rec 290(11):1325–1340

Craven BA, Paterson EG, Settles GS (2010) The fluid dynamics of canine olfaction: unique nasal airflow patterns as an explanation of macrosmia. J R Soc Interface 7(47):933–943

Crimaldi JP, Wiley MB, Koseff JR (2002) The relationship between mean and instantaneous structure in turbulent passive scalar plumes. J Turbul 3(14):1–24

DeBose JL, Nevitt GA (2008) The use of odors at different spatial scales: comparing birds with fish. J Chem Ecol 34(7):867–881

Dell'Ariccia G, Bonadonna F (2013) Back home at night or out until morning? Nycthemeral variations in homing of anosmic Cory's shearwaters in a diurnal colony. J Exp Biol 216(8):1430–1433

Dokter AM, Shamoun-Baranes J, Kemp MU, Tijm S, Holleman I (2013) High altitude bird migration at temperate latitudes: a synoptic perspective on wind assistance. PLoS One 8(1): e52300

Dudley R (2002) The biomechanics of insect flight: form, function, evolution. Princeton University Press, Princeton

Dusenbery DB (1992) Sensory ecology: how organisms acquire and respond to information. WH Freeman, New York

Eikenaar C, Schmaljohann H (2015) Wind conditions experienced during the day predict nocturnal restlessness in a migratory songbird. Ibis 157(1):125–132

Emlen ST (1967) Migratory orientation in the Indigo Bunting, Passerina cyanea: part I: evidence for use of celestial cues. Auk 84(3):309–342

Erni B, Liechti F, Bruderer B (2005) The role of wind in passerine autumn migration between Europe and Africa. Behav Ecol 16(4):732–740

Gagliardo A (2013) Forty years of olfactory navigation in birds. J Exp Biol 216(12):2165–2171

Gagliardo A, Ioale P, Savini M, Dell'Omo G, Bingman VP (2009) Hippocampal-dependent familiar area map supports corrective re-orientation following navigational error during pigeon homing: a GPS-tracking study. Eur J Neurosci 29(12):2389–2400

Gagliardo A, Bried J, Lambardi P, Luschi P, Wikelski M, Bonadonna F (2013) Oceanic navigation in Cory's shearwaters: evidence for a crucial role of olfactory cues for homing after displacement. J Exp Biol 216(15):2798–2805

Garratt JR (1994) The atmospheric boundary layer. Cambridge University Press, Cambridge

Gauthreaux SA (1978) Importance of the daytime flights of nocturnal migrants: redetermined migration following displacement. In: Schmidt-Koenig K, Keeton WT (eds) Animal migration, navigation, and homing. Springer, Berlin, pp 219–227

Gauthreaux SA, Able KP (1970) Wind and the direction of nocturnal songbird migration. Nature 228:476–477

Gibo DL, Pallett MJ (1979) Soaring flight of monarch butterflies, Danaus plexippus (Lepidoptera: Danaidae), during the late summer migration in southern Ontario. Can J Zool 57(7):1393–1401

Gnatzy W, Tautz J (1980) Ultrastructure and mechanical properties of an insect mechanoreceptor: stimulus-transmitting structures and sensory apparatus of the cereal filiform hairs of Gryllus. Cell Tissue Res 213(3):441–463

Gomez G, Atema J (1996) Temporal resolution in olfaction: stimulus integration time of lobster chemoreceptor cells. J Exp Biol 199(8):1771–1779

Greenewalt CH (1975) The flight of birds: the significant dimensions, their departure from the requirements for dimensional similarity, and the effect on flight aerodynamics of that departure. Trans Am Philos Soc 65(4):1–67

Griffin DR (1952) Bird navigation. Biol Rev 27(4):359–390

Grubb TC (1974) Olfactory navigation to the nesting burrow in Leach's petrel (Oceanodroma leucorrhoa). Anim Behav 22(1):192–202

Guerra PA, Reppert SM (2015) Sensory basis of lepidopteran migration: focus on the monarch butterfly. Curr Opin Neurobiol 34:20–28

Guilford T, Åkesson S, Gagliardo A, Holland RA, Mouritsen H, Muheim R, Wiltschko R, Wiltschko W, Bingman VP (2011) Migratory navigation in birds: new opportunities in an era of fast-developing tracking technology. J Exp Biol 214(22):3705–3712

Hein CM, Zapka M, Mouritsen H (2011) Weather significantly influences the migratory behaviour of night-migratory songbirds tested indoors in orientation cages. J Ornithol 152(1):27–35

Henry M, Thomas DW, Vaudry R, Carrier M (2002) Foraging distances and home range of pregnant and lactating little brown bats (Myotis lucifugus). J Mammal 83(3):767–774

Hershberger WA, Jordan JS (1998) The phantom array: a perisaccadic illusion of visual direction. Psychol Rec 48(1):21–32

Ioalé P, Papi F, Fiaschi V, Baldaccini NE (1978) Pigeon navigation: effects upon homing behaviour by reversing wind direction at the loft. J Comp Physiol 128(4):285–295

Ioalè P, Nozzolini M, Papi F (1990) Homing pigeons do extract directional information from olfactory stimuli. Behav Ecol Sociobiol 26(5):301–305

Kerlinger P (1989) Flight strategies of migrating hawks. University of Chicago Press, Chicago

Kerlinger P, Bingman VP, Able KP (1985) Comparative flight behaviour of migrating hawks studied with tracking radar during autumn in central New York. Can J Zool 63(4):755–761

Kishkinev D, Chernetsov N, Heyers D, Mouritsen H (2013) Migratory reed warblers need intact trigeminal nerves to correct for a 1000 km eastward displacement. PLoS One 8(6):e65847

Kumagai T, Shimozawa T, Baba Y (1998) Mobilities of the cercal wind-receptor hairs of the cricket, Gryllus bimaculatus. J Comp Physiol A 183(1):7–21

Liechti F, Bruderer B (1998) The relevance of wind for optimal migration theory. J Avian Biol 29:561–568

Masson C, Mustaparta H (1990) Chemical information processing in the olfactory system of insects. Physiol Rev 70(1):199–245

Matsumoto SG, Hildebrand JG (1981) Olfactory mechanisms in the moth Manduca sexta: response characteristics and morphology of central neurons in the antennal lobes. Proc R Soc Lond B Biol Sci 213(1192):249–277

McKeegan DE (2002) Spontaneous and odour evoked activity in single avian olfactory bulb neurones. Brain Res 929(1):48–58

Metcalfe J, Schmidt KL, Kerr WB, Guglielmo CG, MacDougall-Shackleton SA (2013) White-throated sparrows adjust behaviour in response to manipulations of barometric pressure and temperature. Anim Behav 86(6):1285–1290

Moore P, Crimaldi J (2004) Odor landscapes and animal behavior: tracking odor plumes in different physical worlds. J Mar Syst 49(1):55–64

Mouritsen H, Frost BJ (2002) Virtual migration in tethered flying monarch butterflies reveals their orientation mechanisms. Proc Natl Acad Sci 99(15):10162–10166

Mouritsen H, Derbyshire R, Stalleicken J, Mouritsen OØ, Frost BJ, Norris DR (2013) An experimental displacement and over 50 years of tag-recoveries show that monarch butterflies are not true navigators. Proc Natl Acad Sci 110(18):7348–7353

Murlis J, Elkinton JS, Carde RT (1992) Odor plumes and how insects use them. Annu Rev Entomol 37(1):505–532

Nevitt GA, Veit RR, Kareiva P (1995) Dimethyl sulphide as a foraging cue for Antarctic procellariiform seabirds. Nature 376(6542):680–682

Nevitt GA, Losekoot M, Weimerskirch H (2008) Evidence for olfactory search in wandering albatross, Diomedea exulans. Proc Natl Acad Sci 105(12):4576–4581

Newland PL, Rogers SM, Gaaboub I, Matheson T (2000) Parallel somatotopic maps of gustatory and mechanosensory neurons in the central nervous system of an insect. J Comp Neurol 425(1):82–96

Nisbet ICT (1955) Atmospheric turbulence and bird flight. Br Birds 48:557–559

Nisbet IC (1970) Autumn migration of the Blackpoll Warbler: evidence for long flight provided by regional survey. Bird-Banding 41(3):207–240

O'Keefe J, Nadel L (1978) The hippocampus as a cognitive map. Clarendon Press, Oxford

Papi F, Ioalé P, Fiaschi V, Benvenuti S, Baldaccini NE (1974) Olfactory navigation of pigeons: the effect of treatment with odorous air currents. J Comp Physiol A 94(3):187–193

Patzke N, Manns M, Güntürkün O, Ioale P, Gagliardo A (2010) Navigation-induced ZENK expression in the olfactory system of pigeons (Columba livia). Eur J Neurosci 31(11):2062–2072

Reynolds AM, Reynolds DR (2009) Aphid aerial density profiles are consistent with turbulent advection amplifying flight behaviours: abandoning the epithet 'passive. Proc R Soc B Biol Sci 276(1654):137–143

Reynolds AM, Reynolds DR, Riley JR (2009) Does a 'turbophoretic'effect account for layer concentrations of insects migrating in the stable night-time atmosphere? J R Soc Interface 6(30):87–95

Richardson WJ (1990) Wind and orientation of migrating birds: a review. Experientia 46(4):416–425

Riley JR, Reynolds DR, Smith AD, Edwards AS, Osborne JL, Williams IH, McCartney HA (1999) Compensation for wind drift by bumble-bees. Nature 400(6740):126

Rumbo ER, Kaissling KE (1989) Temporal resolution of odour pulses by three types of pheromone receptor cells in Antheraea polyphemus. J Comp Physiol A 165(3):281–291

Sachs G, Traugott J, Nesterova AP, Bonadonna F (2013) Experimental verification of dynamic soaring in albatrosses. J Exp Biol 216(22):4222–4232

Sapir N, Horvitz N, Dechmann DK, Fahr J, Wikelski M (2014) Commuting fruit bats beneficially modulate their flight in relation to wind. Proc R Soc B Biol Sci 281(1782):20140018

Schmidt-Koenig K (1958) Experimentelle Einflußnahme auf die 24-Stunden-Periodik bei Brieftauben und deren Auswirkungen unter besonderer Berücksichtigung des Heimfindevermögens. Z Tierpsychol 15(3):301–331

Schneider D (1964) Insect antennae. Annu Rev Entomol 9(1):103–122

Shöne H (1984) Spatial orientation: the spatial control of behavior in animals and man. Princeton University Press, Princeton

Srygley RB (2001) Compensation for fluctuations in crosswind drift without stationary landmarks in butterflies migrating over seas. Anim Behav 61(1):191–203

Srygley RB (2003) Wind drift compensation in migrating dragonflies Pantala (Odonata: Libellulidae). J Insect Behav 16(2):217–232

Stefanescu C, Alarcon M, AVila A (2007) Migration of the painted lady butterfly, Vanessa cardui, to north-eastern Spain is aided by African wind currents. J Anim Ecol 76(5):888–898

Steiger SS, Fidler AE, Valcu M, Kempenaers B (2008) Avian olfactory receptor gene repertoires: evidence for a well-developed sense of smell in birds? Proc R Soc B Biol Sci 275(1649):2309–2317

Uchida N, Kepecs A, Mainen ZF (2006) Seeing at a glance, smelling in a whiff: rapid forms of perceptual decision-making. Nat Rev Neurosci 7(6):485–491

Van Doren BM, Sheldon D, Geevarghese J, Hochachka WM, Farnsworth A (2014) Autumn morning flights of migrant songbirds in the northeastern United States are linked to nocturnal migration and winds aloft. Auk 132(1):105–118

Vanrullen R, Thorpe SJ (2001) The time course of visual processing: from early perception to decision-making. J Cogn Neurosci 13(4):454–461

Vickers NJ (2000) Mechanisms of animal navigation in odor plumes. Biol Bull 198(2):203–212

Vogel S (1996) Life in moving fluids: the physical biology of flow. Princeton University Press, Princeton

Wallraff HG (1966) Über die Heimfindeleistungen von Brieftauben nach Haltung in verschiedenartig abgeschirmten Volieren. Z Vgl Physiol 52(3):215–259

Wallraff HG (2004) Avian olfactory navigation: its empirical foundation and conceptual state. Anim Behav 67(2):189–204

Wallraff HG (2005) Avian navigation: pigeon homing as a paradigm. Springer, Berlin

Wallraff HG (2013) Ratios among atmospheric trace gases together with winds imply exploitable information for bird navigation: a model elucidating experimental results. Biogeosciences 10(11):6929–6943

Wallraff HG, Andreae MO (2000) Spatial gradients in ratios of atmospheric trace gases: a study stimulated by experiments on bird navigation. Tellus B 52(4):1138–1157

Walther Y, Bingman VP (1984) Orientierungsverhalten von Trauerschnäppern (Ficedula hyoleuca) während des Frühjahrszuges in Abhängigkeit von Wetterfaktoren. Vogelwarte 32:201–205

Wehner R (1987) 'Matched filters'—neural models of the external world. J Comp Physiol A 161(4):511–531

Wiltschko W, Wiltschko R (1972) Magnetic compass of European robins. Science 176:62–64

Wiltschko R, Wiltschko W (2003) Avian navigation: from historical to modern concepts. Anim Behav 65(2):257–272

Riders on the Wind: The Aeroecology of Insect Migrants

7

Don R. Reynolds, Jason W. Chapman, and V. Alistair Drake

Abstract

Migratory flight close to the Earth's surface (within the so-called flight boundary layer) occurs in some insects, but the vast majority of migrants ascend above this layer and harness the power of the wind for transport. The resulting displacements range from dispersive movements over a few tens of metres to seasonal migrations covering thousands of kilometres. In this chapter, we summarize knowledge of the use of the aerosphere by insects, focusing particularly on longer migrations, in relation to: the height and duration of flight, direction and speed of movement, seasonal and diel patterns, and responses to atmospheric conditions and phenomena. The seasonal mass movements have major ecological consequences in the invaded areas, and these are discussed briefly. We also highlight recent comparisons of insect movement strategies with those of flying vertebrates and mention interactions between these groups in the atmosphere. We conclude with some suggestions for the future development of these topics.

D.R. Reynolds (✉)
Natural Resources Institute, University of Greenwich, Kent, UK
e-mail: D.Reynolds@greenwich.ac.uk

J.W. Chapman
Centre for Ecology and Conservation, and Environment and Sustainability Institute, University of Exeter, Penryn, Cornwall, UK
e-mail: j.chapman2@exeter.ac.uk

V.A. Drake
School of Physical, Environmental and Mathematical Sciences, University of New South Wales at Canberra, Canberra, ACT, Australia

Institute for Applied Ecology, University of Canberra, Canberra, ACT, Australia
e-mail: a.drake@adfa.edu.au

© Springer International Publishing AG, part of Springer Nature 2017
P.B. Chilson et al. (eds.), *Aeroecology*,
https://doi.org/10.1007/978-3-319-68576-2_7

1   Introduction

The discipline of aeroecology is concerned with the interaction of airborne organisms—principally insects, birds, and bats—with the atmosphere and with each other. The important role played by the atmosphere in insect migration and the effects of atmospheric phenomena on insect flight have long been recognized by entomologists (Johnson 1969; Pedgley 1982; Drake and Farrow 1988; Rainey 1989; Isard and Gage 2001). As insects are rather small animals, their self-propelled flight speeds (airspeeds) are comparatively low, and thus their 'appetitive' or foraging flights (e.g. those in search of food, a mate, or an oviposition site) take place very largely within their 'flight boundary layer' (FBL)—the zone extending up from the ground where the ambient wind speed is lower than the insect's air speed (Taylor 1974). Consequently, insects high in the air are very likely to be engaged in *migration*, while birds or bats flying at altitude may be engaged in either foraging (with insects often the prey items) or migratory movement.

In order to achieve generality across the animal kingdom, many migration biologists, particularly entomologists, have concluded that migration is best recognized *behaviourally* (e.g. Kennedy 1985; Dingle and Drake 2007; Dingle 2014; Reynolds et al. 2014; Chapman et al. 2015a; Chapman and Drake 2017). Migratory organisms exhibit a syndrome of morphological, physiological and behavioural traits that bring about persistent, rectilinear movements accompanied by some inhibition of responsiveness to appetitive stimuli. This allows the migrant to move away from its 'home range', and migration is consequently quite distinct from foraging, commuting, or territorial patrolling (Dingle 2014). The distinctive behavioural processes of migration in individuals lead, of course, to a collective outcome—a significant spatial redistribution of the population—and the migration syndrome is understood to have evolved principally in response to spatio-temporal variation in the availability and/or quality of resources, either seasonally or on some shorter time scale (Drake and Gatehouse 1995; Dingle 2014; Chapman and Drake 2017). Aerial movement provides insects with a potentially highly efficient means of transportation, allowing populations to colonize and exploit new habitats, in some cases over considerable distances because the migrants concerned can remain airborne for long periods of time (see below) while they are carried along by the wind.

When conditions are sufficiently warm (see below), insects can generally be found at altitudes up to ~2 km above ground level (a.g.l.) (Fig. 7.1), and occasionally atmospheric temperatures may allow flight up to 4–5 km. Thus, the region of the atmosphere we are concerned with is the 'atmospheric boundary layer' (ABL) and the region of the lower troposphere immediately above it. The ABL is typically convective by day, extending to ~1 km, while at night a shallower 'nocturnal (stable) boundary layer' often forms with a 'residual layer' above it (Fig. 7.2). Our coverage in this chapter will concentrate on migratory movements above the vegetation canopy and will exclude further discussion of:

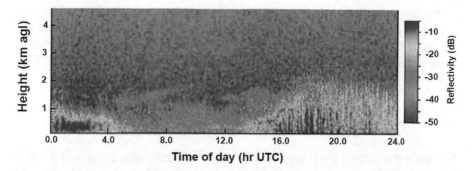

Fig. 7.1 Diurnal variability in intensity and vertical extent of a layer of small insect targets—as shown by reflectivity from a 35-GHz cloud radar on a cloud-free day in Oklahoma, USA. The depth of the insect layer follows the diurnal variation of the convective boundary layer and reaches a maximum in the afternoon when it can extend up to 2 km (From Luke et al. 2008. © American Meteorological Society. Used with permission)

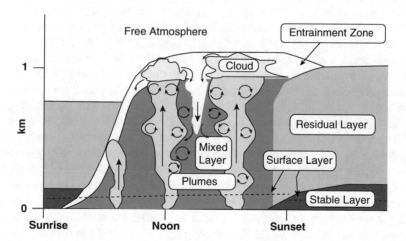

Fig. 7.2 Schematic diagram showing the structure of the ABL during fair weather over flat terrain (Modified from http://lidar.ssec.wisc.edu/papers/akp_thes/node6.html. With permission from Dr Antti Piironen)

(a) Almost all foraging or appetitive flights (except a few that extend to considerable height).

(b) Those migrations, typical of many butterflies, that take place near the ground, within the FBL (although we note some butterflies may employ both within- and above-FBL migratory flight; Stefanescu et al. 2007, 2013).

(c) Movements not employing powered flight, i.e. those undertaken with the aid of silk draglines or relying on minute body size alone (Reynolds et al. 2014).

(d) Very short-range migrations and the 'ranging' movements (Dingle 2014) associated with dispersal.

Much of the material presented here derives from observations made with meteorological and special-purpose entomological radars (see Chaps. 9 and 12). As radar observation of insect migration has recently been the subject of both a review (Chapman et al. 2011a) and a monograph (Drake and Reynolds 2012), we will focus on the more recent literature.

2   The Composition and Incidence of Insects in the Air

An important early contribution to what is now termed aeroecology was made in England shortly after World War II by C.G. Johnson and L.R. Taylor who sampled airborne insects quantitatively and at a series of heights with traps attached to towers and tethered balloons (Johnson 1969). Distributions of small day-flying migrants, particularly aphids, airborne in convective conditions were found to display a linear inverse relation of log density with log height, with the slope of the line varying with the degree of atmospheric stability; a discontinuity in the profile near the surface was probably associated with the insects' FBL (Johnson 1969; Taylor 1974). The gradient of (log) density against (log) height also varied with type of insect: a steep regression line (i.e. one nearly parallel with the height axis) was typical of very small day-flying insects like thrips (Order Thysanoptera) or aphids that are easily carried up by convection. Insects with somewhat stronger flight, e.g. Coccinellid beetles, showed gradients with larger coefficients because a greater proportion of fliers remained at low altitudes. Integration of the density profile over height allowed the estimation of the total number of the particular species in the air (per unit area of the surface). The form of the profile is determined by the processes of insects continually taking off, being circulated around in the atmosphere by convective up- and downdrafts, and eventually settling out again. This work has recently been re-visited (Wainwright et al. 2017); small insect flight in updrafts was found to be unenergetic—individuals rise but at a slower speed than the surrounding air, and their ascent is slowed further if updraft strength increases.

Typical distributions of insect masses estimated by aerial trapping and by X-band entomological radar are illustrated in Fig. 7.3, for a site in southern England at night. The aerial netting samples show a peak in the size distribution at 0.5–1 mg, and in Britain these would comprise mainly aphids, small flies (Diptera), and parasitoid wasps (Hymenoptera particularly Ichneumonoidea and Chalcidoidea) (Chapman et al. 2004). It can be seen (Fig. 7.3) that number concentrations of insects >10 mg are at least two orders of magnitude down from those in the 1-mg range, and for insects of 100 mg or larger, aerial densities may be lower by five orders of magnitude. There is little comparable information from other regions of the world, but our expectation is that the general pattern would be similar.

It should be noted that the current X-band entomological radars (Drake and Reynolds 2012; and see Chap. 9) can detect individual insects weighing ~2 mg at their shortest range (~150 m), but in practice they are not efficient at registering targets of this size (mainly due to numerous small targets interfering with each other), and it is only for insects of ~10 mg and heavier that the radar and netting densities approximately coincide (Fig. 7.3). Furthermore, only individuals above

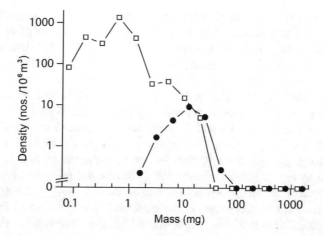

Fig. 7.3 Average densities (numbers per 10^6 m³) for a series of insect mass categories estimated with an X-band vertical-looking entomological radar (circles) and from aerial netting (squares). Aerial catches were taken at ~200 m a.g.l. on several July (i.e. mid-summer) nights at Cardington, Bedfordshire, southern England, and the radar data are for a night in July at a nearby site (Harpenden, Hertfordshire). Only targets detected by the radar at its lowest observing height, which was close to the height at which the net was operated, are included (From Drake and Reynolds 2012)

about 15 mg are detectable over the whole sensed volume of the (vertical-looking) radar, up to ~1 km a.g.l. The vertical-looking X-band systems (VLRs) have, therefore, been used primarily for studies of relatively *large* insects (e.g. larger moths, grasshoppers, and locusts). For smaller species, a shorter radar wavelength is required (Drake and Reynolds 2012); meteorological research 'cloud radars', for example, during cloudless periods in the warm season, will be detecting almost solely insect targets (Luke et al. 2008; Wood et al. 2009; Chandra et al. 2010; Wainwright et al. 2017; and see Fig. 7.1). In the Wood et al. (2009) study, there was good agreement between the target size range detected by a 35 GHz cloud radar and the most commonly occurring insect masses, as identified from aerial netting.

3   Insect Migration-in-Progress 1: Broad-Scale Phenomena

Because aeroecology is the study of organisms while they are in the atmosphere, it is concerned primarily with the *process of migration* rather than its outcomes or its broader function as part of the organisms' life histories and strategies. This section considers those aspects of 'migration-in-progress', such as synchronized emigration, altitudinal layering and orientation patterns, that are manifested over large areas—perhaps several hundred kilometres across. More local and transient phenomena are discussed in section 4.

The numbers of a particular species found high in the air are, of course, primarily dependent on the numbers of that species on the ground in potential source areas and the proportion of individuals whose migration syndrome has prepared them for take-off (Chap. 5 in Johnson 1969). Environmental factors such as previous weather in the source areas may be significant, particularly in small fast-breeding insects like aphids. The daytime emigration curves of *Aphis fabae*, for example, are determined by the rate of moulting into the winged adults, the length of the teneral period from moulting to flight, and the environmental factors inhibiting flight; together these processes result in a temporary decrease in flight-ready individuals during the middle of the day, and the daily take-off pattern is typically bimodal (Johnson 1969, p. 279). More generally, when the migrant is flight mature, various thresholds for flight (particularly light intensity, temperature, and perhaps wind speed) have to be exceeded (Johnson 1969). While the *intensity* of migration is influenced by weather factors, especially temperature, the *timing* of migratory flight is often determined by diel cycles in illumination levels (Lewis and Taylor 1964; Johnson 1969). This is particularly evident in dusk and dawn flights, which change seasonally with the timing of sunset/sunrise or can change from day to day with respect to cloud cover (or even a solar eclipse) (Chap. 10 in Drake and Reynolds 2012).

Although there are exceptions, radar observations reveal little insect activity when surface temperatures are below ~10 °C (Chaps. 10 and 15 in Drake and Reynolds 2012), so this apparently represents a temperature threshold for high-flying insect migrants of many taxa. In northern temperate areas, like the UK, it would seem that temperatures are nearly always suboptimal as the (log) number of insects recorded by entomological radar increases linearly with surface temperature and shows no sign of tailing off (Smith et al. 2000; Chapman et al. 2002a). Where temperatures *are* favourable for take-off and ascent, the diel pattern of insect migratory flight activity revealed by studies in various parts of the world is highly consistent (Reynolds et al. 2008; Chapman et al. 2011a; Drake and Reynolds 2012). A typical sequence, with three fairly distinct emigration periods, viz.: dusk, dawn, and daytime, is shown in Fig. 7.4.

3.1   Dusk Emigration

A mass emigration peak at dusk is one of the most consistently observed temporal features in insect migration studies, being found in virtually all investigations that had the capability to detect it (Chap. 10 in Drake and Reynolds 2012). The dusk emigration often consists of two parts: first, a peak composed of mainly small, crepuscular insects that take off while it is still quite light and secondly a take-off of larger insects, usually a little later when it is becoming dark. Many of the small insects fly for relatively short periods of an hour or so (Riley et al. 1987; Chapman et al. 2004), while the larger species may be deferring their flight to avoid attack by daytime predators. On the displays of X-band scanning entomological radars, the early peak appears as a diffuse echo (of the type produced by multiple small targets within each radar pulse volume) while the later-occurring larger insects produce discrete echoes that can be followed individually over several scans

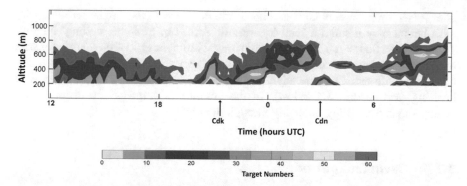

Fig. 7.4 Time/height plot of insect numbers recorded by a vertical-looking entomological radar, in 5-min samples taken every 15 min, from midday on 16 June 2000 to 10:00 h the next day. The color-scale bar refers to the number of resolvable insect targets detected in each of the 15 vertical sampling intervals, during each 5-min sampling period (No data was available for heights below ~180 m above the radar). The figure shows some typical features in the vertical profile, including: the extended daytime flight period ending about 18:00 h, the activity peak at dusk, a post-dusk elevated insect layer (here forming in the evening after about 22:15 h and persisting until dawn), and the activity peak at dawn. Additionally, in this case, a narrow elevated layer formed after dawn at an altitude of about 400 m and this merged into the daytime flight activity which started about 07:30 h. Daytime convection generally disrupts dawn layers but on this occasion layering continued (and, in fact, intensified) at the higher altitude of 600 m, until about 10:00 h. Cdk = civil dusk (21:21 h) and Cdn = civil dawn (02:59 h)

(i.e. for ~10–20 s). After ascent to altitudes of a few hundred metres, migrant insects (both small and large) may maintain flight for several hours or sometimes for the whole night and may become concentrated into narrow altitudinal layers (Reynolds et al. 2005; Chap. 10 in Drake and Reynolds 2012; Rennie 2014; and see below).

Often the two phases of dusk emigration overlap, but on the occasion shown in Fig. 7.4 they appear more distinct—the peak of mainly small insect emigration taking place before civil twilight, while the ascent of larger insects occurred after it. The different faunal composition of these phases is reflected in radar-derived mass estimates: the insects recorded between 20.30 and 22.05 h had a peak in the 5–20-mg category, while those in the layer and recorded between 22.15 h and midnight exhibited a rather broad mass distribution with a median of ~150 mg. There was an invasion of noctuid moths (including *Autographa gamma*) on southerly winds on this night (Chapman et al. 2008b), and these migrants will have contributed to the layer which persisted until dawn. Generally in temperate areas, many of the large nocturnally flying insects will be noctuid moths, but grasshoppers and locusts form an important component of the night-time aerial fauna in the semi-arid subtropics (Chap. 13 in Drake and Reynolds 2012; Drake and Wang 2013).

The development of the vertical profile of aerial density has been quite well studied for larger insects (see Chap. 10 in Drake and Reynolds 2012). Initially, large numbers of targets appear at the lowest altitudes observable with the radar. A little later, when more insects have ascended, the decrease in the insect aerial density with height may assume a semi-logarithmic form (up to some 'ceiling' height). Then, after an hour or so, numbers at the lowest altitudes (100–200 m) usually start to thin out as take-off ceases (and perhaps some very short-duration migrants land), and the profile assumes a relatively stable state which often shows evidence of layering (see section 3.5 below). In the absence of atmospheric disturbances, changes then occur only gradually through the night.

3.2   Dawn Emigration

There is generally a discrete take-off of crepuscular species during the dawn twilight period. This flight tends to be short-lived, although occasionally it continues for some time (Riley et al. 1991, 1994; Irwin and Thresh 1988; Reynolds et al. 2008; Rennie 2014) and leads to the formation of layer concentrations that may persist for extended periods in the stable early-morning atmosphere. Dawn layers sometimes ascend as the morning progresses, and if they last long enough they will usually be disrupted by the growth of daytime convection; their constituent insects then merge with the insect activity that has built up later in the morning in association with the convection. Occasionally, one or more insect layers associated with stable zones in the atmosphere above the convective boundary layer are still present at midday (e.g. Campistron 1975; Lothon et al. 2002).

3.3   Daytime Flight Activity

In fair weather, day-flying migrants take off from mid-morning onward as surface heating promotes boundary layer convection; these insects generally descend as convection dies down before nightfall. Insects flying in the convective ABL will tend to be circulated around in the atmosphere by up- and down-drafts, perhaps taking ~1–2 h to complete an individual flight (Johnson 1969). Continuously operating vertical-pointing meteorological or entomological radars are particularly suited to recording the development of daytime insect activity in the convective boundary layer (e.g. Contreras and Frasier 2008; Luke et al. 2008; Wood et al. 2009; Drake and Reynolds 2012; Bell et al. 2013). Detailed observations of insect targets within fair-weather convection cells will be discussed further below.

Occasionally, in northern temperate regions, small numbers of day-flying species such as aphids remain airborne into the dusk period or later in the night (Irwin and Thresh 1988; Chapman et al. 2004). This phenomenon may be much more common in warmer regions of the world (e.g. high densities of aphids were caught at altitude after midnight in northeast India; Reynolds et al. 1999). The mass invasion of aphids into southern Finland evidently required night flights across the Gulf of Finland/Baltic Sea (Nieminen et al. 2000; Leskinen et al. 2011), but here the over-water trajectories may have forced the aphids to continue in flight, and in any case during the midsummer migration periods there would be no astronomical darkness at these latitudes.

3.4   The 'Transmigration' Phase

Insects that enter the atmosphere above their FBL may not stay there very long—a huge variety of insects, particularly small ones, seem to ascend for relatively short periods at dusk, and sometimes at dawn (Riley et al. 1987; Drake and Reynolds 2012), leading to movements of a few kilometres or perhaps a few tens of kilometres. On the other hand, some highly migratory species, after the initial ascent phase, may maintain flight at heights well above their FBL for many hours before the final descent and alighting. This main period of horizontal translocation or *transmigration* is characterized by continuous steady wing-beating and, at least in the stable nocturnal ABL, rather little change in orientation direction or flight altitude. Displacement in near-geostrophic winds means that long distances can be covered in a night's flight (e.g. up to ~400 km for the Australian plague locust, *Chortoicetes terminifera*) (Drake et al. 2001; Drake and Reynolds 2012; Drake and Wang 2013). Other well-documented examples of this type of transmigration include solitarious desert locusts, *Schistocerca gregaria* (Pedgley 1981), migratory grasshoppers (Chap. 13 in Drake and Reynolds 2012), some moths particularly those from the family Noctuidae (Chapman et al. 2010, 2012; Drake and Reynolds 2012) but also including the tiny diamond-back moth, *Plutella xylostella* (Chapman et al. 2002b), and certain rice planthoppers (Riley et al. 1991; Otuka et al. 2005, 2010).

Other long transmigrations occur when typical day-fliers (e.g. host-alternating aphids) sometimes continue flight after dark (Irwin and Thresh 1988; Reynolds et al. 1999, 2006). If migrants find themselves over the sea at the time of day (or night) when they would normally land, flights may be further extended and result in amazing traverses (e.g. migrations of the wandering glider dragonfly, *Pantala flavescens*, from India across the Indian Ocean via the Maldives, Chagos, and Seychelles to east Africa; Anderson 2009).

3.5   Insect Layering

The relatively stable insect density profile that typically becomes established an hour or so after the dusk emigration is representative of the 'transmigration' phase of migratory activity. Sometimes densities may be relatively constant up to a flight 'ceiling', but often the migrants concentrate around particular altitudes forming one or more layers of very broad horizontal extent (Drake 2014; Rennie 2014; Chap. 10 in Drake and Reynolds 2012). The insect layers so formed may become highly pronounced with densities one or two orders of magnitude greater than those just above or below the layer. The layering is likely to be a reflection of the stratified character of the nocturnal ABL, with migrants reacting to physical features of their atmospheric environment.

Many layers form at maxima in the temperature profile, particularly near the top of the nocturnal surface inversion (Drake and Farrow 1988; Wood et al. 2010; Chaps. 10 and 13 in Drake and Reynolds 2012; Rennie 2014; Boulanger et al. 2017), suggesting a response of the migrants to air temperature. In warm conditions where temperatures are less limiting, insects may fly far above the top of the surface inversion, in the residual layer. Elevated layers may form here associated with higher-altitude isothermal zones or inversions (e.g. a capping inversion left from the daytime atmospheric boundary layer) (Reynolds et al. 2005). In other cases, a layer may form at a flight ceiling, or there may be a gradual increase in density up to a distinct upper boundary at the ceiling. An example from China involved the brown planthopper, *Nilaparvata lugens*, and here the well-defined ceiling corresponded to a known temperature threshold for sustained flight in this species (Riley et al. 1991).

In other situations, the observed insect layers do not correspond to any temperature feature or known threshold, but they are associated with a local maximum in wind speed (see Chap. 10 in Drake and Reynolds 2012; Qi et al. 2014). This raises the question of what wind-related cues are being used by migrants in the selection of height of flight, and this is discussed further in the section on 'Orientation behaviour'. Migration at the top of the nocturnal surface inversion would frequently locate the migrants at a boundary-layer wind speed maximum (i.e. in the low-level jet), so it may be difficult to separate the effects of temperature and wind-related cues in this case.

Insect layers have also been observed to form in regions of wind shear at the interface between a lower and an upper airflow. This may occur following the passage of a 'gravity current' (see Fig. 7.5) if insects are not retained within the head of the current but are swept back over its upper surface, forming a layer that may extend for many tens of kilometres (Schaefer 1976). In this case, the insects in the layer may be merely avoiding colder air in the density current itself, but there may also be responses to turbulence or Kelvin–Helmholtz waves at the current's upper surface. In some circumstances, the insects seemed to be concentrated at an altitude where wind shear was greatest (Schaefer 1976; Hobbs and Wolf 1989), but in other situations layers were located in regions of reduced shear, e.g. migrants flying preferentially near the centreline of a nocturnal low-level jet (Wolf et al. 1986). Complex behavioural responses by the insects to the shear-zone turbulence, which may contribute to the formation of these layers (Reynolds et al. 2009), are of relevance to possible cues used by migrants to sense wind velocity at altitude and thus take-up wind-related orientation directions (see below); radar observations indicate that well-developed layering and the common orientation of migrants often occur together (e.g. Hobbs and Wolf 1989; Aralimarad et al. 2011).

Insect layers may also occur during the day. Apart from layering in the stable early-morning atmosphere (mentioned above in connection with dawn take-off), layers of stable air may occur later in the day above the ABL (Campistron 1975; Chap. 10 in Drake and Reynolds 2012); for example, in warm layered airstreams advected above a cooler surface flow (Browning et al. 2011). It seems that day-flying insects (particularly small species that depend on convective upcurrents to assist in their ascent) will be confined by *any* neutral or stable layers aloft and accumulate near the base of these features leading to highly stratified density profiles (Isard et al. 1990; Lothon et al. 2002).

There are occasional radar reports of large insects in layers at particularly high altitudes (~3 km a.g.l.) during daytime (Drake and Farrow 1985; Leskinen et al. 2012). These observations were made by experienced observers (so the echoes concerned were not likely to have been confused with those from birds), and they are of interest because flight was occurring at very low temperatures (~5 °C), and because the insects were probably noctuid moths and these might be expected to land at daybreak. We must conclude that these large moths [the case described by Leskinen et al. (2012) involved in mass migration of underwings *Catocala* spp.] generate enough internal heat by wing-beating that they are able to continue flying in low ambient temperatures.

Lastly, in conditions of high directional wind shear, it is conceivable that layers could arise from differing take-off numbers in different source regions, or from differing flight durations in the two airflows, one of which is closer to a favoured displacement direction (Chapman et al. 2010)—though only if there is little ascent or descent of migrants from one airflow to the other.

3.6   The High-Altitude Orientation Behaviour of Migrants

It might be thought that high-flying nocturnal insect migrants would be swept along by strong high-altitude winds, orientating at random in what amounts to a 'sensory vacuum', but from the earliest days of radar entomology it has been clear that large insects, at least, often show a high degree of common alignment (e.g. Schaefer 1976). A recent advance has been the revelation of the degree of sophistication of some orientation strategies (see review by Reynolds et al. 2016) and accumulating evidence that these would significantly enhance migration success, i.e. the observed orientations are *ecologically adaptive* (Chapman et al. 2012). The best documented case is that of *A. gamma* (silver Y moth) which shows the following elements: (i) migrations are only initiated (or sustained for any length of time) on nights when the winds at altitude will carry the moths approximately in the seasonally favourable direction; (ii) there is a seasonal reversal of the preferred direction between spring and autumn (but the nature of the internal compass sense is not known); (iii) flight is often at the altitude of the fastest winds; and (iv) headings are adopted that partially counteract crosswind drift from the preferred migration direction (Chapman et al. 2008a, b, 2010, 2013, 2015b).

Some smaller migrants use a simpler strategy—i.e. they adopt a more-or-less downwind heading thus adding their airspeed to the wind speed and maximizing the displacement distance in a given time. Current evidence indicates that high-altitude nocturnal orientations are maintained by some intrinsic property of the wind, rather than (or as well as) an optomotor-type visual reaction to relative ground movement (Reynolds et al. 2016). In fact, the night-flying migrants seem to be using turbulent velocity and acceleration cues to determine the mean wind direction, and this suggestion is supported by a consistent tendency for headings to be offset on average by ~20° to the right of the downwind direction—the bias expected (in the

Northern Hemisphere) from theory (Reynolds et al. 2010, 2016; Aralimarad et al. 2011). The day-flying windborne migrant fauna has been rather neglected in radar entomological studies, but recent analyses (Hu et al. 2016a) found that adaptive tailwind selectivity and wind-related orientations were extremely common in medium-sized (~10–70 mg) and large (70–500 mg) insects migrating during the daytime, and in the period around sunset. Observations of insect layers with powerful Doppler meteorological radars show that echo patterns due to common orientation extend over very large areas (e.g. Lang et al. 2004; Rennie 2014), and this is further evidence that behavioural mechanisms are dependent on broad-scale, rather than localized, cues.

Lastly, we note that in very small migrant insects, headings at altitude may be random (Riley et al. 1991; Qi et al. 2014), or at least unrelated to facilitating movement in any ecologically adaptive migration direction. This is not surprising as, due to their low airspeeds, these insects will have very little influence on their migration direction when flying above their FBL. In the case of aphid clonal migrants, such large numbers of offspring are produced that a strategy of flight in winds from all directions is quite sustainable.

3.7   Aerial Densities and Migration Rates of Insects Aloft

Aerial densities and other quantitative measures of intensity of insect flight activity may provide information useful in many fields of basic and applied ecology. In the context of aeroecology, they may be of particular interest in connection with the feeding resources of aerial insectivores (McCracken et al. 2008; Kelly et al. 2013; Krauel et al. 2015). To mention one example, large numbers of Brazilian free-tailed bats (*Tadarida brasiliensis*) are known to forage at high altitude over Texas and perform a valuable ecological service for farmers by preying on migrant pest moths such as *Helicoverpa zea* (Cleveland et al. 2006; McCracken et al. 2012). Calculations of insect migration intensity from radar-derived data can be used to assess the ecological significance of a migration (Hu et al. 2016b) and to determine the food resources available to the bats (see p. 230 of Drake and Reynolds 2012).

Density values recorded with radar during peak emigration periods, when insect flight activity is likely to be most intense, are shown in Table 7.1; the values span almost six orders of magnitude (1×10^{-7}–1×10^{-1} m^{-3}). As might be expected, species in which outbreaks occur (locusts, and some grasshoppers and moths) exhibit high aerial densities on evenings immediately after a population has developed into its adult (i.e. flight-capable) form (e.g. Riley et al. 1983; Rose et al. 1985). Area densities of largish insects are, however, dwarfed by those of small species such as the rice brown planthopper (Table 7.1). To put these values in context, aphids and other small insects near the ground, and swarms of the desert locust, may sometimes attain densities of ~10 m^{-3} (Johnson 1969), though typical values would be much lower (Table 11.1 in Drake and Reynolds 2012).

One measure of the outcome of a migration is the increase in population in the destination region. Radar observations provide a proxy measure for this in the form

Table 7.1 Examples of volume and area densities of insects flying at altitude, recorded in radar studies mainly during the dusk emigration peak (data extracted from Table 10.2 in Drake and Reynolds 2012)

Taxon	Volume density (insects per m³)	Area density[a] (insects per hectare)
Grasshoppers, e.g. *Oedaleus senegalensis*, *Aiolopus simulatrix*, and Australian plague locusts (*Chortoicetes terminifera*)	2×10^{-7}–5×10^{-3}	~5–6000
Medium/large moths (~100–250 mg), e.g. *Spodoptera exempta*, *S. frugiperda*, *Helicoverpa zea*	2×10^{-7}–6×10^{-3}	~10–25,000[b]
Small moths (8–40 mg), e.g. *Cnaphalocrocis medinalis*, *Choristoneura fumiferana*	5×10^{-5}–3×10^{-3}	Up to 15,000
Small insects: rice planthoppers, e.g. *Nilaparvata lugens*	9×10^{-4}–8×10^{-2}	Up to 120,000
Small insects including aphids, e.g. *Rhopalosiphum maidis*	1×10^{-3}–1×10^{-2}	–

[a]Area density is estimated by summing volume densities, weighted by their associated height intervals for all (or most) heights of flight. Because it is independent of the height distribution of the migrants, it is a more representative measure of migration intensity at a particular location and time than is volume density

[b]This high area density was estimated from the nightly emergence of African armyworm moths at an outbreak site, but as virtually all moths emigrate in this situation, the values give some indication of the densities during the dusk emigration

Note: The volume and area density measurements for a species were not necessarily measured on the same occasion

of the 'passage rate' or the 'total overflight' (over a line drawn perpendicular to the direction of motion; see Chap. 9). Once again, published values (see Table 10.3 in Drake and Reynolds 2012) are likely to be associated with significant migration events and thus tend towards the maximal rather than the typical. Migrations of large insects like common noctuid moths or night-flying grasshoppers/locusts have recorded total overflights ranging from a few tens to a few thousands per meter per night, while for small insects like the brown planthopper overflights as high as 1.2 million per meter in just 2.5 h have been estimated. The overflight quantity can be accumulated over extended periods where appropriate, e.g. for the duration of fledging/emergence for a particular subpopulation (Chen et al. 1989), or throughout the migration period for the seasonal redistribution of a species between different climate zones (Chapman et al. 2012; Hu et al. 2016b).

3.8   Insect Movement in Relation to Large-Scale Weather Systems

The relation of long-range insect migrants to synoptic airflows is discussed by Johnson (1969), Pedgley (1982), Drake and Farrow (1988), Johnson (1995) and Pedgley et al. (1995). In middle latitudes, weather conditions and wind directions are predominantly determined by a succession of large-scale pressure systems (depressions and their associated fronts, and anticyclones) which move generally eastwards, each one taking a few days to pass over any particular locality (see Chap. 2). Depressions are associated with cloud, rain, and strong winds while conditions during the passage of an anticyclone are characteristically settled, with usually clear skies. Winds circulate around both types of system (in opposite directions), so that a particular locality experiences changing wind directions as the system passes over. Johnson (1969, pp. 477–481) presents some idealized trajectories that take account of the ambient wind, the movement of a depression, and the orientation and air speed of an insect. In practice, of course, trajectories are likely to be disturbed by weather associated with the depression, particularly if/when the insect encounters cold air or falling rain.

Winds from lower latitudes will generally be warm, and these winds provide transport for insect migrants to move polewards in spring, when cooler regions become seasonally favourable for breeding; numerous examples are mentioned in the references cited in the previous paragraph. In north-west Europe, some mass arrivals are from the east rather than the south, from sources in western Russia or even further afield, on warm south-east winds blowing around large anticyclones centred over north-western Russia (Mikkola 1986; Pedgley et al. 1995). Selective movement on warmer winds could arise without a specific adaptation, if the threshold temperature for flight is only exceeded when these are blowing. However, as temperatures fall in autumn, many migrant species need to return to warmer winter-rainfall regions, and this movement will often require a cued response, with flight in warmer conditions suppressed as it would now take the migrants in exactly the wrong direction (McNeil 1987; Chapman et al. 2012).

Radar observations in southeastern Australia identified warm northerly and northwesterly winds ahead of a cold front as a vehicle for major southward (i.e. poleward) movements of a variety of moth species in spring (Drake et al. 1981; Drake and Farrow 1985). In contrast, a northward movement of locusts that persisted for several days in early summer occurred in southerly winds on the western side of an almost stationary anticyclone (Drake and Farrow 1983). In the southern USA, southward movements of moths in autumn occurred on northerly airflows immediately following the passage of a cold front (Pair et al. 1987; Beerwinkle et al. 1994; Krauel et al. 2015). Similar examples have been observed in East Asia (Ming et al. 1993; Riley et al. 1995; Feng et al. 2009). In other cases, the suitability of winds for equator-ward 'return' migrations cannot be assessed from simple meteorological factors such as temperature, and migrants seemingly use an internal compass sense to assess whether the direction of windborne displacement is close enough to an innate seasonally preferred direction (e.g. in A. gamma in northern Europe).

In the tropics, climate and weather are dominated by the seasonal movements of the Intertropical Convergence Zone (ITCZ) and its associated monsoonal winds, which bring alternating wet and dry seasons (see Chap. 2). In the savanna and steppe regions in West Africa, for example, several species of grasshoppers, heteropteran bugs, moths, and flies are able to take advantage of the constancy of the seasonal reversal in the monsoon/trade wind system to move between complementary breeding (or breeding and diapausing) areas (Pedgley et al. 1995). As the ITCZ moves northwards, rain falls at progressively higher latitudes, and windborne migrants move into areas where vast seasonal resources are becoming available (e.g. new growth of semi-arid vegetation or newly flowing waterways). Correspondingly in autumn, equatorward migrations predominate on the re-established trade winds in the wake of the retreating ITCZ. Some populations may move far to the south where the rains are late enough to allow breeding, while others move less far and persist in dry-season refuges. We note that the putative world's longest insect migration—that of the above-mentioned *P. flavescens* dragonflies from India to Africa across the western Indian Ocean—takes place on north-easterly winds at altitudes above 1000 m behind the southward-moving ITCZ (Anderson 2009).

Another remarkable migration is the movement of pest planthoppers and moths from central and southern China to Korea and Japan (e.g. Otuka et al. 2005; Tojo et al. 2013), which occurs in June–July every year on low-level jets associated with the quasi-stationary *Bai-u* front (East Asian subtropical rain-band)—the boundary separating the hot humid tropical air mass to the south from the cooler polar air mass to the north (see Chap. 2 on the transport of biota on Asian monsoonal winds). Because of the economic implications of these pest invasions, the planthopper movements, particularly, have been intensively modelled using numerical weather prediction programmes combined with submodels to take account of atmospheric dispersion and insect flight parameters (e.g. Otuka et al. 2005). In North America, migrations of moths of the fall armyworm *Spodoptera frugiperda*, a serious crop pest, have been the subject of a comparably comprehensive computer modelling effort (Westbrook et al. 2016).

The subtropics are the locations of extensive deserts, where insect breeding is limited by rainfall. The classic long-distance desert migrant is the desert locust, *S. gregaria*, whose downwind movements are able to span the Sahara, allowing breeding on monsoon rains associated with the ITCZ in summer, and on cyclonic rains associated with midlatitude westerly airflows in the Mediterranean in winter (Pedgley 1981). It has recently been shown that migrations of the painted lady butterfly *Vanessa cardui* extend from northern Europe to both the Maghreb region north of the Sahara and the Sahel zone to its south and therefore involve a desert crossing (Stefanescu et al. 2016; Talavera and Vila 2017).

4   Insect Migration-in-Progress 2: Localized Aggregations of Insects Aloft

Aggregations of insects in the air can arise from a variety of causes (Chap. 11 in Drake and Reynolds 2012). In this section, we will be mainly concerned with the effects on flying insects of small-scale wind systems, or of precipitation, but first three other causes of localized concentrations seen on radar displays will be mentioned.

If there is non-uniformity in the spatial distribution of ground populations around the radar site, this may be reflected in patchiness of echoes from emigrating insects displayed on the radar PPI screen. These patches can be plumes of insects streaming away from what is virtually a point source (a single tree, perhaps) or patches of dimensions of several hundred meters such as crop fields (Chap. 11 in Drake and Reynolds 2012) up to extensive areas (thousands of square kilometres) observable with operational weather radars (Westbrook 2008; Rennie 2012; Westbrook et al. 2014; Boulanger et al. 2017; Westbrook and Eyster 2017). Secondly, migrating insects may show an active behavioural response to a *topographical feature* (hill, coastline, river, etc.). For example, Russell and Wilson (1996, 2001) found radar evidence of a tendency in insects aloft to resist being carried over a coastline by (light) off-shore winds, and this resulted in a temporary accumulation of insects along the coast. Thirdly, there may be behavioural responses to other individuals, of the same or a different species. The classic example here is the locust swarm, where cohesion is maintained through mutual interactions between the gregarious individuals comprising the swarm (pp. 257–258 in Drake and Reynolds 2012). Other examples of 'swarms' observed on radar include: honeybee *drone congregation areas* (pp. 352–354 in Drake and Reynolds 2012) and the nuptial and oviposition swarms of mayflies (Ephemeroptera) above rivers and lakes (pp. 350–352 in Drake and Reynolds 2012). In the latter case, the enormous swarms of mayflies on the Upper Mississippi valley regularly produce spectacular displays on the WSR-88D radars of the US National Weather Service (http://www.crh.noaa.gov/arx/?n=mayflygeneral). It should be noted that these phenomena involve appetitive flights rather than migration. An example of another taxon, namely bats, causing an effect on concentrations of large migrating insects (e.g. grasshoppers and moths) is manifested by distinctive echo-free lacunae that persist for a few tens of seconds on PPI displays of entomological radars (pp. 258–260 in Drake and Reynolds 2012). These 'blank patches' are the result of the insects making evasive flight manoeuvres (power diving, stalling, turning away) in response to the ultrasonic emissions from an approaching hunting bat.

4.1 Concentration by Small-Scale Meteorological Disturbances

The relatively steady movement of migrant insects in broad-scale wind flows (described above) may be disrupted by disturbances on horizontal scales of a few tens of meters up to ~200 km (i.e. more or less the micro-α, meso-γ, and meso-β scales in standard meteorological parlance, see Orlanski 1975). These disturbances comprise *terrain-induced phenomena* (organized daytime convection, land/sea breezes, slope winds, lee waves, and vortices) and those (e.g. convective storms, squall lines) *induced by instabilities* in moving synoptic-scale systems. Most of these wind perturbations will have associated convergence zones within which airborne insects will tend to become concentrated. The insects are carried horizontally towards the zone of convergence by the low-altitude wind flow; they then resist (to a greater or lesser extent) being carried upward in the rising air at the convergence and thus tend to accumulate there, forming distinctive echo patterns on entomological and meteorological scanning radars. At a certain altitude, air temperatures will drop below the insects' flight threshold, and the migrants would be physiologically unable to fly; before this point is reached, however, it appears that migrants often respond *behaviourally*, by ceasing flight or actively flying downward (Achtemeier 1991, 1992; Geerts and Miao 2005; Browning et al. 2011). Some such mechanism has presumably evolved to prevent insects being carried involuntarily to great heights in the violent updrafts associated with thunderstorms (Browning et al. 2011), though this does occasionally happen—as evidenced by small insects forming the core of hailstones (Browning 1981).

In daytime fair weather, mesoscale shallow (1–2 km deep) convection may take the form of polygonal cells or parallel bands in the ABL, and these patterns are visible on radar due to insect echoes. The smallest-scale insect concentrations detected by radar include portions of the walls of individual convective cells (Schaefer 1976), and a series of studies with different types of radar confirm that the regions of high reflectivity are located in updrafts (Chandra et al. 2010, pp. 261–264 in Drake and Reynolds 2012); the emptier centres of the cells are broader regions of downdraft, which are relatively free of insects. Occasionally, large areas can be covered by patterns of these 5–10 km diameter 'Bénard-like' cells (Hardy and Ottersten 1969; Rennie 2012). Under certain conditions of surface-layer heat flux and wind shear, the cellular pattern changes to 'streets' of counter-rotating convective rolls aligned downwind (Weckwerth et al. 1999; Koch 2006). In fine weather, day-flying insects ranging from aphids to swarming locusts can take advantage of updrafts produced in unstable convective atmospheres in order to reduce the energy needed to remain aloft, so staying within a region of uplift in a convection cell is clearly adaptive. In fact, on days (or at times of day) when atmospheric convection is absent, insects do not show this 'cumuliform' vertical distribution; instead, aerial density profiles are stratified with aphids, for example, being restricted to within a few tens of metres of the surface (Johnson 1969), or confined by stable or neutral layers at higher altitude (Isard et al. 1990). However, as far as we know, even large insects like swarming desert locusts do not appear to actively seek out thermals (pp. 37–38 in Rainey 1989) in the way that soaring birds like vultures and storks do (see Chaps. 8 and 9).

Gravity Currents Another common insect-concentrating mechanism is the atmospheric *gravity current* (also known as a 'density current' or 'buoyancy current'), which occurs when a higher density (cooler) airflow intrudes into and displaces lower density (warmer) air. The leading edge of the advancing gravity current forms a distinct 'front' where it undercuts the ambient warmer air and both airstreams (mostly) ascend. The gravity current is often slightly deeper immediately behind the front—in other words, it forms a 'head' on the top surface of which there may be a zone of intense mixing, with breaking waves (Kelvin–Helmholtz waves) streaming back over the upper surface of the current (Simpson 1997; and see Fig. 7.5). The passage of the leading edge of a gravity current is typically marked by a gust line with a wind-shift, turbulence, a pressure jump and a temperature fall.

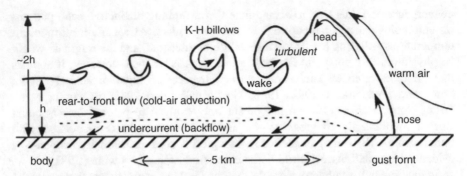

Fig. 7.5 Schematic view of a gravity current which is advancing towards the right (From Geerts et al. 2006. © American Meteorological Society. Used with permission)

Insects carried into the head from the rear tend to resist being taken up by the rising air and thus accumulate there. In addition, any insects that are carried upwards in the rising ambient air just ahead of the front may be entrained by the zone of strong turbulent mixing around the head and be incorporated into the concentration there.

Various manifestations of the atmospheric gravity current revealed by radar returns from insects include the following: storm outflows (Lothon et al. 2011), sea or lake breezes (Simpson 1994; Sauvageot and Despaux 1996; Russell and Wilson 1997; Isard et al. 2001), katabatic flows (Schaefer 1976), drylines (sharp boundaries separating moist and dry air masses) (Weiss et al. 2006), mesoscale cold fronts (Geerts et al. 2006), and prefrontal wind-shift lines (Bluestein et al. 2014); see Chap. 11 of Drake and Reynolds (2012) for a unified treatment of these phenomena. One striking example is shown in Fig. 7.6—a radially spreading storm outflow observed in Niger in the West African Sahel (Lothon et al. 2011).

The insect density in the concentrations may be 1–2 orders of magnitude greater than in surrounding areas, and volume densities may approach $10^{-1}\,\mathrm{m}^{-3}$ for small day-flying insects. These concentrations may form a valuable feeding resource for aerial insectivores, and, indeed, sea-breeze fronts were delineated on early air-traffic control radars by swifts feeding along the front (p.135 in Simpson 1994). The aerial concentration of pest insects, if these land while in a convergence or soon after encountering one, may lead to high ground densities and serious localized outbreaks. In the case of migratory grasshoppers such as *Oedaleus senegalensis* or *Aiolopus simulatrix* in the Sahel, the flight-capable (i.e. adult) stage can cause immediate damage to crops, and Joyce (1983) considered that *A. simulatrix* is a problem *only* when grasshoppers have been concentrated in convergence zones. In other cases, such as with the African armyworm moth (*Spodoptera exempta*), the damage is caused by caterpillars developing after mass oviposition by the initial migrants. There is a well-attested case in this insect for the initial low-density moth populations being brought together by wind convergence associated with rainstorms, so that the resulting caterpillars occur as very dense infestations of the 'gregarious' phase which can cause widespread and serious

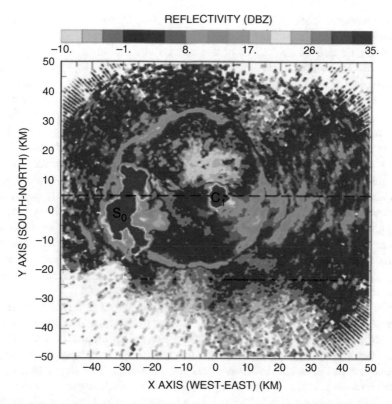

Fig. 7.6 Reflectivity pattern in a horizontal cross section at 600 m height, using data from a C-band Doppler radar at Niamey, Niger, in late afternoon (17.20 UTC) of 10 July 2006. A westward-moving storm system (denoted S_0) has produced the circular and radially expanding gust front made visible by insect scatterers. A second, smaller, convective cell (denoted C_1) has formed at the east side of the circular outflow apparently caused by the interaction of the outflow with a low-altitude wave train previously generated by S_0. Further observation indicated the circular gust front dissipated shortly after 2100 UTC, about 60 or 70 km away from the radar (From Lothon et al. 2011. © American Meteorological Society. Used with permission)

damage to crops (Rose et al. 2000; Chap. 12 in Drake and Reynolds 2012). In fact, satellite-derived information on rainstorm distribution is useful in predicting potential armyworm outbreak sites (Tucker 1997).

There can be little doubt that gravity currents perturb the flight trajectories of migrants flying at lower altitudes, and the flight orientations of insects trapped in the head of the current are certainly disrupted (Greenbank et al. 1980). In the case of storm outflows, modifications to migration flight paths would appear to be limited in duration and usually rather random in direction and timing, but gravity currents associated with large-scale topographic features, such as sea breezes and katabatic flows, may produce more systematic effects. For example, the regular penetration of sea breezes deep inland has significant effects on population distribution of the spruce budworm moth (*Choristoneura fumiferana*) in New Brunswick, Canada (Neumann and Mukammal 1981).

A dramatic detrimental effect of the coastal circulation on small insects was highlighted in a study made with a polarimetric K_a-band Doppler radar on the Atlantic coast of France (Sauvageot and Despaux 1996). On days when the land/sea breeze cell became well established, small insects were seen to be carried out to sea on the 'return flow' (between ~500 and 1000 m altitude) but were brought back towards land when they descended into the sea breeze layer (below 500 m). By the early evening, however, the diurnal veering of the low-altitude wind meant that the insects were no longer being returned by the onshore flow; instead they accumulated in a band located between a few kilometres and ~50 km out to sea. Eventually, the bands dissipated *in situ*, as the insects were apparently deposited in the ocean. This phenomenon has seemingly not been reported from elsewhere, so it is not clear how widespread its occurrence is; indeed, there is radar evidence from elsewhere indicating that insects sometimes resist being taken over coastlines (see above). Experiments with tethered individuals suggest that desert locusts, at least when flying at low altitudes, can detect the polarized reflections from large bodies of water and thus avoid moving over them (Shashar et al. 2005).

Lake breezes are atmospheric circulations similar to (but generally weaker than) sea breezes that form along the shorelines of large bodies of water such as the Great Lakes of North America. Isard et al. (2001) review information on a variety of insect taxa with particular reference to the mechanisms causing the movement between the Great Lakes' coastlines and the inland, and the deposition of high concentrations of (living) insects along the lake shore. They suggest that the insects are initially collected and lifted up at a lake breeze front and are transported in the counter-flow out over the lake. If they then descend to near the water surface, they will be carried in the surface flow back to the coastline, where retardation of the air flow by dunes and trees slows the wind and allows the insects to land (Isard et al. 2001). Clearly, other outcomes are possible: if the lake breeze is strong, insects at the coastline may be merely blown inland. Interpretation of insect movement over land–water boundaries may also be complicated by interactions between the wind flows of the lake/sea breeze circulation, and the insects' optomotor reactions to the water or land surface, which may lead to so-called 'rectification'— i.e. the boundary facilitates transport in one direction, but hinders it in another (for details, see pp. 134–136 in Johnson 1969).

Bores and Waves Another group of phenomena that may perturb the steady horizontal migration of insects comprises atmospheric *bores* (propagating *hydraulic jumps* indicated by an abrupt change in the height of an atmospheric layer) and *waves* (e.g. solitary waves, Kelvin–Helmholtz waves, lee waves) (Chap. 11 in Drake and Reynolds 2012). A bore may take the form of a smooth wave front followed by a train of waves (an *undular bore*). These atmospheric *internal bores* can be initiated by colliding gravity currents or by an advancing gravity current intruding into an existing stable layer. Once generated, a bore can propagate far ahead of the initiating gravity current, and its passage is typically attended by a wind shift and a small but sustained increase in surface pressure, but (unlike a gravity current) there is no rapid drop in surface temperature. The bore may then evolve further into a *solitary wave*—a stable 'wave of elevation' (occurring either individually or as an amplitude-ordered short wave train) which can propagate for long distances with little loss of form (Knupp 2006). Bore and wave disturbances generally seem to displace migrants upwards and downwards, rather than increase aerial densities, but concentration may occur in closed horizontal vortices which develop in some large amplitude waves. Insects need to remain entrained in the closed circulations for some time, however, before there will be any significant effects on the migrants' flight trajectories (Drake 1985). Bore-type disturbances can be quite common in certain locations /seasons, most famously as the 'Morning Glory' cloud formation over the Gulf of Carpentaria in northern Australia, but radar observations have recently documented examples in other parts of Australia (Rennie 2012) and in the USA (Knupp 2006). Figure 7.7 shows an example observed on a Doppler C-band weather radar in northern Victoria, Australia (Rennie 2012). Rennie commented that waves often preceded or sometimes

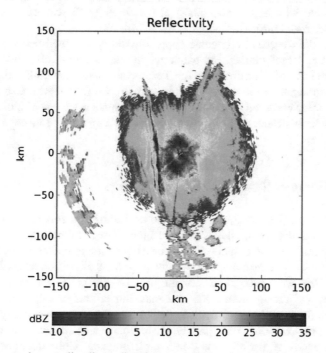

Fig. 7.7 Insect echo revealing linear disturbance, probably an atmospheric solitary wave train, shown in a reflectivity scan from a Doppler C-band weather radar at Yarrawonga, N.E. Victoria, Australia (00.12 h LST, 25 Nov 2010; elev. ~1°). The wave train—three lines are visible, the first being the most prominent—is approaching from the west, ahead of rain echo. In this example, the line echo was not associated with any substantial change in the insect velocities measured by the radar (not shown), indicating a wave disturbance rather than a gravity current (From Rennie 2012. © 2012 CSIRO and the Bureau of Meteorology)

followed precipitation, suggesting their origin in a storm outflow or gust front; insect migration often coincides with unsettled convective weather in this part of inland Australia, rendering these disturbances detectable with the weather radar.

Lee waves are stationary waves on the downwind side of a topographic barrier such as a mountain ridge, and if the waves are of large amplitude they can contain stationary closed horizontal vortices or 'rotors'. The effect of rotors and other topographic eddies on flying African armyworm moths has been documented in the Rift Valley of Kenya (Chap. 11 in Drake and Reynolds 2012). Rotor circulations may occur repeatedly at particular locations and so potentially have more significant effects than travelling waves on migrant insect populations transiting or developing in the vicinity.

Tornadic Thunderstorms and Hurricane Eyes 'Hook echoes' seen on Doppler weather radars are typical of supercell thunderstorms that give rise to tornadoes, and these echoes are likely to be regions of very strong inflow. The Differential Doppler velocity (DDV) field can exhibit very large values of both signs at the ends of 'hooks' and this, as well as other *polarimetric* radar variables, indicates these areas are filled with highly non-spherical (i.e. biological) scatterers—most likely trapped insects and birds (Melnikov et al. 2014a, b).

Insects and birds can also become trapped in the eyes of well-developed tropical cyclones, e.g. large numbers of black witch moths (*Ascalapha odorata*) were present in the eye of a hurricane on the Texas coast (Freeman 2003). Early reports of this phenomenon are reviewed by Van den Broeke (2013) who also documents recent examples observed by polarimetric radar where the areas of bioscatter were revealed by high differential reflectivity combined with a very low copolar correlation coefficient.

4.2 Effects of Rain on Migrating Insects

In many areas of the world, outside of the arid/semi-arid zones, one of the most tangible agents disrupting steady long-distance insect migration is rainfall. The extensive nature of some synoptic rain-bearing systems makes them difficult to avoid, and these systems may have significant general effects on migratory activity in insects. However, radar studies of the precise effects of rain on migration are hindered by the strong attenuation and masking effects of raindrops at X-band frequencies, thus making it difficult to detect insect targets in anything other than the lightest precipitation. Consequently, the effects of rain are less certain compared to the influences of various fine-weather phenomena. Complications also arise because there are several mechanisms associated with rainy weather, other than direct impaction by raindrops, that may cause migrants to descend or land. These include: descent due to ambient temperatures falling below the flight threshold; descent due to cessation of convection (which many day-flying migrants require to remain aloft); and insects being forced down by strong downdraughts associated with convective rainstorms.

A recent laboratory study has shown that a relatively small flying insect (an *Anopheles* mosquito) flying in heavy rain is likely to be struck by raindrops but that its small mass, and the fact that it can be displaced downwards (unlike an insect on a solid surface), allows it to survive these high-speed impacts (Dickerson et al. 2012). Nevertheless, the insect experiences enormous accelerative forces (100–300 g) and its trajectory is greatly disrupted, so the rain may be best avoided.

Spruce budworm moths (*C. fumiferana*) have been observed to continue flying in very light rain associated with a synoptic warm sector (Greenbank et al. 1980). In cases like this, the air temperature evidently remained above the flight threshold, and the probability of raindrop impact may be so low that migrants are unaware of the rain. In heavier rain, most radar evidence points to a termination of migratory flight (see Chap. 11 in Drake and Reynolds 2012). In the case of scattered convective rainfall cells, radar observations have shown that either the heavy rain itself or associated cold-air downdrafts locally interrupted migratory flight (Greenbank et al. 1980). The rain shafts falling from higher altitudes were vividly envisaged as puncturing a layer of migrating moths producing a high-density strip of deposited moths on the ground; however, flight continued outside the actual rain cells. In the case of synoptic systems (e.g. midlatitude warm fronts) that produce widespread moderate-intensity rainfall over an extended period, one can envisage that migration may well be terminated for the night over an extensive area.

In arid and semi-arid areas of the world, where rainfall is erratic or highly seasonal, deposition (whether active or passive) in the vicinity of rainstorms may be highly adaptive. Here migrants such as locusts and African armyworm moths (see above) need to arrive in locations where rain has recently fallen, as this will provide the flushes of new vegetation and other resources needed for survival and breeding (Pedgley 1981; Rose et al. 2000).

5   Ecosystem Consequences of Long-Range Mass Movements of Insects

While aeroecology is concerned primarily with the processes of animal movement within the atmosphere, much of the work described here has been inspired and driven by concerns about its consequences, i.e. with the outcomes of the migrations. Quantitative measures of intensity and directional transport of migrant insects may provide information useful in the fields of population ecology and ecological genetics, the ecology of vectored microbes, pest and disease management, and conservation biology (Chapman et al. 2015a; Bauer et al. 2017). Clearly, the immense seasonal shifts of insect biomass to and from higher latitudes in response to seasonal temperature changes (Hu et al. 2016b), or tracking erratic or seasonal rainfall in arid lands (Drake et al. 2001), will have huge ecological impact as a result of trophic and transport effects (see reviews by Holdo et al. 2011; Bauer and Hoye 2014; Landry and Parrott 2016). The effects of insect migration on ecosystem functioning include *direct* 'trophic' effects of a mass arrival of consumers, competitors, predators or prey—for example, competition with concomitant migrant or resident species for shared resources, or interactions mediated by shared predators (apparent competition). For example, an investigation of a dual-migrant (noctuid moth and bat) system in Texas found that spikes in local abundances of moths were associated with synoptic weather (e.g. passage of cold fronts), and this in turn influenced behaviour and resource acquisition of Brazilian free-tailed bats (Krauel et al. 2015) (Chap. 15). The *indirect* 'transport' effects include: the transport and recycling of nutrients and energy (via feeding, excretion, reproduction, and decomposition), the 'vectoring' of parasites, pathogenic agents and symbionts, and the spread of genetic material (by means of the migrants themselves or by the carriage of seeds and pollen).

Many effects of insect migrants will be benign, e.g. delivery of ecosystem services such as pollination (Ssymank et al. 2008), biological control of pests (Chapman et al. 2006; Jeffries et al. 2013), maintaining susceptibility to insecticides by diluting the pool of resistant genes (Brévault et al. 2008), and provision of food to higher trophic levels (e.g. the above-mentioned aerial insectivores, grizzly bears feeding on the army cutworm (*Euxoa auxiliaris*) (Bauer and Hoye 2014), or the entire suite of predators sustained by bogong moths (*Agrotis infusa*) (Green 2011; Warrant et al. 2016)). Other effects will be harmful, however, e.g. infesting crops with pests that cause major yield losses (Chen et al. 1989), spreading infectious plant and animal diseases (Reynolds et al. 2006; Otuka 2013), and the spreading of genotypes associated with resistance to insecticides (Scott et al. 2005) and overcoming insect-resistance in crop varieties (Tanaka and Matsumura 2000). Elucidating the diverse impacts of a mass arrival of insect migrants on ecosystem function remains an extremely challenging prospect, but the first stage is to accurately quantify it, both in terms of the numbers of insects involved and their biomass. For some classic invasive agricultural pests (locusts, aphids, armyworm moths), a major objective of determining the *population trajectory* of the population (or its various subpopulations) through space and time has been largely achieved [e.g. for swarming populations of the desert locust (*Schistocerca gregaria*) (Pedgley 1981) and Australian plague locust (Deveson and Walker 2005), or for aphid crop pests in the UK (Taylor 1985)], although there may be current concerns as to whether these trajectories will alter with climate change and whether current operational forecasting capabilities can be sustained into the future. Operational applications of radar in insect pest forecasting and management have already been implemented in a few instances (e.g. Leskinen et al. 2011; Drake and Wang 2013), and there is current interest in developing further systems, e.g. in China and the USA.

6   Contrasting the Airborne Migration Strategies in Large Insects and Vertebrates

Here, we draw attention to the opportunities provided by modern observing techniques for making comparisons of migration physiology, behaviour and ecology in distantly related taxa (Hein et al. 2012); migration in a fluid medium (air or water) where movement is affected, to a greater or lesser extent, by the flow of the medium itself appears of particular interest (Chapman et al. 2011b). Highlighting similarities and differences among diverse taxa in orientation and navigation mechanisms, the energetics of migratory flight, and mortality rates during migration (Dudley 2000; Dingle 2014; Chapman et al. 2012, 2015a, b; Reynolds et al. 2016) will surely lead to insights into the evolutionary drivers of animal migration strategies.

For example, a comparative radar study of passerine songbirds (Old World warblers, flycatchers and chats) and noctuid moths (*A. gamma*) migrating over north-western Europe showed that the moths, despite having much slower self-propelled airspeeds, achieved virtually the same ground speeds and track directions as the birds (Alerstam et al. 2011; Chapman et al. 2015b, 2016a). The primary reason for this surprising result was that the moths took greater advantage of wind assistance (but only flew when the wind direction was favourable), whereas the birds were able to migrate on winds from a variety of directions but consequently received less assistance. There were also interesting differences in the types of orientation behavior exhibited by the two groups of migrants (Chapman et al. 2015b, 2016a). Overall, the birds' strategies allowed more temporal and spatial control over their flight paths, and they were able to be more risk-averse than were the moths. The high fecundity of many migratory noctuids (Chapman et al. 2012) would, of course, enable them to bear large migration losses.

Quantification of mortality during migration is particularly problematic—there is very little empirical data on survival during *regular* seasonal migrations of insects, such as are becoming available for birds (Sillett and Holmes 2002; Menu et al. 2005; Klaassen et al. 2014) (but note Ward et al. 1998; Chapman et al. 2012), and it is difficult to assess the effect of enormous losses at sea which are occasionally reported (e.g. Farrow 1975; Sauvageot and Despaux 1996). The advent of continuous monitoring of insect movement over large areas by weather surveillance radar networks (see below) should help in this respect.

Incidentally, the recent comparison of migration distances among taxa (Hein et al. 2012) showed how truly spectacular insect migrations are, when scaled for body size; the number of body lengths covered by moths and dragonflies, for example, was roughly 25-times that travelled by geese.

7 Prospects and Future Challenges

There is no shortage of specific unresolved aeroecological questions that can be investigated by radar and complementary techniques (see Chaps. 9 and 12). For example, tracking radar could be used to identify individual high-altitude 'hawking' targets, i.e. types of bats or birds, while at the same time, information could be obtained on their insect prey by means of vertical-looking entomological radars (VLRs). In the case of bats such as the Brazilian free-tailed bat (*Tadarida brasiliensis*) that prey on quite large insects, the prey items (e.g. noctuid moths) could be identified with current X-band VLRs, but if the aerial predators consume small insects, a higher frequency VLR system would be needed, in order to reliably detect micro-insects (mass ~1 mg) over all their flight heights.

Future prospects for the remote sensing of airborne migrants over *extensive* spatial scales is entering an exciting new phase due to (a) the expansion of networks of specialized biological radars and (b) their integration with weather surveillance radars in order to monitor continent-wide movement of birds, bats, and insects (Dokter et al. 2011; Chilson et al. 2012a, b; Kelly et al. 2012; Shamoun-Baranes et al. 2014). In Europe, this initiative is being implemented under the ENRAM (European Network for the Radar surveillance of Animal Movement) research network (Shamoun-Baranes et al. 2014), and the entomological aspects of the program include pilot studies to create a platform for intensive cross-calibration measurements, involving the simultaneous deployment of multiple remote sensing devices (e.g. air traffic surveillance radar, dedicated ornithological and entomological radars, thermal imaging equipment, and opto-electronic systems) all within the sensed volume of one or more operational weather radars. The ultimate objective is to combine real-time insect altitude profiles into a multi-radar mosaic so as to produce visualizations of insect migration on a regional- or continental scale. Such radar-based monitoring networks will provide essential data on seasonal numbers, timings, and routes of insect mass migrations—there is, for example, a pressing need for more information on the overwintering distributions for a range of insects that invade northern temperate areas each summer. We note that case studies exist for quite a few long-range migratory insects, particularly pest species (Drake and Reynolds 2012), and the challenge is now to extend these to ecosystem-wide monitoring over regional scales (Kelly and Horton 2016). Wide-area monitoring networks will also help the operational detection (and perhaps 'nowcasting') of irregular incursions due to irruptive outbreaks of particular pest species (c.f. Nieminen et al. 2000; Leskinen et al. 2011); similarly, fully automated vertical-beam entomological radars located in a region prone to locust outbreaks have been envisaged as fulfilling a 'sentinel' role (Drake and Wang 2013).

In conclusion, we note that in over four decades of radar entomological research, much of the 'phenomenology' of high-altitude insect interactions with wind and weather are reasonably well understood. The outstanding uncertainties concern the mechanisms for detection of wind velocity at attitude, and the maintenance of seasonal orientation directions, in nocturnal insect migrants. There is also a need for more radar-derived information on the termination stages of migration: there have been very few observations of migrants in the process of descent and landing (Chap. 10 in Drake and Reynolds 2012). Progress is currently being made in quantifying the interaction of insects with other animals (particularly insectivores) high in the air, as indicated above. Finally, just as the first deployment of scanning radars for entomological purposes nearly 50 years ago revealed previously unsuspected phenomena, new perspectives on insect movement can be expected to emerge through the application of novel radar, optical, and data-logging technologies that are now becoming available (see Chap. 9).

Acknowledgements We acknowledge the support provided by COST—European Cooperation in Science and Technology through the Action ES1305 'European Network for the Radar Surveillance of Animal Movement' (ENRAM). JWC acknowledges support of Rothamsted Research, where he was employed during much of the drafting process; Rothamsted Research is a national institute of bioscience strategically funded by the UK Biotechnology and Biological Sciences Research Council (BBSRC).

References

Achtemeier GL (1991) The use of insects as tracers for 'clear-air' boundary-layer studies by Doppler radar. J Atmos Ocean Technol 8:746–765

Achtemeier GL (1992) Grasshopper response to rapid vertical displacements within a 'clear air' boundary layer as observed by Doppler radar. Environ Entomol 21:921–938

Alerstam T, Chapman JW, Bäckman J et al (2011) Convergent patterns of long-distance nocturnal migration in noctuid moths and passerine birds. Proc R Soc B Biol Sci 278:3074–3080

Anderson RC (2009) Do dragonflies migrate across the western Indian Ocean? J Trop Ecol 25:347–348

Aralimarad P, Reynolds AM, Lim KS, Reynolds DR, Chapman JW (2011) Flight altitude selection increases orientation performance in high-flying nocturnal insect migrants. Anim Behav 82:1221–1225

Bauer S, Hoye BJ (2014) Migratory animals couple biodiversity and ecosystem functioning worldwide. Science 344:1242552. https://doi.org/10.1126/science.1242552

Bauer S, Chapman JW, Reynolds DR et al (2017) From agricultural benefits to aviation safety: realizing the potential of continent-wide radar networks. BioScience 67(10):912–918

Beerwinkle KR, Lopez JD, Witz JA, Schleider PG, Eyster RS, Lingren PD (1994) Seasonal radar and meteorological observations associated with nocturnal insect flight at altitudes to 900 meters. Environ Entomol 23:676–683

Bell JR, Aralimarad P, Lim K-S, Chapman JW (2013) Predicting insect migration density and speed in the daytime convective boundary layer. PLoS One 8(1):e54202

Bluestein HB, Snyder JC, Thiem KJ, Wienhoff ZB, Reif D, Turner D (2014) Doppler-radar observations of a prefrontal wind-shift line in the Southern Plains of the U.S. Poster MES. P01 presented at ERAD2014: Eighth European Conference on Radar in Meteorology and Hydrology, 1–5 Sept 2014, Garmisch-Partenkirchen, Germany. DLR- Institut für Physik der Atmosphäre, Oberpfaffenhofen, Germany (Extended abstract)

Boulanger Y, Fabry F, Kilambi A et al (2017) The use of weather surveillance radar and high-resolution three dimensional weather data to monitor a spruce budworm mass exodus flight. Agric For Meteorol 234:127–135

Brévault T, Achaleke J, Sougnabé SP, Vaissayre M (2008) Tracking pyrethroid resistance in the polyphagous bollworm, *Helicoverpa armigera* (Lepidoptera: Noctuidae) in the shifting landscape of a cotton growing area. Bull Entomol Res 98:565–573

Browning KA (1981) Ingestion of insects by intense convective updraughts. Antenna 5:14–17

Browning KA, Nicol JC, Marsham JH, Rogberg P, Norton EG (2011) Layers of insect echoes near a thunderstorm and implications for the interpretation of radar data in terms of airflow. Q J R Meteorol Soc 137:723–735

Campistron B (1975) Characteristic distributions of angel echoes in the lower atmosphere and their meteorological implications. Bound Layer Meteorol 9:411–426

Chandra AS, Kollias P, Giangrande SE, Klein SA (2010) Long-term observations of the convective boundary layer using insect radar returns at the SGP ARM climate research facility. J Climate 23:5699–5714

Chapman JW, Reynolds DR, Smith AD, Riley JR, Pedgley DE, Woiwod IP (2002a) High-altitude migration of the diamondback moth *Plutella xylostella* to the U.K.: a study using radar, aerial netting and ground trapping. Ecol Entomol 27:641–650

Chapman JW, Smith AD, Woiwod IP, Reynolds DR, Riley JR (2002b) Development of vertical-looking radar technology for monitoring insect migration. Comput Electron Agric 35:95–110

Chapman JW, Reynolds DR, Smith AD, Smith ET, Woiwod IP (2004) An aerial netting study of insects migrating at high-altitude over England. Bull Entomol Res 94:123–136

Chapman JW, Reynolds DR, Brooks SJ, Smith AD, Woiwod IP (2006) Seasonal variation in the migration strategies of the green lacewing *Chrysoperla carnea* species complex. Ecol Entomol 31:378–388

Chapman JW, Reynolds DR, Hill JK, Sivell D, Smith AD, Woiwod IP (2008a) A seasonal switch in compass orientation in a high-flying migrant moth. Curr Biol 18:R908–R909

Chapman JW, Reynolds DR, Mouritsen H et al (2008b) Wind selection and drift compensation optimize migratory pathways in a high-flying moth. Curr Biol 18:514–518

Chapman JW, Nesbit RL, Burgin LE, Reynolds DR, Smith AD, Middleton DR, Hill JK (2010) Flight orientation behaviors promote optimal migration trajectories in high-flying insects. Science 327:682–685

Chapman JW, Drake VA (2017) Insect migration. Elsevier, In Reference Module in Life Sciences. Oxford. http://www.sciencedirect.com/science/article/pii/B9780128096338012486

Chapman JW, Drake VA, Reynolds DR (2011a) Recent insights from radar studies of insect flight. Annu Rev Entomol 56:337–356

Chapman JW, Klaassen RHG, Drake VA et al (2011b) Animal orientation strategies for movement in flows. Curr Biol 21:R861–R870

Chapman JW, Bell JR, Burgin LE, Reynolds DR, Pettersson LB, Hill JK, Bonsall MB, Thomas JA (2012) Seasonal migration to high latitudes results in major reproductive benefits in an insect. Proc Natl Acad Sci USA 109(37):14924–14929

Chapman JW, Lim KS, Reynolds DR (2013) The significance of midsummer movements of *Autographa gamma*: Implications for a mechanistic understanding of orientation behavior in a migrant moth. Curr Zool 59:360–370

Chapman JW, Nilsson C, Lim KS, Bäckman J, Reynolds DR, Alerstam T, Reynolds AM (2015a) Detection of flow direction in high-flying insect and songbird migrants. Curr Biol 25(17): R733–R752

Chapman JW, Reynolds DR, Wilson K (2015b) Long-range seasonal migration in insects: mechanisms, evolutionary drivers and ecological consequences. Ecol Lett 18:287–302

Chapman JW, Nilsson C, Lim KS, Bäckman J, Reynolds DR, Alerstam T (2016) Adaptive strategies in nocturnally migrating insects and songbirds: contrasting responses to wind. J Anim Ecol 85:115–124

Chen R-L, Bao X-Z, Drake VA, Farrow RA, Wang S-Y, Sun Y-J, Zhai B-P (1989) Radar observations of the spring migration into northeastern China of the oriental armyworm moth, *Mythimna separata* and other insects. Ecol Entomol 14:149–162

Chilson PB, Bridge E, Frick WF, Chapman JW, Kelly JF (2012a) Radar aeroecology: exploring the movements of aerial fauna through radio-wave remote sensing. Biol Lett 8:698–701

Chilson PB, Frick WF, Kelly JF, Howard KW, Larkin RP, Diehl RH, Westbrook JK, Kelly TA, Kunz TH (2012b) Partly cloudy with a chance of migration: weather, radars, and aeroecology. Bull Am Meteorol Soc 93:669–686

Cleveland CJ, Betke M, Federico P et al (2006) Economic value of the pest control service provided by Brazilian free-tailed bats in south-central Texas. Front Ecol Environ 4:238–243

Contreras RF, Frasier SJ (2008) High-resolution observations of insects in the atmospheric boundary layer. J Atmos Ocean Technol 25:2176–2187

Deveson ED, Walker PW (2005) Not a one-way trip: historical distribution data for Australian plague locusts supports frequent seasonal exchange migrations. J Orthop Res 14:95–109

Dickerson AK, Shankles PG, Madhavan NM, Hu DL (2012) Mosquitoes survive raindrop collisions by virtue of their low mass. Proc Natl Acad Sci USA 109(25):9822–9827

Dingle H (2014) Migration: the biology of life on the move, 2nd edn. Oxford University Press, Oxford

Dingle H, Drake VA (2007) What is migration? BioScience 57:113–121

Dokter AM, Liechti F, Stark H, Delobbe L, Tabary P, Holleman I (2011) Bird migration flight altitudes studied by a network of operational weather radars. J R Soc Interface 8:30–43

Drake VA (1985) Solitary wave disturbances of the nocturnal boundary layer revealed by radar observations of migrating insects. Bound Layer Meteorol (3):269–286

Drake VA (2014) Estimation of unbiased insect densities and density profiles with vertically pointing entomological radars. Int J Remote Sens 35(13):4630–4654

Drake VA, Farrow RA (1983) The nocturnal migration of the Australian plague locust, *Chortoicetes terminifera* (Walker) (Orthoptera: Acrididae): quantitative radar observations of a series of northward flights. Bull Entomol Res 73:567–585

Drake VA, Farrow RA (1985) A radar and aerial-trapping study of an early spring migration of moths (Lepidoptera) in inland New South Wales. Aust J Ecol:10223–10235

Drake VA, Farrow RA (1988) The influence of atmospheric structure and motions on insect migration. Annu Rev Entomol 33:183–210

Drake VA, Gatehouse AG (eds) (1995) Insect migration: tracking resources through space and time. Cambridge University Press, Cambridge

Drake VA, Reynolds DR (2012) Radar entomology: observing insect flight and migration. CABI, Wallingford

Drake VA, Wang HK (2013) Recognition and characterization of migratory movements of Australian plague locusts, *Chortoicetes terminifera*, with an insect monitoring radar. J Appl Remote Sens 7(1):075095, 17pp

Drake VA, Helm KF, Readshaw JL, Reid DG (1981) Insect migration across Bass Strait during spring: a radar study. Bull. Entomol. Res.71:449–466

Drake VA, Gregg PC, Harman IT, Wang H-K, Deveson ED, Hunter, DM, Rochester WA (2001) Characterizing insect migration systems in inland Australia with novel and traditional methodologies. In: Woiwod IP, Reynolds DR, Thomas CD (eds) Insect movement: mechanisms and consequences. CAB International, Wallingford, pp 207–233

Dudley R (2000) The biomechanics of insect flight: form, function, evolution. Princeton University Press, Princeton, NJ

Farrow RA (1975) Offshore migration and the collapse of outbreaks of the Australian plague locust *Chortoicetes terminifera* Walk. in south-east Australia. Aust J Zool 234:569–595

Feng H-Q, Wu X-F, Wu B, Wu K-M (2009) Seasonal migration of *Helicoverpa armigera* (Lepidoptera: Noctuidae) over the Bohai Sea. J Econ Entomol 102:95–104

Freeman B (2003) A fallout of black witches (*Ascalapha odorata*) associated with Hurricane Claudette. News Lepid Soc 43(3):71

Geerts B, Miao Q (2005) The use of millimeter Doppler radar echoes to estimate vertical air velocities in the fair-weather convective boundary layer. J Atmos Ocean Technol 22:225–246

Geerts B, Damiani R, Haimov S (2006) Finescale vertical structure of a cold front as revealed by an airborne Doppler radar. Mon Weather Rev 134:251–271

Green K (2011) The transport of nutrients and energy into the Australian Snowy Mountains by migrating Bogong moths *Agrotis infusa*. Austral Ecol 36:25–34

Greenbank DO, Schaefer GW, Rainey RC (1980) Spruce budworm (Lepidoptera: Tortricidae) moth flight and dispersal: new understanding from canopy observations, radar, and aircraft. Mem Entomol Soc Can 110:1–49

Hardy KR, Ottersten H (1969) Radar investigations of convective patterns in the clear atmosphere. J Atmos Sci 26:666–672

Hein AM, Hou C, Gillooly JF (2012) Energetic and biomechanical constraints on animal migration distance. Ecol Lett 15:104–110

Hobbs SE, Wolf WW (1989) An airborne radar technique for studying insect migration. Bull Entomol Res 79:693–704

Holdo RM, Holt RD, Sinclair ARE, Godley BJ, Thirgood S (2011) Migration impacts on communities and ecosystems: empirical evidence and theoretical insights. In: Milner-Gulland EJ, Fryxell JM, Sinclair ARE (eds) Animal migrations: a synthesis. Oxford University Press, Oxford, pp 131–143

Hu G, Lim KS, Reynolds DR, Reynolds AM, Chapman JW (2016a) Wind-related orientation patterns in diurnal, crepuscular and nocturnal high-altitude insect migrants. Front Behav Neurosci 10: article 32, 8pp

Hu G, Lim KS, Horvitz N, Clark SJ, Reynolds DR, Sapir N and Chapman JW (2016b) Mass seasonal bioflows of high-flying insect migrants. Science 354: 1584–1587

Isard SA, Gage SH (2001) Flow of life in the atmosphere: an airscape approach to understanding invasive organisms. Michigan State University Press, East Lansing

Irwin ME, Thresh JM (1988) Long range aerial dispersal of cereal aphids as virus vectors in North America. Philos Trans R Soc Lond B 321:421–446

Isard SA, Irwin ME, Hollinger SE (1990) Vertical distribution of aphids (Homoptera: Aphididae) in the planetary boundary layer. Environ Entomol 19:1473–1484

Isard SA, Kristovich DAR, Gage SH et al (2001) Atmospheric motion systems that influence the redistribution and accumulation of insects on the beaches of the Great Lakes in North America. Aerobiologia 17:275–291

Jeffries DL, Chapman J, Roy HE, Humphries S, Harrington R et al (2013) Characteristics and drivers of high-altitude ladybird flight: insights from vertical-looking entomological radar. PLoS One 8(12):e82278

Johnson CG (1969) Migration and dispersal of insects by flight. Methuen, London

Johnson SJ (1995) Insect migration in North America: synoptic-scale transport in a highly seasonal environment. In: Drake VA, Gatehouse AG (eds) Insect migration: tracking resources through space and time. Cambridge University Press, Cambridge, pp 31–66

Joyce RJV (1983) Aerial transport of pests and pest outbreaks. EPPO Bull 13:111–119

Kelly JF, Horton KG (2016) Toward a predictive macrosystems framework for migration ecology. Glob Ecol Biogeogr 25(10):1159–1165. https://doi.org/10.1111/geb.12473

Kelly JF, Shipley JR, Chilson PB, Howard KW, Frick WF, Kunz TH (2012) Quantifying animal phenology in the continental scale using NEXRAD weather radars. Ecosphere 3:article 16. https://doi.org/10.1890/ES11-00257.1

Kelly JF, Bridge ES, Frick WF, Chilson PB (2013) Ecological energetics of an abundant aerial insectivore, the purple martin. PLoS One 8(9):e76616

Kennedy JS (1985) Migration: behavioral and ecological. In: Rankin MA (ed) Migration: mechanisms and adaptive significance. Contributions in Marine Science 27 (Supplement). Marine Science Institute, University of Texas at Austin, Port Aransas, TX, pp 5–26

Klaassen RH, Hake M, Strandberg R et al (2014) When and where does mortality occur in migratory birds? Direct evidence from long-term satellite tracking of raptors. J Anim Ecol 83(1):176–184

Knupp KR (2006) Observational analysis of a gust front to bore to solitary wave transition within an evolving nocturnal boundary layer. J Atmos Sci 63:2016–2035

Koch GJ (2006) Using a Doppler light detection and ranging (lidar) system to characterize an atmospheric thermal providing lift for soaring raptors. J Field Ornithol 77:315–318

Krauel JJ, Westbrook JK, McCracken GF (2015) Weather-driven dynamics in a dual-migrant system: moths and bats. J Anim Ecol 84:604–614

Landry JS, Parrott L (2016) Could the lateral transfer of nutrients by outbreaking insects lead to consequential landscape-scale effects? Ecosphere 7:e01265

Lang TJ, Rutledge SA, Stith JL (2004) Observations of quasi-symmetric echo patterns in clear air with the CSU-CHILL polarimetric radar. J Atmos Ocean Technol 21:1182–1189

Leskinen M, Markkula I, Koistinen J, Pylkkö P, Ooperi S, Siljamo P, Ojanen H, Raiskio S, Tiilikkala K (2011) Pest insect immigration warning by an atmospheric dispersion model, weather radars and traps. J Appl Entomol 135:55–67

Leskinen M, Rojas L, Mikkola K (2012) Weather radar observations of Underwing moth immigrations to Finland. Poster at ERAD 2012—the Seventh European Conference on Radar in Meteorology and Hydrology, Toulouse, France, 24–29 June 2012. Météo France, Toulouse. http://www.meteo.fr/cic/meetings/2012/ERAD/extended_abs/NMUR_215_ext_abs.pdf

Lewis T, Taylor LR (1964) Diurnal periodicity of flight by insects. Trans R Entomol Soc Lond 116:393–479

Lothon M, Campistron B, Jacoby-Koaly S, Bénech B, Lohou F, Girard-Ardhuin F (2002) Comparison of radar reflectivity and vertical velocity observed with a scannable C-band radar and two UHF profilers in the lower troposphere. J Atmos Ocean Technol 19:899–910

Lothon M, Campistron B, Chong M, Couvreux F, Guichard F, Rio C, Williams E (2011) Life cycle of a mesoscale circular gust front observed by a C-Band Doppler radar in West Africa. Mon Weather Rev 139:1370–1388

Luke EP, Kollias P, Johnson KL, Clothiaux EE (2008) A technique for the automatic detection of insect clutter in cloud radar returns. J Atmos Ocean Technol 25:1498–1513

McCracken GF, Gillam EH, Westbrook JK, Lee Y, Jensen M, Balsley B (2008) Brazilian free-tailed bats (*Tadarida brasiliensis*: Molossidae, Chiroptera) at high altitude: links to migratory insect populations. Integr Comp Biol 48:107–118

McCracken GF, Westbrook JK, Brown VA et al (2012) Bats track and exploit changes in insect pest populations. PLoS One 7(8):e43839

McNeil JN (1987) The true armyworm, *Pseudaletia unipuncta*—a victim of the pied piper or a seasonal migrant. Insect Sci Appl 8:591–597

Melnikov V, Leskinen M, Koistinen J (2014a) Doppler velocities at orthogonal polarizations in radar echoes from insects and birds. IEEE Geosci Remote Sens Lett 11(3):592–596

Melnikov V, Zrnic D, Burgess D, Mansell E (2014b) Observations of hail cores of tornadic thunderstorms with three polarimetric radars. Presentation at 94th American Meteorological Society Annual Meeting, 2–6 Feb, 2014, Atlanta, 16pp. https://ams.confex.com/ams/94Annual/webprogram/Paper233038.html

Menu S, Gauthier G, Reed A (2005) Survival of young greater snow geese (*Chen caerulescens atlantica*) during fall migration. Auk 122:479–496

Mikkola K (1986) Direction of insect migrations in relation to the wind. In: Danthanarayana W (ed) Insect flight: dispersal and migration. Springer-Verlag, Berlin, pp 152–171

Ming J-G, Jin H, Riley JR et al (1993) Autumn southward 'return' migration of the mosquito *Culex tritaeniorhynchus* in China. Med Vet Entomol 7:323–327

Neumann HH, Mukammal EI (1981) Incidence of mesoscale convergence lines as input to spruce budworm control strategies. Int J Biometeorol 25:175–187

Nieminen M, Leskinen M, Helenius J (2000) Doppler radar detection of exceptional mass migration of aphids into Finland. Int J Biometeorol 44:172–181

Orlanski I (1975) A rational subdivision of scales for atmospheric processes. Bull Am Meteorol Soc 56:527–530

Otuka A (2013) Migration of rice planthoppers and their vectored re-emerging and novel rice viruses in East Asia. Front Microbiol 4:article 309. https://doi.org/10.3389/fmicb.2013.00309.

Otuka A, Dudhia J, Watanabe T, Furuno A (2005) A new trajectory analysis method for migratory planthoppers, *Sogatella furcifera* (Horváth) (Homoptera: Delphacidae) and *Nilaparvata lugens* (Stål), using an advanced weather forecast model. Agric For Entomol 7:1–9

Otuka A, Matsumura M, Sanada-Morimura S, Takeuchi H, Watanabe T, Ohtsu R, Inoue H (2010) The 2008 overseas mass migration of the small brown planthopper, *Laodelphax striatellus*, and subsequent outbreak of rice stripe disease in western Japan. Appl Entomol Zool 45:259–266

Pair SD, Raulston JR, Westbrook JR, Wolf WW, Sparks AN, Schuster MF (1987) Development and production of corn earworm and fall armyworm in the Texas High Plains: evidence for the reverse fall migration. Southwestern Entomologist 12:89–99

Pedgley DE (ed) (1981) Desert Locust forecasting manual, vols 1 and 2. Centre for Overseas Pest Research, London

Pedgley DE (1982) Windborne pests and diseases: Meteorology of airborne organisms. Ellis Horwood, Chichester

Pedgley DE, Reynolds DR, Tatchell GM (1995) Long-range insect migration in relation to climate and weather: Africa and Europe. In: Drake VA, Gatehouse AG (eds) Insect migration: tracking resources through space and time. Cambridge University Press, Cambridge, pp 3–29

Qi H, Jiang C, Zhang Y, Yang X, Cheng D (2014) Radar observations of the seasonal migration of brown planthopper (*Nilaparvata lugens* Stal) in Southern China. Bull Entomol Res 104:731–741

Rainey RC (1989) Migration and meteorology: flight behaviour and the atmospheric environment of locusts and other migrant pests. Oxford University Press, Oxford

Rennie SJ (2012) Doppler weather radar in Australia. CAWCR (Centre for Australian Weather and Climate Research) Technical Report 055. Bureau of Meteorology, Melbourne

Rennie SJ (2014) Common orientation and layering of migrating insects in southeastern Australia observed with a Doppler weather radar. Meteorol Appl 21:218–229

Reynolds DR, Mukhopadhyay S, Riley JR, Das BK, Nath PS, Mandal SK (1999) Seasonal variation in the windborne movement of insect pests over northeast India. Int J Pest Manag 45:195–205

Reynolds DR, Chapman JW, Edwards AS, Smith AD, Wood CR, Barlow JF et al (2005) Radar studies of the vertical distribution of insects migrating over southern Britain: the influence of temperature inversions on nocturnal layer concentrations. Bull Entomol Res 95:259–274

Reynolds DR, Chapman JW, Harrington R (2006) The migration of insect vectors of plant and animal viruses. Adv Virus Res 67:453–517

Reynolds DR, Smith AD, Chapman JW (2008) A radar study of emigratory flight and layer formation by insects at dawn over southern Britain. Bull Entomol Res 98:35–52

Reynolds AM, Reynolds DR, Riley JR (2009) Does a 'turbophoretic' effect account for layer concentrations of insects migrating in the stable night-time atmosphere? J R Soc Interface 6:87–95

Reynolds AM, Reynolds DR, Smith AD, Chapman JW (2010) A single wind-mediated mechanism explains high-altitude 'non-goal oriented' headings and layering of nocturnally migrating insects. Proc R Soc B Biol Sci 277:765–772

Reynolds DR, Reynolds AM, Chapman JW (2014) Non-volant modes of migration in terrestrial arthropods. Anim Migr 2:8–28

Reynolds AM, Reynolds DR, Sane SP, Hu G, Chapman JW (2016) Orientation in high-flying migrant insects in relation to flows: mechanisms and strategies. Philos Trans R Soc B: 371 (1704):20150392

Riley JR, Reynolds DR, Farmery MJ (1983) Observations of the flight behaviour of the armyworm moth, *Spodoptera exempta*, at an emergence site using radar and infra-red optical techniques. Ecol Entomol 8:395–418

Riley JR, Reynolds DR, Farrow RA (1987) The migration of *Nilaparvata lugens* (Stal) (Delphacidae) and other Hemiptera associated with rice during the dry season in the Philippines: a study using radar, visual observations, aerial netting and ground trapping. Bull Entomol Res 77:145–169

Riley JR, Cheng X-N, Zhang X-X, Reynolds DR, Xu G-M, Smith AD, Cheng J-Y, Bao A-D, Zhai B-P (1991) The long-distance migration of *Nilaparvata lugens* (Stal) (Delphacidae) in China: radar observations of mass return flight in the autumn. Ecol Entomol 16:471–489

Riley JR, Reynolds DR, Smith AD, Rosenberg LJ, Cheng X-N, Zhang X-X, Xu G-M, Cheng J-Y, Bao A-D, Zhai B-P, Wang H-K (1994) Observations on the autumn migration of *Nilaparvata lugens* (Homoptera: Delphacidae) and other pests in east central China. Bull Entomol Res 84:389–402

Riley JR, Reynolds DR, Smith AD, Edwards AS, Zhang X-X, Cheng X-N, Wang H-K, Cheng J-Y, Zhai B-P (1995) Observations of the autumn migration of the rice leaf roller *Cnaphalocrocis medinalis* (Lepidoptera: Pyralidae) and other moths in eastern China. Bull Entomol Res 85:397–414

Rose DJW, Page WW, Dewhurst CF et al (1985) Downwind migration of the African armyworm moth, *Spodoptera exempta*, studied by mark-and-recapture and by radar. Ecol Entomol 10:299–313

Rose DJW, Dewhurst CF, Page WW (2000) The African armyworm handbook: the status, biology, ecology, epidemiology and management of *Spodoptera exempta* (Lepidotera: Noctuidae), 2nd edn. Natural Resources Institute, Chatham

Russell RW, Wilson JW (1996) Aerial plankton detected by radar. Nature 381:200–201

Russell RW, Wilson JW (1997) Radar-observed 'fine lines' in the optically clear boundary layer: reflectivity contribution from aerial plankton and its predators. Bound Layer Meteorol 82:235–262

Russell RW, Wilson JW (2001) Spatial dispersion of aerial plankton over east-central Florida: aeolian transport and coastline concentrations. Int J Remote Sens 22:2071–2082

Sauvageot H, Despaux G (1996) The clear-air coastal vespertine radar bands. Bull Am Meteorol Soc 77:673–681

Schaefer GW (1976) Radar observations of insect flight. In: Rainey RC (ed) Insect flight (Symposia of the Royal Entomological Society no. 7). Blackwell Scientific, Oxford, pp 157–197

Scott KD, Lawrence N, Lange CL, Scott LJ, Wilkinson KS, Merritt MA, Miles M, Murray D, Graham GC (2005) Assessing moth migration and population structuring in *Helicoverpa armigera* (Lepidoptera: Noctuidae) at the regional scale: example from the Darling Downs, Australia. J Econ Entomol 98:2210–2219

Shamoun-Baranes J, Alves JA, Bauer S et al. (2014) Continental-scale radar monitoring of the aerial movements of animals. Mov Ecol 2: article 9, 6pp

Shashar N, Sabbah S, Aharoni N (2005) Migrating locusts can detect polarized reflections to avoid flying over the sea. Biol Lett 1:472–475

Sillett TS, Holmes RT (2002) Variation in survivorship of a migratory songbird throughout its annual cycle. J Anim Ecol 71:296–308

Simpson JE (1994) Sea breeze and local winds. Cambridge University Press, Cambridge

Simpson JE (1997) Gravity currents in the environment and the laboratory, 2nd edn. Cambridge University Press, Cambridge

Smith AD, Reynolds DR, Riley JR (2000) The use of vertical-looking radar to continuously monitor the insect fauna flying at altitude over southern England. Bull Entomol Res 90:265–277

Ssymank A, Kearns CA, Pape T, Thompson FC (2008) Pollinating flies (Diptera): a major contribution to plant diversity and agricultural production. Biodiversity 9(1–2):86–89

Stefanescu C, Alarcón M, Àvila A (2007) Migration of the painted lady butterfly, *Vanessa cardui*, to north-eastern Spain is aided by African wind currents. J Anim Ecol 76:888–898

Stefanescu C, Páramo F, Åkesson S et al (2013) Multi-generational long-distance migration of insects: studying the painted lady butterfly in the Western Palaearctic. Ecography 36:474–486

Stefanescu C, Soto DX, Talavera G, Vila R, Hobson KA (2016) Long-distance autumn migration across the Sahara by painted lady butterflies: exploiting resources in the tropical savannah. Biol Lett 12:20160561

Talavera G, Vila R (2017) Discovery of mass migration and breeding of the painted lady butterfly *Vanessa cardui* in the Sub-Sahara: the Europe–Africa migration revisited. Biol J Linn Soc 120:274–285

Tanaka K, Matsumura M (2000) Development of virulence to resistant rice varieties in the brown planthopper, *Nilaparvata lugens* (Homoptera: Delphacidae), immigrating into Japan. Appl Entomol Zool 35:529–533

Taylor LR (1974) Insect migration, flight periodicity and the boundary layer. J Anim Ecol 43:225–238

Taylor LR (1985) An international standard for the synoptic monitoring and dynamic mapping of migrant pest aphid populations. In: MacKenzie DR, Barfield CS, Kennedy GC, Berger RD, Taranto DJ (eds) The movement and dispersal of agriculturally important biotic agents. Claitor's Publishing Division, Baton Rouge, pp 337–380

Tojo S, Ryuda M, Fukuda T, Matsunaga T, Choi D-R, Otuka A (2013) Overseas migration of the common cutworm, *Spodoptera litura* (Lepidoptera: Noctuidae), from May to mid-July in East Asia. Appl Entomol Zool 48:131–140

Tucker MR (1997) Satellite-derived rainstorm distribution as an aid to forecasting African armyworm outbreaks. Weather 52:204–212

Van Den Broeke MS (2013) Polarimetric Radar observations of biological scatterers in Hurricanes Irene (2011) and Sandy (2012). J Atmos Ocean Technol 30:2754–2767

Wainwright CE, Stepanian PM, Reynolds DR, Reynolds AM (2017) The movement of small insects in the convective boundary layer: linking patterns to processes. Sci Rep 7(1):article 5438

Ward SA, Leather SR, Pickup J, Harrington R (1998) Mortality during dispersal and the cost of host-specificity in parasites: how many aphids find hosts? J Anim Ecol 67:763–773

Warrant E, Frost B, Green K, Mouritsen H, Dreyer D, Adden A, Brauburger K, Heinze S (2016) The Australian bogong moth *Agrotis infusa*: a long-distance nocturnal navigator. Front Behav Neurosci 10:77

Weckwerth T, Horst TW, Wilson JW (1999) An observational study of the evolution of horizontal convective rolls. Mon Weather Rev 127:2160–2179

Weiss CC, Bluestein HB, Pazmany AL (2006) Finescale radar observations of the 22 May 2002 dryline during the International H_2O Project (IHOP). Mon Weather Rev 134:273–293

Westbrook JK (2008) Noctuid migration in Texas within the nocturnal aeroecological boundary layer. Integr Comp Biol 48:99–106

Westbrook JK, Eyster RS (2017) Doppler weather radar detects emigratory flights of noctuids during a major pest outbreak. Remote Sensing Applications: Society and Environment 8:64–70

Westbrook JK, Eyster RS, Wolf WW (2014) WSR-88D Doppler radar detection of corn earworm moth migration. Int J Biometeorol 58:931–940

Westbrook JK, Nagoshi RN, Meagher RL, Fleischer SJ, Jairam S (2016) Modeling seasonal migration of fall armyworm moths. Int J Biometeorol 60:255–267

Wolf WW, Westbrook JK, Sparks AN (1986) Relationship between radar entomological measurements and atmospheric structure in south Texas during March and April 1982. In: Sparks AN (ed) Long-range migration of moths of agronomic importance to the United States and Canada: Specific examples of occurrence and synoptic weather patterns conducive to migration. ARS-43. United States Department of Agriculture, Agricultural Research Service, Washington, DC, pp 84–97

Wood CR, O'Connor EJ, Hurley RA, Reynolds DR, Illingworth AJ (2009) Cloud-radar observations of insects in the UK convective boundary layer. Meteorol Appl 16:491–500

Wood CR, Clark SJ, Barlow JF, Chapman JW (2010) Layers of nocturnal insect migrants at high-altitude: the influence of atmospheric conditions on their formation. Agric For Entomol 12:113–121

Facing the Wind: The Aeroecology of Vertebrate Migrants

8

Felix Liechti and Liam P. McGuire

Abstract

The aerosphere is an essential part of the habitat of flying vertebrates. Birds and bats make use of the airspace for daily activities like foraging, commuting, mating, and seasonal movements including migration. In this chapter, we focus on how the properties of the aerosphere affect migration and a few other regular large-scale movements. For animals moving between seasonally favourable habitats across hundreds or thousands of kilometres, the conditions of the aerosphere have a substantial impact on energy and time demands, on orientation and navigation, and finally on survival. Although bats and birds often make similar use of the aerosphere, there is a huge difference in our actual knowledge of these interactions. There are about 4000 migratory bird species comprising 100–150 billion individuals undertaking regular seasonal movements within and between continents and across the oceans. Our understanding of bat migration is much more limited and such estimates are not available, but many bat species and many millions of individuals show similar kinds of migratory behaviour. Atmospheric conditions vary across time and space, including variation in air flow and air temperature, humidity, and density. In this chapter, we emphasize the importance of wind and precipitation as the main factors driving behaviour and evolutionary adaptations. Flying within a moving air space, bats and birds can make use of regular seasonal wind fields, like the trade and anti-trade winds, but they also must deal with irregular events, like heavy storms. In combination with the distribution of their preferred habitats, large-scale atmospheric conditions guide their flight routes and shape their migratory strategies. The

F. Liechti (✉)
Swiss Ornithological Institute, Sempach, Switzerland
e-mail: felix.liechti@vogelwarte.ch

L.P. McGuire
Department of Biological Sciences, Texas Tech University, Lubbock, TX, USA
e-mail: liam.mcguire@ttu.edu

© Springer International Publishing AG, part of Springer Nature 2017
P.B. Chilson et al. (eds.), *Aeroecology*,
https://doi.org/10.1007/978-3-319-68576-2_8

179

timing of individual flight stages is directed by weather conditions, mainly wind and precipitation. Once aloft, individuals may select among varying wind conditions at different flight altitudes to achieve beneficial wind conditions en route. These behavioural patterns have a strong effect on the time needed to move between suitable habitats, but probably more importantly on overall energy demand and thus foraging time/cost and survival. While birds are known to explore heights up to 8000 m asl during migration, bats are generally restricted to heights below 3000 m asl, possibly due to differences in lung morphology. On the other hand, many bats can withstand harsh weather conditions by using torpor, while birds may have to leave or starve. Birds and bats are confronted with the regular occurrence of both predictable (e.g. wind support deviations from target) and unpredictable (e.g. storms) displacements and are therefore equipped with excellent orientation capabilities (Chap. 6). This chapter provides the background of what we really know about the role of the aerosphere for the migration of birds and bats and where we still marvel.

1   Introduction

Vertebrates interact with the aerosphere on very different spatial and temporal scales (Frick et al. 2013). One extreme are gliding vertebrates which are airborne only for seconds covering some tens of metres. This strategy is relatively common, including flying fishes (Davenport 1994) and at least 30 independent evolutionary lineages among terrestrial vertebrates including amphibians, reptiles, and mammals (Dudley et al. 2007). By comparison, flapping flight is relatively uncommon, only having evolved in three vertebrate lineages—the extinct pterosaurs and the extant birds and bats. Airborne journeys range from foraging flights up to migratory movements among and across continents (Alerstam 1990; Fleming and Eby 2003; Newton 2010). The temporal and spatial scales at which flying vertebrates encounter the aerosphere can reach extremes. Barnacle geese (*Branta leucopsis*) may spend only ~20 min in flight per day in non-migratory periods but may fly ~20 h per day as they travel 2500 km during migration (Portugal et al. 2012). Swifts (*Tachymarptis melba, Apus apus*) can stay airborne continuously for hundreds of days covering thousands of kilometres (Liechti et al. 2013; Hedenström et al. 2016). Bar-tailed godwits (*Limosa lapponica baueri*) may fly continuously for >7 days while migrating >10,000 km in a single flight (Gill et al. 2005; Gill et al. 2009). Flight altitudes up to 8000 m above sea level occur occasionally above lowlands (Liechti and Schaller 1999) but regularly above high mountain ranges (Bishop et al. 2015). Thus, there is great diversity in the spatial and temporal scales, and behavioural and ecological contexts under which vertebrates interact with the aerosphere. In this chapter, we refrain from covering all of these aspects and, instead, focus on migratory movements of birds and bats, the only extant flying vertebrates, across the aerosphere. There is already an extensive body of literature regarding bird migration (e.g. Alerstam 1990; Berthold 2001; Greenberg and Marra

2005; Newton 2010; Shamoun-Baranes et al. 2017), whereas much less is published on bat migration (Fleming and Eby 2003; McGuire and Guglielmo 2009; Krauel and McCracken 2013). Specifically, we examine the interactions of vertebrate migrants and the aerosphere in terms of short- and long-term adaptations to the spatial and temporal conditions of the atmosphere, referring to both bat and bird literature where possible.

Why is the aerosphere such an attractive space for large-scale movements? Flying allows animals to move quickly between distant places. Compared to running, flying is faster and energetically cheaper per unit distance—compared to swimming, flying is faster, but more expensive (cf. Chap. 5). For example, a swallow weighing ~20 g could fly from Zurich to Paris with a fat store ~10% of body mass (Delingat et al. 2008), while a similar sized mouse would achieve only about a tenth of the distance with the same amount of energy within a time of probably some days (Alexander 2006), which makes it virtually impossible for small walkers to travel long distances. Flight also allows an animal to travel in a straight line between two sites, avoiding any detours. No wonder, flying animals have taken these advantages to explore habitats all around the world, regardless of whether these are on land or at sea. The extraordinary mobility allows them individually to exploit distant habitats with short-term peaks in seasonal food abundance.

Mass movements of geese, waders, and other large migratory birds have fascinated people all around the world. However, more than 90% of migratory birds fly across sea and land without notice by the public. Most of them are small passerines travelling at night (Suter 1957) or flying during the day at heights where they are hardly detected by visual observation (Bruderer 1971). There are very few estimates of the number of individual migrants. About 2.1 billion birds are expected to cross the Sahara on their autumn migration from the European breeding grounds to the sub-Saharan winter quarters (Hahn et al. 2009). However, such estimates are very difficult to calculate and based on simplifications and assumptions. No wonder, there are no scientific estimates of individuals involved in any other migratory flyway. We only can assume that similar numbers cross from North to South America and considerably more move within Asia and Oceania. The global bird population has been estimated at 200–400 billion individuals (Gaston and Blackburn 1997), and of ~10,000 bird species, about 40% undertake regular seasonal movements within and between continents and across the oceans. Thus, we assume that worldwide about 100–150 billion individual birds migrate along distinct flyways every year.

Similar estimates are nearly impossible for bats. Although most of the ~1300 bat species are comparatively sedentary, spending the entirety of their lives in the same region (Fleming and Eby 2003; Hutterer et al. 2005), migration has evolved repeatedly (Bisson et al. 2009) and members of at least nine bat families are known to migrate (McGuire and Ratcliffe 2011). Estimating the number of migrating bats is a near impossible task given their cryptic, nocturnal nature and the challenges of estimating population size for even the most common species (e.g. O'Shea et al. 2003). However, examples such as the millions of Brazilian free-tailed bats (*Tadarida brasiliensis*) migrating between Mexico and the United States (Cockrum 1969; Horn and Kunz 2008), the millions of migratory Straw-Coloured fruit bats (*Eidolon helvum*) in many African countries (Richter and Cumming 2006; Byng et al. 2009), and the hundreds of thousands of migrating bats killed by wind turbines (Arnett and Baerwald 2013; Arnett et al. 2016; Frick et al. 2017) all indicate the large numbers of bats that travel long distances through the aerosphere. The numbers of migrating birds and bats underline the significance of the aerosphere for the transfer of a huge biomass of aerial vertebrates between distant habitats.

The aerosphere is a highly dynamic space in the truest sense of the word. It is formed by a constant exchange of air masses around the globe (cf. Chap. 2), driven by the seasonality of solar radiation and the rotation of the earth. The exchange of warm and cold air including more or less precipitation results in a regular annual pattern of temporal and spatial food abundance for migrants. The air masses, in which flying animals move, are constantly in motion. Most flights take place within the planetary boundary layer (Chap. 2), where wind speeds are of the same order of magnitude as the airspeed of birds and bats (5–25 m/s; birds: Bruderer and Boldt 2001; Alerstam et al. 2007; Pennycuick et al. 2013; bats: Williams et al. 1973; McCracken et al. 2008; Parsons et al. 2008; McCracken et al. 2016). Therefore, any flight trajectory is the combined result of self-propelled animal movement and the flow of the air. Although the aerosphere is for most species "only" a medium to go from one place to the other, it has a substantial impact on energy and time demands, on orientation and navigation, and ultimately on survival. For Black-Throated Blue Warblers (*Setophaga caerulescens*), migration accounted for 85% of annual mortality (Sillett and Holmes 2002), but Grüebler et al. (2014) did find similar survival during breeding and non-breeding periods in barn swallows (*Hirundo rustica*). Migration mortality is estimated to account for 44–57% of annual mortality in Leisler's bats (Giavi et al. 2014).

There are three distinct aspects by which the aerosphere affects long-distance movements. First, the more or less regular flow of air around the globe has moulded the general *migratory flyways* of numerous migratory species. Secondly, the seasonality of temperature and rain affects the abundance of food and, hence, the *timing of migration*. Thirdly, the high spatial and temporal variability in weather conditions has induced adaptations in *flight behaviour* (Fig. 8.1). The aerosphere has and still is shaping migration from the local to the global scale and from the individual to the population and species level.

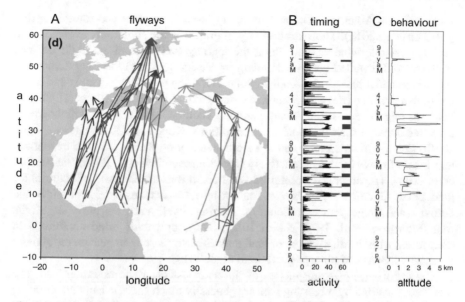

Fig. 8.1 Examples for the three main topics treated in this chapter: the impact of the aerosphere on general flyways (**a**), the timing of flight stages (**b**), and flight behavior (**c**). The graphs refer to spring migration of great reed warblers (*Acrocephalus arundinaceus*). (**a**) Spring flyways of six different populations (reproduced with permission from Koleček et al. 2016), (**b**) acceleration logger data from a single individual during 20 days of spring migration, red bars indicate stages of continuous flight activity, and (**c**) altitude data (km above sea level) for the same individual in panel B (**b** and **c** unpublished data from FL)

2   Migratory Flyways

Many species have stable breeding ranges (where they foster their young) and non-breeding residencies governed by the seasonal pattern of food abundance (e.g. Fleming et al. 1993; Richter and Cumming 2006). However, in many cases migratory routes do not follow the shortest flyway between these sites. The evolution of migratory routes is moulded by multiple factors, e.g. by historical, ecological, geographical, and physical factors (Alerstam et al. 2003). Atmospheric conditions and the distribution of possible stopover sites make some migratory routes more favourable to follow than others (e.g. Erni et al. 2005). Although atmospheric conditions are highly variable, they are predictable at least to some extent. Apart from the long-term seasonal cycles of weather conditions, passages of low- and high-pressure systems follow specific rules, allowing migrants to consider prevailing winds (Kranstauber et al. 2015) or precipitation for their journey (Chap. 2). Therefore, many migrants do not follow a great circle route, the shortest distance between two points on earth, but take advantage of supportive wind conditions (Tottrup et al. 2017). Blackpoll warblers (*Setophaga striata*) migrate from northeastern North America to the northern coast of South America flying south over the Atlantic Ocean and returning north by an overland route in spring (DeLuca et al. 2015). This extraordinary migratory route could only evolve due to the very reliable wind support along the journey (Williams et al. 1977; DeLuca et al. 2015). Bar-tailed godwits breeding in Alaska fly directly to New Zealand to overwinter and return via the Asian mainland (Battley et al. 2012). This round trip is facilitated by the prevailing hemispheric wind conditions (Gill et al. 2014). Seabirds migrate across the Atlantic and the Pacific Ocean following flight routes favoured by large-scale seasonal wind conditions (Kopp et al. 2011; Egevang et al. 2010). Soaring birds on land and sea are dependent on the specific use of thermal updrafts or the sea breeze, respectively (e.g. Weimerskirch et al. 2016; Vansteelant et al. 2017). The absence of preferred wind conditions at sea (thermal updrafts for land birds) or on land (sea breeze for sea birds) is responsible for considerable detours and huge concentrations of soaring land and sea birds at straits (e.g. Finlayson et al. 1976; Smith 1980). Thus, prevailing wind conditions in connection with the distribution of land masses have a strong impact on the general migratory flyways of birds (e.g. Nourani et al. 2016).

The importance of atmospheric conditions is less clear for migrating bats. The distances travelled by migrating bats are generally much shorter than for migrating birds (Fleming and Eby 2003; Bisson et al. 2009). Few bat species travel >1000 km during migration, and regional migrations of <500 km are common (Hutterer et al. 2005; Bisson et al. 2009). Over these relatively short distances, deviating from the great circle distance for optimal atmospheric conditions may not outweigh the benefits of other route selection criteria. For example, nectarivorous bats may follow the phenology of flowering plants (Fleming et al. 1993). Similarly, patterns of fatalities at wind turbines have found that migrating bats concentrate along the foothills of the Rocky Mountains where potential roost trees are more abundant than in the adjacent prairies, suggesting migratory route is influenced by the availability of roosting habitat (Baerwald and Barclay 2009). In Europe, migrating Nathusius' pipistrelles (*Pipistrellus nathusii*) concentrate along the coast of the Baltic Sea during migration (Pētersons 2004). Despite these specific examples, general migratory flyways have not been identified for bats as they have been for birds, which may indicate a more broad front migration in bats (as observed in Rydell et al. 2014), or the greater importance of smaller spatial scale patterns over the generally shorter migration distances. Understanding the patterns and drivers of spatial distribution during migratory periods remains a research priority (Arnett and Baerwald 2013).

Although migratory flyways determined by atmospheric conditions are not known for migratory bats, the aerosphere certainly has a large influence on bat migration. Several studies have attempted to understand migration patterns by considering atmospheric conditions. Several atmospheric factors including wind speed, wind direction and atmospheric pressure have been shown to influence migratory activity (e.g. Cryan and Brown 2007; Baerwald and Barclay 2011; Dechmann et al. 2017; Pettit and O'Keefe 2017). The arrival of Hoary Bats (*Lasiurus cinereus*) on an offshore island was associated with low barometric pressure (Cryan and Brown 2007), as was activity of Hoary Bats at a wind energy facility (Baerwald and Barclay 2011). In North America, low barometric pressure is associated with autumn storm fronts, suggesting that bats may take advantage of these storm fronts as they move south. These same autumn storm fronts indirectly influence migratory bat systems. To deposit the fat store required to fuel their migration, Brazilian free-tailed bats (*T. brasiliensis*) prey on waves of migrating moths following autumn cold front passages (Krauel et al. 2015). Thus, although the patterns are perhaps not as clear as for migratory birds, large-scale weather patterns influence migration in bats both directly and indirectly.

The seasonal regularity of such large-scale atmospheric patterns drives population-specific selection processes for innate changes in flight directions and optimal detours. The reliance of long-distance migrants on such patterns is also expressed by rare but regular weather-related mass-mortality events in migrants. Unseasonable cold weather, heavy precipitation, and/or wind conditions can kill thousands of migrants within a few days (reviewed by Newton 2007). More regularly, birds and bats are blown off their course by strong unfavourable winds, appearing as vagrants at sites far off their normal range. Birds are more conspicuous than bats, and with many observers storm-displaced birds are regularly documented. Although less frequently documented due to their cryptic nature, storm displacement of migrating bats appears to occur regularly (e.g. Van Gelder and Wingate 1961). In fact, storm displacements during migratory seasons may have contributed to the colonization of remote oceanic islands (Bonaccorso and McGuire 2013). Like normal migratory routes, vagrancy often follows directional patterns along well-known storm passages (Newton 2010).

3   Timing of Migration

Atmospheric conditions have a strong impact not only on the route but also on the timing of migration. Migration timing has a considerable influence on reproductive success and survival (Bauer et al. 2016). Inclement weather can cause heavy mortality during flight, displacement to unfavourable or unfamiliar habitats, and much more frequently delays in departure for migration. Delays occur not only through poor flight conditions but also due to a lack of food (Newton 2007).

The main task for a migrant is to be at the right time at the right place. This is an ambitious task because the temporal pattern and abundance of seasonal food resources exploited by migratory animals fluctuates among years. Variability in food resources makes it difficult to establish optimal strategies for an individual's annual cycle. In all migrants, physiological limits combine with behavioural responses and extrinsic factors to produce observed migration patterns (Bowlin et al. 2010). Mainly for short-lived species, like small passerines and some bats, behavioural patterns controlled by the annual innate program must account for variable actual environmental conditions. In general, long-distance migrants rely on an innate program as the main driver for the timing of annual cycle events, such as breeding, migration, and moult, while phenotypic flexibility allows short-distance migrants (and residents) to react to the actual environmental conditions (Tøttrup et al. 2008).

An important driver for this flexibility in behavioural traits is the dynamic nature of the aerosphere. The range and the frequency of specific atmospheric conditions (wind, precipitation, etc.) impair not only survival during migration but also future reproductive success. Therefore, the stochasticity of atmospheric conditions within and among years forces migratory birds and bats to have some flexibility in their timing for migration. The major question arising from this fact is: How can selection mould migratory strategies within a population by taking into account the stochasticity in the aerosphere? For short-lived species with only one or two reproductive cycles, the impact of weather conditions during migration can vary considerably between subsequent generations; e.g. arriving at the breeding site early might be favourable in 1 year but disastrous in the next. Therefore, many short-lived migratory birds show a high flexibility in the timing of migration, which seems to be strongly related to atmospheric conditions (Hüppop and Hüppop 2003). For example, within three subpopulations of barn swallows (*Hirundo rustica*) by far most of the variance in the timing of migration was explained by the year effect (Liechti et al. 2015).

Individuals must make departure and arrival decisions based on their innate program and informed by environmental conditions encountered en route (e.g. Tøttrup et al. 2008; Schmaljohann and Naef-Daenzer 2011; Eikenaar and Schmaljohann 2015). To understand the decision-making process, we need to know what an individual can sense and the specific decision rules. This prior knowledge is moulded by selection processes; e.g. individuals avoiding heavy rain or opposing winds will most likely have lower mortality risks. On the other hand, being very selective for favourable winds might delay departure with negative consequences on survival and reproduction. In 1975, an early, cold, and wet September reduced food abundance in western central Europe dramatically and consistently for several weeks. Thousands of barn swallows were not able to move southward and many of them starved (Bruderer 1975). Similarly, temperate bats may be limited by temperature and the associated availability of insect prey (Rydell et al. 2014). However, the use of torpor (reduced metabolic rate and body temperature) to save energy may allow some bats (and possibly some birds) to mitigate the consequences of unfavourable weather encountered en route (Geiser and Brigham 2012; McGuire et al. 2014). If bats use torpor to minimize the consequences of weather conditions experienced en route, migration phenology may be relatively consistent among years (Pētersons 2004; McGuire et al. 2012). Among migratory bats in the Baltic and North Sea regions of Europe, some species and sites varied in median migration date by only 1–3 days between years, while other species and site combinations differed by up to 15 days between years (Rydell et al. 2014).

Although bird migration has been studied intensively, only a few simple rules of thumb have been established as reaction norms up to now. Obviously, migratory soaring land birds need thermal updrafts and thus sunshine, while the huge proportion of nocturnal migrants prefers calm winds or tailwinds for migration (reviewed by Liechti 2006). Most migrants interrupt migration during heavy rain and avoid flying against strong headwinds (at least land birds over land; Richardson 1990). Headwinds can similarly affect migrating bats, potentially stalling migration for several nights at a time (e.g. Pētersons 2004). In a study of Common Noctules during spring migration, the likelihood of departure was greater with strong tailwinds (or the absence of strong headwinds when atmospheric pressure was low) (Dechmann et al. 2017). A study of American Redstarts (*Setophaga ruticilla*) and yellow-rumped Warblers (*Setophaga coronata*) identified tailwind assistance as an important predictor of departure decisions from stopover sites (Dossman et al. 2016). However, in the same region a study of Silver-haired bats (*Lasionycteris noctivagans*) did not find any influence of wind speed or wind direction on departure from the site (although sample size was relatively small; McGuire et al. 2012). Rather, departure from the stopover site was limited by rain, likely because the energetic costs of flight are doubled in bats that become wet due to rain (Voigt et al. 2011). This aspect of atmospheric influence on migration is specific to bats, rather than birds, as fur becomes wet and holds moisture while feathers shed raindrops.

Thus, it appears that the absence of headwinds and precipitation represents a simple rule of thumb for migration. However, reality is much more complex and simple reaction norms can easily be overruled or at least modified in response to geographic position, local food abundance, physiological state, and seasonal timing. Migrants must not only choose the right time period for migration, the conditions encountered *en route* must also be considered, as well as predictions of food abundance at forthcoming stopover sites and current energy stores. Departure decisions for a barrier crossing must be taken with much more care with respect to atmospheric conditions (Sjöberg et al. 2015; Dossman et al. 2016) than when crossing more favourable habitat with the opportunity to interrupt flight at any time.

The optimal timing of migration in relation to atmospheric conditions (mainly wind) and quality of stopover sites has been modelled in various studies (e.g. Weber and Hedenström 2000; Erni et al. 2003), but it is only recently that predictions have been tested with real data from individual tracks (McLaren et al. 2016). Long-lived species can accumulate experiences over several annual cycles and, thus, could adapt their behaviour correspondingly, e.g. how to avoid or compensate for wind drift (Alerstam 1979; Liechti 2006; Horvitz et al. 2014). Currently, there is little information available about the individual learning process in migratory birds and bats, but with increasing individual lifetime tracking much more data will be available soon to investigate how migrants adjust or learn to adjust their timing to atmospheric conditions (Mitchell et al. 2015).

4   Flight Behaviour

Flight only accounts for a fraction of the total time and energy costs of migration (Hedenström and Alerstam 1997; Wikelski et al. 2003). Nonetheless, the rate of energy consumption per distance flown has a strong and direct impact on the amount of refuelling time needed at subsequent stopover sites and the number stopovers required (Shamoun-Baranes et al. 2010). Therefore, optimizing flights with respect to atmospheric conditions not only includes the selection of the best temporal (timing) pattern and large-scale spatial factors (migratory route) but also selecting the optimal flight altitude with the best wind support. Within the planetary boundary layer where almost all migration takes place, winds are of the same order of magnitude as the airspeeds of birds or bats. Wind flow therefore has a strong influence on flight costs for migrants. Wind conditions aloft can easily double (headwinds) or halve (tailwinds) flight costs and time needed for a specific flight stage. A migrant must consider the impact of wind on ground speed and track direction when choosing a flight altitude (Alerstam 1979). An individual migrant can either compensate for lateral drift or allow partial or full drift. In theory, all of these behavioural patterns can be profitable within a given frame of assumptions (reviewed by Alerstam 2011). Several studies have shown that birds only partially compensate for wind drift but select specific altitudinal layers with the most profitable wind support (reviewed by Liechti 2006; Kemp et al. 2013; Dokter et al. 2013; Safi et al. 2013; Horton et al. 2016a). Selecting an optimal flight altitude is often a trade-off between maximizing ground speed and minimizing lateral drift. Wind, and not temperature, relative humidity, or water content, explains the height distribution of diurnal (Mateos-Rodriguez and Liechti 2012) and nocturnal bird migration (e.g. Bruderer et al. 1995; Liechti et al. 2000; Schmaljohann et al. 2009; Kahlert et al. 2012; Dokter et al. 2013; Horton et al. 2016b) in most studies (but see Kemp et al. 2013). Furthermore, selection for flight altitude happens even during a single nocturnal flight (Bowlin et al. 2015). Although migrating birds often select lower altitudes (<1500 m) to benefit from tailwinds, many birds are observed migrating at altitudes up to 5000 m (Schmaljohann et al. 2009; Bruderer and Peter 2017). The classic example of high-altitude migration is bar-headed geese (*Anser indicus*) that migrate over the Himalayas (Hawkes et al. 2011), at a maximum recorded altitude of 7290 m (Hawkes et al. 2013). However, while physically able to fly at extreme altitudes, geese follow a route that is longer than the great circle distance, do not select flight altitudes to benefit from favourable tailwinds, and generally do not appear to fly higher than required by the mountainous terrain (Hawkes et al. 2013). Nonetheless, observations of bird migrants crossing the Negev desert (Israel) at heights up to 8000 m asl and taking advantage of a low-level jet stream (Liechti and Schaller 1999) suggest that physiological adaptations to very high altitudes are within the range of several bird species.

While rarely considered in bats, the capacity for high-altitude flight may be an aspect in which migrating birds and bats interact differently with the aerosphere. Bats never reach the highest altitudes observed for migrating birds, and there are few records of bats flying at even moderate altitudes. Among the few recorded observations, Brazilian free-tailed bats (*Tadarida brasiliensis*) forage on moths migrating at high altitude, up to ~1000 m (McCracken et al. 2008), and have been observed on radar at altitudes up to 3000 m (Williams et al. 1973). Based on aircraft strikes in Australia, the highest observed altitude for flying foxes (Pteropodidae) was 1524 m and most strikes occurred below 300 m (Parsons et al. 2008). With limited observations and generally lower flight altitudes, one might hypothesize that a morphological or physiological difference between bats and birds allows birds access to a greater range of the aerosphere than is typically available to bats.

Lung morphology and function is a key difference between bats and birds that may influence how these groups of migrants interact with the aerosphere. The mammalian lung is pliable with bidirectional flow, while the avian lung is more rigid with unidirectional flow (Maina 2000). However, phenotypic flexibility is an intriguing possibility. While the size of many organs changes during migration for birds (e.g. Piersma and Gill 1998; Bauchinger et al. 2005; McWilliams and Karasov 2005), the lungs are one organ that has not been observed to vary seasonally. By comparison, increased lung mass was observed in migrating hoary bats (*Lasiurus cinereus*), which may be related to compensation for lower oxygen concentration that would be encountered by bats if they fly at higher altitudes during migration (McGuire et al. 2013a). This idea needs to be further explored or tested.

Regardless of flight altitude, migrants must adjust their heading and airspeed to compensate for lateral wind drift (Liechti 2006). The extent to which lateral drift should be fully or partially compensated cannot be understood without knowing the final goal of an individual migrant (Alerstam 1979). As long as the expected goal (breeding, residency, or stopover site) of the tracked individual is unknown, the behavioural patterns can only be interpreted qualitatively (e.g. Green et al. 2004) or theoretically (McLaren et al. 2012). Adapting heading and airspeed has important consequences for the migratory route, the number of flight stages, and flight energetic costs. Nonetheless, birds and bats undertake flights under varying atmospheric conditions and seem to be able to compensate for suboptimal atmospheric conditions en route to a much greater extent than other migratory groups such as insects (Sapir et al. 2014; Chapman et al. 2016; Chap. 7).

5   Sensing Atmospheric Conditions

Behavioural adaptations to atmospheric conditions are constrained by the ability of migrants to perceive relevant stimuli. Numerous studies have clearly shown that migratory birds respond to atmospheric conditions (e.g. Bruderer 1971; Kerlinger and Moore 1989; Richardson 1990; Schaub et al. 2004; Breuner et al. 2013), but there is still very little known about the relevant sensory systems and the sensitivity of these systems. Sensitivity to changes in barometric pressure has been examined in detail only in pigeons (Kreithen and Keeton 1974), and functional receptors for both bats and birds are assumed to be in the inner ear (Vitali 1911; Paige 1995; von Bartheld and Giannessi 2011). Nowadays, it is widely accepted that most birds can sense even small changes in air pressure (O'Neil 2013) and that they do react to approaching cold fronts probably due to decreasing air pressure (reviewed by Richardson 1990).

Short-term flights of nocturnal long-distance migrants in the beginning of the night indicate that birds explore the air column to decide whether or not to embark for a barrier crossing (Schmaljohann et al. 2010). These flights are assumed to check wind conditions aloft. However, for an animal moving within the air flow, the detection of the speed and the direction of the flow is a difficult task (Liechti 2006). Mechanoreceptors near feather follicles (Brown and Fedde 1993) and on bat wing surface (Sterbing-D'Angelo et al. 2011) provide information of air flow around the body. For birds, these sensors allow for measuring wind conditions while still sitting on the ground. Whether or not this also allows measuring small-scale turbulences related to height levels with changing winds (Elkins 1988) aloft is unknown. Anecdotal observations of nocturnal migrants moving backwards above an extensive fog layer (pers. obs.) indicate that birds need ground features to estimate wind displacement.

Apparently birds and bats can sense temperature and relative humidity. Both parameters have little direct effect on flight costs but strongly affect water balance (Torre-Bueno 1978; Engel et al. 2006). A challenge faced by many migrants is maintaining osmotic homeostasis, coping with low humidity in the aerosphere (Gerson and Guglielmo 2011). Migrating birds and bats alike meet the energetic demands of migration by relying on stored fat to fuel endurance flights (Jenni and Jenni-Eiermann 1998; McGuire and Guglielmo 2009; McGuire et al. 2013a, b; cf. Chap. 5). Indeed, the common selective pressures of vertebrate flight are expected to result in evolutionary convergence among these groups (McGuire and Guglielmo 2009). Yet, basic differences in physiology and morphology will affect these groups differently. For birds flying under conditions where water balance would be compromised (e.g. over deserts, possibly long-distance trans-oceanic flights), protein catabolism is an important source of endogenous water production (Gerson and Guglielmo 2011). The protein-for-water strategy is an effective method for migrating birds to deal with the challenging conditions in the aerosphere because most birds are generally uricotelic (Wright 1995), expelling nitrogenous waste as uric acid with little associated water requirement. Conversely, mammals are generally ureotelic, producing urea which requires large amounts of water to excrete. Therefore, protein-for-water is likely a strategy only available to birds, and bats may be more selective of atmospheric layers based on relative humidity.

Based on physiological models, it has been assumed that water balance might be a limiting factor for migratory birds (Klaassen 1995), but there is little empirical evidence for flight strategies governed by water loss (Landys et al. 2000; Liechti et al. 2000). Birds migrating across the Sahara desert select flight altitudes that would increase dehydration rates but provide favourable tailwinds (Schmaljohann et al. 2009). However, differences in the number of birds performing early morning flights when crossing the Sahara between autumn and spring may indicate restricted flight times due to constraints in water balance (Schmaljohann et al. 2009; Adamík et al. 2016). The ability to increase protein catabolism to maintain water balance in these dehydrating conditions allows birds to take advantage of a range of options in the aerosphere that migrating bats would not have access to. While first consideration of the aerosphere may suggest a relatively homogenous and open space for vertebrate migrants, ambient humidity conditions are an important factor that may be a barrier to bats and require physiological adjustment by migrating birds.

6   Conclusions

There are many important similarities but also key differences in the way that atmospheric conditions affect the aeroecology of migrating birds and bats. Wind direction and wind speed affect birds and bats similarly, providing tailwind assistance, headwind challenges, and the need to compensate or account for drift. But the scale of migrations may be an important factor. The periodic seasonal and spatial pattern of large-scale atmospheric conditions shapes migratory flyways, mainly of long-distance bird migrants benefiting from regular tailwinds. For migrating bats that travel relatively shorter distances, optimal wind conditions may not offset the necessary deviations from the shorter great circle distances, or other important factors (e.g. stopover habitat en route). Short-term weather conditions have a fundamental influence on the timing of individual flight stages, which can lead to peak migration events with millions of birds and/or bats moving at the same time. Similarly, storms may regularly displace migrating bats and birds, far off their intended course. For optimizing wind support, birds appear to select from a larger altitude range than bats and may select winds that would be otherwise unavailable to migrating bats due to humidity challenges. On the other hand, bats can endure harsh weather conditions better by using torpor to save energy. Clearly, there are important similarities and differences in the way that these two groups of vertebrate migrants interact with the aerosphere. With the limited knowledge of bat migration further comparison is difficult at present, but as migratory bats receive further research attention we anticipate many novel and interesting findings.

To further understand migration with respect to atmospheric conditions, long-term and large-scale studies are needed. Continent-wide radar surveillance networks allow us to follow, in almost real time, the mass movements of millions of migrants in context with weather conditions (Kelly et al. 2012; Alves et al. 2016). Lightweight tracking devices give more and more insight into the individual behaviour of even small passerines during migration, and the establishment of continental-scale telemetry networks (e.g. motus-wts.org; Taylor et al. 2017) promises to provide many novel insights. Due to the stochasticity of weather conditions, it will take multiple years before we better understand how birds and bats "know" when to move and when to stay.

Continued investigation of the aeroecology of vertebrate migrants promises valuable insights, both basic and applied. From an applied perspective, wind energy is an excellent renewable resource provided by the aerosphere, but it has become clear that very large numbers of migrants (especially bats) are killed by turbines every year (Lehnert et al. 2014; Arnett and Baerwald 2013, cf. Chap. 18). Wind turbine mortality presents a major conservation concern but may also provide information about the aeroecology of migrants that are otherwise rarely encountered. From a more fundamental perspective, one of the most interesting questions might be to understand how evolutionary processes mould optimal behaviour within this highly dynamic but not stochastic system. How do experience and reaction norms account for adaptations, particularly in the context of short- and long-lived species? Flying vertebrate migrants present a study system where animal behaviour and physiology are strongly influenced by environmental conditions in the aerosphere, affecting fitness and thus leading to adaptation and evolution. Thus, the aeroecology of vertebrate migrants presents an exciting opportunity for integrative investigations, from behaviour and physiology through to ecology and evolution.

Migration biology remains a topic of great interest to biologists, and the already extensive body of literature continues to grow. As previously understudied areas receive more attention (e.g. bat migration) and more researchers explicitly consider their studies in the context of aeroecology, we will continue to develop novel insights. As for many systems, technological advancements will play a role in novel investigations. Miniaturized tracking devices or improved tracking systems will generate improved lifetime track data that is now only sparsely available. Current studies are necessarily limited to large bodied animals, or brief snapshots of long migratory journeys. With the dynamic nature of the aerosphere and the diversity of vertebrate migrants, long-term studies (both within and among seasons, and multi-year studies) will be highly valuable contributions. Ultimately, the combination of increasing efforts in understudied areas, integration among scientific disciplines, improved technology, and a greater set of long-term data will lead to greater understanding, though we will surely continue to marvel at the fantastic feats of flying vertebrate migrants.

Acknowledgements We thank P. Chilson for the initiative to compile this book, and J. Kelly and W. Frick for revising our manuscript. We are also grateful for the many insightful and stimulating conversations with colleagues and collaborators that have contributed to the ideas for this chapter.

References

Adamík P, Emmenegger T, Briedis M, Gustafsson L, Henshaw I, Krist M, Laaksonen T, Liechti F, Procházka P, Salewski V, Hahn S (2016) Barrier crossing in small avian migrants: individual tracking reveals prolonged nocturnal flights into the day as a common migratory strategy. Sci Rep 6:21560

Alerstam T (1979) Wind as a selective agent in bird migration. Ornis Scand 10:76–93

Alerstam T (1990) Bird migration. Cambridge University Press, Cambridge

Alerstam T (2011) Optimal bird migration revisited. J Ornithol 152:5–23

Alerstam T, Hedenström A, Åkesson S (2003) Long-distance migration: evolution and determinants. Oikos 103:247–260

Alerstam T, Rosen M, Bäckman J, Ericson PGP, Hellgren O (2007) Flight speeds among bird species: allometric and phylogenetic effects. PLoS Biol 5:e197

Alexander RM (2006) Principles of animal locomotion, 2nd edn. University Press, Princeton

Alves JA, Shamoun-Baranes J, Desmet P, Dokter A, Bauer S, Hüppop O et al (2016) Monitoring continent - wide aerial patterns of bird movements usi ng weather radars. In: Proceedings of the BOU's 2015 Annual Conference, #BOU2015 2015, pp 1–5

Arnett EB, Baerwald EF (2013) Impacts of wind energy development on bats: implications for conservation. In: Adams RA, Pedersen SC (eds) Bat evolution, ecology, and conservation. Springer, New York, pp 435–456

Arnett EB, Baerwald EF, Matthews F, Rodrigues L, Rodríguez-Durán A, Rydell J, Villegas-Patraca R, Voigt CC (2016) Impacts of wind energy development on bats: a global perspective.

In: Voigt CC, Kingston T (eds) Bats in the Anthropocene: conservation in a Changing World. Springer, Cham, pp 295–324

Baerwald EF, Barclay RMR (2009) Geographic variation in activity and fatality of migratory bats at wind energy facilities. J Mammal 90:1341–1349

Baerwald EF, Barclay RMR (2011) Patterns of activity and fatality of migratory bats at a wind energy facility in Alberta, Canada. J Wildl Manag 75:1103–1114

Battley PF, Warnock N, Tibbitts TL, Gill RE, Piersma T, Hassell CJ, Douglas DC, Mulcahy DM, Gartrell BD, Schuckard R, Melville DS, Riegen AC (2012) Contrasting extreme long-distance migration patterns in bar-tailed godwits Limosa lapponica. J Avian Biol 43:21–32

Bauchinger U, Wohlmann A, Biebach H (2005) Flexible remodeling of organ size during spring migration of the garden warbler (*Sylvia borin*). Zoology 108:97–106

Bauer S, Lisovski S, Hahn S (2016) Timing is crucial for consequences of migratory connectivity. Oikos 125:605–612

Berthold P (2001) Bird migration—a general survey, 2nd edn. Oxford University Press, Oxford

Bishop CM, Spivey RJ, Hawkes LA, Batbayar N, Chua B, Frappell PB, Milsom WK, Natsagdorj T, Newman SH, Scott GR, Takekawa JY, Wikelski M, Butler PJ (2015) The roller coaster flight strategy of bar-headed geese conserves energy during Himalayan migrations. Science 347:250–254

Bisson I-A, Safi K, Holland RA (2009) Evidence for repeated independent evolution of migration in the largest family of bats. PLoS One 4:e7504

Bonaccorso FJ, McGuire LP (2013) Modeling the colonization of Hawaii by hoary bats (*Lasiurus cinereus*). In: Adams RA, Pedersen SC (eds) Bat ecology, evolution, and conservation. Springer, New York, pp 187–206

Bowlin MS, Bisson I-A, Shamoun-Baranes J, Reichard JD, Sapir N, Marra PP, Kunz TH, Wilcove DS, Hedenström A, Guglielmo CG, Åkesson S, Ramenofsky M, Wikelski M (2010) Grand challenges in migration biology. Integr Comp Biol 50:261–279

Bowlin MS, Enstrom DA, Murphy BJ, Plaza E, Jurich P, Cochran J (2015) Unexplained altitude changes in a migrating thrush: long-flight altitude data from radio-telemetry. Auk 132:808–816

Breuner CW, Sprague RS, Patterson SH, Woods HA (2013) Environment, behavior and physiology: do birds use barometric pressure to predict storms? J Exp Biol 216:1982–1990

Brown RE, Fedde MR (1993) Airflow sensors in the avian wing. J Exp Biol 179:13–30

Bruderer B (1971) Radarbeobachtungen über den Frühlingszug im Schweizerischen Mittelland. (Ein Beitrag zum Problem der Witterungsabhängigkeit des Vogelzugs). Ornithol Beob 68:89–158

Bruderer B (1975) Zur Schwalbenkatastrophe im Herbst 1974. Tierwelt 4–6:1–20

Bruderer B, Boldt A (2001) Flight characteristics of birds: I. Radar measurements of speeds. Ibis 143:178–204

Bruderer B, Underhill LG, Liechti F (1995) Altitude choice of night migrants in a desert area predicted by meteorological factors. Ibis 137:44–55

Bruderer B, Peter D (2017) Windprofit favouring extreme altitudes of bird migration. Ornithologische Beobachter 114:73–86

Byng JW, Racey PA, Swaine MD (2009) The ecological impacts of a migratory bat aggregation on its seasonal roost in Kasanka National Park, Zambia. Afr J Ecol 48:29–36

Chapman JW, Nilsson C, Lim KS, Bäckman J, Reynolds DR, Alerstam T (2016) Adaptive strategies in nocturnally migrating insects and songbirds: contrasting responses to wind. J Anim Ecol 85:115–124

Cockrum EL (1969) Migration in the guano bat, *Tadarida brasiliensis*. Misc Pub Univ Kansas Museum Nat Hist 51:303–336

Cryan PM, Brown AC (2007) Migration of bats past a remote island offers clues toward the problem of bat fatalities at wind turbines. Biol Conserv 139:1–11

Davenport J (1994) How and why do flying fish fly? Rev Fish Biol Fish 4:184–214

Dechmann DKN, Wikelski M, Ellis-Soto D, Safi K, Teague O'Mara M (2017) Determinants of spring migration departure decision in a bat. Biol Lett 13(9):20170395

Delingat J, Bairlein F, Hedenström A (2008) Obligatory barrier crossing and adaptive fuel management in migratory birds: the case of the Atlantic crossing in Northern Wheatears (*Oenanthe oenanthe*). Behav Ecol Sociobiol 62:1069–1078

DeLuca WV, Woodworth BK, Rimmer CC, Marra PP, Taylor PD, McFarland KP, Mackenzie SA, Norris DR (2015) Transoceanic migration by a 12 g songbird. Biol Lett 11:20141045

Dokter AM, Shamoun-Baranes J, Kemp MU, Tijm S, Holleman I (2013) High altitude bird migration at temperate latitudes: a synoptic perspective on wind assistance. PLoS One 8:e52300

Dossman BC, Mitchell GW, Norris DR, Taylor PD, Guglielmo CG, Matthews SN, Rodewald PG (2016) The effects of wind and fuel stores on stopover departure behavior across a migratory barrier. Behav Ecol 27:567–574

Dudley R, Byrnes G, Yanoviak SP, Borrell B, Brown RM, McGuire JA (2007) Gliding and the functional origins of flight: biomechanical novelty or necessity? Annu Rev Ecol Evol Syst 38:179–201

Egevang C, Stenhouse IJ, Phillips RA, Petersen A, Fox JW, Silk JRD (2010) Tracking of Arctic terns *Sterna paradisaea* reveals longest animal migration. Proc Natl Acad Sci USA 107:2078–2081

Eikenaar C, Schmaljohann H (2015) Wind conditions experienced during the day predict nocturnal restlessness in a migratory songbird. Ibis 157:125–132

Elkins N (1988) Can high-altitude migrants recognize optimum flight levels? Ibis 130:562–563

Engel S, Herbert B, Visser GH (2006) Metabolic costs of avian flight in relation to flight velocity: a study in Rose Coloured Starlings (Sturnus roseus, Linnaeus). J Comp Physiol B 176:415–427

Erni B, Liechti F, Bruderer B (2003) How does a first year passerine migrant find its way? Simulating migration mechanisms and behavioural adaptations. Oikos 103:333–340

Erni B, Liechti F, Bruderer B (2005) The role of wind in passerine autumn migration between Europe and Africa. Behav Ecol 16:732–740

Finlayson JC, Garcia EFJ, Mosquera MA, Bourne WRP (1976) Raptor migration across the Strait of Gibraltar. Br Birds 69:77–87

Fleming TH, Eby P (2003) Ecology of bat migration. In: Kunz TH, Fenton MB (eds) Bat ecology. University of Chicago Press, Chicago, pp 156–208

Fleming TH, Nuñez RA, Sternberg LSL (1993) Seasonal changes in the diets of migrant and non-migrant nectarivorous bats as revealed by carbon stable isotope analysis. Oecologia 74:72–75

Frick WF, Baerwald EF, Pollock JF, Barclay RMR, Szymanski JA, Weller TJ, Russell AL, Loeb SC, Medellin RA, McGuire LP (2017) Fatalities at wind turbines may threaten population viability of a migratory bat. Biol Conserv 209:172–177

Frick WF, Chilson PB, Fuller NW, Bridge ES, Kunz TH (2013) Aeroecology. In: Adams RA, Pedersen SC (eds) Bat ecology, evolution, and conservation. Springer, New York, pp 149–168

Gaston KJ, Blackburn TM (1997) How many birds are there? Biodivers Conserv 6:615–625

Geiser F, Brigham RM (2012) The Other Functions of Torpor. In: Ruf T, Bieber C, Arnold W, Millesi E (eds) Living in a Seasonal World. Springer, Heidelberg

Gerson AR, Guglielmo CG (2011) House sparrows (Passer domesticus) increase protein catabolism in response to water restriction. Am J Phys 300:R925–R930

Giavi S, Moretti M, Bontadina F, Zambelli N, Schaub M (2014) Seasonal survival probabilities suggest low migration mortality in migrating bats. PLoS One 9:e85628

Gill RE, Piersma T, Hufford G, Servranckx R, Riegen A (2005) Crossing the ultimate ecological barrier: evidence for an 11000-km-long nonstop flight from Alaska to New Zealand and Eastern Australia by bar-tailed godwits. Condor 107:1–20

Gill RE Jr, Tibbitts TL, Douglas DC, Handel CM, Mulcahy DM, Gottschalck JC, Warnock N, McCaffery BJ, Battley PF, Piersma T (2009) Extreme endurance flights by landbirds crossing the Pacific Ocean: ecological corridor rather than barrier? Proc R Soc B 276:447–457

Gill J, Douglas DC, Handel CM, Tibbitts TL, Hufford G, Piersma T (2014) Hemispheric-scale wind selection facilitates bar-tailed godwit circum-migration of the Pacific. Anim Behav 90:117–130

Green M, Alerstam T, Gudmundsson GA, Hedenström A, Piersma T (2004) Do Arctic waders use adaptive wind drift? J Avian Biol 35:305–315

Greenberg R, Marra PP (2005) Birds of two worlds: the ecology and evolution of migration. The John Hopkins University Press, Baltimore

Grüebler MU, Korner-Nievergelt F, Naef-Daenzer B (2014) Equal nonbreeding period survival in adults and juveniles of a long-distant migrant bird. Ecol Evol 4:756–765

Hahn S, Bauer S, Liechti F (2009) The natural link between Europe and Africa – 2.1 billion birds on migration. Oikos 118:624–626

Hawkes LA, Balachandran S, Batbayar N, Butler PJ, Frappell PB, Milsom WK, Tseveenmyadag N, Newman SH, Scott GR, Sathiyaselvam P, Takekawa JY, Wikelski M, Bishop CM (2011) The trans-Himalayan flights of bar-headed geese (Anser indicus). Proc Natl Acad Sci 108 (23):9516–9519

Hawkes LA, Balachandran S, Batbayar N, Butler PJ, Chua B, Douglas DC, Frappell PB, Hou Y, Milsom WK, Newman SH, Prosser DJ, Sathiyaselvam P, Scott GR, Takekawa JY, Natsagdorj T, Wikelski M, Witt MJ, Yan B, Bishop CM (2013) The paradox of extreme high-altitude migration in bar-headed geese *Anser indicus*. Proc R Soc B 280:20122114

Hedenström A, Alerstam T (1997) Optimum fuel loads in migratory birds: distinguishing between time and energy minimization. J Theor Biol 189:227–234

Hedenström A, Norevik G, Warfvinge K, Andersson A, Bäckman J, Åkesson S (2016) Annual 10-Month Aerial Life Phase in the Common Swift Apus apus. Curr Biol 26(22):3066–3070

Horn JW, Kunz TH (2008) Analyzing NEXRAD Doppler radar images to assess nightly dispersal patterns and population trends in Brazilian free-tailed bats (*Tadarida brasiliensis*). Integr Comp Biol 48:24–39

Horton KG, Van Doren BM, Stepanian PM, Farnsworth A, Kelly JF (2016a) Where in the air? Aerial habitat use of nocturnally migrating birds. Biol Lett 12(11):20160591

Horton KG, Van Doren BM, Stepanian PM, Hochachka WM, Farnsworth A, Kelly JF (2016b) Nocturnally migrating songbirds drift when they can and compensate when they must. Sci Rep 6(1)

Horvitz N, Sapir N, Liechti F, Avissar R, Mahrer I, Nathan R (2014) The gliding speed of migrating birds: slow and safe or fast and risky? Ecol Lett 17:670–679

Hüppop O, Hüppop K (2003) North Atlantic Oscillation and timing of spring migration in birds. Proc R Soc B Biol Sci 270(1512):233–240

Hutterer R, Teodora I, Meyer-Cords C, Rodrigues L (2005) Bat migrations in Europe. Naturschutz und Biologische Vielfalt, Bonn

Jenni L, Jenni-Eiermann S (1998) Fuel supply and metabolic constraints in migrating birds. J Avian Biol 29:521–528

Kahlert J, Leito A, Laubek B, Luigujõe L, Kuresoo A, Aaen K, Luud A (2012) Factors affecting the flight altitude of migrating waterbirds in Western Estonia. Ornis Fenn 89:241–253

Kelly JF, Ryan Shipley J, Chilson PB, Howard KW, Frick WF, Kunz TH (2012) Quantifying animal phenology in the aerosphere at a continental scale using NEXRAD weather radars. Ecosphere 3(2):art16

Kemp MU, Shamoun-Baranes J, Dokter AM, van Loon E, Bouten W (2013) The influence of weather on the flight altitude of nocturnal migrants in mid-latitudes. Ibis 155:734–749

Kerlinger P, Moore FR (1989) Atmospheric structure and avian migration. In: Power DM (ed) Current ornithology. Plenum Press, New York, pp 109–142

Klaassen M (1995) Water and energy limitations on flight range. Auk 112:260–262

Koleček J, Procházka P, El-Arabany N, Tarka M, Ilieva M, Hahn S, Honza M, de la Puente J, Bermejo A, Gürsoy A, BenschS, Zehtindjiev P, Hasselquist D, Hansson B (2016) Cross-continental migratory connectivity and spatiotemporal migratory patterns in the great reed warbler. J Avian Biol. https://doi.org/10.1111/jav/00929

Kopp M, Peter HU, Mustafa O, Lisovski S, Ritz MS, Phillips RA, Hahn S (2011) South polar skuas from a single breeding population overwinter in different oceans though show similar migration patterns. Mar Ecol Prog Ser 435:263–267

Kranstauber B, Weinzierl R, Wikelski M, Safi K (2015) Global aerial flyways allow efficient travelling. Ecol Lett 18:1338–1345

Krauel JJ, McCracken GF (2013) Recent advances in bat migration research. In: Adams RA, Pedersen SC (eds) Bat ecology, evolution, and conservation. Springer, New York, pp 293–314

Krauel JJ, Westbrook JK, McCracken GF (2015) Weather-driven dynamics in a dual-migrant system: moths and bats. J Anim Ecol 84:604–614

Kreithen ML, Keeton WT (1974) Detection of atmospheric pressure by the homing pigeon, *Columba liviia*. J Comp Physiol 89:73–82

Landys MM, Piersma T, Visser GH, Jukema J, Wijker A (2000) Water balance during real and simulated long-distance migratory flight in the bar-tailed godwit. Condor 102:645–652

Lehnert LS, Kramer-Schadt S, Schönborn S, Lindecke O, Niermann I, Voigt CC (2014) Wind farm facilities in Germany kill noctule bats from near and far. PLoS One 9:e103106

Liechti F (2006) Birds: blowin' by the wind? J Ornithol 147:202–211

Liechti F, Schaller E (1999) The use of low-level jets by migrating birds. Naturwissenschaften 86:549–551

Liechti F, Klaassen M, Bruderer B (2000) Predicting migratory flight altitudes by physiological migration models. Auk 117:205–214

Liechti F, Witvliet W, Weber R, Bächler E (2013) First evidence of a 200-day non-stop flight in a bird. Nat Commun 4:2554

Liechti F, Scandolara C, Rubolini D, Ambrosini R, Korner-Nievergelt F, Hahn S, Lardelli R, Romano M, Caprioli M, Romano A, Sicurella B, Saino N (2015) Timing of migration and residence areas during the non-breeding period of barn swallows Hirundo rustica in relation to sex and population. J Avian Biol 46:254–265

Maina JN (2000) What it takes to fly: the structural and functional respiratory refinements in birds and bats. J Exp Biol 203:3045–3064

Mateos-Rodriguez M, Liechti F (2012) How do diurnal long-distance migrants select flight altitude in relation to wind? Behav Ecol 23:403–409

McCracken GF, Gillam EH, Westbrook JK, Lee Y-F, Jensen ML, Balsley BB (2008) Brazilian free-tailed bats (*Tadarida brasiliensis*: Molossidae, Chiroptera) at high altitude: links to migratory insect populations. Integr Comp Biol 48:107–118

McCracken GF, Safi K, Kunz TH, Dechmann DKN, Swartz SM, Wikelski M (2016) Airplane tracking documents the fastest flight speeds recorded for bats. R Soc Open Sci 3(11):160398

McGuire LP, Guglielmo CG (2009) What can birds tell us about the migration physiology of bats? J Mammal 90:1290–1297

McGuire LP, Ratcliffe JM (2011) Light enough to travel: migratory bats have smaller brains, but not larger hippocampi, than sedentary species. Biol Lett 7:233–236

McGuire LP, Guglielmo CG, Mackenzie SA, Taylor PD (2012) Migratory stopover in the long-distance migrant silver-haired bat, *Lasionycteris noctivagans*. J Anim Ecol 81:377–385

McGuire LP, Fenton MB, Guglielmo CG (2013a) Phenotypic flexibility in migrating bats: seasonal variation in body composition, organ sizes and fatty acid profiles. J Exp Biol 216:800–808

McGuire LP, Fenton MB, Guglielmo CG (2013b) Seasonal upregulation of catabolic enzymes and fatty acid transporters in the flight muscle of migrating hoary bats, *Lasiurus cinereus*. Comp Biochem Physiol B 165:138–143

McGuire LP, Jonasson KA, Guglielmo CG (2014) Bats on a budget: torpor-assisted migration saves time and energy. PLoS One 9:e115724

McLaren JD, Shamoun-Baranes J, Bouten W (2012) Wind selectivity and partial compensation for wind drift among nocturnally migrating passerines. Behav Ecol 23:1089–1101

McLaren JD, Shamoun-Baranes J, Camphuysen CJ, Bouten W (2016) Directed flight and optimal airspeeds: homeward-bound gulls react flexibly to wind yet fly slower than predicted. J Avian Biol. https://doi.org/10.1111/jav.00828

McWilliams SR, Karasov WH (2005) Migration takes guts – digestive physiology of migratory birds and its ecological significance. In: Greenberg R, Marra PP (eds) Birds of two worlds – the ecology and evolution of migration. Johns Hopkins University Press, Baltimore, pp 67–78

Mitchell GW, Woodworth BK, Taylor PD, Norris DR (2015) Automated telemetry reveals age specific differences in flight duration and speed are driven by wind conditions in a migratory songbird. Mov Ecol 3:1–13

Newton I (2007) Weather-related mass-mortality events in migrants. Ibis 149:453–467

Newton I (2010) Bird migration. Harper Collins, London

Nourani E, Yamaguchi NM, Manda A, Higuchi H (2016) Wind conditions facilitate the seasonal water-crossing behaviour of Oriental Honey-buzzards Pernis ptilorhynchus over the East China Sea. Ibis 158:506–518

O'Neill P (2013) Magnetoreception and baroreception in birds. Dev Growth Differ 55:188–197

O'Shea TJ, Bogan MA, Ellison LE (2003) Monitoring trends in bat populations of the United States and territories: status of the science and recommendations for the future. Wildlife Soc B 31:16–29

Paige KN (1995) Bats and barometric pressure: conserving limited energy and tracking insects from the roost. Funct Ecol 9:463–467

Parsons JG, Blair D, Luly J, Robson SKA (2008) Flying-fox (Megachiroptera: Pteropodidae) flight altitudes determined via an unusual sampling method: aircraft strikes in Australia. Acta Chiropterol 10:377–379

Pennycuick CJ, Åkesson S, Hedenström A (2013) Air speeds of migrating birds observed by ornithodolite and compared with predictions from flight theory. J R Soc Interface 10:20130419

Pētersons G (2004) Seasonal migrations of north-eastern populations of Nathusius' bat *Pipistrellus nathusii* (Chiroptera). Myotis 41–42:29–56

Pettit JL, O'Keefe JM (2017) Day of year, temperature, wind, and precipitation predict timing of bat migration. J Mammal 98(5):1236–1248

Piersma T, Gill RE Jr (1998) Guts don't fly: small digestive organs in obese bar-tailed godwits. Auk 115:196–203

Portugal SJ, Green JA, White CR, Giullemette M, Butler PJ (2012) Wild geese do not increase flight behaviour prior to migration. Biol Lett 8:469–472

Richardson WJ (1990) Timing of bird migration in relation to weather: updated review. In: Gwinner E (ed) Bird migration. Springer, Berlin, pp 78–101

Richter HV, Cumming GS (2006) Food availability and annual migration of the straw-colored fruit bat (*Eidolon helvum*). J Zool 268:35–44

Rydell J, Bach L, Bach P, Diaz LG, Furmankiewicz J, Hagner-Wahlsten N, Kyheröinen E-M, Lilley T, Masing M, Meyer MM, Pētersons G, Šuba J, Vasko V, Vintulis V, Hedenström A (2014) Phenology of migratory bat activity across the Baltic Sea and south-eastern North Sea. Acta Chiropterol 16:139–147

Safi K, Kranstauber B, Weinzierl R, Griffin L, Rees EC, Cabot D, Cruz S, Proaño C, Takekawa JY, Newman SH, Waldenström J, Bengtsson D, Kays R, Wikelski M, Bohrer G (2013) Flying with the wind: scale dependency of speed and direction measurements in modelling wind support in avian flight. Mov Ecol 1:1–13

Sapir N, Horvitz N, Dechmann DKN, Fahr J, Wikelski M (2014) Commuting fruit bats beneficially modulate their flight in relation to wind. Proc R Soc B 281:20140018

Schaub M, Liechti F, Jenni L (2004) Departure of migrating European robins, *Erithacus rubecula*, from a stopover site in relation to wind and rain. Anim Behav 67:229–237

Schmaljohann H, Naef-Daenzer B (2011) Body condition and wind support initiate the shift of migratory direction and timing of nocturnal departure in a songbird. J Anim Ecol 80:1115–1122

Schmaljohann H, Liechti F, Bruderer B (2009) Trans-Sahara migrants select flight altitudes to minimize energy costs rather than water loss. Behav Ecol Sociobiol 63:1609–1619

Schmaljohann H, Becker PJJ, Karaardic H, Liechti F, Naef-Daenzer B, Grandío JM (2010) Nocturnal exploratory flights, departure time, and direction in a migratory songbird. J Ornithol 152:439–452

Shamoun-Baranes J, Leyrer J, van Loon E, Bocher P, Robin F, Meunier F, Piersma T (2010) Stochastic atmospheric assistance and the use of emergency staging sites by migrants. Proc R Soc B 277:1505–1511

Shamoun-Baranes J, Liechti F, Vansteelant WMG (2017) Atmospheric conditions create freeways, detours and tailbacks for migrating birds. J Comp Physiol A 203(6–7):509–529

Sillett TS, Holmes RT (2002) Variation in survivorship of a migratory songbird throughout its annual cycle. J Anim Ecol 71:296–308

Sjöberg S, Alerstam T, Åkesson S, Schulz A, Weidauer A, Coppack T, Muheim R (2015) Weather and fuel reserves determine departure and flight decisions in passerines migrating across the Baltic Sea. Anim Behav 104:59–68

Smith NG (1980) Hawk and vulture migrations in the Neotropics. In: Keast A, Morton ES (eds) Migrant birds in the Neotropics. Smithsonian Institution Press, Washington, pp 51–65

Sterbing-D'Angelo S, Chadha M, Chiu C, Falk B, Xian W, Barcelo J, Zook JM, Moss CF (2011) Bat wing sensors support flight control. Proc Natl Acad Sci USA 108:11291–11296

Suter E (1957) Radar Beobachtungen über den Verlauf des nächtlichen Vogelzuges. Rev Suisse Zool 64:294–303

Taylor PD, Crewe TL, Mackenzie SA, Lepage D, Aubry Y, Crysler Z, Finney G, Francis CM, Guglielmo CG, Hamilton DJ, Holberton RL, Loring PH, Mitchell GW, Ryan Norris D, Paquet J, Ronconi RA, Smetzer JR, Smith PA, Welch LJ, Woodworth BK (2017) The Motus Wildlife Tracking System: a collaborative research network to enhance the understanding of wildlife movement. Avian Conservation and Ecology 12(1):8

Torre-Bueno JR (1978) Evaporative cooling and evaporative water loss in the flying birds. J Exp Biol 75:231–236

Tøttrup AP, Pedersen L, Onrubia A, Klaassen RHG, Thorup K (2017) Migration of red-backed shrikes from the Iberian Peninsula: optimal or sub-optimal detour? J Avian Biol 48(1):149–154

Tøttrup AP, Thorup K, Rainio K, Yosef R, Lehikoinen E, Rahbek C (2008) Avian migrants adjust migration in response to environmental conditions en roue. Biol Lett 4:685–688

Van Gelder RG, Wingate DB (1961) The taxonomy and status of bats in Bermuda. Am Mus Novit 2029:1–9

Vansteelant WMG, Shamoun-Baranes J, McLaren J, van Diermen J, Bouten W (2017) Soaring across continents: decision-making of a soaring migrant under changing atmospheric conditions along an entire flyway. J Avian Biol 48(6):887–896

Vitali G (1911) Di un interessante deivato della prima fessura branchiale nel passero. Anat Anz 39:219–224

Voigt CC, Schneeberger K, Voigt-Heucke SL, Lewanzik D (2011) Rain increases the energy cost of bat flight. Biol Lett 7:793–795

von Bartheld CS, Giannessi F (2011) The paratympanic organ: a barometer and altimeter in the middle ear of birds? J Exp Zool 316B:402–408

Weber TP, Hedenström A (2000) Optimal stopover decisions under wind influence: the effects of correlated winds. J Theor Biol 205:95–104

Weimerskirch H, Bishop C, Jeanniard-du-Dot T, Prudor A, Sachs G (2016) Frigate birds track atmospheric conditions over months-long transoceanic flights. Science 353:74–78

Wikelski M, Tarlow EM, Raim A, Diehl RH, Larkin RP, Visser GH (2003) Costs of migration in free-flying songbirds. Nature 423:704

Williams TC, Williams JM, Ireland LC, Teal JM (1977) Autumnal bird migration over the western North Atlantic Ocean. Am Birds 31:251–267

Williams TC, Ireland LC, Williams JM (1973) High altitude flights of the free-tailed bat, *Tadarida brasiliensis*, Observed with Radar. J Mammal 54(4):807–821

Wright PA (1995) Nitrogen excretion: three end products, many physiological roles. J Exp Biol 198:273–281

Part II

Methods of Observation

Aeroecological Observation Methods

9

V. Alistair Drake and Bruno Bruderer

Abstract

Observation of animals flying in the atmosphere is the core empirical process of aeroecology. For species that are small, or that fly by night or at high altitudes, this presents a considerable challenge. Even for the more visible species and for flights near the ground, recording the animals' movements requires specialised techniques. Fortunately, continuing rapid advances in radio and optical technologies, electronics, and computing are providing numerous opportunities for developing new and improved observing capabilities. The larger, more complete, and more precise observational datasets that these new technologies are providing underlie the current wave of discovery and growth in this novel discipline. This chapter is mainly concerned with methods for detecting and studying insects, birds, and bats flying in the open air, i.e. above the vegetation layer. Detection of these animals, and estimation of their numbers, can be achieved through in-flight capture or by remote sensing, with the latter comprising visual observation (including technologies for augmenting human sight), aural monitoring, radar, and laser/lidar. Remotely sensed animals can be identified, though sometimes only to a group of species, from characteristic features of the signals or images received. Information about the animals' activities—their mode of flight, orientation, etc.—can be obtained either by remote sensing or from sensors mounted on the animals. The latter method, which relies on radio-telemetry or archival logging to record the acquired data, may also be used to monitor the animal's physiological state, the environment it is moving in, and its

V. Alistair Drake (✉)
School of Physical, Environmental and Mathematical Sciences, University of New South Wales, Canberra, ACT, Australia

Institute for Applied Ecology, University of Canberra, Canberra, ACT, Australia
e-mail: a.drake@adfa.edu.au

B. Bruderer
Swiss Ornithological Institute, Sempach, Switzerland

© Springer International Publishing AG, part of Springer Nature 2017
P.B. Chilson et al. (eds.), *Aeroecology*,
https://doi.org/10.1007/978-3-319-68576-2_9

trajectory. The chapter also examines how information about the timing and geographical extent of movements, and the environmental conditions the animals are experiencing, can be obtained. Finally, the particular challenges of observational aeroecology are identified, the multidisciplinary nature of the observing task is recognised, and some possible developments are proposed.

1   Introduction

Aeroecology is concerned with the activity of living organisms within the atmosphere, especially with the active flight of winged animals such as insects, birds, and bats. This chapter deals with the methods that have been employed to study the presence and behaviour of such organisms in the atmosphere. Probably the most common function of these flights is migration, and the examples presented here are predominantly concerned with this aspect of aeroecology.

We first briefly consider the low-technology methods used to study the general timing and geographical extent of animal movements. Then we focus on *remote-sensing* techniques that provide more detailed information on the temporal and spatial distribution of animals in the air and their relation to weather conditions. The next step of methodological development was the study of flight behaviour, such as relating track direction to heading, groundspeed to airspeed, and flight altitude to the vertical variation of atmospheric conditions. As a complement to these behavioural studies, techniques for discriminating remotely sensed targets became available: e.g. insects *vs* birds *vs* bats, or passerine birds *vs* swifts *vs* shorebirds, or even more precise taxonomic groupings. More recently, *electronic tags* have been developed that allow trajectories for entire migration paths to be recorded for some species; the trajectories can then be related to the large-scale conditions along the routes. The latest tags can measure and record physiological functions and behaviours of a flying individual as well as the physical features of its immediate environment, and have potential to reveal new perspectives on migratory flight.

Insights about aeroecological phenomena, and especially animal behaviours, can be investigated through laboratory experiments (e.g. Reynolds and Riley 2002; Wiltschko and Wiltschko 2003; Engel et al. 2010; Guerra et al. 2014). Such methods have the advantage of allowing manipulation of the test animal's environment. This chapter will, however, be confined to methods applicable in the field.

2   Timing and Geographical Extent of Movements

Information about where animal movements start and finish, and their timing, is of particular value to population biologists, pest managers, dispersal ecologists, and conservation scientists; the latter especially will also need to know the general form of the route taken and critical stopover points. For aeroecologists, such information allows observations made at a particular locality to be related to large-scale migratory journeys, thus providing insights into the function of the observed flight activities. Knowing the overall timing of migration, the general directions and distances, and the environmental constraints en route facilitates interpretation of the observations of flight activities, in particular with respect to their adaptive value.

Various methods are available for determining that an individual animal, or a population, has moved from one location or region to another, and observations of this type were among the earliest contributions to migration and dispersal biology (e.g. Dorst 1962). *Observations of presence or absence* in different regions at different times of year, and transitory appearances at intermediate points, are often sufficient to reveal the annual pattern of seasonal movements between breeding and non-breeding areas (Alerstam 1990; Stefanescu et al. 2013). Internet-based cooperative recording systems such as *eBird* (http://ebird.org/), EuroBirdPortal (http://www.eurobirdportal.org/), and *eButterfly* (http://www.e-but terfly.org/) encourage contributions from "citizen scientists" and make them readily available for analyses, e.g. for recognising migration trajectories (La Sorte et al. 2016) and changing migration phenologies (Kelly et al. 2016). *Trapping* and recording of casualties (e.g. at man-made structures or following storms) provide additional information about the passage of inconspicuous species. Analysing the physiological state (e.g. weight, fat score, muscle score) of birds trapped before and after extended barrier crossings informs about the amount of fuel consumed per unit distance or time and thus about the significance of the barrier involved (Salewski et al. 2009). In the case of most insects, trapping (e.g. Harrington and Woiwod 2007) will usually replace field observation for establishing presence and absence; the interpretation of movement patterns for insect migrants, which have only limited control over their flight trajectories and whose life cycle often comprises more than one generation per year, is particularly challenging (e.g. Taylor 1986). *Marking* in one form or another (Bub 1995; Hagler and Jackson 2001) has been widely used to disclose point-to-point movements by individuals; cooperative schemes for marking birds with individually numbered leg bands are particularly well developed. Occasionally natural markers such as attached pollen grains on insects (e.g. Gregg et al. 2001) or biochemical markers in the tissue of migrant animals may be used as well. Stable isotopes of some abundant elements (e.g. hydrogen, carbon, nitrogen, strontium) vary across broad geographical regions or between habitats and food types (Hobson 1999; Pekarsky et al. 2015). If the geographical incidence of the isotopes is sufficiently known, their occurrence in wings (insects) and feathers or claws (birds) may inform about the general location of the individual animals during growth (e.g. Webster et al. 2002; Hahn et al. 2014; Stefanescu et al. 2016). For birds, these methods are perhaps being superseded by the use of electronic tags (see below) which provide more precise route and timing information.

3 Detection and Quantification of Animals in Flight

The first requirement for aerobiological observation is simply detection, i.e. determining that animals are present in the atmosphere, and—if possible—estimating their distribution and numbers. This can be achieved either by capturing specimens or by remote sensing (including visual observation). An important requirement is to discriminate and quantify the main participants of the aerial fauna, i.e. to distinguish between insects, birds, and bats. Identification to broad taxonomic group, to species, or even to sex and age cohort are more challenging aims. For precise identifications, capture will generally be more effective than remote sensing, but automated remote-sensing methods are much more practicable for longer-term observation campaigns, environmental monitoring programmes, and operational applications (e.g. detecting crop pest invasions or bird hazards to aviation) and for these identification to species may not be essential.

3.1 Capture of Animals in Flight

Capture of animals in flight well above the surface has been widely used for insects, which have been sampled using traps mounted on masts or towers, mountain tops, tethered balloons, kites, and aircraft (Reynolds et al. 1997). Traps are usually of either the suction or the interception types, as these do not depend on a behavioural response from the insects. Light traps attract insects and thus modify their behaviour. Nevertheless, light traps mounted on mountain-top towers (Gregg et al. 1993) have proved valuable for obtaining samples of larger species, which are both less numerous and stronger flying than the microinsects that represent the majority of the catch from suction and interception traps (e.g. Chap. 7). Powerful lamps producing narrow vertically pointing beams attract insects overflying at heights up to a few hundred metres, as demonstrated by major differences between the composition of the catches in such a "searchlight trap" and a conventional light trap operating nearby (Feng et al. 2004). Suction traps operated with their apertures more than 10 m above a crop sample dispersing or migrating insects and their catches are thus representative of the wider area rather than just the trap's immediate surroundings (e.g. Riley et al. 1987; Harrington and Woiwod 2007).

A notable early contribution to aeroecology was made by Glick (1939) who used aircraft-borne insect traps in flights over Louisiana, USA; the insect samples he obtained at heights up to 4.5 km established the extraordinary diversity and abundance of insects and spiders in the air well above the surface. Intensive studies of the height distribution of insects over England, with simultaneous sampling at a series of heights using suction traps on towers and tethered balloons, established the power-law decrease of microinsect numbers with height that typically applies for small windborne migrants (Johnson 1969; Chap. 7). Large surface-mounted barrier traps, with interceptions from opposing directions kept separate, have proved effective for quantifying seasonal migrations of low-flying butterflies (Walker 2001). Aerial trapping is still employed (e.g. Hua et al. 2002; Chapman et al.

2004; Reynolds et al. 2013), although its application is constrained by operating costs (especially of personnel) and, for all but the lowest altitudes, aviation safety restrictions. Contemporary insect trapping programmes usually either complement entomological-radar observations to provide identifications for the radar targets (e.g. Chapman et al. 2006) or are undertaken to monitor pest species buildup and immigration (e.g. Harrington and Woiwod 2007; Otuka 2013).

In-flight trapping is less common for birds and bats than for insects, mainly because these animals are much less numerous in the air than insects. However, when migrating birds and bats are funnelled into passes by high mountain ridges functioning as leading lines (close to the principal direction of migration), they can be caught during active migratory flight. On Col de Bretolet (1923 m.a.s.l.) in the Swiss Alps, thousands of birds have been caught each autumn since 1960. Trapping over 24 h during at least 3 months per autumn has provided detailed insights into the seasonal and diurnal migratory patterns of long- and short-distance migrants (Dorka 1966; Jenni 1984). Direct comparison of captures and radar counts proved that nocturnal captures are strongly correlated with radar-recorded migratory passage (Komenda-Zehnder et al. 2010). Nocturnally migrating bats are caught as relatively rare "by-catch" during such bird-trapping activities.

3.2   Visual Observation and Electro-optical Imaging

Visual observation is essentially a form of remote sensing in which the detecting component is the human eye (and brain). In its basic form, in which the eye is augmented only with binoculars or a telescope, it is primarily suited for daytime observation. It is usually carried out by observers on the ground, but airborne observations of larger species in flight, and locust swarms, have been made from powered aircraft, motorised gliders, and remotely controlled drones (Pennycuick et al. 1979; Rainey 1989, pp. 62–87; Leshem and Yom-Tov 1996; Leshem and Bahat 1999). Single- and multiple-camera digital video systems have been developed for detecting and tracking flying animals (Riley 1994; Chap. 10), and technological advances in recent years have made automated track retrieval possible. While the current emphasis of this work appears to be on tracking animals within a quite small volume of space near the surface, telescope-based video systems for observing flight at altitude (the principal focus of this chapter) appear feasible. Telescope-based systems with simple non-imaging electro-optical detection are also an option (e.g. Brydegaard et al. 2016b).

Large soaring birds can be observed, and often identified, at altitudes up to several hundred metres. Observation is most efficient at bottleneck points, such as narrow land bridges or mountain passes (e.g. Therrien et al. 2012). A comprehensive scheme has been established to quantify soaring bird migration through Israel, where migration from western Asia and eastern Europe funnels towards Africa (Alon et al. 2004). Counts of small birds are possible when these are migrating near the surface at mountain passes, along a coast or near similar guideline features. The passage of low-flying passerine migrants through various mountain passes in

Switzerland has been monitored visually over many years (e.g. Jenni 1984). Standardised visual observations from a station in the Jura Mountains revealed the seasonal and diurnal pattern of 15 diurnally migrating passerine species (Korner-Nievergelt et al. 2007). Ornithologists refer to these types of daytime movements as "visible migration"; they recognise that these flights may be outnumbered by high-altitude and night-time movements that cannot be seen from the ground and that radar observations (see below) often show to have quite different forms (Wilcock 1964; Alerstam and Ulfstrand 1972). Seabirds can be observed at prominent headlands, such as at Tarifa, the southern-most point of Europe in the Strait of Gibraltar, where visual and radar observations were combined to characterise the migratory behaviour of the most numerous seabird species (Mateos and Bruderer 2010). Large insects, such as butterflies and dragonflies, sometimes make highly visible migratory flights just above the ground, and these can be counted by an observer (Williams 1958; Baker 1978; Russell et al. 1998); while such movements can be concentrated along landscape features, they also often occur over a broad front in open terrain.

At night, conventional observation becomes impossible, but both insects and birds can be detected through a telescope as they move across the face of the moon (Lowery 1951; Pruess and Pruess 1971; Liechti et al. 1995). Counts can be corrected for the geometrical biases arising from the moon's phase and its changing position in the sky; for birds, identification to the family-level appears sometimes possible (Peter et al. 1999). This "moon-watching" method is demanding of observer effort and possible only on occasions when the moon is above the horizon, its disc nearly full, and the sky clear. As the method is cheap and easily applicable by amateurs, coordinated observations over large areas can provide useful information on local and regional differences in migration traffic rates (see below) and flight directions (Liechti et al. 1996). Automated recording has proven to be practicable (Peter et al. 1999).

Other approaches to night-time observation of high-flying animals include use of a searchlight beam combined with a telescope (Able and Gauthreaux 1975). An electro-optical variant of this method, with heights of the detected animals determined by the parallax effect (i.e. use of two or more separated imagers to determine distance stereoscopically), has been demonstrated (Bolshakov et al. 2010); this system claims to reveal wing-beat frequencies and, for a fraction of the detections, heading direction and size. Illumination with visible light attracts insects (see above) and may also disturb birds (Bruderer et al. 1999). This problem can be avoided by using near-infrared (IR) illumination and observing with either night-vision goggles (Lingren et al. 1995) or an IR-sensitive video camera (Riley 1994). Passive thermal-IR imaging systems do not require illumination and can operate at long range, allowing observation of birds to heights of ~3 km (Liechti et al. 1995; Zehnder et al. 2001; Gauthreaux and Livingstone 2006; Weisshaupt et al. 2016). Thermal imaging depends on temperature differences between the animal and the background sky and can be used by day as well as by night, though it is most effective when the sky is clear and the air cold. It can be expected to work less well for insects, which are poikilotherms, than for birds and bats, but as active flight produces significant body warming (Dudley 2000, pp. 196–199), a detectable temperature contrast against a clear sky can be expected for larger insects. These methods have been effective but have not been widely adopted; in the case of thermal imaging, this is likely to be due in part to the high cost of the equipment. In all these cases, limited identification capability and lack of proper range information makes quantification difficult.

3.3   Aural Monitoring

Many bird species make calls when in flight and if the bird is not too high these can be heard at the surface and identified by a skilled listener (Farnsworth 2005). The method can help to reveal some components of night-time bird migrations. In order to detect calls from greater heights, a directional microphone and an amplifier are usually employed (Evans and Mellinger 1999). The microphone is best mounted on a rooftop or mast so that it is clear of vegetation and less impaired by background noise from anthropogenic or natural sources. Sounds can be automatically recorded and observations continued for long periods at modest cost. Computerised analyses have been developed to detect calls in the recordings, calculate and display spectrograms, and identify species (or species groups) using a database of identified flight calls (e.g. Evans and O'Brien 2002); however, the use of expert analysts, although very time-consuming, improves the quality of the extracted data (Sanders and Mennill 2014). With multiple microphones, some on towers, it is possible to determine the height of the calling bird from the different arrival times of the calls at the various microphone locations (Stepanian et al. 2016).

In North America, calling rates for species or species groups with identifiable calls showed significant correlation with catches of migrants at nearby stopover locations (Sanders and Mennill 2014), but there were only weak correlations between calling rates and estimates of bird density from weather surveillance radar observations (Farnsworth et al. 2004). The latter result may be accounted for by birds calling more at certain times of night and in particular weather conditions (with some species not calling at all), by the differing height coverages of the two techniques, and by variation in atmospheric attenuation of the calls with temperature and humidity (Horton ct al. 2015). Flight calls have been used to relate songbird migration intensity to wind direction (Smith et al. 2014). Calling during migratory flight is particularly prevalent among Nearctic birds (Farnsworth 2005). In Europe, most nocturnal passerine migrants (mainly Sylviidae and Muscicapidae) are silent during undisturbed migration. Even at an Alpine pass with concentrated migration, only one nocturnally migrating bunting (*Emberiza hortulana*) and some thrushes and larks (mainly *Alauda arvensis*) can be heard more or less regularly (Dorka 1966). Increased calling occurs when the birds are flying in disturbed conditions (e.g. in the light dome above a town under low clouds). Hüppop and Hilgerloh (2012) automatically recorded the calls of three thrush species at an illuminated platform in the North Sea and concluded that the frequency of calls did not really reflect migratory intensity but rather deteriorating flight conditions.

Shorebirds and herons can regularly be heard (in spite of their low passage numbers). In contrast to European passerines migrating at night, their diurnal counterparts call frequently (Dorka 1966; B. Bruderer, unpublished observations). While aural monitoring does not seem to be a reliable method to reveal the normal course of nocturnal migration in Europe, aural identification of diurnal migrants may be used as a complement to radar observations (e.g. Gehring 1963).

Bats are highly vocal in flight, producing calls with communication functions similar to those of birds and a variety of short, frequent, and often high-intensity emissions that are used for echolocation (Griffin 1958; Schnitzler and Kalko 2001). All vocalisations are at high frequencies and predominantly in the ultrasound range (i.e. >20 kHz). Bat detectors, electronic devices that record the high frequencies and convert them into the audible range, are now commercially available from a number of sources (Adams et al. 2012); systems that store the recordings for detailed analysis in the laboratory are also available (Frick 2013). Variations in the sensitivity, directionality, and frequency coverage of the recording systems, and also in the signal-processing methods used, lead to differences in effectiveness, both among the systems and depending on the application. Interspecific differences in calling frequencies and modulations make identification to species, or at least to species group, possible in well-studied faunas (Ahlén and Baagøe 1999). Automated identification has also been developed (e.g. Jennings et al. 2008).

Electronic bat detection and recording is commonly used for species identification (biodiversity studies) and research on echolocation itself, but systems have also been used for monitoring bat activity (Frick 2013). A limitation, however, is that acoustic bat detection is possible only over ranges of a few tens of metres, because high-frequency sounds are strongly absorbed by the atmosphere (Lawrence and Simmons 1982). For example, at 50 kHz, a typical frequency for echolocation pulses, attenuation occurs at a rate of ~0.6–1.7 dB (i.e. 13–32% drop in intensity, depending on the humidity) for each metre (NPL 1995). Attenuation adds to the reduction in intensity arising from the usual inverse-square law spreading of the sound power and at these levels becomes more significant than spreading over distances of ~10 m or greater. Lower-frequency components of the pulses are less affected but are usually emitted less strongly by the bats. For these reasons, detection of bats at high altitude with ground-based detectors is essentially unachievable. Jensen and Miller (1999) mounted three microphones up a 15-m mast and were able to estimate the height and distance of the bats from the differing arrival times of their calls at the microphones. For observations at higher altitudes, microphones can be carried aloft on kites, tethered balloons, or free-flying constant volume balloons ("tetroons") that drift along at an approximately constant (and selectable) height (Fenton and Griffin 1997; McCracken et al. 2008); logistical, regulatory, and safety constraints, plus cost, make these latter methods unsuitable for routine use.

Tiger moths (Arctiidae) produce loud clicks in flight that can be recorded with bat detectors (Barber and Conner 2006), but as these sounds are emitted in response to bat attacks their value for studying the moths' flight is probably limited to this one activity type. At the opposite end of the sound spectrum, in some insect groups wing beating produces audible buzzes or whines that have potential value for monitoring and identification (Raman et al. 2007), but detection ranges are generally low and again there appears to be little scope for utilising these in aeroecological investigations.

3.4 Radar

Radar has probably been the most widely used and productive observation technique in aeroecological research, starting in the 1950s with the naissance of "radar ornithology" (Eastwood 1967) and continuing to the present (e.g. Bruderer et al. 2012; Drake and Reynolds 2012). As such, it has its own chapter in this book (Chap. 12), which the reader should consult for the basics of radar theory, descriptions of the principal radar types, and discussion of the effectiveness of animals at reflecting radio waves (i.e. of their *radar cross sections*, RCS). Aeroecological observations with radar fall into two broad categories: those made with large surveillance radars that have been deployed primarily for other purposes (air defence, air traffic control, and weather forecasting) and small units operated by researchers specifically for observing animals in flight. The first category (especially weather radars, which in their modern forms are very effective at detecting high-flying insects, birds, and bats) is covered in Chap. 12, while in this section the design and use of the small research units is described.

An important distinction between the two categories is that the large radars, which observe at long ranges (of order 100 km), receive echo from a large volume that almost invariably contains numerous biological targets; in contrast, the small radars observe at short ranges (order 1 km) and the echo they receive often originates from a single animal. Thus, the echo intensity for small radars is given by equation (1) of Chap. 12 rather than Eqs. (3) and (5), which apply to the multiple-target echo received by the large radars. As Eq. (1) shows, the intensity of the echo from a single target decreases as the fourth power of the target's range R (R^{-4} dependence). In practice, this means that biological targets detected at short range may include smaller types like ladybugs and moths, while at longer ranges only larger types like birds and locusts will appear in the target sample. Recognising and accounting for this rapid decrease of detection performance with range is critical for the effective use of all of these small radar types. Another distinction is that the large radars generally transmit at ~5 cm ("C band") or longer ("S band" or "L band" at ~10 cm and ~23 cm, respectively) wavelength, while the small units operate at "X band" (~3 cm) or, in a few cases where very small insects are being studied, K_a band (~8 mm wavelength). Small units based on X-band civil-marine technology are relatively economical to procure and operate, and many are trailer mounted and easily relocated from one study site to the next. While their range is limited, they provide much finer spatial resolution than the large surveillance units. Generally, small and large radars have been used to explore different aspects of animal activity in the atmosphere. As technologies develop, however, the limitations of each type have been reduced and their potential application areas may be starting to overlap.

The earliest examples of short-range radars employed for aeroecological observations were units built to control the firing of anti-aircraft guns; these were capable of tracking, scanning, and fixed-beam operation (Schaefer 1968; Bruderer 1971). With modernisation, they remain in use by radar ornithologists today (Bruderer 2007; Bruderer et al. 2012). A different approach was taken in 1968 when civil marine radar technology was used as the basis for a unit purpose built for insect observation; this was deployed for several weeks in west Africa, with remarkable success (Roffey 1972). Several entomological radars of a similar form were operated over the following two decades, and the more modern types that have now replaced them also rely on civil marine transceivers. During the last decade, a number of manufacturers have started commercial production of "bird/bat radars" in response to a growing need to monitor flights of these taxa at existing or proposed airfields, wind farms, and some industrial sites. Harmonic entomological radars, which require the insect being tracked to carry a diode tag and which are used in studies of short-distance flights near the ground, are described in Sect. 4.5.

3.4.1 Entomological Radars

Scanning entomological radars are usually operated with the PPI display (Chap. 12) set to show a maximum range of about 1.5 km. Observations are made at a series of beam elevation angles ranging from near horizontal to an elevation angle of about 60°, to provide almost complete coverage up to about 1200 m. Vegetation, buildings, and terrain features produce "clutter" echoes and shadowing that can obscure low-flying insects, and observation below ~50 m is possible only in flat and treeless locations. Observations of the surveillance type can also be made with these units, out to ranges of ~50 km; these can reveal insect concentrations (Chap. 7) and nearby rainstorms.

Scanning radars can be used to observe localised phenomena such as plumes of insects taking off from particular emergence or roosting sites (Drake and Reynolds 2012, Chap. 11). More usually, however, they have been employed to study migration at altitude throughout the day or the night. The PPI may show both distributed echo (due to multiple targets) and stronger point (or "dot") echoes. The former is produced by (usually smaller) insects at high densities and the latter by individual larger insects (most often dragonflies, grasshoppers, locusts, butterflies, or moths). The distributed echo usually disappears soon after nightfall, leaving only targets of the point type. These can be counted and then divided by the effective scanning volume (calculated from the radar parameters and a typical radar cross section for the targets; Drake and Reynolds 2012, Chap. 7; Drake et al. 2017) to provide an estimate of the number density. By repeating the procedure at a series of beam elevation angles, a density profile can be obtained. The effective width of the beam, which determines the effective scanning volume for a target of a particular size, at first increases with range (i.e. the beam broadens) and then narrows as the maximum detection range for the target size is approached.

The scan rate of these radars is quite rapid—20 revolutions per minute (r.p.m.) is normal—so individual insects appear on several successive PPI images and their speeds and directions of motion can be determined. The speed and density values can be combined to produce estimates of migration flux. The density and flux estimates from different heights can be integrated to produce two measures of the total flight activity at the location: the area density and the migration passage (or traffic) rate (either the vector sum of the fluxes at all heights or the component of this in a defined direction). If the insects have some degree of common orientation (as is often the case, at least at night; see Chap. 7), this manifests as a symmetric pattern of echoes on the PPI (because an elongated insect viewed sideways-on has a much larger RCS than one that is pointing towards or away from the radar). An axis of alignment can be estimated quite easily from the pattern, though there are still two possibilities, 180° apart, for the heading direction (or *orientation*).

These methods allowed a quantitative approach to the study of insect migration to be developed as early as the 1970s (Schaefer 1976). A limitation is that there is often uncertainty about the identity of the detected targets and the number of different species within the target population. Modern signal processing techniques and digital recording of the data make estimation of target sizes from the observations (Cheng et al. 2002) more feasible than previously, but vertical-beam radars (see below) can do this better and have largely superseded scanning types in entomological research. It is worth noting that both types effectively provide point estimates of the various migration quantities: the 3-km-diameter area examined by a scanning radar amounts to only a few percent of the distance covered in a typical overnight migration.

Fixed-beam entomological radars. A fixed beam brings advantages of simplicity, makes automation easier, and reduces procurement and operating costs. When a beam remains still, flying animals are detected as they fly through it instead of as the beam passes over them. This represents a fundamentally different type of observation: while scanning radars measure target *density*, stationary beam types actually measure *flux* (Drake and Reynolds 2012, pp. 129–131). To obtain an estimate of density from a flux measurement (and vice versa), the speed of the targets must be determined.

Fixed-beam entomological radars almost always employ a vertical beam. This is necessary for observing smaller insects because these can only be detected at shorter ranges. It also has two important advantages that apply to large as well as small targets: a consistent, approximately ventral, beam-target aspect (provided the insects are maintaining steady flight) and (at least potentially) symmetry with respect to flight direction. However, observations below ~100 m may be obscured by clutter, or lost due to a minimum-range limitation of the magnetron-type transceivers employed in most current units. The narrowness of the beam at short ranges will also reduce detection rates. Insects flying through a fixed beam usually remain detectable for a few seconds, and this allows their wing beating to be observed. Speeds can also be obtained from the rate at which the echo intensity increases and then decreases during the transit. For detection ability to be independent of the insect's movement direction and body alignment, the vertical beam must have a circular form and the beam must either be circularly polarised or employ linear polarisation that rotates rapidly enough to complete one or more cycles before the target has left the beam. The latter approach allows the insect's alignment to be determined, though as with the scanning radar there is a 180° ambiguity.

The most sophisticated vertical-beam entomological radars employ the "ZLC configuration" in which a narrow-angle conical scan is added to the rotating polarisation; it is then possible to estimate the target's direction of travel as well as its speed and alignment (Drake and Reynolds 2012, pp. 90–93, 151–159). This last type also provides an estimate of the target's RCS and two measures of its shape; these, along with the wing-beat frequency, are valuable for helping to establish the target's identity (Drake 2016; Drake et al. 2017). The extra information provided by the scan enables more accurate flux and density estimation, but the additional signal modulations reduce detection rates and complicate the data analysis procedure (Drake 2013, 2014). In compensation for these difficulties, ZLC radars can operate autonomously and therefore provide a practicable means of obtaining datasets extending over several years. They have become the mainstay of radar-based entomological research and have formed the basis for major investigations of insect migratory behaviour (e.g. Chapman et al. 2010; Wood et al. 2010), "bioflows" (Hu et al. 2016), and comparative aeroecology (Alerstam et al. 2011; Chapman et al. 2016) as well as finding a role in operational locust forecasting (Drake and Wang 2013).

Tracking radars have only exceptionally been used in entomology. Gehring (1967) and Bruderer (1969) used military tracking radars (see below) to track insects for up to 1 km and found much smaller modulations of the echo signal than for birds and bats. Together with important differences in speed, this allowed discrimination of insect and vertebrate targets (Bruderer and Steidinger 1972). Recent studies on bird migration across the Sahara incorporated tracking of locusts, which were followed over distances of up to 1200 m and at heights of 500–1000 m; these targets flew at speeds of 23–27 km h^{-1} (6.4–7.5 m s^{-1}) and produced relative signal fluctuations of ±1.5 to ±5 dB (Zaugg et al. 2008; B. Bruderer, unpublished data). Insects have also been tracked with X-band radars in the USA, in one case with a unit developed specifically for that purpose (Drake and Reynolds 2012, pp. 97–98); individuals were followed for 0.5–1 km and wing-beat frequencies could not be detected. The fraction of tracked targets with weak modulations is used by radar ornithologists as a measure of the proportion of (large) insects in their target samples.

3.4.2   Ornithological Radars

Research on bird migration that uses small radars has continued to rely mainly on ex-military fire-control units as these allow operation in scanning, fixed-beam, and tracking mode, all with the same pencil beam. When X-band transmission is employed, insect echoes are often numerous so that birds usually constitute only a minority of the detected targets. A sensitivity-time control (STC) circuit, which reduces the detection threshold with range so as to compensate for the R^{-4} dependence, is used to eliminate the smaller targets. Use of S-band wavelengths considerably reduces this problem, because most insects then fall into the Rayleigh-scattering region (Chap. 12) and produce echoes so weak as to be undetectable. However, only a few ornithological studies (e.g. Schaefer 1968) have used S-band;

one reason for this is that concentration of these longer waves into a narrow pencil beam requires very large antennas.

The continuous application of the Swiss-made military fire-control radar "Superfledermaus" (Contraves, Zurich-Oerlikon) for ornithological research over more than 40 years has seen data recording methods progress from an observer reading the radar's instrument panel to fully automatic digital data acquisition. This development illustrates some specific problems and caveats that have to be considered when applying such equipment. The radar itself (3.3 cm wavelength, vertical polarization, pencil beam with nominal width of 2.2°, 0.3 µs pulse length, 150-kW peak pulse power, 90 m minimum observing range) has remained unmodified except that the stability of the receiver was improved in the mid-1990s by replacing the amplifiers with solid-state units. Development has been focused on the recording system, which became increasingly computerised from 1980 onwards. In 2003, a study on bird migration across the Sahara required the mounting of the core components of a "Superfledermaus" (without tracking capabilities) on an off-road truck. This was followed by the development of a specifically designed bird radar that combined a commercially available civil marine radar with a parabolic dish antenna to produce a pencil beam that can be directed at a range of azimuth and elevation angles (Bruderer 2007; Bruderer et al. 2012).

Counting birds by radar. First attempts to quantify nocturnal bird migration by means of short-range high-precision X-band radar (with either fixed or scanning beams) go back to the late 1950s and early 1960s (Eastwood 1967, Chap. 11). Systematic migration studies commenced in 1968 using the "Superfledermaus" to measure the migration traffic rate (MTR, i.e. the migration flux through a stationary counting plane) and later also the density of migration (i.e. the number of birds flying per unit volume) at a series of heights. A narrow, clearly defined pencil beam is optimal for quantitative recording of MTR. Knowing the features and limitations of the measuring system is also essential and in the case of the Swiss system a program to determine the counting effectiveness of the unit was initiated at the very beginning (Bruderer 1971; Bruderer and Steidinger 1972). This has included measurements of radar cross sections of various birds and determination of their detection ranges, estimation of the detection volume for various bird sizes, screening of ground clutter, and elimination of non-bird echoes (small insects by means of STC and larger targets by their echo signatures). Observations comprised a pair of measurements, one taken with the pencil beam directed vertically and the other with it directed at a low elevation angle perpendicular to the principal direction of migration, in order to provide coverage of heights between 30 and 4000 m.

Vertical scanning of the pencil beam perpendicular to the direction of migration has potential to increase the counting surface. A trial of this was carried out at an Alpine pass, with the elevations from 0 to 90° covered in three steps as the automated scan extended only over 30°. These tests showed that the fast movements of the beam reduce the detection range considerably (due to the reduction of hits per target); furthermore, the interpretation of the photos taken from the range-height indicator (RHI; Chap. 12) was problematic (due to unknown rates of decrease of echo detectability with distance and elevation). As a consequence, the method was abandoned (Bruderer 1980).

Cross-calibration of the radar with a long-range thermal-IR imaging device and with moon watching led to the conclusion that within the range of the optical systems the operational opening angle of the radar beam was approximately double the nominal beam width of 2.2° (Liechti et al. 1995). Electronic signal recording directly from the solid-state amplifier (introduced in the mid-1990s) allowed averaging of consecutive samples obtained at the radar's pulse repetition frequency of 2083 Hz. Averaging every 16 samples, which reduced the effective sampling rate to ~130 Hz, led to a much improved signal-to-noise ratio and increased the detection range by about one-third.

A method for automatically discriminating birds from other targets, using echo signatures identified by experts to train a classification algorithm, has been developed (Zaugg et al. 2008). Use of this algorithm substantially reduces analysis time for large samples and is an important step towards real-time bird identification by pencil-beam radar. A major problem (that is rarely considered in ornithological studies) is the detectability of differently sized targets in different parts of the radar beam. Schmaljohann et al. (2008) showed that after exclusion of roughly 85% of non-bird targets, the size of bird echoes over the Sahara (scaled by the R^{-4} law to a standard range of 3 km) was between -90 and -50 dBm. As the smallest birds could be detected up to 3 km, the STC compensation for the R^{-4} law was set to work up to this distance, with the consequence that the surveyed volume increased up to this point according to the effective opening angle of the beam. Beyond 3 km the detection probability for small birds started to decrease, while for large targets the reduction of the effective beam width began beyond about 5 km.

In order to increase the space under surveillance, conical scanning at various elevation angles together with electronic echo sampling along the beam was introduced in the late 1980s. Various difficulties with the analysis of the recorded data were identified and methods to overcome the problems designed (Bruderer et al. 1995a). Due to the conical movement of the beam, the detection probability is reduced in comparison with the fixed beam at all distances and particularly beyond 3 km. Corrections were needed (a) to eliminate echoes of targets smaller than the smallest birds (achieved by applying STC at ranges up to 3 km), (b) to compensate for the decreasing detection probability beyond this point, (c) to compensate for increasing detection probability with increasing scanning elevation (due to more hits per target when the beam scans smaller volumes at higher elevations), and (d) to account for the aspect dependence of bird RCSs (in low elevation beams, birds present small RCSs in head- and tail-on views and large RCSs when seen from the side). With these corrections, the method is practicable provided the density of insects isn't so high that the resulting composite echoes exceed the STC-modified detection threshold.

Taking into account all these precautions, the method is still in use today to obtain data on the spatial distribution of birds within a hemisphere of radius ~4 km. With careful selection of the radar site, bird targets can be detected down to about 30 m above the ground. The radar is usually positioned in a shallow hollow for ground clutter reduction. Combining the data gathered from several elevation angles allows calculation of the density of migration for height zones 200 m deep. Height distributions of birds above the Arava Valley in Israel were compared with simultaneous measurements of the altitudinal profiles of wind, temperature, pressure, and humidity; it was found that birds carefully select flight levels according to wind conditions aloft (Bruderer et al. 1995b; Liechti et al. 2000).

The caveats described above for vertically and conically scanning pencil beams apply also to scanning fan beams, and in fact the problems are then greater because the distribution of energy in a fan beam lacks the circular symmetry of a pencil beam and the wide vertical opening angles increase the rate of multiple-target echoes. For this reason, fan beams have generally not been favoured by ornithological researchers; they are however useful for general monitoring of bird and bat activity and feature in several commercial radar units produced for that purpose (see below).

Ornithological tracking radars. Military tracking radars are designed to detect flying objects and to track selected targets automatically. In bird studies, tracking is used principally to observe flight behaviour and determine how it varies with environmental conditions and height. The normal procedure comprises searching, detection, and selection of a target to be tracked. Since the mid-1990s, these target-selection processes have become increasingly automated.

Once an echo is selected and locked on, tracking is achieved by the beam scanning rapidly around the axis of the antenna with an offset of about 1°, keeping the target close to the axis and within a short "range gate" (Bruderer 1997). Tracking pilot balloons with the same radar at intervals between the bird observations provides complementary wind data. The Superfledermaus was used to track identified birds (e.g. with a tiny light attached at night), allowing flight speeds of 139 species to be established (Bruderer and Boldt 2001). Combined tracking and recording with a cine camera led to a catalogue of wing-beat patterns for 155 species (Bruderer et al. 2010). This radar type has also been used to follow released bats, in order to record their wing-beat patterns and flight speeds (Bruderer and Popa-Lisseanu 2005). Observations of the flight paths of migrating birds were used to analyse flight behaviour in relation to topographical features at several locations in the Alps and the Mediterranean region. Flight behaviour of diurnal (mainly soaring birds) and nocturnal migrants in relation to meteorological conditions was studied in all these areas and in the Sahara (e.g. Bruderer 2007; Bruderer et al. 2012; Bruderer and Peter 2017; Bruderer et al. 2018). Other ornithological tracking radars have been employed by researchers investigating migration directions in polar regions where there is no surveillance-radar coverage (e.g. Alerstam et al. 2007), the responses of migrant birds to side winds (Green et al. 2004) and coastlines (Nilsson et al. 2014), adjustment of migration directions following take-off (Sjöberg and Nilsson 2015), and the comparative migration strategies of insects and birds (Chapman et al. 2016).

3.4.3 Commercial Bird/Bat Radars

Radars that have become available commercially over the last 10–15 years for monitoring bird and bat activity around airfields and at industrial sites (Nohara et al.

Table 9.1 Examples of commercially available bird/bat radars

Manufacturer (Country) Reference	Model[a]	Wavelength	Scan type	Transmitter/receiver
Accipiter (Canada, USA) http://www.accipiterradar.com/	AR-2	X	Dual pencil beam, two elevation angles	Non-coherent (magnetron)
DeTect (USA) http://www.detect-inc.com/avian.html	Merlin ABAR	S	Dual fan beam (horizontal, vertical)	Solid-state, coherent, pulse-Doppler
Robin (Netherlands) http://www.robinradar.com/	3D Flex	S	Mechanically steered two-axis beam	Frequency-modulated continuous wave (FMCW)
SRC http://www.srcinc.com/	BSTAR-3	L	Electronically steered two-axis beam	Coherent Doppler
Swiss-Birdradar http://www.swiss-birdradar.com/	BirdScan MV1	X	Wide-angle vertical beam	Non-coherent (magnetron)

[a]Models selected to indicate the variety of types available. Most manufacturers offer a range of models and/or options; models in the table may no longer be current

2007; Beason et al. 2013) are almost all of the scanning type. Technical details of some examples of recently available models are presented in Table 9.1. All but one of these are based on civil marine radar components and transmit in the S- and X-bands, with some systems operating dual units, one at each wavelength, simultaneously. The longer S-band wavelength is less affected by insect and rain clutter but produces a broader beam. A variety of antenna types and scanning configurations are employed. The most basic units employ a vertical fan beam turning azimuthally, as in a marine radar, though sometimes the fan is tilted upwards to reduce ground clutter and increase height coverage. This is often supplemented with a second fan beam that is oriented laterally and scans vertically. Some types use one or more dish antennas to produce pencil beams that are rather broader than those used by radar entomologists and ornithologists and that are directed at various fixed (often low) elevation angles. At least one top-level model has full azimuth-elevation control and incorporates, as a supplementary operating mode, the ability to follow a target like an ornithological tracking radar.

As the market has developed, these radars have become more sophisticated, both by adopting the latest civil-marine technology, such as frequency-modulated continuous wave (FMCW), pulse compression, and Doppler signal processing, and through incorporation of specific-purpose target-recognition and tracking systems. The BSTAR unit, which is based on military rather than civil-marine technology, carries this trend towards sophistication furthest. The information stream from avian radars at airports is now being integrated into the operational air-traffic control system (FAA 2010), and these units must meet the very high standards of reliability, accuracy, and prompt information delivery that are critical for maintaining flight schedules and safety. Specifications for units used for surveys may be less demanding, and analysis of the observations can be done offline and at a later date. The BirdScan MV1 radar differs from the other units in Table 9.1 as it has a more restricted function: to support wind-turbine operations. It produces estimates of the rate of bird movements at different (rather low) heights and can be integrated into the wind farm control system to bring the turbines to a stop when migrating birds are likely to collide with the moving blades.

As with all modern radars, the displays are generated digitally, and this provides opportunities for developing systems that are fine-tuned to the users' needs. In the case of bird/bat radars, this typically means distinguishing bird and bat targets from all other sources of echo, distinguishing individuals from flocks, tracking multiple targets simultaneously, issuing warnings when targets appear likely to enter specified regions, and archiving the target tracks for later re-examination and to accumulate statistics. Here tracking is achieved by associating the echoes detected at nearby locations on successive scans to form trajectories for individual targets. Systems are now available that can calculate the tracks of hundreds of birds or bats simultaneously (Beason et al. 2013).

Applications of these radars fall generally into two categories: (a) operational use by air-traffic controllers and the managers of wind farms and some industrial and mining sites where there is a need to respond promptly to any bird or bat flight activity that conflicts with their operations (Klope et al. 2009; Coates et al. 2011) and (b) accumulating data on bird and bat flight patterns as part of the environmental impact assessment process for proposed developments that might affect, or be affected by, bird or bat flights (Plonczkier and Simms 2012). These units have also been used in more general studies of bird migration and its relation to weather (e.g. Contreras 2013), for which they appear very suitable.

3.4.4 Meteorological Research Radars and Operational Wind Profilers

A variety of radars are used in meteorological research, often of types very different from those designed for routine weather forecasting (and of which the aeroecological applications are described in Chap. 12). Some of these research radars, especially types designed for studies of the lower troposphere (including the atmospheric boundary layer) or of cloud droplets, routinely detect flying animals (Drake and Reynolds 2012, Chap. 15). Data from these units, if accessible, are a potentially valuable resource for aeroecological studies. The meteorological researchers may also have complementary observing systems operating at the same site, providing the wind and weather data needed to interpret the animal observations. Recent examples using vertical-beam radars operating in the Ka or W bands (i.e. at millimetre wavelengths), have been concerned with insects flying in daytime convection (Wood et al. 2009; Wainwright et al. 2017) and with the role of the Great Plains low-level jet (see Chap. 2) in assisting or hindering migrants (Wainwright et al. 2016). An airborne W-band radar, with both nadir- and zenith-pointing beams, has also been used to observe daytime thermal convection (revealed by echo from entrained insects;

Geerts and Miao 2005); this study made the interesting aeroecological finding that insects in ascending plumes were flying actively downwards.

In addition to the weather surveillance radars now widely exploited for observation of bird and bat flight (see Chap. 12), many weather forecasting services make operational use of radar wind profilers. These have fixed beams directed at high elevation angles and, as they are designed to detect reflections from atmospheric turbulence, typically operate at wavelengths longer than those of the radars considered elsewhere in this chapter. Birds passing through the beams of these radars produce echoes that contaminate the meteorological data if not filtered out (Wilczak et al. 1995). It has recently been demonstrated that, for one commercially available model operating at L-band (23 cm wavelength), bird echoes can be reliably identified and bird detection is possible to heights of 4 km (Weisshaupt et al. 2017). This raises the interesting possibility that existing operational networks of these profilers may, without impinging on their primary meteorological function, be able to provide data on both bird migration and the winds in which the birds are flying.

3.5 Laser and Lidar

Remote sensing of flying animals is possible with light as well as radio waves (Drake and Reynolds 2012, pp. 19–20) and a *lidar* is a device essentially analogous to a radar but operating at frequencies in the visible spectrum or at nearby infrared or ultraviolet wavelengths (Weitkamp 2005). Lidars usually operate with laser illumination (and the name "laser radar" is then sometimes preferred). Laser light is monochromatic and a narrow-band optical filter in the receiving system eliminates most ambient light and makes operation during daytime possible. Performance can be improved by the use of a black termination screen, but this requires a fixed (and approximately horizontal) beam which obviously presents a serious constraint for aeroecological applications. Lasers produce beams that are very narrow and well collimated (so that they broaden with distance only very gradually), and these would only infrequently detect well-separated targets like insects or birds in the atmosphere. Incorporation of a beam expander increases the width to ~10 cm, but this is still ~100 times less than with a pencil-beam radar. Laser and lidar systems may therefore be most useful for studying small insects near the ground, where densities are high (Kirkeby et al. 2016). Ranging by means of the Scheimpflug principle works well in such contexts and allows construction of effective systems at a moderate cost (Brydegaard et al. 2014, 2016a). If the aim is simply to monitor activity near the surface, a ranging capability may not be needed; a continuous laser beam directed horizontally and relatively simple telescope-mounted opto-electronic detectors may then be sufficient (Brydegaard et al. 2016b).

Use of laser and lidar systems for detecting and studying flying animals has so far comprised mainly trials with insects. Systems capable of detecting flying honeybees at distances of ~100 m, and of measuring their wing-beat modulations, have been demonstrated (Carlsten et al. 2011; Tauc et al. 2017). Observation at kilometre ranges has been achieved (Brydegaard et al. 2016a). As laser light is monochromatic, little information on the target's colour is available, but a system incorporating illumination with ultraviolet light and a receiving system able to detect natural fluorescent emissions of visible light has been demonstrated, indicating that there is scope for obtaining additional information about the target's identity and, for some species, even its age and sex (Brydegaard et al. 2009, 2010; Guan et al. 2010). A vertical beam lidar transmitting three light colours simultaneously has detected aerofauna to heights of 600 m and demonstrated differences in coloration (as indicated by reflecting effectiveness at the three wavelengths) between targets (Jansson et al. 2017).

4   Tracking and Monitoring Using Electronic Tags

Electronic tags mounted on flying animals are designed to send information from the tagged individual to remote receivers or to store such information locally for later downloading (reviews by Kenward 2001; Fuller et al. 2005; Rutz and Hays 2009; Tomkiewicz et al. 2010; Bridge et al. 2011; Sokolov 2011; Kissling et al. 2014; Kays et al. 2015). In the simplest case, there is just a simple signal that is used to locate the animal at distances of up to a few kilometres by *radio-telemetry* or worldwide by *satellite telemetry*. *Archival loggers* store sensed data along with the precise time and may operate autonomously for a year until the animal is recaptured and the logger removed. More sophisticated tags may monitor the animal's physiology and behaviour and/or a range of environmental variables from its immediate surroundings; such data can be stored in a logger or relayed to external receivers (e.g. on satellites). Aeroecological topics studied by means of electronic tags comprise: timing of dispersal and migration movements; duration of flights and of stopovers; migratory routes, speeds, and altitudes; energetics of flight; and orientation, navigation, and homing. An extensive list of examples and applications is given by Sokolov (2011), and Vardanis et al. (2016) provide a recent example that illustrates the depth of information obtainable.

Whenever tags are mounted on animals, care has to be taken in regard to additional weight, drag or other impediment of movement, and the animal's general health (including possible toxic effect of glues). These constraints are particularly severe for flying animals for which the weight of the devices and their energy sources should not exceed 2–5% of the animals' body weight (Bowlin et al. 2010; Cooke et al. 2004; Cochran 1980; Naef-Daenzer et al. 2005). Transmitters weighing as little as 0.1 g are now available (Naef-Daenzer 2013); these allow large insects to be tracked. The light batteries required for small animals have a low power output and short lives and solar cells have sometimes been used to prolong their operation. Transmitters may be glued to the pronotum of a locust (Fischer and Ebert 1999) or to the back of a bat (after trimming the fur; Sapir et al. 2014a); on birds, they are usually mounted dorsally by means of a harness (Naef-Daenzer et al. 2005; Naef-Daenzer 2007). For relatively large birds (ducks and geese), implantation into the body cavity (with a percutaneous antenna) is possible; for small implanted devices, no negative effects on survival and breeding success were found (Guillemette et al.

2002), but larger units led to an increase in mortality of the surgically treated birds, particularly during the first 10 days after release (Sexon et al. 2014).

4.1   Radio-telemetry and Cellular Systems

When radio-telemetry is employed, transmitters attached to the study animals broadcast pulsed radio signals, typically in the very high-frequency (VHF) band. These allow an individual's presence to be monitored or the animal to be located and followed over short or medium distances with directional antennas. Receivers range from simple hand-held equipment to computerized and automated systems with receiving antennas mounted on towers to increase the range of operation. Receivers may also be mounted on vehicles or aircraft (Sapir et al. 2014b; Wikelski et al. 2006) which then follow the animal as it moves over an extended distance, or search for it periodically (e.g. once a day) to determine its location. A specifically designed array of antennas has been used to study age-specific differences in departure, flight duration, and speed of passerine migrants along ~100 km of the northern-most part of the US east coast (Mitchell et al. 2015). The Motus Wildlife Tracking system of Bird Studies Canada incorporates more than 300 receiving stations, with detecting ranges of up to 15 km, covering mainly the eastern half of Canada and the USA (Taylor et al. 2017).

Ground-based studies are, however, often constrained by a low number of receiving stations. Cellular phone technology may have the potential to overcome this limitation. In the simplest case, a tag on an animal would send an identifiable signal to a nearby cellular tower, which would cause it to be recorded as within the tower's range. More advanced tags would incorporate a Global Positioning System (GPS) unit and transmit their coordinates to the Global System for Mobile Communications (GSM) network. In a recent study, GPS/GSM data revealed age-specific differences in the migration routes of Golden Eagles *Aquila chrysaetos* in eastern North America (Miller et al. 2016). Logging devices may store not only the GPS locations but additional data about the animal itself and its environment and relay these to the Internet periodically or when contact to a receiver is established. However, combined GPS/GSM devices are still relatively heavy, and significant miniaturization is needed to apply them on birds lighter than ~500 g (Bridge et al. 2011). Bouten et al. (2013) have developed a GPS tracking unit weighing only 12 g that incorporates 4 MB of memory for data storage, a solar panel, a tri-axial accelerometer, and a radio transceiver for bidirectional communication with one or more ground-based antennas. The radio link can be used both to download the logged data and to alter the measurement schedule after the animal carrying the unit has been released.

4.2   Satellite Telemetry

Satellite telemetry allows worldwide tracking and doesn't require an observer to physically follow the tagged animal. It became a feasible option when the ARGOS system (http://www.argos-system.org/), originally designed to track oceanographic buoys and upload data from them, became available to biologists. The system calculates locations by measuring the Doppler effect on an exact frequency (401.650 MHz ± 30 kHz) sent from a Platform Terminal Transmitter (PTT) on the animal. Each of several Argos satellites orbits the earth 14 times a day, detecting PTTs within about 5000 km. At the equator, a PTT can communicate with a satellite about three times daily. Data are transmitted from the satellites to an Argos ground station and from there to the customer (Fancy et al. 1988). While tracking of animals was initially limited by the mass of the PTTs, transmitters have now been reduced to about 5 g and solar cells provide essentially unlimited battery life. The precision of the Doppler-based locations is often rather coarse (150–1000 m) and a Global Positioning System (GPS) is sometimes incorporated into the PTT to improve the precision to a few metres, the coordinates being transmitted via the Argos messaging facility (Tomkiewicz et al. 2010).

The main application of satellite telemetry has been the determination of the routes taken by long-distance migrants, e.g. the non-stop flights of Alaskan Bar-tailed Godwits (*Limosa lapponica*), covering 11,700 km across the Pacific to New Zealand (Gill et al. 2009), and the trans-Himalaya flights of Bar-headed Geese (*Anser indicus*), climbing from sea level to passes at 4000–6000 m within 7–8 h (Hawkes et al. 2011).

A plan is currently being developed, under the acronym ICARUS (International Cooperation for Animal Research Using Space, http://icarusinitiative.org/; Wikelski et al. 2007), to establish a remote-sensing platform on the International Space Station (ISS) that would allow global tracking of hundreds of medium-sized to small animals. It is proposed to record both their routes and physiological and behavioural data on loggers and then transmit these data in packages when the ISS passes nearby; the data would then be transmitted to ground stations and distributed to researchers.

4.3   Archival Loggers

Archival loggers are designed to store data on the environment of the animal carrying the logger or on the behaviour or physiological state of the animal itself. The smallest types do not use satellite- or radio-telemetry but require recapture of the animal to download the data; this disadvantage is offset by the minimal mass of the loggers (0.3–1 g), which allows them to be used with animals too small to carry transmitting units. Use of low-power and data-compression technologies enables recording to continue over long periods (up to several years).

Loggers have proved particularly effective when used as *geolocators* to track migratory birds between breeding and wintering grounds (overview in McKinnon et al. 2013). Geolocators incorporate an electronic light sensor and record light levels over time. The duration of day length (the time between dawn and dusk) is used to determine latitude, while the mid-time between dawn and dusk is used to determine longitude. The location data are not as accurate as those from GPS or ARGOS, and around the equinoxes latitude can only be estimated by interpolation of positions from earlier and later dates. A famous result from geolocator studies is the demonstration that Pacific Golden Plovers (*Pluvialis fulva*) make round trips of 16,000–27,000 km from wintering grounds on south-western Pacific islands via Japan to Alaska and then back south across the open ocean (Johnson et al. 2012). Geolocators of 1.4 g on Arctic Terns (*Sterna paradisea*) revealed migratory distances of 80,000 km or more (Egevang et al. 2010; Fijn et al. 2013). Among small passerines, the longest geolocator-recorded migration is that of Northern Wheatears (*Oenanthe oenanthe*) flying from Alaska to East Africa via Siberia (Bairlein et al. 2012). Geolocator recordings have established that nocturnal migrants sometimes extend their flight into the day when crossing a desert or a sea barrier (Adamik et al. 2016). Geolocators have also been used to study migratory activity in bats (Weller et al. 2016).

4.4   Monitoring In-Flight Activities

In addition to just following the animals, it may be desirable to record data on their behaviour and physiological state and on the environment in which they are flying. Data can be stored on the tag until the animal is recovered, or transmitted to a ground-based or satellite-borne receiving system. The increasing importance of such studies in aeroecology is reflected by the recent introduction of the term "biologging" (Rutz and Hays 2009). Various methods and research applications have been suggested, *viz.*: (a) accelerometers (measuring acceleration in up to three dimensions), to monitor energy expenditure or activity patterns (Wilson et al. 2008; Harel et al. 2016); (b) heart-rate loggers, with similar aims (Cochran and Wikelski 2005); (c) mini-neuro-loggers recording neuronal activity in the brain (electroencephalograph) and combining this with GPS data to study orientation behaviour (Vyssotski et al. 2006); and (d) animal-borne video tags (with a deployment mass of <14 g and VHF radio transmission) to provide a bird's-eye view of resource use and social interactions along a movement trajectory (Bluff and Rutz 2008).

Cochrane and Wikelski (2005) obtained data on heart rate, wing-beat frequency, and respiration rate of thrushes in flight by means of electrodes placed under the skin. The electrode potential modulated an 1800-Hz tone that was transmitted to a radio receiver. Liechti et al. (2013) mounted 1.5-g data loggers on 100-g Alpine Swifts (*Tachymarptis melba*) to record light intensity, pitch (inclination of body axis), and vertical acceleration (revealing wing beats). These data demonstrated that the birds stayed airborne over more than 6 months both during migration and in their African winter quarters. Lightweight transmitters on flying locusts provided electromyograms from which wing-beat frequencies were determined (Fischer and Ebert 1999).

4.5   Harmonic Radar

Harmonic tags incorporate an antenna and a diode and return a signal at double the frequency of an incoming radio wave (Drake and Reynolds 2012, Chaps. 8 and 14). As they have no battery, they can be made very light (down to 1 mg), which makes them suitable for use with insects. The tag is illuminated, and its echo detected, with a special scanning entomological radar that transmits at X-band and receives at the harmonic frequency; this unit is capable of detecting tags to ~900 m. The frequency change makes it possible to detect a tag-carrying animal flying at heights of only a few metres, where a normal radar echo would be obscured by ground clutter. Current harmonic radars work only with insects that fly quite close to the surface. They have been used mainly in studies of insect behaviour and cognition, but aeroecological investigations include a study of honeybee flight relative to wind (which at these heights is straightforward to measure with recording anemometers) (Riley et al. 1999) and of moth flight responses to wind-borne pheromones (Riley et al. 1998).

5   Environmental Observations

The activities undertaken by flying animals are influenced by the environmental conditions the animals are encountering or have previously encountered. Interpreting observed activities, and inferring their adaptive value, will therefore require information on environmental conditions. The list of potentially important parameters includes a range of weather variables, visibility, measures of illumination (night and day), geomagnetism, the positions of the sun and moon, and the moon phase. As temperature and wind have particular significance for aeroecology, we will focus on these parameters.

Weather conditions at the surface and throughout the troposphere can be obtained from national meteorological services. They obtain information from a worldwide network of weather stations that provide surface and radiosonde (upper air) measurements as well as from weather radars and satellites. For aeroecologists, the original meteorological data provide general information about the wind directions and speeds likely to be encountered by flying animals, including the availability of "transport" for wind-borne species; they appear particularly valuable for understanding how long-distance migrations are achieved (e.g. Symmons 1986; Erni et al. 2005). Numerical weather prediction (NWP) models integrate current and recent previous measurements to produce analyses of the large-scale weather systems that extend throughout the troposphere and to predict their development (Coiffier 2011). Reanalyses are available for past periods, with worldwide or continental-scale coverage at a series of altitudes (e.g. https://www.ncdc.noaa. gov/data-access/model-data/model-datasets/reanalysis). Wind and temperature values for the actual locations and times of the aeroecological observations can be obtained from the NWP outputs, but these are estimates and therefore of limited value for direct comparison with observed animal movements. NWP outputs can also be used to calculate trajectories of wind-borne (insect) migrants and infer source and destination regions of migrant populations (e.g. Otuka et al. 2005; Stefanescu et al. 2007); the flight behaviour and airspeed contribution of the flying animals is sometimes incorporated into the trajectory calculation (Chapman et al. 2010). Use of regional NWP models that incorporate more detailed information about the local terrain and the frictional effects of the surface on boundary-layer airflows (e.g. Hurley 2008) may be beneficial as aeroecological observations are very often of animals flying at heights of no more than ~1 km.

Observation, rather than calculation, of air-parcel trajectories has been achieved through the use of constant-volume balloons ("tetroons"), which drift at altitudes of constant air density (Westbrook et al. 1995). The tetroons were ballasted to float at the height at which the species of interest (a moth) was expected to fly and after rising to this height remained close to it for the duration of their flight. A radio transmitter was carried, and the tetroon's position recorded continually by observers in a pursuit vehicle. The method is not suitable for routine use but validated calculated trajectories for single-night movements and assisted interpretation of captures of marked moths. Direct observation of the temperature of the air an animal is flying through would appear straightforward in studies that involve fitting animals with a backpack archival logger (see above). GPS observations from backpacks on vultures in soaring flight provided detailed information about the form and strength of the thermals the birds were exploiting (Harel et al. 2016; Treep et al. 2016).

Interpretation of the behaviour of flying animals—e.g. their response to drift (Chapman et al. 2011)—generally requires precise wind speed and direction data obtained from the same location and height as the animal and at the same time. This need for accuracy arises in part from the relatively low airspeeds of the animals which, at least for insects, can be comparable to the magnitude of any discrepancies between NWP-estimated and actual wind vectors. Moreover, individual small-scale weather phenomena such as thunderstorms and their associated lower-atmosphere wind systems (Chap. 2) do not appear in NWP outputs at all but certainly affect animal flight (Chap. 7).

Surface conditions at a field site can readily be recorded with commercially available automatic weather stations, and regional data are available from weather services. More difficult to obtain, but particularly important with respect to flying animals, are data on conditions at heights above ~10 m. A traditional method to obtain wind data is tracking pilot balloons with theodolites or radar (e.g. Bruderer and Steidinger 1972). Such measurements are time-consuming and compete with biological observations or require additional field personnel. For studies of bird migration, measurement every 4 h (e.g. Bruderer et al. 1995b) is considered desirable, but more often balloons are tracked just before the beginning of nocturnal migration, at midnight, and around dawn; if soaring birds are being studied, an additional balloon at midday is added. Most birds can thus be allocated to a wind measurement within ±3 h, and if no weather front passes through the area, this is usually sufficient. If there are noticeable weather changes, or insects (with their lower air speeds) are being studied, the time frame for detailed comparison with flight vectors may be reduced to 1–2 h (e.g. Feng et al. 2004). Near-continuous observations with vertical-beam entomological radars often show a gradual variation of migration direction through the night (Drake and Reynolds 2012, Fig. 10.3), and this can provide an indication of how long a wind measurement can be expected to remain valid.

If temperature, humidity, and pressure values are required, radiosondes can be launched, though this incurs further costs and fieldwork effort. Radiosondes have routinely been applied for the study and prediction of the altitudinal distribution of nocturnal bird migration above the deserts of Israel (Bruderer et al. 1995b) and the Sahara (Schmaljohann et al. 2009), as well as to forecast the flight altitudes and flight behaviour of raptors (Spaar and Bruderer 1996). Simpler temperature sondes have been used in studies of insect migration (e.g. Drake and Reynolds 2012, Fig. 10.8). Remote-sensing systems for determining wind and temperature profiles through the lower atmosphere, using sound, ultra-high-frequency (UHF) electromagnetic waves, or (for temperature) both (May et al. 2002; Bianco 2010), are now commercially available. These appear well suited for continuous operation at long-term aeroecological observing sites, but so far this has hardly been done, probably mainly because of the cost.

When observations are made from an accompanying powered aircraft or glider, instrumentation for recording temperature, ascent rate, etc., can be carried to provide measurements of the conditions the flying animals are experiencing (e.g. Pennycuick et al. 1979; Leshem and Yom-Tov 1996). An aircraft fitted with a Doppler navigation system, temperature sensors, and an airborne entomological radar provided accurate estimates of wind speed and direction along the flight transect and demonstrated that line concentrations of airborne insects were associated with mesoscale convergence zones (Dickison et al. 1990).

6   Challenges and Prospects

The various methods described in this chapter have revealed numerous aeroecological phenomena and in many cases allowed quantitative investigations of them to be undertaken. With their aid, a body of knowledge about the use of the lower atmosphere by flying animals has accumulated and brought aeroecological enquiry to a half-mature state, in which much is understood and yet many questions are unresolved. While further application of existing observing methods can be expected to produce some additional progress, more significant advances are likely to arise through improved observing capabilities. In this section, we identify some of the shortcomings of existing technologies and describe some recent developments that show promise for observational research in the quite near future.

Some obvious limitations of current methods evident from the preceding sections include labour intensiveness, capital costs, need for specialist maintenance and development support, inadequate sample sizes, poor coverage (horizontal or vertical), insufficiently precise target identification, inability to measure all relevant parameters (including environmental factors), and unsuitability of some methods for use with smaller animals. Of course, different research projects are affected in different ways and to different extents; identification precision, for example, is not an issue in trapping studies. Technical advances, usually through adoption and adaptation of new technologies developed in other contexts, are most likely to increase observational effectiveness.

Advances in computing technology have allowed development of automated observing systems that have both reduced the need for field personnel and increased sample sizes. In computer-based observing all data is recorded in digital form, which eliminates the time-consuming initial (data-entry) stages of the analysis process. Losses of data quality common in previous photographic recording methods are also avoided. However, development and maintenance of automated systems usually requires significant commitment of personnel with engineering, instrumentation, and computing skills; equipment and component costs may also be considerable. Aeroecological research may best prosper through interdisciplinary collaborations drawing on skills and resources already available in different research groups and institutions. Three continent-wide collaborations relating to diverse aspects of aeroecology or aeroecological observation have been developed over the last decade: the Movebank animal-tracking archive (Kranstauber et al. 2011), the European Network for Radar surveillance of Animal Movement (ENRAM) (Shamoun-Baranes et al. 2014), and the Motus Wildlife Tracking system (see above); technological advances together with data exchange mediated through the Internet (Bridge et al. 2011) have played a critical role in each case.

Aeroecological researchers have drawn heavily on remote-sensing observations up to now and this seems likely to continue. Radar has proved a particularly effective observing tool over several decades and there is scope for further improvement. In the case of weather surveillance radars, further deployment of sensitive Doppler and Doppler-polarimetric units will allow monitoring of flying animals over a wider area, and systems for identifying insect, bird, and bat echo and disseminating observations of these to potential users are already being developed (Chap. 12). For small radars, Dopplerization would provide an estimate of the radial velocity component, which would perhaps be most useful for vertical-beam units where it represents ascent and descent, a component of flight that has not received as much attention as it perhaps deserves. Units employing shorter wavelengths, perhaps even W-band (~3 mm wavelength), would allow studies to be extended to smaller insect species (e.g. Wang et al. 2017), while use of FMCW transmission may improve the coverage of vertical-beam units near the surface where pulse units have a blind zone. Lidar and electro-optical observing techniques are developing rapidly at present and show considerable promise, but so far appear best suited for studies of the numerous insects that fly near the surface; they seem more likely to complement radar than to replace it. Operation of a number of aeroecological and meteorological remote-sensing instruments at the same site, ideally within the coverage of a Doppler-polarimetric weather surveillance radar, would allow numerous comparisons and correlations to be explored. Developing one such site in each continent, in a region where animal flight is prevalent, and maintaining these in operation over several years, might provide an appropriate focal task for the aeroecology research community during the next decade.

The relatively recent, and continuing, development of miniature telemetry and data-logging systems that can be attached to birds and bats without burdening them too greatly provides an alternative observational capability with strengths and limitations that differ significantly from those of the various remote-sensing methods. The ability to record flight paths over a full year has provided many novel insights for migration science. These new technologies have led to major advances in knowledge of bird migrations and have been embraced enthusiastically by researchers. These methods appear to have considerable promise as they allow continuous monitoring of location, physiology, behaviour, and the environmental conditions the animal is experiencing. For smaller species, size and weight constraints are very limiting and at present preclude use of telemetry, so that recapture of the animal is necessary to recover the logged data. Any further miniaturization of electronic and battery technologies will allow the technique to be extended to yet smaller animals, perhaps even larger insects, and enable collection of more comprehensive datasets over longer periods.

Findings from different observing methods generally complement each other. Remote sensing provides information about where and when animals fly in the atmosphere, and the environmental conditions in which the flights occur. Laboratory experimentation allows the behavioural responses that produce these flights, and that determine their particular timing and form, to be identified. Data logging can contribute, though with limitations (perhaps especially of sample size), to both of these types of information. Observation of animals in source, stopover, and destination areas, and physiological examination of captured individuals, elucidates the function of the flight activity and its adaptive value. Broader insights arise especially when observations of different types, and perhaps originating from different researcher perspectives, are combined (Shamoun-Baranes et al. 2017).

Acknowledgements We gratefully acknowledge contributions from F. Liechti and D.R. Reynolds on some of the topics dealt with in this chapter.

References

Able KP, Gauthreaux SA Jr (1975) Quantification of nocturnal passerine migration with a portable ceilometer. Condor 77:92–96

Adamík P, Emmenegger T, Briedis M, Gustafsson L, Henshaw I, Krist M, Laaksonen T, Liechti F, Procházka P, Salewski V, Hahn S (2016) Barrier crossing in small avian migrants: individual tracking reveals prolonged nocturnal flights into the day as a common migratory strategy. Sci Rep 6:21560

Adams AM, Jantzen MK, Hamilton RM, Fenton MB (2012) Do you hear what I hear? Implications of detector selection for acoustic monitoring of bats. Methods Ecol Evol 3:992–998

Ahlén I, Baagøe H-J (1999) Use of ultrasound detectors for bat studies in Europe: experiences from field identification, surveys, and monitoring. Acta Chiropterol 1:137–150

Alerstam T (1990) Bird migration. Cambridge University Press, Cambridge, 420 pp

Alerstam T, Ulfstrand S (1972) Radar and field observations of diurnal bird migration in south Sweden, autumn 1971. Ornis Scand 3:99–139

Alerstam T, Bäckman J, Gudmundsson GA, Hedenström A, Henningsson SA, Karlsson A, Rosén M, Strandberg R (2007) A polar system of intercontinental bird migration. Proc R Soc B 274:2523–2530

Alerstam T, Chapman JW, Bäckman J, Smith AD, Karlsson H, Nilsson C, Reynolds DR, Klaassen RHG, Hill J (2011) Convergent patterns of long-distance nocturnal migration in noctuid moths and passerine birds. Proc R Soc B 278:3074–3080

Alon D, Granit B, Shamoun-Baranes J, Leshem Y, Kirwan GM, Shirihai H (2004) Soaring-bird migration over northern Israel in autumn. Br Birds 97:160–182

Bairlein F, Norris DR, Nagel R, Bulte M, Voigt CC, Fox JW, Hussell DJT, Schmaljohann H (2012) Cross-hemisphere migration of a 25 g songbird. Biol Lett 8:505–507

Baker RR (1978) The evolutionary ecology of animal migration. Hodder and Stoughton, Sevenoaks, UK. 1012 pp

Barber JR, Conner WE (2006) Tiger moth responses to a simulated bat attack: timing and duty cycle. J Exp Biol 209:2637–2650

Beason RC, Nohara TJ, Weber P (2013) Beware the Boojum: caveats and strengths of avian radar. Human-Wildlife Interactions 7:16–46

Bianco L (2010) Introduction to SODAR and RASS wind-profiler systems. In: Cimini D, Marzano FS, Visconti G (eds) Integrated ground-based observing systems. Applications for climate, meteorology, and civil protection. Springer, Heidelberg, pp 89–108

Bluff LA, Rutz C (2008) A quick guide to video-tracking birds. Biol Lett 4:319–322

Bolshakov CV, Vorotkov MV, Sinelschikova A, Bulyuk VN, Griffiths M (2010) Application of the optical-electronic device for the study of specific aspects of nocturnal passerine migration. Avian Ecol Behav 18:23–51

Bouten W, Baaij EW, Shamoun-Baranes J, Camphuysen KCJ (2013) A flexible GPS tracking system for studying bird behaviour at multiple scales. J Ornithol 154:571–580

Bowlin MS, Henningson P, Muijres FT, Vleugels RHE, Liechti F, Hedenström A (2010) The effects of geolocator drag and weight on the flight ranges of small migrants. Methods Ecol Evol 1:1–5

Bridge ES, Thorup K, Bowlin MS, Chilson PB, Diehls RH, Flérons RW, Hartl P, Kays R, Kelly JF, Robinson WD, Wikelski M (2011) Technology on the move: recent and forthcoming innovations for tracking migratory birds. Bioscience 61:689–698

Bruderer B (1969) Zur Registrierung und Interpretation von Echosignaturen an einem 3-cm-Zielverfolgungsradar. Ornithol Beob 66:70–88

Bruderer B (1971) Radarbeobachtungen über den Frühlingszug im Schweizerischen Mittelland. (Ein Beitrag zum Problem der Witterungsabhängigkeit des Vogelzugs). Ornithol Beob 68:89–158

Bruderer B (1980) Vogelzugforschung unter Einsatz von Radargeräten. Medizinische Informatik und Statistik 17:144–154

Bruderer B (1997) The study of bird migration by radar. Part 1: the technical basis. Naturwissenschaften 84:1–8

Bruderer B (2007) Adapting a military tracking radar for ornithological research—The case of the "Superfledermaus". In: Ruth JM (ed) Applying radar technology to migratory bird conservation and management: strengthening and expanding a collaborative. Open-File Report 2007–1361. US Geological Service, Fort Collins, CO, USA, pp 32–37

Bruderer B, Boldt A (2001) Flight characteristics of birds: 1. radar measurements of speeds. Ibis 143:178–204

Bruderer B, Peter D (2017) Windprofit als Ursache extremer Zughöhen. Ornithol Beob 114:73–86

Bruderer B, Popa-Lisseanu AG (2005) Radar data on wing-beat frequencies and flight speeds of two bat species. Acta Chiropterol 7:73–82

Bruderer B, Steidinger P (1972) Methods of quantitative and qualitative analysis of bird migration with a tracking radar. In: Galler SR, Schmidt-Koenig K, Jacobs GJ, Belleville RE (eds)

Animal orientation and navigation. NASA SP-262. US Government Printing Office, Washington, DC, USA, pp 141–167

Bruderer B, Steuri T, Baumgartner M (1995a) Short-range high-precision surveillance of nocturnal migration and tracking of single targets. Israel J Zool 41:207–220

Bruderer B, Underhill LG, Liechti F (1995b) Altitude choice of night migrants in a desert area predicted by meteorological factors. Ibis 137:44–55

Bruderer B, Peter D, Korner-Nievergelt F (2018) Vertical distribution of bird migration between the Baltic Sea and the Sahara. J Ornithol 159. https://doi.org/10.1007/s10336-017-1506-z

Bruderer B, Peter D, Steuri T (1999) Behaviour of migrating birds exposed to X-band radar and a bright light beam. J Exp Biol 202:1015–1022

Bruderer B, Peter D, Boldt A, Liechti F (2010) Wing-beat characteristics of birds recorded with tracking radar and cine camera. Ibis 152:272–291

Bruderer B, Steuri T, Aschwanden J, Liechti F (2012) Vom militärischen Zielfolgeradar zum Vogelradar. Ornithol Beob 109:157–176

Brydegaard M, Guan ZG, Wellenreuther M, Svanberg S (2009) Insect monitoring with fluorescence lidar techniques: feasibility study. Appl Opt 48:5668–5677

Brydegaard M, Lundin P, Guan Z, Runemark A, Åkesson A, Svanberg S (2010) Feasibility study: fluorescence lidar for remote bird classification. Appl Opt 49:4531–4544

Brydegaard M, Gebru A, Svanberg S (2014) Super resolution laser radar with blinking atmospheric particles—application to interacting flying insects. Prog Electromagn Res 147:141–151

Brydegaard M, Gebru A, Kirkeby C, Åkesson S, Smith H (2016a) Daily evolution of the insect biomass spectrum in an agricultural landscape accessed with lidar. EPJ Web of Conferences 119:22004

Brydegaard M, Merdasa A, Gebru A, Jayaweera H, Svanberg S (2016b) Realistic instrumentation platform for active and passive optical remote sensing. Appl Spectrosc 70:372–385

Bub H (1995) Bird trapping and bird banding. A handbook for trapping methods all over the world. Cornell University Press, Ithaca, NY, USA. 330 pp

Carlsten ES, Wicks GR, Repasky KS, Carlsten JL, Bromenshenk JJ, Henderson CB (2011) Field demonstration of a scanning lidar and detection algorithm for spatially mapping honeybees for biological detection of land mines. Appl Opt 50:2112–2123

Chapman JW, Reynolds DR, Smith AD, Smith ET, Woiwod IP (2004) An aerial netting study of insects migrating at high altitude over England. Bull Entomol Res 94:123–136

Chapman JW, Reynolds DR, Brooks SJ, Smith AD, Woiwod IP (2006) Seasonal variation in the migration strategies of the green lacewing Chrysoperla carnea species complex. Ecol Entomol 31:378–388

Chapman JW, Nesbit RL, Burgin LE, Reynolds DR, Smith AD, Middleton DR, Hill JK (2010) Flight orientation behaviors promote optimal migration trajectories in high-flying insects. Science 327:682–685

Chapman JW, Klaassen RHG, Drake VA, Fossette S, Hays GC, Metcalfe JD, Reynolds AM, Reynolds DR, Alerstam T (2011) Animal orientation strategies for movement in flows. Curr Biol 21:R861–R870

Chapman JW, Nilsson C, Lim KS, Bäckman J, Reynolds DR, Alerstam T (2016) Adaptive strategies in nocturnally migrating insects and songbirds: contrasting responses to wind. J Anim Ecol 85:115–124

Cheng DF, Wu KM, Tian Z, Wen LP, Shen ZR (2002) Acquisition and analysis of migration data from the digitised display of a scanning entomological radar. Comput Electron Agric 35:63–75

Coates PS, Casazza ML, Halstead BJ, Fleskes JP, Laughlin JA (2011) Using avian radar to examine relationships among avian activity, bird strikes, and meteorological factors. Human–Wildlife Interactions 5:249–268

Cochran WW (1980) Wildlife telemetry. In: Schemnitz S (ed) Wildlife management techniques manual. The Wildlife Society, Washington, DC, USA, pp 507–520

Cochran WW, Wikelski M (2005) Individual migratory tactics of New World *Catharus* thrushes. In: Greenberg R, Marra PP (eds) Birds of two worlds. The ecology and evolution of migration. The Johns Hopkins University Press, Baltimore, Maryland, pp 274–289

Coiffier J (2011) Fundamentals of numerical weather prediction. Cambridge University Press, Cambridge, 368 pp

Contreras S (2013) Temporal and spatial patterns of bird migration along the lower Texas coast. Ph.D. dissertation, Texas A&M University-Kingsville. 97 pp

Cooke SJ, Hinch SG, Wikelski M, Andrews RD, Kuchel LJ, Wolcott TG, Butler PJ (2004) Biotelemetry: a mechanistic approach to ecology. Trends Ecol Evol 19:334–343

Dickison RBB, Mason PJ, Browning KA, Lunnon RW, Pedgley DE, Riley JR, Joyce RJV (1990) Detection of mesoscale synoptic features associated with dispersal of spruce budworm moths in eastern Canada. Philos Trans R Soc Lond B 328:607–617

Dorka V (1966) Das jahres- und tageszeitliche Zugmuster von Kurz- und Langstreckenziehern nach Beobachtungen auf den Alpenpässen Cou/Bretolet (Wallis). Ornithol Beob 63:165–223

Dorst J (1962) The migrations of birds. Houghton Mifflin, Boston, MA, USA. 476 pp

Drake VA (2013) Signal processing for ZLC-configuration insect-monitoring radars: yields and sample biases. In: Radar 2013. International conference on radar, Adelaide, September 9–12. IEEE, Piscataway, NJ, pp 298–303

Drake VA (2014) Estimation of unbiased insect densities and density profiles with vertically pointing entomological radars. Int J Remote Sens 35:4630–4654

Drake VA (2016) Distinguishing target classes in observations from vertically pointing entomological radars. Int J Remote Sens 37:3811–3835

Drake VA, Reynolds DR (2012) Radar entomology: observing insect flight and migration. CABI, Wallingford, UK. 496 pp

Drake VA, Wang HK (2013) Recognition and characterization of migratory movements of Australian Plague Locusts, *Chortoicetes terminifera*, with an Insect Monitoring Radar. J Appl Remote Sens 7:075095

Drake VA, Chapman JW, Lim KS, Reynolds DR, Riley JR, Smith AD (2017) Ventral-aspect radar cross sections and polarization patterns of insects at X band and their relation to size and form. Int J Remote Sens 38:5022–5044

Dudley R (2000) The biomechanics of insect flight: form, function, evolution. Princeton University Press, Princeton, NJ, USA. 476 pp

Eastwood E (1967) Radar ornithology. Methuen, London, p 278

Egevang C, Stenhouse IJ, Phillips RA, Petersen A, Fox JW, Silk JRE (2010) Tracking of Arctic terns *Sterna paradisea* reveals longest animal migration. Proc Natl Acad Sci 107:2078–2081

Engel S, Bowlin MS, Hedenström A (2010) The role of wind-tunnel studies in integrative research on migration biology. Integr Comp Biol 50:323–335

Erni B, Liechti F, Bruderer B (2005) The role of wind in passerine autumn migration between Europe and Africa. Behav Ecol 16:732–740

Evans WR, Mellinger DK (1999) Monitoring grassland birds in nocturnal migration. Stud Avian Biol 19:219–229

Evans WR, O'Brien M (2002) Flight calls of migratory birds. Eastern North American landbirds. Old Bird Inc, Ithaca, NY, USA. [CD-ROM]

FAA (2010) Airport avian radar systems. Advisory circular 150/5220-25. US Department of Transportation Federal Aviation Administration, Washington, DC, USA. http://www.faa.gov/documentLibrary/media/Advisory_Circular/150_5220_25.pdf

Fancy SG, Pank LF, Douglas DC, Curby CH, Garner GW, Amstrup SC, Regelin WL (1988) Satellite telemetry: a new tool for wildlife research and management. U.S. Fish and Wildlife Service, Washington, DC, USA. 54 pp

Farnsworth A (2005) Flight calls and their value for future ornithological studies and conservation research. Auk 122:733–746

Farnsworth A, Gauthreaux SA Jr, van Blaricom DJ (2004) A comparison of nocturnal call counts of migrating birds and reflectivity measurements on Doppler radar. J Avian Biol 35:365–369

Feng H-Q, Wu K-M, Cheng D-F, Guo Y-Y (2004) Northward migration of *Helicoverpa armigera* (Lepidoptera: Noctuidae) and other moths in early summer observed with radar in northern China. J Econ Entomol 97:1874–1883

Fenton MB, Griffin DR (1997) High-altitude pursuit of insects by echolocating bats. J Mammal 78:247–250

Fijn RC, Hiemstra D, Phillips RA, van der Winden J (2013) Arctic terns *Sterna paradisea* from the Netherlands migrate record distances across three oceans to Wilkes Land, East Antarctica. Ardea 101:3–12

Fischer H, Ebert E (1999) Tegula function during free locust flight in relation to motor pattern, flight speed and aerodynamic output. J Exp Biol 202:711–721

Frick WF (2013) Acoustic monitoring of bats, considerations of options for long-term monitoring. Therya 4:69–78

Fuller MR, Millspaugh JJ, Church KE, Kenward RE (2005) Wildlife radiotelemetry. In: Braun CE (ed) Techniques for wildlife investigations and management. The Wildlife Society, Bethesda, MD, USA, pp 377–417

Gauthreaux SA Jr, Livingston JW (2006) Monitoring bird migration with a fixed-beam radar and a thermal-imaging camera. J Field Ornithol 77:319–328

Geerts B, Miao Q (2005) The use of millimeter Doppler radar echoes to estimate vertical air velocities in the fair-weather convective boundary layer. J Atmospheric Ocean Technol 22:225–246

Gehring W (1963) Radar- und Feldbeobachtungen über den Verlauf des Vogelzuges im Schweizerischen Mittelland: Der Tagzug im Herbst (1957-1961). Ornithol Beob 60:35–68

Gehring W (1967) Analyse der Radarechos von Vögeln und Insekten. Ornithol Beob 64:145–151

Gill RE, Tibbitts TL, Douglas DC, Handel CM, Mulcahy DM, Gottschalck JC, Warnock N, McCaffrey BJ, Battley PF, Piersma T (2009) Extreme endurance flights by landbirds crossing the Pacific Ocean: ecological corridor rather than barrier? Proc R Soc B 276:447–457

Glick PA (1939) The distribution of insects, spiders and mites in the air. Technical Bulletin No. 673. United States Department of Agriculture, Washington, DC, USA. 150 pp

Green M, Alerstam T, Gudmundsson GA, Hedenström A, Piersma T (2004) Do Arctic waders use adaptive wind drift? J Avian Biol 35:305–315

Gregg PC, Fitt GP, Coombs M, Henderson GS (1993) Migrating moths (Lepidoptera) collected in tower-mounted light traps in northern New South Wales, Australia: species composition and seasonal abundance. Bull Entomol Res 83:563–578

Gregg PC, Del Socorro AP, Rochester RA (2001) Field test of a model of migration of moths (Lepidoptera: Noctuidae) in inland Australia. Aust J Entomol 40:249–256

Griffin DR (1958) Listening in the dark. The acoustic orientation of bats and men. Yale University Press, New Haven, CT, 413 pp

Guan Z, Brydegaard M, Lundin P, Wellenreuther M, Runemark A, Svensson EI, Svanberg S (2010) Insect monitoring with fluorescence lidar techniques: field experiments. Appl Opt 49:5133–5142

Guerra PA, Gegear RJ, Reppert SM (2014) A magnetic compass aids monarch butterfly migration. Nat Commun 5:4164

Guillemette M, Woakes AJ, Flagstad A, Butler PJ (2002) Effects of data-loggers implanted for a full year in female Common Eiders. Condor 104:448–452

Hagler JR, Jackson CG (2001) Methods for marking insects: current techniques and future prospects. Annu Rev Entomol 46:511–543

Hahn S, Dimitrov D, Rehse S, Yohannes E, Jenni L (2014) Avian claw morphometry and growth determine the temporal pattern of archived stable isotopes. J Avian Biol 45:202–207

Harel R, Horvitz N, Nathan R (2016) Adult vultures outperform juveniles in challenging thermal soaring conditions. Sci Rep 6:27865

Harrington R, Woiwod I (2007) Foresight from hindsight: the Rothamsted Insect Survey. Outlooks Pest Manage 18:9–14

Hawkes LA, Balachandran S, Batbayar N, Butler PJ, Frapell PB, Milsom WK, Tseveenmyadag N, Newman SH, Scott GR, Sathiyaselvam P, Takekawa JT, Wikelski M, Bishop CM (2011) The trans-Himalayan flights of bar-headed geese (*Anser indicus*). Proc Natl Acad Sci 108:9516–9519

Hobson KA (1999) Tracing origins and migration of wildlife using stable isotopes: a review. Oecologia 120:314–326

Horton KG, Stepanian PM, Wainwright CE, Tegeler AK (2015) Influence of atmospheric properties on detection of wood-warbler nocturnal flight calls. Int J Biometeorol 59:1385–1394

Hu G, Lim KS, Horvitz N, Clark SJ, Reynolds DR, Sapir N, Chapman JW (2016) Mass seasonal bioflows of high-flying insect migrants. Science 354:1584–1587

Hua H-X, Deng W-X, Li R-H (2002) Trajectory analysis on the summer immigrant brown planthoppers *Nilaparvata lugens* in the middle reaches of the Yangtze River captured by aerial net. Acta Entomol Sin 45:68–74

Hüppop O, Hilgerloh G (2012) Flight call rates of migrating thrushes: effect of wind conditions, humidity and time of day at an illuminated platform. J Avian Biol 43:85–90

Hurley P (2008) The development and verification of TAPM. In: Borego C, Miranda AI (eds) Air pollution modelling and its application XIX. Springer, Dordrecht, Netherlands, pp 208–216

Jansson S, Brydegaard M, Papayannis A, Tsaknakis G, Åkesson S (2017) Exploitation of an atmospheric lidar network node in single-shot mode for the classification of aerofauna. J Appl Remote Sens 11:036009

Jenni L (1984) Herbstzugmuster von Vögeln auf dem Col de Bretolet unter besonderer Berücksichtigung nachbrutzeitlicher Bewegungen. Ornithol Beob 81:183–213

Jennings N, Parsons S, Pocock MJO (2008) Human vs. machine: identification of bat species from their echolocation calls by humans and by artificial neural networks. Can J Zool 86:371–377

Jensen ME, Miller LA (1999) Echolocation signals of the bat *Eptesicus serotinus* recorded using a vertical microphone array: effect of flight altitude on searching signals. Behav Ecol Sociobiol 47:60–69

Johnson CG (1969) Migration and dispersal of insects by flight. Methuen, London, 763 pp

Johnson OW, Fielding L, Fisher JF, Gold RS, Goodwill RH, Bruner AE, Furey JF, Brusseau PA, Brusseau NH, Johnson PM, Jukema J, Prince LL, Tenney MJ, Fox JW (2012) New insight concerning transoceanic migratory pathways of Pacific Golden Plover (*Pluvialis fulva*): the Japan stopover and other linkages as revealed by geolocators. Wader Stud Group Bull 119:1–8

Kays R, Crofoot MC, Jetz W, Wikelski M (2015) Terrestrial animal tracking as an eye on life and planet. Science 348:aaa2478

Kelly JF, Horton KG, Stepanian PM, de Beurs KM, Fagin T, Bridge EF, Chilson PB (2016) Novel measures of continental-scale avian migration phenology related to proximate environmental cues. Ecosphere 7:e01434

Kenward RE (2001) A manual for wildlife radio tagging. Academic Press, London, 350 pp

Kirkeby C, Wellenreuther M, Brydegaard M (2016) Observations of movement dynamics of flying insects using high resolution lidar. Sci Rep 6:29083

Kissling WD, Pattemore DE, Hagen M (2014) Challenges and prospects in the telemetry of insects. Biol Rev 89:511–530

Klope MW, Beason RC, Nohara TJ, Begier MJ (2009) Role of near-miss bird strikes in assessing hazards. Human–Wildlife Conflicts 3:208–215

Komenda-Zehnder S, Jenni L, Liechti F (2010) Do bird captures reflect migration intensity?—Trapping numbers on an Alpine pass compared with radar counts. J Avian Biol 41:434–444

Korner-Nievergelt F, Korner-Nievergelt P, Baader E, Fischer L, Schaffner W, Kestenholz M (2007) Jahres- und tageszeitliches Auftreten von Singvögeln auf dem Herbstzug im Jura (Ulmethöchi, Kanton Basel-Landschaft). Ornithol Beob 104:101–130

Kranstauber B, Cameron A, Weinzierl R, Fountain T, Tilak S, Wikelski M, Kays R (2011) The Movebank data model for animal tracking. Environ Model Softw 26:834–835

La Sorte FA, Fink D, Hochachka WM, Kelling S (2016) Convergence of broad-scale migration strategies in terrestrial birds. Proc R Soc B 283:20152588

Lawrence BD, Simmons JA (1982) Measurement of atmospheric attenuation at ultrasonic frequencies and the significance for echolocation by bats. J Acoust Soc Am 71:585–590

Leshem Y, Bahat O (1999) Flying with the birds. Chemed Books, Tel Aviv, Israel. 264 pp

Leshem Y, Yom-Tov Y (1996) The use of thermals by soaring migrants. Ibis 138:667–674

Liechti F, Bruderer B, Paproth H (1995) Quantification of nocturnal bird migration by moonwatching: comparison with radar and infrared observations. Journal of Field Ornithology 66:457–468

Liechti F, Peter D, Lardelli R, Bruderer B (1996) Herbstlicher Vogelzug im Alpenraum nach Mondbeobachtungen—Topographie und Wind beeinflussen den Zugverlauf. Ornithol Beob 93:131–152

Liechti F, Klaassen M, Bruderer B (2000) Predicting migratory flight altitudes by physiological migration models. Auk 117:205–214

Liechti F, Witvliet W, Weber R, Bächler E (2013) First evidence of a 200-day non-stop flight in a bird. Nat Commun 4:2554

Lingren PD, Raulston JR, Popham TW, Wolf WW, Lingren PS, Esquivel JF (1995) Flight behaviour of corn earworm (Lepidoptera: Noctuidae) moths under low wind speed conditions. Environ Entomol 24:851–860

Lowery GH (1951) A quantitative study of the nocturnal migration of birds. Univ Kans Publ Mus Nat Hist 3:361–472

Mateos M, Bruderer B (2010) Anwendung von Radar für das Studium des Zuges von Meeresvögeln durch die Strasse von Gibraltar. Ornithol Beob 107:179–190

May PT, Cummings F, Koutsovasilis J, Jones R, Shaw D (2002) The Australian Bureau of Meteorology 1280-MHz wind profiler. J Atmos Ocean Technol 19:911–923

McCracken GF, Gillam EH, Westbrook JK, Lee Y-F, Jensen ML, Balsley BB (2008) Brazilian free-tailed bats (Tadarida brasiliensis: Molossidae, Chiroptera) at high altitude: Links to migratory insect populations. Integr Comp Biol 48:107–118

McKinnon EA, Fraser KC, Stutchbury BJM (2013) New discoveries in landbird migration using geolocators, and a flight plan for the future. Auk 130:211–222

Miller TA, Brooks RP, Lanzone MJ, Brandes D, Cooper J, Trembley JA, Wilhelm J, Duerr A, Katzner TE (2016) Limitations and mechanisms influencing the migratory performance of soaring birds. Ibis 158:116–134

Mitchell GW, Woodworth BK, Taylor PD, Norris DR (2015) Automated telemetry reveals age specific differences in flight duration and speed are driven by wind conditions in a migratory songbird. Mov Ecol 3:1–13

Naef-Daenzer B (2007) An allometric function to fit leg-loop harnesses to terrestrial birds. J Avian Biol 38:404–407

Naef-Daenzer B (2013) Entwicklungen in der Telemetrie und ihre Bedeutung für die ornithologische Forschung. Ornithol Beob 110:307–318

Naef-Daenzer B, Früh D, Stalder M, Wetli P, Weise E (2005) Miniaturization (0.2 g) and evaluation of attachment techniques of telemetry transmitters. J Exp Biol 208:4063–4068

Nilsson C, Bäckman J, Alerstam T (2014) Are flight paths of nocturnal songbird migrants influenced by local coastlines at a peninsula? Curr Zool 60:660–669

Nohara TJ, Weber P, Ukrainec A, Premji B, Jones G (2007) An overview of avian radar developments—past, present and future. In: Proceedings, bird strike 2007 conference, September 10–13, 2007. Kingston, Ontario, Canada

NPL (1995) Tables of physical & chemical constants, 16th edition. 2.4.1 speed and attenuation of sound. Kaye & Laby online, Version 1.0. National Physical Laboratory, Teddington, UK. http://www.kayelaby.npl.co.uk/general_physics/2_4/2_4_1.html

Otuka A (2013) Migration of rice planthoppers and their vectored re-emerging and novel rice viruses in East Asia. Front Microbiol 4:309

Otuka A, Dudhia J, Watanabe T, Furuno A (2005) A new trajectory analysis method for migratory planthoppers, *Sogatella furcifera* (Horváth) (Homoptera: Delphacidae) and *Nilaparvata lugens* (Stål), using an advanced weather forecast model. Agric For Entomol 7:1–9

Pekarsky S, Angert A, Haese B, Werner M, Hobson KA, Nathan R (2015) Enriching the isotopic toolbox for migratory connectivity analysis: a new approach for migratory species breeding in remote or unexplored areas. Divers Distrib 21:416–427

Pennycuick CJ, Alerstam T, Larsson B (1979) Soaring migration of the Common Crane *Grus grus* observed by radar and from an aircraft. Ornis Scand 10:241–251

Peter D, Trösch B, Lücker L (1999) Intensiver Vogelzug im Spätherbst als Folge einer Stauentladung. Ornithol Beob 96:285–292

Plonczkier P, Simms IC (2012) Radar monitoring of migrating pink-footed geese: behavioural responses to offshore wind farm development. J Appl Ecol 49:1187–1194

Pruess KP, Pruess NC (1971) Telescopic observation of the moon as a means for observing migration of the army cutworm, *Chorizagrotis auxiliaris* (Lepidoptera: Noctuidae). Ecology 52:999–1007

Rainey RC (1989) Migration and meteorology. Flight behaviour and the atmospheric environment of locusts and other migrant pests. Oxford University Press, Oxford, 344 pp

Raman DR, Gerhardt RR, Wilkerson JB (2007) Detecting insect flight sounds in the field: implications for acoustical counting of mosquitoes. Trans Am Soc Agric Biol Eng 50:1481–1485

Reynolds DR, Riley JR (2002) Remote-sensing, telemetric and computer-based technologies for investigating insect movement: a survey of existing and potential techniques. Comput Electron Agric 35:271–307

Reynolds DR, Riley JR, Armes NJ, Cooter RJ, Tucker MR, Colvin J (1997) Techniques for quantifying insect migration. In: Dent DR, Walton MP (eds) Methods in ecological and agricultural entomology. CAB International, Wallingford, UK, pp 111–145

Reynolds DR, Nau BS, Chapman JW (2013) High-altitude migration of Heteroptera in Britain. Eur J Entomol 110:483–492

Riley JR (1994) Flying insects in the field. In: Wratten SD (ed) Video techniques in animal ecology and behaviour. Chapman and Hall, London, pp 1–15

Riley JR, Reynolds DR, Farrow RA (1987) The migration of *Nilaparvata lugens* (Stål) (Delphacidae) and other Hemiptera associated with rice during the dry season in the Philippines: a study using radar, visual observations, aerial netting and ground trapping. Bull Entomol Res 77:145–169

Riley JR, Valeur P, Smith AD, Reynolds DR, Poppy G, Löfsted C (1998) Harmonic radar as a means of tracking the pheromone-finding and pheromone following flight of male moths. J Insect Behav 11:287–296

Riley JR, Reynolds DR, Smith AD, Edwards AS, Osborne JL, Williams IH, McCartney HA (1999) Compensation for wind drift by bumble-bees. Nature 400:126

Roffey J (1972) Radar studies of insects. PANS 18:303–309

Russell RW, May ML, Soltesz KL, Fitzpatrick JW (1998) Massive swarm migrations of dragonflies (Odonata) in eastern North America. Am Midl Nat 140:325–342

Rutz C, Hays GC (2009) New frontiers in biologging science. Biol Lett 5:289–292

Salewski V, Kéry M, Herremans M, Liechti L, Jenni L (2009) Estimating fat and protein fuel from fat and muscle scores in passerines. Ibis 151:640–653

Sanders CE, Mennill DJ (2014) Acoustic monitoring of nocturnally migrating birds accurately assesses the timing and magnitude of migration through the Great Lakes. Condor 116:371–383

Sapir N, Horvitz N, Dechmann DKN, Fahr J, Wikelski M (2014a) Commuting fruit bats beneficially modulate their flight in relation to wind. Proc R Soc B 281:20140018

Sapir N, Horvitz N, Wikelski M, Avissar R, Nathan R (2014b) Compensation for lateral drift due to crosswind in migrating European Bee-eaters. J Ornithol 155:745–753

Schaefer GW (1968) Bird recognition by radar. A study in quantitative radar ornithology. In: Murton RK (ed) The problems of birds as pests. Academic Press, New York, NY, USA, pp 53–86

Schaefer GW (1976) Radar observations of insect flight. In: Rainey RC (ed) Insect flight. Symposia of the Royal Entomological Society no. 7. Blackwell Scientific, Oxford, UK, pp 157–197

Schmaljohann H, Liechti F, Bächler E, Steuri T, Bruderer B (2008) Quantification of bird migration by radar—a detection probability problem. Ibis 150:342–355

Schmaljohann H, Liechti F, Bruderer B (2009) Trans-Sahara migrants select flight altitudes to minimize energy costs rather than water loss. Behav Ecol Sociobiol 63:1609–1619

Schnitzler H-U, Kalko EKV (2001) Echolocation by insect-eating bats. Bioscience 51:557–569

Sexon MG, Mulcahy DM, Spriggs M, Myres GE (2014) Factors influencing immediate post-release survival of Spectacled Eiders following surgical implantation of transmitters with percutaneous antennae. J Wildl Manag 78:550–560

Shamoun-Baranes J, Alves JA, Bauer S, Dokter AM, Hüppop O, Koistinen J, Leijnse H, Liechti F, van Gasteren H, Chapman JW (2014) Continental-scale radar monitoring of the aerial movements of animals. Mov Ecol 2:9

Shamoun-Baranes J, Liechti F, Vansteelant WMG (2017) Atmospheric conditions create freeways, detours and tailbacks for migrating birds. J Comp Physiol A 203:509–529

Sjöberg S, Nilsson C (2015) Nocturnal migratory songbirds adjust their travelling direction aloft: evidence from a radiotelemetry and radar study. Biol Lett 11:20150337

Smith AD, Paton PWC, McWilliams SR (2014) Using nocturnal flight calls to assess the fall migration of warblers and sparrows along a coastal ecological barrier. PLoS One 9:e92218

Sokolov LV (2011) Modern telemetry: new possibilities in ornithology. Biol Bull 38:885–904

Spaar R, Bruderer B (1996) Soaring migration of Steppe Eagles *Aquila nipalensis* in southern Israel: flight behaviour under various wind and thermal conditions. J Avian Biol 27:289–301

Stefanescu C, Alarcón M, Àvila A (2007) Migration of the painted lady butterfly, *Vanessa cardui*, to north-eastern Spain is aided by African wind currents. J Anim Ecol 76:888–898

Stefanescu C, Páramo F, Åkesson S, Alarcón M, Ávila A, Brereton T, Carnicer J, Cassar LF, Fox R, Heliölä J, Hill JK, Hirneisen N, Kjellén N, Kühn E, Kuussaari M, Leskinen M, Liechti F, Musche M, Regan EC, Reynolds DR, Roy DB, Ryrholm N, Schmaljohann H, Settele J, Thomas CD, van Swaay C, Chapman JW (2013) Multi-generational long-distance migration of insects: studying the painted lady butterfly in the Western Palearctic. Ecography 36:474–486

Stefanescu C, Soto DX, Talavera G, Vila R, Hobson KA (2016) Long-distance autumn migration across the Sahara by painted lady butterflies: exploiting resources in the tropical savannah. Biol Lett 12:20160561

Stepanian PM, Horton KG, Hille DC, Wainwright CE, Chilson PB, Kelly JF (2016) Extending bioacoustic monitoring of birds aloft through flight call localization with a three-dimensional microphone array. Ecol Evol 6(19):7039–7046

Symmons PM (1986) Locust displacing winds in eastern Australia. Int J Biometeorol 30:53–64

Tauc MJ, Fristrup KM, Shaw JA (2017) Development of a wing-beat-modulation scanning lidar system for insect studies. In: Singh UN (ed) Lidar remote sensing for environmental monitoring 2017, Proc SPIE 10406:104060G

Taylor LR (1986) Synoptic dynamics, migration and the Rothamsted Insect Survey: presidential address to the British Ecological Society, December 1984. J Anim Ecol 55:1–38

Taylor PD, Crewe TL, Mackenzie SA, Lepage D, Aubry Y, Crysler Z, Finney G, Francis CM, Guglielmo CG, Hamilton DJ, Holberton RL, Loring PH, Mitchell GW, Norris D, Paquet J, Ronconi RA, Smetzer J, Smith PA, Welch LJ, Woodworth BK (2017) The Motus Wildlife Tracking System: a collaborative research network to enhance the understanding of wildlife movement. Avian Cons Ecol 12(1):8

Therrien J-F, Goodrich LJ, Barber DR, Bildstein KL (2012) A long-term database on raptor migration at Hawk Mountain Sanctuary, northeastern United States. Ecology 93:1979

Tomkiewicz SM, Fuller MR, Kie JG, Bates KK (2010) Global positioning system and associated technologies in animal behaviour and ecological research. Philos Trans R Soc B 365:2163–2176

Treep J, Bohrer G, Shamoun-Baranes J, Duriez O, Frasson RPdM, Bouten W (2016) Using high-resolution GPS tracking data of bird flight for meteorological observations. Bull Am Meteorol Soc 97:951–961

Vardanis Y, Nilsson J-A, Klaassen RHG, Strandberg R, Alerstam T (2016) Consistency in long-distance bird migration: contrasting patterns in time and space for two raptors. Anim Behav 113:177–187

Vyssotski AL, Serkov AN, Itskov PM, Dell'Omo G, Latanov AV, Wolfer DP, Lipp H-P (2006) Miniature neurologgers for flying pigeons: multichannel EEG and action and field potentials in combination with GPS recording. J Neurophysiol 95:1263–1273

Wainwright CE, Stepanian PM, Horton KG (2016) The role of the US Great Plains low-level jet in nocturnal migrant behavior. Int J Biometeorol 60:1531–1542

Wainwright CE, Stepanian PM, Reynolds DR, Reynolds AM (2017) The movement of small insects in the convective boundary layer: linking patterns to processes. Sci Rep 7:5438

Walker TJ (2001) Butterfly migrations in Florida: seasonal patterns and long-term changes. Environ Entomol 30:1052–1060

Wang R, Hu C, Fu X, Long T, Zeng T (2017) Micro-Doppler measurement of insect wing-beat frequencies with W-band coherent radar. Sci Rep 7:1396

Webster MS, Marra PP, Haig SM, Bensch S, Holmes RT (2002) Links between worlds: unraveling migratory connectivity. Trends Ecol Evol 17:76–83

Weisshaupt N, Maruri M, Arizaga J (2016) Nocturnal bird migration in the Bay of Biscay as observed by a thermal-imaging camera. Bird Study 63:533–542

Weisshaupt N, Lehmann V, Arizaga J, Maruri M, Freckleton R (2017) Radar wind profilers and avian migration: a qualitative and quantitative assessment verified by thermal imaging and moon watching. Methods Ecol Evol 8:1133–1145

Weitkamp C (2005) Lidar: range-resolved optical remote sensing of the atmosphere. Springer, New York, 456 pp

Weller TJ, Castle KT, Liechti F, Hein CD, Schirmacher MR, Cryan PM (2016) First direct evidence of long-distance seasonal movements and hibernation in a migratory bat. Sci Rep 6:34585

Westbrook JK, Eyster RS, Wolf WW, Lingren PD, Raulston JR (1995) Migration pathways of corn earworm (Lepidoptera: Noctuidae) indicated by tetroon trajectories. Agric For Meteorol 73:67–87

Wikelski M, Moskowitz D, Adelman JS, Cochran J, Wilcove DS, May ML (2006) Simple rules guide dragonfly migration. Biol Lett 2:325–329

Wikelski M, Kays RW, Kasdin NJ, Thorup K, Smith JA, Swenson GW Jr (2007) Going wild: what a global small-animal tracking system could do for experimental biologists. J Exp Biol 210:181–186

Wilcock J (1964) Radar and visible migration in Norfolk, England: a comparison. Ibis 106:101–109

Wilczak JM, Strauch RG, Martin FM, Weber BL, Meritt DA, Jordan JR, Wolfe DE, Lewis LK, Wuertz DB, Gaynor JE, McLaughlin SA, Rogers RR, Riddle AC, Dye TS (1995) Contamination of wind profiler data by migrating birds: characteristics of corrupted data and potential solutions. J Atmos Ocean Technol 12:449–467

Williams CB (1958) Insect migration. Collins, London, 237 pp

Wilson RP, Shepard EL, Liebsch N (2008) Prying into the intimate details of animal lives: use of a daily diary on animals. Endanger Species Res 4:123–137

Wiltschko R, Wiltschko W (2003) Mechanism of orientation and navigation in migratory birds. In: Berthold P, Gwinner E, Sonnenschein E (eds) Avian migration. Springer, Berlin, Germany, pp 433–456

Wood CR, O'Connor EJ, Hurley RA, Reynolds DR, Illingworth AJ (2009) Cloud-radar observations of insects in the UK convective boundary layer. Meteorol Appl 16:491–500

Wood CR, Clarke SJ, Barlow JF, Chapman JW (2010) Layers of nocturnal insect migrants at high-altitude: the influence of atmospheric conditions on their formation. Agric For Entomol 12:113–121

Zaugg S, Saporta G, van Loon E, Schmaljohann H, Liechti F (2008) Automatic identification of bird targets with radar via patterns produced by wing flapping. J R Soc Interface 5:1041–1053

Zehnder S, Åkesson S, Liechti F, Bruderer B (2001) Nocturnal autumn bird migration at Falsterbo, South Sweden. J Avian Biol 32:239–248

Multi-Camera Videography Methods for Aeroecology

10

Margrit Betke, Tyson Hedrick, and Diane Theriault

Abstract

The ability to study the behavior of flying animals in their natural environment has been dramatically improved by recent advances in imaging technologies and video analysis tools. Stereo videography, in particular, has taken aeroecology to a new level. We here describe methods for estimating the three-dimensional flight paths of animals from image data recorded simultaneously by multiple visible-light or infrared cameras. We explain how the accuracy of these estimates is influenced by camera properties and placements, as well as calibration procedures. We include a description of potential pitfalls drawn from field experience with various bat and bird species and show some example stereoscopic reconstructions of bat and bird flight.

1 Introduction and Related Work

From Single-Camera to Multi-Camera Videography When the first Aeroecology Symposium was organized by Kunz et al. in 2008, the imaging technologies used at the time to locate free-ranging animals in the aerosphere included radar, radio transmission, and single-camera videography. As described in the preceding chapter, camera-based measurement methods have an extensive history in the study of animals in flight. By 2008, single-camera videography had been used to study free-ranging bats, birds, and insects (Balch et al. 2001; Betke et al. 2007, 2008; Hallam et al. 2010; Hristov et al. 2010; Khan et al. 2004; Kunz et al. 2009;

M. Betke (✉) • D. Theriault
Boston University, Boston, MA, USA
e-mail: betke@bu.edu; dht@bu.edu

T. Hedrick (✉)
University of North Carolina at Chapel Hill, Chapel Hill, NC, USA
e-mail: thedrick@bio.unc.edu

© Springer International Publishing AG, part of Springer Nature 2017
P.B. Chilson et al. (eds.), *Aeroecology*,
https://doi.org/10.1007/978-3-319-68576-2_10

Tweed and Calway 2002; Veeraraghavan et al. 2008). Census data obtained from single-camera videography, for example, had yielded new insights into the pest services provided by bats (Cleveland et al. 2006; Federico et al. 2008). While multi-camera videography had been used successfully in laboratory studies of animal flight (e.g., Hedrick et al. 2004; Riskin et al. 2008; Erwin et al. 2001), the first attempts to measure the three-dimensional (3D) positions of free-ranging animals in the field encountered various challenges (Lee et al. 2004; Premerlani 2007; Cavagna et al. 2008; Hristov et al. 2008). These challenges have been resolved only in recent years because of improvements to the camera hardware and analysis software available for the task. In particular, new insights into calibration (Hedrick 2008; Towne et al. 2012; Theriault et al. 2014) and the development of three-dimensional tracking and pose-estimation software (Wu et al. 2009a, b, 2011, 2012, 2014; Attanasi et al. 2013; Breslav et al. 2014) are responsible for multi-camera approaches becoming feasible and popular, providing precise 3D measurements over large volumes.

Compared to other aeroecology measurement techniques, such as radar or GPS tags carried by individual animals, multi-camera videography provides precise measurements over a small volume and for a limited amount of time. In general, video provides the most precise measurements but imposes the strictest limits on measurement volume and recording time. Radar extends to a greater volume but at lower resolution. Animal-borne GPS or inertial sensing tags are in principle unlimited in their range but cannot be easily applied to more than a few individuals in a group. They also demand a minimum body size of the animals studied.

Spatial scale limitations in multi-camera measurements spring from the requirement that the animals of interest be at least one pixel and preferably at least ten pixels in image size. For current-generation camera hardware recording 2336×1728 pixel images with a 16.35×12.10 mm sensor and 20 mm lens, this implies a practical detection and measurement volume of approximately 680,000 m^3 (or 250,000 body lengths^3) for small birds with a 20 cm characteristic length, assuming the bird images must be at least 3 pixels in size for effective tracking. In addition to being a function of the camera and lens, the image size of an animal depends on its body size and scales with l^2 where l is a characteristic body length. Due to the physics of flight and lift production, the typical flight speed of animals also increases with body size but scales with $l^{0.5}$. Thus, small animals will move faster through a measurement volume based on minimum observable size, and it is therefore easier to make multi-camera aeroecology measurements of larger animals such as birds or bats than insects. The measurement volume and precision of multi-camera methods are strongly influenced by camera placement decisions often made early in the course of a study; these considerations are examined in detail in later sections of this chapter.

Multi-camera techniques require a set of two or more cameras recording high-resolution movies with precise time synchronization among cameras to within a single video frame and preferably to the shutter opening event itself. This degree of time synchronization is typically the domain of costly video systems intended for the research and industrial marketplace, but these systems historically save only short duration videos. In contrast, consumer video cameras can often produce recordings many hours in length but do not facilitate time synchronization among cameras. This gap is being

Fig. 10.1 Three-step workflow for multi-camera videography of animals in the aerosphere: (1) Planning the field experiment with simulations informs how to set up the cameras in the field in order to achieve observational objectives (details in Sect. 2). (2) The multi-camera system is calibrated in a multistep process that uses both videos of calibration objects and of the animals themselves (details in Sect. 3). (3) Three-dimensional flight trajectories are estimated from a tracking process (Sect. 4)

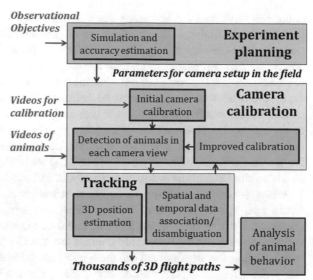

closed by progress from both ends; research and industrial cameras are increasingly capable of long duration recordings, while recording frame rates and audiovideo synchronization are now of sufficient quality in some consumer video cameras to permit post-recording synchronization to less than a single frame. In all cases, the pixel resolution, light sensitivity, and dynamic range of digital cameras are rapidly increasing. On the analysis software side, standardized tools and workflows are emerging for placing cameras in the field, recording video data, calibrating the cameras for stereo reconstruction and 3D measurement, extracting animal movement from the video images, and merging movement tracks from several cameras to a single 3D track. While previously multi-camera methods were the province of research groups with specialists devoted to image and video analysis techniques, this is no longer necessary, and description of the appropriate methods is a major focus of this chapter.

Although multi-camera videography workflows are simple to describe (Fig. 10.1), implementation is another matter and prospective users should be aware of subtleties in camera placement (Sect. 2), camera calibration (Sect. 3), and track reconstruction (Sect. 4) that are dealt with in the subsequent sections of this chapter. These expand and update material covered by Theriault et al. (2014). The chapter also describes examples of the stereo videography approach applied to reconstructing the flight paths of free-ranging cliff swallows (*Petrochelidon pyrrhonota*), chimney swifts (*Chaetura pelagica*), and Brazilian free-tailed bats (*Tadarida brasiliensis*) (Sect. 5).

2   Planning Field Experiments with Multiple Cameras

When the goal of 3D videography is to quantify the kinematics of airborne animals and facilitate the study of their behavior, it is important to estimate the level of calibration and reconstruction accuracy that may be achieved in a field experiment. Any uncertainty in the estimation of the 3D position of the animals affects the uncertainty in derived calculations like velocity and acceleration which may be of direct biological interest. The protocol we recommend for field work with multiple cameras therefore includes an important planning phase. Prior to any camera setup and recordings, a preliminary plan for the location of the recording space, choice of cameras and lenses, and placement of the cameras should be made. Simulation experiments should be performed to estimate the uncertainty in 3D location measurements that this plan entails. In planning the field experiment, it is important to formulate the observational objectives, which, for example, are the level of uncertainty in position estimation that can be tolerated or the requirements on image resolution (number of pixels per animal) and image contrast (to distinguish animals from backgrounds) that video analysis software may have.

When selecting a multi-camera system, scientists should consider whether the frame rate, spatial resolution, field of view, and synchronization ability of the cameras are appropriate for the size and speed of the study organisms. Two obser- vational objectives that commonly conflict are the size of the volume of space in which the animals are observed and the spatial resolution at which they are observed. When designing an experiment for such studies, Theriault et al. (2014) suggested to impose a lower bound on the size of animals in the image, so that they are not recorded at sizes that will make post-experiment analysis difficult. Based on a pinhole camera model, the bound x_{max} on the pixel span of the animal in the image can only be guaranteed if the observation distance between animals and each camera is at most $D_{max} = \frac{fX}{x_{max}p}$, where X is the length of the animal, f is the focal length of the camera, and p is the physical width of a pixel. For our studies, in the interest of observing the flight paths of the animals over a large distance, we chose to allow a small image size. A 10-pixel nose-to-tail span of a 10-cm-long bat in an image was ensured for animals that flew at distances smaller than $D_{max} = (25 \text{ mm} \times 10 \text{ cm})/(10 \text{ pixels} \times 18 \text{ µm/pixel}) = 13.8 \text{ m}$ from our thermal cameras.

As Theriault et al. (2014) showed, an inescapable source of uncertainty in the stereoscopic reconstruction of a 3D point is the quantization of intensity measurements (light or thermal radiation) into an array of discrete pixels. Each pixel, projected into space, defines a pyramidal frustum expanding outward from the camera. The location of the 3D point resides somewhere in the intersection of the frustums defined by pixels in each image. For any camera configuration, this uncertainty can be estimated for every 3D point observed by at least two cameras via simulation.

Theriault et al. (2014) provided the "easyCamera" software which first projects the 3D point onto the image plane of each camera and quantizes the location of each projection according to the pixel grid of each camera. The discrete pixel coordinates of the image points in each camera can then be used to reconstruct a 3D position via triangulation. The software computes the reconstruction uncertainty as the difference between the original and reconstructed positions of the point. The software easyCamera was written in MATLAB, is available online (http://hdl.handle.net/ 2144/8456), and comes with usage instructions, video tutorials, and sample data.

The size and shape of the observation volumes and the uncertainty due to quantization within these volumes can differ significantly depending on the number of cameras and their placement (Fig. 10.2). Ensuring that the angle between the

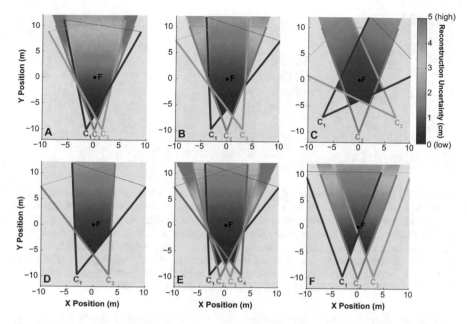

Fig. 10.2 Reconstruction uncertainty due to quantization effects is shown for six hypothetical camera configurations (Theriault et al. 2014). The cameras were simulated to have a pixel width of 18 μm and field-of-view angle 40.5° and be positioned at a fixed height Z and aimed at a common, equidistant fixation point $F = (0, 0, Z)$. Horizontal cuts of the 3D view frustums of the cameras at height Z and distance $D_{max} = 20$ m are shown from above at the level and elevation angle of the cameras (cuts at a different level or angle would show slightly greater reconstruction uncertainty but similar trends). Placing the cameras further apart reduces reconstruction uncertainty (**a** versus **b**). If the cameras are placed too far apart (**c**), however, the view volume is "closed," and there are unobservable regions of space where the cameras will be looking past each other. If the distance between the outermost cameras is held constant, adding cameras may not decrease the uncertainty due to image quantization in the common observable region (**d** versus **e**). When the image planes of the cameras are parallel (**f**), the common view volume is smaller and further away from the cameras than in the other configurations. Among the configurations shown, configuration **b** may be most desirable

optical axes is not wider than the field-of-view angle of the cameras leads to "open" intersection volumes that extend infinitely far away from the cameras (all examples in Fig. 10.2 except **c**), which is desirable because it facilitates recording even if the animals appear in an unexpected location. The level of uncertainty increases with the distance from the cameras because the volume of the intersection of the pixel frustums also increases with this distance.

Theriault et al. (2014) also considered the reconstruction uncertainty that arises from the difficulty in identifying the location of an animal in an image (Fig. 10.3). The location of an animal is often thought of as a single point, e.g., at the center of its body. Localization accuracy of this ill-defined point depends on the resolution of the animal in the image (Fig. 10.3b). To estimate uncertainty in localization, Theriault et al. (2014) included a stochastic element to the simulation procedure described above by adding noise to the 2D projections before quantization and computing the

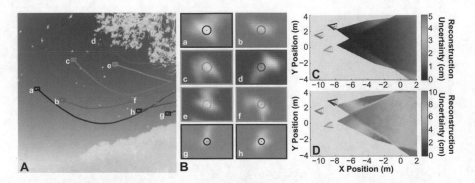

Fig. 10.3 Reconstruction uncertainty due to quantization and resolution issues (Theriault et al. 2014). In a video frame obtained for a flight study (**a**), the automatically detected locations of the animals may not be at their centers (*colored dots* in **b**). In the estimation of reconstruction uncertainty (**c, d**), the effect was included by corrupting the image projections of simulated world points, generated throughout the whole space, with Gaussian noise where the standard deviation is one-sixth of the calculated apparent size of an animal at that location (*circles* in **b**). When estimating the reconstruction uncertainty, including image location ambiguity (**d**) increases the estimated uncertainty more than threefold over image quantization alone (**c**) (note the change in *color scale*)

root-mean-squared (RMS) distance between the original and reconstructed scene points (Fig. 10.3d).

Reconstruction error occurs when image locations corresponding to different animals are mistakenly used to reconstruct 3D positions. These "data association" errors are commonly made by automated tracking methods, especially if the animals appear similar and small in the images. Camera selection and placement can reduce the potential occurrence of data association errors by imposing appropriate geometric constraints on the triangulation (Fig. 10.4). Theriault et al. (2014) recommended use of three or more cameras and a noncollinear camera placement that ensures that the image planes are not parallel (avoiding the configuration in Fig. 10.2f).

3   Camera Calibration

Image observations from multiple cameras can be used to "triangulate" or "reconstruct" the 3D coordinates of scene points that give rise to these observations. The triangulation process relies on having the relative position and orientation of the cameras, the "extrinsic parameters," and their focal lengths, principal points, and nonlinear distortion coefficients, the "intrinsic parameters." The process by which these parameters are estimated is known as "camera calibration" and typically involves matching points on a calibration object across camera views.

A popular calibration procedure is the direct linear transformation (DLT) (Abdel-Aziz and Karara 1971), with some researchers also using a camera calibration toolbox for MATLAB (Bouguet 1999). When using DLT, it is important to obtain calibration points throughout the volume of interest; otherwise reconstruction accuracy may be reduced (Hedrick 2008). Previous authors who have used DLT in a field

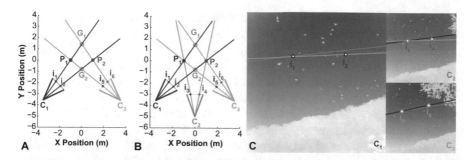

Fig. 10.4 Using three appropriately placed cameras instead of two leads to reduced data association errors (Theriault et al. 2014). A data association error may occur with a two-camera system (**a**)
if image points i_1 and i_6 are used to reconstruct "ghost" point G_1 and i_2 and i_5 to reconstruct "ghost"
point G_2. The correct association would be to match image points i_1 and i_5 to localize P_1 and image
points i_2 and i_6 to localize P_2. If a third camera is introduced (**b**), this error cannot be made because
the image points i_3 and i_4 can only be projections of P_1 and P_2. A similar three-camera configuration
was used for the bat recordings. It facilitated the process of matching corresponding image points
across camera views: The ray through i_3 and P_1 (*yellow*) and the ray through i_5 and P_1 (*green*)
appear as "epipolar" *lines* in the image of camera C_1 (large image in panel **c**), which intersect in
image point i_1 (*red*). Similarly, the projections into camera C_2 of the (*red*) ray through i_1 and P_1 and
the (*green*) ray through i_5 and P_1 at image point i_3 in camera C_2 (*yellow* in panel **c**); the projections
into camera C_3 intersect at i_5 (*green*). These "epipolar constraints" are only satisfied if the image
points i_1, i_3, and i_5 are matched to reconstruct the position P_1. A collinear camera placement
(Fig. 10.2f) would result in parallel epipolar lines, making the matching of corresponding image
points difficult

setting have constructed a large physical calibration object at the field site (Clark
2009; Munk 2011), limiting the size of the calibration volume. Others have carefully
measured the extrinsic parameters by hand (Cavagna et al. 2008), relying on a
semipermanent placement of their cameras in a sheltered location. Theriault et al.
(2014) proposed a different calibration approach that is particularly useful for field
settings where the volume of interest may be tens of thousands of cubic meters or
where cameras cannot be left in place for multiple sessions.

The approach by Theriault et al. (2014) combines a sequence of algorithms for
estimating the intrinsic and extrinsic parameters of the cameras. Its implementations
(a version with a graphical user interface and a command-line Python version) were
published as "easyWand calibration software" online (http://hdl.handle.net/2144/
8456), are licensed under the GNU public license version 3, and come with usage
instructions, video tutorials, and sample data. The first, most time-consuming step of
the calibration procedure is to manually or automatically digitize (annotate) a large
number of image locations of objects recorded in all views. The second step is to
produce preliminary estimates of the relative positions and orientations of the
cameras and the 3D positions of the calibration objects. This step uses the focal
lengths and principal points obtained directly from the lenses and image sensors as
preliminary estimates of the intrinsic camera parameters and the 8-point algorithm
(Hartley and Zisserman 2004). The third step involves the "sparse bundle adjustment
(SBA) algorithm" (Lourakis and Argyros 2009) to obtain refined estimates for all

calibration parameters. This algorithm minimizes the difference between the observed and ideal locations of the calibration points in each camera view. Bundle adjustment had been chosen previously by biologists for its 3D reconstruction accuracy (Walker et al. 2009) but lacked wide use because of the absence of easily accessible software implementations that were directly applicable to analysis of field data. In the last step, the easyWand software converts the estimates of the intrinsic and extrinsic parameters in the form of the DLT coefficients (this allows smooth integration into previously existing calibration workflows).

Wands as calibration objects are convenient for their mobility and as a means to measure scene scale and conduct additional error checking. None of the algorithms in the easyWand software pipeline explicitly requires use of a wand or other calibration objects. Other sources of matched scene points can be used as input. Theriault et al. (2014) proposed to use points on calibration objects tossed through-out the air space of interest as well as on the animals flying through this space. In practice, as shown in Fig. 10.1, animal points detected using the initial wand calibration can be passed back into the easyWand camera calibration routine as background points to iteratively improve the calibration. Note that because our methods do not require use of a calibration object, they are not only useful for field work but also in laboratory situations where introducing a calibration object into the volume to be measured is difficult or infeasible.

To enable evaluation of calibration inaccuracy, the easyWand software offers three measures: (1) The "reprojection error," measured in pixels for each camera, is the RMS distance between the original and reprojected image points of each calibration point, where the "reprojected" image points are computed using the estimated 3D position of the calibration point and the estimated camera parameters. (2) Inaccuracy can be described by the ratio of the standard deviation of wand-length estimates to their mean. A large ratio may indicate problems with the calibration, for example, unidentified lens distortion. (3) The average uncertainty in the position of each wand tip can be estimated from the known distance between the two tips.

3.1 Calibrating Consumer-Grade Cameras

Calibration of consumer-grade cameras for aeroecology data collection raises two primary issues. First, many consumer-grade cameras have exceptionally high distortion lenses. These may actually be desirable because they typically correspond to a very wide field of view but also make calibration difficult. In principle, the SBA algorithm can estimate nonlinear distortion parameters from the scene information itself, but nonlinear estimation is subject to substantial uncertainty, and it typically takes upward of 10,000 image points for a 4th-order radial distortion estimate to stabilize if the initial estimate is of no distortion. Thus, a high-quality initial estimate to either be refined by SBA or used without modification is essential. Fortunately, the radial and tangential distortion correction scheme Heikkila and Silvén (1997) used in SBA is a de facto standard and the distortion coefficients for a given lens or camera can be measured separately using software more appropriate to the task including the Camera Calibration Toolbox for MATLAB (Bouguet 2004) or the OpenCV environment (Bradski 2000). Since distortion coefficients are constant for non-zoom lenses and can be estimated from only one camera, this process is not difficult in practice.

The second difficulty presented by the use of consumer-grade hardware is the absence of support for synchronization of multiple cameras. Since multi-camera videography depends on having images of the same scene from multiple viewpoints, this absence is particularly troublesome. In most cases, the cameras can be synchronized to ±0.5 frames by a signal introduced into the camera audio recording channel. Note that the speed of sound is slow enough that users synchronizing via the audio channel should distribute the sound to each camera using electronic transmission methods; Walkie Talkie attention tones work well. A small, centrally controlled lamp mounted at the periphery of each camera lens can also provide synchronization information if the audio channel is unavailable or not perfectly synchronized with the video recording. The expected uncertainty from the ±0.5 frame offset can be estimated by adding the appropriate pixel noise factor to the easyCamera experiment planning package of Theriault et al. (2014). Once 3D tracks have been determined, the underlying 2D pixel location sequences can be interpolated to remove even the ±0.5 frame offset. Note that the wand calibration routines are exceptionally sensitive to frame offsets and often cannot cope with even the ±0.5 uncertainty produced by audio synchronization. Wand calibration users should briefly hold the wand in position for each of the 10–60 wand images or make extensive use of static background scenery for calibration.

4   Methods for Computing Three-Dimensional Flight Paths

Automated methods for computing 3D flight trajectories from calibrated multi-view video sequences rely on automated detection and tracking algorithms (Fig. 10.1). To compute the 2D locations of animals automatically in each video frame, our detection algorithm first subtracts a "background model" of typical intensity values in the scene from each video frame and then determines the animals to be in the regions with large changes in intensity values (Fig. 10.5). The background model is initialized by averaging the image intensities of the first few frames of the video (e.g., 25 frames). Each pixel value of the background model B_t is then recursively updated by adding a fraction α of the current image intensity I_t and $(1 - \alpha)$ of the value in the background model B_{t-1} for the previous frame: $B_t = \alpha I_t + (1 - \alpha)B_{t-1}$.

Our method uses the "reconstruction-tracking approach" by Wu et al. (2009b) to first triangulate the 3D positions of the animals from their 2D image locations, based on the estimated camera calibration parameters, and then to "track" each animal. "Tracking" means a sequence of 3D positions of the flying animal is estimated. For each animal, our method employs Kalman smoothing (Brown and Hwang 1997) to assign a newly triangulated 3D position to a previously established animal flight path. This assignment is called "temporal data association." We also need to solve

Original Image Background Model Pixels with Changes of Thermal Values

Fig. 10.5 Automated detection of bats in thermal videos

the "spatial data association," i.e., the problem of identifying the same animal across camera views. The epipolar constraints shown in Fig. 10.4 support this task.

Disambiguating associations for all animals in a scene may not be possible within one time step (processing one frame) due to various reasons, e.g., the calibration is inaccurate, the animals fly in large groups, have similar appearance, and occlude each other. Uncertainties in the current time step may be resolved when evidence for or against a hypothesized association has been collected in subsequent frames. This technique is called "look-ahead" or "deferred-logic tracking." The reconstruction-tracking approach by Wu et al. (2009b) uses the classic method for this technique, multiple hypothesis tracking (MHT; Reid 1979), with a sliding time window during which hypotheses can be resolved. The number of hypotheses that the algorithm has to check grows exponentially with the number of animals that appear in the scene during this time window. The method by Wu et al. (2009b) therefore uses a "greedy approximation algorithm" to disambiguate associations and append the reconstructed flight path of each animal with the appropriate newly computed 3D position of the animal.

5 Example Results for Birds and Bats

Estimating the 3D Flight Paths of Bats For capturing video of Brazilian free-tailed bats (*Tadarida brasiliensis*), we used three thermal infrared cameras (FLIR SC8000, FLIR Systems, Inc., Wilsonville, OR) with variable-focus 25-mm lenses and a pixel width of 18 μm, providing a 40.5° field of view. The 14-bit per pixel video has a frame size of 1024 × 1024 pixels and frame rate of 131.5 Hz. We used hardware synchronization to ensure accurate temporal alignment of frames across cameras.

We recorded video of Brazilian free-tailed bats during their evening emergence from Davis Blowout Cave in Blanco County, Texas (Fig. 10.6). The camera placements we selected for our bat field experiments (Fig. 10.6a) were similar to the configuration shown in Fig. 10.2b. With our simulation, we were able to

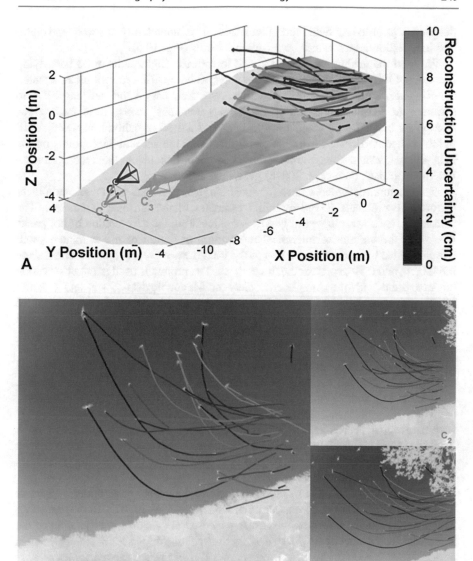

Fig. 10.6 3D flight trajectories of 28 Brazilian free-tailed bats during a 1-s interval (**a**), estimated from video captured with three synchronized, high-speed thermal infrared cameras (**b**). The tracks are shown in the context of the spatially varying reconstruction uncertainty arising due to both image quantization and image localization ambiguity. The *colored pyramids* represent the three cameras, while the colored lines each mark the trajectory of an individual bat. The observation distance between cameras and bats was approximately 10 m, chosen so that the nose-to-tail span of a bat in an image was at least 10 pixels. The baseline distance between the outermost cameras was approximately 6 m, chosen so that the expected uncertainty in reconstructed 3D positions at the observation distance due to image quantization and image localization ambiguity was <10 cm, the length of a bat. The RMS reconstruction uncertainty for the 1656 estimated 3D positions shown was 7.8 cm. The trajectories were smoothed with Kalman filtering

determine, prior to any field work, that the levels of uncertainty due to quantization and localization issues would be acceptable for us (Fig. 10.6).

We used the methods described above to estimate flight paths of 28 bats flying through a 1400-m^3 volume during a 1-s interval with a 7.8-cm root mean squared (RMS) uncertainty in their 3D positions. We used easyWand and easySBA to estimate the extrinsic and intrinsic camera parameters, using, from each view, 226 points manually digitized from the ends of a 1.56-m calibration wand, 2010 points manually digitized from hot packs thrown in the air, and 7135 points on the bats flying through the volume of interest identified using the reconstruction-tracking approach (Wu et al. 2009b) described above.

Manufacturer-provided values for focal lengths, principal points, and nonlinear distortion coefficients served as initial estimates of the intrinsic parameters. The calibrated space was aligned to gravity by calculating the acceleration of hot packs thrown in the volume of interest. The standard deviation of the estimated wand length divided by its mean was 0.046 (4.6%). The respective reprojection errors were 0.63, 0.74, and 0.59 pixels for the three views. The protocols used in bat observation are consistent with the American Society of Mammalogists (Sikes and Gannon 2011) and were approved by the Institutional Animal Care and Use Committee (IACUC) of Boston University and the Texas Parks and Wildlife Department (Permit Number: SPR-0610-100).

Estimating the 3D Flight Paths of Birds For capturing video of cliff swallows, we used three high-speed cameras (N5r, Integrated Design Tools, Inc., Tallahassee, Florida, USA) with 20-mm lenses (AF NIKKOR 20 mm f/2.8D, Nikon Inc., Melville, NY), recording 2336 × 1728 10-bit grayscale pixel images at 100 Hz with hardware synchronization. For recording chimney swifts, we used three Canon EOS 6D (Canon U.S.A. Inc., Melville, New Jersey, USA) cameras equipped with 35-mm lenses (Canon EF 35 mm f/1.4L) recording 1920 × 1080 24-bit color images at 29.97 Hz with audio synchronization.

A cliff swallow flock was recorded adjacent to the colony roost under a highway bridge in Chatham County, North Carolina, at 35° 49' 42" N, 78° 57' 51" W (Fig. 10.7). The proposed stereo videography process was used to support the estimation of the flight paths of 12 birds flying through a 7000-m^3 volume during a 2.3-s interval with a 5.9-cm RMS uncertainty in their 3D positions. The extrinsic camera parameters were estimated by easyWand and easySBA using, from each view, 58 points from the ends of a 1.0-m wand tossed through the scene and 4946 points acquired from swallows flying through the volume of interest via automated processing of the video sequences (Wu et al. 2009a). Manufacturer-provided values were used as the intrinsic parameters. The calibrated space was aligned to gravity by measuring the acceleration of a rock thrown through the volume of interest. The standard deviation of the estimated wand length divided by its mean was 0.0056 (0.56%); the respective RMS reprojection errors were 1.16, 1.58, and 1.17 pixels for the three cameras. The swallow observation protocol was approved by the University of North Carolina IACUC.

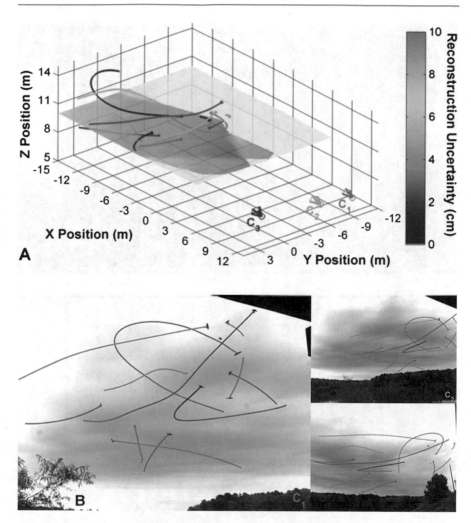

Fig. 10.7 The flight paths of 12 cliff swallows during a 2.3-s interval (**a**), estimated from video that was captured using three synchronized, high-speed visible light cameras (**b**). The tracks are shown in the context of the spatially varying reconstruction uncertainty arising due to both image quantization and image localization ambiguity. At an observation distance of approximately 20 m, the birds, which are approximately 13 cm long, were imaged at an average length of 18 pixels. The baseline distance between the outermost cameras was approximately 11 m. The RMS reconstruction uncertainty for the 2796 estimated 3D points shown was 5.9 cm, less than half the length of a bird. The 3D trajectories were smoothed using a smoothing spline based on this uncertainty

A chimney swift group, Fig. 10.8, was recorded in the city of Raleigh, North Carolina, during migratory roost flocking (Evangelista et al. 2017); the protocols used here permitted estimation of flight paths for a roosting group as large as 5000 birds for 30 min per evening in a 1,250,000 m^3 volume. The intrinsic parameters of the lenses were measured in lab using OpenCV (Bradski 2000) and the extrinsic

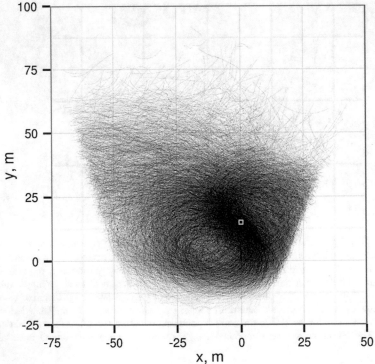

Fig. 10.8 The chimney swift roosting flock, detected using background subtraction and marked white in one of the original video frames (**a**), along with an overhead view of the flight paths of ca. 1500 birds during a 590 s interval (**b**). (**a**) From early in the evening while the flock is small and (**b**) from paths recorded as the flock is beginning to enter the chimney, shown as a white square near the center of the plot. The characteristic figure-eight or double torus flight pattern used by the birds is beginning to take shape. The edges of the imaging volume are also clearly visible on the left and right sides of the figure

parameters acquired from landmarks on buildings in the scene using easyWand and easySBA. The calibration was eventually upgraded using 40,000 points acquired from swifts flying through the volume of interest. The calibrated space was aligned to gravity using buildings in the scene. The median RMS reprojection errors for the animals were 0.47 pixels among the three cameras, corresponding to a dimensional RMS uncertainty of 0.28 m in 3D position. The swift observation protocol was approved by the University of North Carolina IACUC.

6   Discussion and Conclusions

As our results show, we balanced our two conflicting observational objectives of recording sufficiently long flight paths in a large volume of interest and recording the animals so that they appeared sufficiently large in all camera views. The resulting camera placement yielded a level of uncertainty in estimated 3D locations of the animals that was less than the length of a bat, half the length of a swallow, and the wingspan of a swift. The values of our two measures of calibration inaccuracy were sufficiently small to indicate accurate calibrations without any errors in determining corresponding calibration points.

Our recordings were part of extensive multiday experiments. Camera setup in the field can be done easily because no field measurements of camera pose or distances to the animals of study are needed. The only measurements required are references for scene scale and orientation. In our experiments, the known length of a calibration wand moved through the scene provided a scale reference, and gravitational acceleration, estimated from the ballistic trajectories of thrown objects, provided a reference for scene orientation. We typically recorded our calibration videos before obtaining study videos of bats and birds, but this order can be reversed. Camera placement and calibration recordings required 45 min at the beginning of each daily recording bout. This short setup time may be critical when the study organism or group is on the move and must be followed, or when site location, daily weather patterns, tides, or safety considerations dictate.

We suggest that the methods and implementations provided here will substantially aid biologists seeking to make quantitative measures of animal movements and behavior in field settings, as they have done for our own work on bat and bird flight. While previous studies have gathered similar data, they required heroic attempts at calibration frame construction in the field or carefully controlled field environments. We believe that the ease of setup and accuracy of calibration afforded by our methods opens up a wide range of previously unachievable studies and plan to continue refining the publicly available software implementations to fit a variety of needs.

As our results demonstrate, a stereo videography workflow that uses our pre-experiment planning software and post-experiment calibration and analysis software yields accurate reconstructions of 3D flight paths of bats and birds in field settings. The workflow has been applied in various other experiments to estimate 3D flight paths that were then used to analyze bat and bird behavior

(Boardman et al. 2013; Breslav et al. 2012; Kong et al. 2013, 2014; Sheltom et al. 2014; Theriault et al. 2010; Yang et al. 2013).

The next chapter, Chap. 11, discusses how to scale from individuals to populations. For single-camera videography, there are successful examples how "small" spatial and temporal resolution data may be incorporated into larger scale measurements and models. Thermal video analysis of the cave emergence flights of entire colonies has revealed significantly smaller Brazilian free-tailed bat colonies than previously estimated (Betke et al. 2008). Census data obtained from single-camera videography have been used to estimate the agro-economic value of pest services provided by entire bat populations (Cleveland et al. 2006; Federico et al. 2008). Ongoing efforts in applying multi-camera videography to analyze the behavior of groups of flying bats and birds have shown much promise in yielding large-scale results (Theriault et al. 2013; Attanasi et al. 2014).

Acknowledgements We wish to thank N. Fuller, B. Jackson, E. Bluhm, D. Evangelista, Z. Wu, B. Borucki, A. Banks, A. Froschauer, and K. Swift for their research assistance. We thank D. Davis and D. Bamberger for property access and the Department of Texas Parks and Wildlife for permitting assistance. This work was partially funded by the Office of Naval Research [N000141010952 to M.B. and T.H.], the National Science Foundation [0910908 and 0855065 to M. B. and 1253276 to T. H.], and the Air Force Office of Scientific Research [FA9550-07-1-0540 to M. B.].

References

Abdel-Aziz YI, Karara HM (1971) Direct linear transformation from comparator coordinates into object space coordinates in close-range photogrammetry. In: Proceedings of the symposium on close range photogrammetry, Direct linear transformation into object space coordinates in close-range photogrammetry. American Society of Photogrammetry, Falls Church, VA, pp 1–18

Attanasi A, Cavagna A, Castello LD, Giardina I, Jelic A, Melillo S, Parisi L, Pellacini F, Shen E, Silvestri E, Viale M (2013) Tracking in three dimensions via recursive multi-path branching. http://arxiv.org/abs/1305.1495

Attanasi A, Cavagna A, Castello LD, Giardina I, Jelic A, Melillo S, Parisi L, Pohl O, Shen E, Viale M (2014) Emergence of collective changes in travel direction of starling flocks from individual birds fluctuations. http://arxiv.org/abs/1410.3330

Balch T, Khan Z, Veloso M (2001) Automatically tracking and analyzing the behavior of live insect colonies. In: Proceedings of the fifth international conference on autonomous agents, Montreal, Canada, pp 521–528

Betke M, Hirsh DE, Bagchi A, Hristov NI, Makris NC, Kunz TH (2007) Tracking large variable numbers of objects in clutter. In: Proceedings of the IEEE computer society conference on computer vision and pattern recognition (CVPR), Minneapolis, MN, 8 p. https://doi.org/10.1109/CVPR.2007.382994

Betke M, Hirsh DE, Makris NC, McCracken GF, Procopio M, Hristov NI, Tang S, Bagchi A, Reichard JD, Horn JW, Crampton S, Cleveland CJ, Kunz TH (2008) Thermal imaging reveals significantly smaller Brazilian free-tailed bat colonies than previously estimated. J Mammal 89 (1):18–24. https://academic.oup.com/jmammal/article/89/1/18/1021254 [Also discussed in Nature 452, Research highlights, p 507]

Boardman BL, Hedrick TL, Theriault DH, Fuller NW, Betke M, Morgansen KA (2013) Collision avoidance in biological systems using collision cones. In: Proceedings of the 2013 American control conference, 8 p. https://doi.org/10.1109/ACC.2013.6580285

Bouguet JY (1999) Visual methods for three-dimensional modeling. PhD thesis, California Institute of Technology

Bouguet JY (2004) Camera calibration toolbox for Matlab. http://www.vision.caltech.edu/bouguetj/calib_doc/

Bradski G (2000) The OpenCV library. Doctor. Dobbs J 25(11):120–126

Breslav M, Fuller NW, Betke M (2012) Vision system for wingbeat analysis of bats in the wild. In: Proceedings of the workshop on visual observation and analysis of animal and insect behavior (VAIB 2012), held in conjunction with the 21st international conference on pattern recognition (ICPR 2012) Tsukuba, Japan, p 4. http://homepages.inf.ed.ac.uk/rbf/VAIB12PAPERS/breslav.pdf

Breslav M, Fuller N, Sclaroff S, Betke M (2014) 3d pose estimation of bats in the wild. In: Proceedings of the IEEE winter conference on applications of computer vision, Steamboat Springs, CO, 8 p. https://doi.org/10.1109/WACV.2014.6836114, http://ieeexplore.ieee.org/document/6836114/

Brown RG, Hwang PYC (1997) Introduction to random signals and applied Kalman filtering. Wiley, New York

Cavagna A, Giardina I, Orlandi A, Parisi G, Procaccini A, Viale M, Zdravkovic V (2008) The STARFLAG handbook on collective animal behaviour: 1. Empirical methods. Anim Behav 76 (1):217–236

Clark CJ (2009) Courtship dives of Anna's Hummingbird offer insights into flight performance limits. Proc R Soc B Biol Sci 276(1670):3047–3052

Cleveland CJ, Betke M, Federico P, Frank JD, Hallam TG, Horn J, Kunz TH, López JD, McCracken GF, Medellín RA, Moreno-Valdez A, Sansone CG, Westbrook JK (2006) Economic value of pest control services provided by Brazilian free-tailed bats in south-central Texas. Front Ecol Environ 4:238–243. https://doi.org/10.1890/1540-9295(2006)004[0238:EVOTPC]2.0.CO;2

Erwin HR, Wilson WW, Moss CF (2001) A computational sensorimotor model of bat echolocation. J Acoust Soc Am 110(2):1176–1187

Evangelista JD, Ray DD, Raja SK, Hedrick TL (2017) Three-dimensional trajectories and network analyses of group behaviour within chimney swift flocks during approaches to the roost. Proc R Soc B 284(1849):20162602. https://doi.org/10.1098/rspb.2016.2602, http://rspb.royalsocietypublishing.org/content/royprsb/284/1849/20162602.full.pdf

Federico P, Hallam TG, McCracken GF, Purucker S, Grant W, Sandoval AN, Westbrook J, Medellín R, Cleveland C, Sansone CG, López JD Jr, Betke M, Moreno-Valdez A, Kunz TH (2008) Brazilian free-tailed bats (Tadarida brasiliensis) as insect pest regulators in transgenic and conventional cotton crops. Ecol Appl 18(4):826–837. http://onlinelibrary.wiley.com/doi/10.1890/07-0556.1/epdf

Hallam TG, Raghavan A, Kolli H, Dimitrov D, Federico P, Qi H, McCracken GF, Betke M, Westbrook JK, Kennard K, Kunz TH (2010) Dense and sparse aggregations in complex motion: video coupled with simulation modeling. Ecol Complex 7(1):69–75. https://doi.org/10.1016/j.ecocom.2009.05.012

Hartley R, Zisserman A (2004) Multiple view geometry in computer vision, 2nd edn. Cambridge University Press, Cambridge

Hedrick TL (2008) Software techniques for two- and three-dimensional kinematic measurements of biological and biomimetic systems. Bioinspir Biomim 3:034001. http://iopscience.iop.org/article/10.1088/1748-3182/3/3/034001/pdf

Hedrick T, Usherwood J, Biewener A (2004) Wing inertia and whole-body acceleration: an analysis of instantaneous aerodynamic force production in cockatiels Nymphicus hollandicus flying across a range of speeds. J Exp Biol 207:1689–1702

Heikkila J, Silvén O (1997) A four-step camera calibration procedure with implicit image correction. In: Computer vision and pattern recognition, 1997. Proceedings, 1997 I.E. computer society conference on, IEEE, pp 1106–1112

Hristov NI, Betke M, Kunz TH (2008) Applications of thermal infrared imaging for research in aeroecology. Integr Comp Biol 48(1):50–59. https://academic.oup.com/icb/article/48/1/50/628016

Hristov NI, Betke M, Hirsh D, Bagchi A, Kunz TH (2010) Seasonal variation in colony size of Brazilian free-tailed bats at Carlsbad Caverns using thermal imaging. J Mammal 91(1):183–192. https://doi.org/10.1644/08-MAMM-A-391R.1

Khan Z, Balch T, Dellaert F (2004) A Rao-Blackwellized particle filter for eigentracking. In: IEEE conference on computer vision and pattern recognition, Washington, DC, pp 980–986

Kong Z, Özcimder K, Fuller N, Greco A, Theriault D, Wu Z, Kunz T, Betke M, Baillieul J (2013) Optical flow sensing and the inverse perception problem for flying bats. In: The 2013 I.E. conference on decision and control (CDC), Florence, Italy, 8 p. https://doi.org/10.1109/CDC.2013.6760112, http://ieeexplore.ieee.org/document/6760112/

Kong Z, Özcimder K, Fuller N, Theriault D, Betke M, Baillieul J (2014) Perception and steering control in paired bat flight. In: The 19th world congress of the international federation of automatic control (IFAC), South Africa, 7 p

Kunz T, Gauthreaux SJ, Hristov N, Horn J, Jones G, Kalko E, Larkin R, McCracken G, Swartz S, Srygley R, Dudley R, Westbrook J, Wikelski M (2008) Aeroecology: probing and modeling the aerosphere. Integr Comp Biol 48(1):1–11

Kunz TH, Betke M, Hristov NI, Vonhof M (2009) Methods for assessing colony size, population size, and relative abundance of bats. In: Kunz TH, Parsons S (eds) Ecological and behavioral methods for the study of bats, 2nd edn. Johns Hopkins University Press, Baltimore, MD, pp 133–157

Lee EY, Betke M, Kunz TH (2004) Bats in motion: stereo object recognition and trajectory analysis of flying bats. In: Proceedings of the 34th Annual North American symposium on bat research (NASBR), Salt Lake City, UT, p 61

Lourakis MA, Argyros A (2009) SBA: a software package for generic sparse bundle adjustment. ACM Trans Math Softw 36(1):1–30. https://doi.org/10.1145/1486525.1486527

Munk JD (2011) The descent of ant. PhD thesis, University of California, Berkeley. http://rspb.royalsocietypublishing.org/content/royprsb/284/1849/20162602.full.pdf

Premerlani LB (2007) Stereoscopic reconstruction and analysis of infrared video of bats. Master's thesis, Boston University

Reid DB (1979) An algorithm for tracking multiple targets. IEEE Trans Autom Control 24:843–854

Riskin D, Willis D, Iriarte-Díaz J, Hedrick T, Kostandov M, Chen J, Laidlaw D, Breuer K, Swartz S (2008) Quantifying the complexity of bat wing kinematics. J Theor Biol 254(3):604–615

Sheltom RM, Jackson BE, Hedrick TL (2014) The mechanics and behavior of cliff swallows during tandem flights. J Exp Biol 217: 2717–2725. https://doi.org/10.1242/jeb.101329, http://jeb.biologists.org/content/jexbio/217/15/2717.full.pdf?with-ds=yes

Sikes R, Gannon W (2011) Animal Care and Use Committee of the American Society of Mammalogists guidelines of the American Society of Mammalogists for the use of wild mammals in research. J Mammal 92:235–253

Theriault DH, Wu Z, Hristov NI, Swartz SM, Breuer KS, Kunz TH, Betke M (2010) Reconstruction and analysis of 3D trajectories of Brazilian free-tailed bats in flight. In: Workshop on visual observation and analysis of animal and insect behavior, held in conjunction with the 20th international conference on pattern recognition, August, 2010, Istanbul, Turkey, 4 p. http://homepages.inf.ed.ac.uk/rbf/VAIB10PAPERS/theriault-icpr-20pdf

Theriault DH, Fuller N, Betke M (2013) Understanding collective behavior in *Tadarida brasiliensis* using computer vision and multi-target tracking. In: Proceedings of the 16th international bat research conference, San José, Costa Rica

Theriault DH, Fuller NW, Jackson BE, Bluhm E, Evangelista D, Wu Z, Betke M, Hedrick TL (2014) A protocol and calibration method for accurate multi-camera field videography. J Exp Biol 217:1843–1848. https://doi.org/10.1242/jeb.100529, http://jeb.biologists.org/content/217/11/1843.full.pdf

Towne G, Theriault DH, Wu Z, Fuller NW, Kunz TH, Betke M (2012) Error analysis and design considerations for stereo vision systems used to analyze animal behavior. In: Proceedings of the workshop on visual observation and analysis of animal and insect behavior (VAIB 2012), held in conjunction with the 21st international conference on pattern recognition (ICPR 2012) Tsukuba, Japan, 4 p. http://www.cs.bu.edu/fac/betke/papers/VAIB-Towne-etal-2012.pdf

Tweed D, Calway A (2002) Tracking many objects using subordinated CONDENSATION. In: Proceedings of the British machine vision conference, Cardiff, UK, pp 283–292

Veeraraghavan A, Chellappa R, Srinivasan M (2008) Shape-and-behavior-encoded tracking of bee dances. IEEE Trans Pattern Anal Mach Intell 30(3):463–476

Walker SM, Thomas AL, Taylor GK (2009) Photogrammetric reconstruction of high-resolution surface topographies and deformable wing kinematics of tethered locusts and free-flying hoverflies. J R Soc Interface 6(33):351–366

Wu Z, Hristov NI, Hedrick TL, Kunz TH, Betke M (2009a) Tracking a large number of objects from multiple views. In: Proceedings of the international conference on computer vision (ICCV), Kyoto, Japan, 8 p. https://doi.org/10.1109/ICCV.2009.5459274, http://ieeexplore.ieee.org/document/5459274/

Wu Z, Hristov NI, Kunz TH, Betke M (2009b) Tracking-reconstruction or reconstruction-tracking? Comparison of two multiple hypothesis tracking approaches to interpret 3D object motion from several camera views. In: Proceedings of the IEEE workshop on motion and video computing (WMVC), Snowbird, Utah, 8 p. https://doi.org/10.1109/WMVC.2009.5399245, http://ieeexplore.ieee.org/document/5399245/

Wu Z, Kunz TH, Betke M (2011) Efficient track linking methods for track graphs using network-flow and set-cover techniques. In: Proceedings of the IEEE conference on computer vision and pattern recognition (CVPR), Colorado Springs, pp 1185–1192. https://doi.org/10.1109/CVPR.2011.5995515, http://ieeexplore.ieee.org/document/5995515/

Wu Z, Thangali A, Sclaroff S, Betke M (2012) Coupling detection and data association for multiple object tracking. In: Proceedings of the IEEE conference on computer vision and pattern recognition (CVPR), Providence, RI, 8 p. https://doi.org/10.1109/CVPR.2012.6247896, http://ieeexplore.ieee.org/document/6247896/

Wu Z, Fuller N, Theriault D, Betke M (2014) A thermal infrared video benchmark for visual analysis. In: Proceedings of the 10th IEEE workshop on perception beyond the visible spectrum (PBVS), Columbus, OH, 8 p. https://doi.org/10.1109/CVPRW.2014.39, http://ieeexplore.ieee.org/document/6909984/

Yang X, Schaaf C, Strahler A, Kunz T, Fuller N, Betke M, Zu WZ, Theriault D, Culvenor D, Jupp D, Newnham G, Lovell J (2013) Study of bat flight behavior by combining thermal image analysis with a LiDAR forest reconstruction. Can J Remote Sens 39(S1):S112–S125. https://doi.org/10.5589/m13-034

Using Agent-Based Models to Scale from Individuals to Populations

11

Eli S. Bridge, Jeremy D. Ross, Andrea J. Contina, and Jeffrey F. Kelly

Abstract

All aggregate biological phenomena in the aerosphere are due to the behaviors of unique individuals acting according to their own rules of behavior and perceived external stimuli. An appealing characteristic of aeroecology is that we can observe both these aggregate behaviors of large groups (using tools such as radar and observational networks) as well as the behavior of individual animals (by employing animal tracking technology). Traditional population modeling efforts focus on equations that mimic natural populations in terms of overall population size and/or mean population parameters, often discounting mechanisms operating on the individual level that give rise to overall population dynamics. Concurrent advancements in computing capacity and animal tracking methodologies provide us with the opportunity to examine how the actions of individuals scale up to give rise to population-level phenomena in the aerosphere. More specifically, we can now model populations as a collection of individuals that behave independently, and we can validate the inferences from these agent-based models by tracking actual animals over the course of their annual cycle. In this chapter, we provide an example of how an agent-based model can be used to predict migration behavior across a species range based on a small set of actual migration tracks. The example provides a generalizable framework for using agent-based models as a link between data from individuals and broad-scale phenomena.

E.S. Bridge (✉) · J.D. Ross · A.J. Contina
Oklahoma Biological Survey, University of Oklahoma, Norman, OK, USA
e-mail: ebridge@ou.edu; rossjd@ou.edu; andrea.contina@ou.edu

J.F. Kelly
Oklahoma Biological Survey and Department of Biology, University of Oklahoma, Norman, USA
e-mail: jkelly@ou.edu

© Springer International Publishing AG, part of Springer Nature 2017
P.B. Chilson et al. (eds.), *Aeroecology*,
https://doi.org/10.1007/978-3-319-68576-2_11

259

1   Introduction: Agent-Based Modeling in Aeroecology

An appealing characteristic of aeroecology is that current methodologies allow us to observe aggregate behaviors of large groups of animals as well as individual behaviors. Tools such as weather radar and observational networks like eBird allow us to make large-scale quantifications of animal populations that use the aerosphere (see Chaps. 4, 12, 13 and 16), whereas tracking devices like satellite transmitters and geolocators let us determine the locations and behaviors of individual animals (see Chap. 9). However, when we try to apply these data sources to address fundamental questions about what governs individual or aggregate behaviors, we often find that the specificity of the broad-scale data is lacking and that the generalizability of individual-based tracking data is limited. This dilemma is analogous to the more general problem of understanding animal behavior and population biology from top-down and bottom-up approaches. We can represent the gross dynamics of a population as a simple mathematical formulation (e.g., a Lotka–Volterra equation; Yorke and Anderson 1973), but this approach fails to offer detailed explanations for the observed population dynamics, especially when applied to systems with dynamically interacting heterogeneous components (Kaul and Ventikos 2015). Yet, a focus on the behavior of a few individual animals or even many animals within a single population may be biased toward the peculiarities of the studied population. Moreover, expanding a study to include a large number of individuals and populations is often economically or logistically impossible.

In this chapter, we argue that agent-based models (ABMs) can serve as a unifying link between knowledge about individuals and observations of broad, population-level phenomena. The theoretical basis for ABMs assumes that all aggregate biological phenomena are due to the behaviors of unique individuals acting according to their own rules of behavior and perceived external stimuli (Marceau 2008). If we can generate a realistic simulation of how individuals interact with their environment and each other, then we can simulate entire populations of individuals that demonstrate the same population-level characteristics of actual populations. Basic concepts of population growth and demography can still be applied, but there is the added benefit of knowing how individual-level behaviors relate to the overall population dynamics. Moreover, we can experimentally change these individual parameters and observe the aggregate outcome. This key capacity in agent-based modeling provides insight into the evolutionary forces that shape both individual and, consequently, aggregate behavior. Even further insight is possible given that we can follow the histories of individual agents to see how their particular decisions led to particular fates. For example, Kanarek et al. (2008) sought to explain the use of traditional foraging grounds by Barnacle Geese (*Branta leucopsis*) by generating agents (geese) that could choose unoccupied sites according to social rank, memories from past seasons, past reproductive success, and an inherited genetic propensity for site faithfulness. They found that year-to-year site fidelity is highly dependent upon the principle that increased familiarity with a site leads to increased food intake.

Hence, site fidelity is largely due to memory and prior experience, which suggests that geese may not adapt readily to rapid redistributions of resources.

ABMs generally seek at least one of two forms of insight. The first is large-scale emergent behaviors or phenomena not explicitly described by the behavior of the components of the system that may be nonintuitive, context dependent, or otherwise too complex to predict with knowledge only of how individual agents behave. Discovery of emergent behaviors was a goal of what is arguably the first agent-based model, that of Botkin et al. (1972). This model, called JABOWA, used known measures of growth as well as shade sensitivity, water availability, and temperature effects to simulate forest dynamics and succession within a mixed-species, mixed-age expanse of trees. The model was successful in achieving realistic simulations of interactions among trees of different sizes and different species, and the study helped to explain the maintenance of species diversity in old-age forests.

A second general goal of ABMs is to determine a simple suite of individual behavioral traits that can give rise to a known aggregate behavior. An example that sought this goal was a model by Parrish et al. (2002), which synthesized several mathematical models to generate a simple set of individual behavioral traits that gave rise to stereotypical schooling behavior in collections of computer-generated fish. Their ABM was capable of generating schooling behavior among individuals whose movement decisions amounted to a sum of locomotory, social, and random forces acting on each individual, with no inherent drive to match each others' behavior. They found that social forces (attraction to conspecific groups) were the driving force of school formation, but physiological and physical forces (i.e., drag and/or swimming speed) determined the type of schooling behavior (e.g., polarized swimming) that emerged. The now famous "Boids" program was an even simpler means of simulating a natural phenomenon. In this model, Reynolds (1987) was able to generate realistic flocking behavior among virtual birds that "flew" amidst each other as well as various objects. The agents' paths were determined by only three simple rules: (1) maintain a minimum distance from any object/bird, (2) match the speed of nearby birds, and (3) move toward the center of the flock. Although this model is a simplification of the behavioral rules that govern real flight behavior, the emergent behavior of a group of "boids" closely resembles the flight behavior of real flocks.

At present, ABMs are used for everything from economics to oncology (Tesfatsion 2002; Wang et al. 2015), and they have progressed toward ever greater levels of realism. The earliest ABMs were often confined to a theoretical landscape. However, increased availability of geographic information has given rise to a generation of agent-based models that uses representations of a real-world landscape as an arena for agents to interact with their environment and each other. Extending this paradigm further are ABMs that predict evolutionary responses to environmental change by altering the current digital representations of landscapes to match predictions of future changes (Coss-Custard and Stillman 2008). The ecological scope of ABMs is also expanding. Of particular interest is a recent methodological convergence by ecologists and epidemiologists in the use of ABMs to investigate the role of individual behavior and social networks in disease spread (see Chap. 17). Despite these advances over the last 40 years, ABMs are not a mainstream form of analysis. At present, ABMs are biased toward addressing pragmatic problems (e.g., management of threatened populations), but they also are increasingly used as a tool for addressing more general questions (DeAngelis and Grimm 2014).

We argue that in fields like aeroecology, where the gap between our understanding of the individual and that of broad-scale phenomena (e.g., transcontinental migration) is particularly wide, ABMs offer a much-needed link between individuals and populations. Although they cannot be applied to every situation, ABMs work well when individuals respond strongly to a minimal set of environmental cues or spatiotemporal features in the landscape. An excellent example is a simulation by Erni et al. (2003) that explores the possible migration strategies used by inexperienced (first year) Garden Warblers to navigate from their natal locations in Scandinavia to potential wintering sites in West Africa. The modeling effort compared a variety of instinctive orientation strategies (i.e., flying in a fixed direction) and incorporation of simple movement decisions wherein birds followed coastlines that led toward the wintering grounds. They found that survival, as estimated from the availability of refueling habitat along a route, was enhanced when birds followed coastlines (thus avoiding long-distance water crossings), even though these routes were not as efficient as those that relied upon fixed flight headings. Another example comes from Dennhardt et al. (2015), who present an agent-based model of Golden Eagle (*Aquila chrysaetos*) movements in the Northeastern USA that is founded on the premise that these birds respond strongly to updrafts as they soar southward during their fall migration. Dennhardt et al. used a real-world physical model of updrafts to predict the migratory movements of eagles with regard to both space and time, and their model results corresponded closely with historical raptor monitoring data as well as monitoring in new sites selected as a result of the modeling output. The following chapter section offers a similar but more generalized approach to using ABMs to predict migration behavior.

2 Extending an Existing Model to Predict Bird Migration Routes

2.1 Background

As a more general example of agent-based modeling, we present an extension of a previous optimization exercise that attempted to determine if a molt-migrant songbird, the Painted Bunting (*Passerina ciris*), migrated in a manner that optimized its exposure to primary productivity. Molt migrants are so called because they molt their flight feathers during fall migration as opposed to immediately before or after, often pausing at traditional molting sites each year en route between breeding and wintering areas. Among North American songbirds, many molt migrants fly to the North American Monsoon Region in Northwestern Mexico to molt their feathers before moving farther south to their wintering grounds. Some birds, including many Painted Buntings, follow a markedly indirect path between their breeding and wintering grounds to molt their feathers in the monsoon region (Contina et al. 2013; Jahn et al. 2013), and this behavior has been attributed to the flush of primary productivity that is typical of this region in early fall (Leu and Thompson 2002; Rohwer et al. 2005). Bridge et al. (2016) created an agent-based model wherein virtual birds, with movement capabilities similar to real birds, were allowed to move upon a dynamic map of real-world primary productivity in North America that represented the amount of new plant growth (based on a 28-day differential of EVI) within each grid point (pixel) for every day of the fall migration period. Through comparisons among millions of simulations generated by several different search algorithms, this modeling effort yielded sets of migration paths that optimized exposure to primary productivity during the fall molt migration. These optimal paths agreed strongly with the actual migration tracks of real Painted Buntings from southwestern Oklahoma (Fig. 11.1; Contina et al. 2013), which suggests that real Painted Bunting migration paths can be predicted by finding movement paths that maximize exposure to primary productivity as the breeding season ends.

We used this observation as the basis for predicting the migration paths of Painted Buntings from locations throughout their breeding range. Interestingly, Painted Buntings have a split geographic distribution among breeding populations. A large western population in the South-Central USA is separated from a smaller breeding population near the Atlantic Coast by a ~500 km east-west expanse of coastal plain that lacks Painted Buntings (Fig. 11.1). The eastern population does

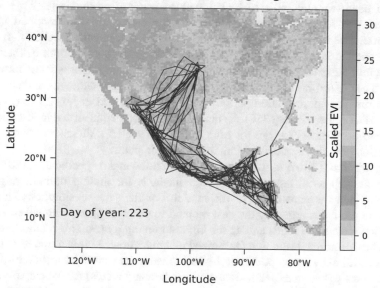

Fig. 11.1 Migration tracks of 12 real Painted Buntings (blue) and the best ten tracks (red) from the "random-optimized" simulation method in Bridge et al. (2016). Tracks are shown on a background map of the scaled EVI values that served as the basis for selection of the best simulated migration tracks

not consist of molt migrants. Rather, these birds molt on the breeding grounds in a manner typical of other eastern songbirds. Hence, our modeling effort not only addresses migration routes but also general molt-migration strategies.

2.2   Model Description

For full details about the ABM used here, refer to Bridge et al. (2016). Briefly, the model prescribes realistic movements to individual agents by having the agents draw from distributions of movement distances based on the movements of real birds. For each day, agents were first assigned to a migratory or nonmigratory state for a particular day with a probability of 0.93 for the nonmigratory state. Birds that entered the nonmigratory state either drew from a distribution of short, local movement distances or did not move at all. Agents in the migratory state traveled at a random heading for a given distance that was drawn from a gamma distribution of distances reflecting the capabilities of real birds making migratory flights (see Contina et al. 2013). For the next 5 days after entering the migratory state, there was a probability of 0.2 that long-distance movement would continue, and if that happened, the heading would be maintained within 20° of the original heading. Except for the constrained headings associated with the migratory state, travel direction was a randomly chosen heading from 0 to 359°.

Millions of movement paths were generated with this process with the intent of exploring a wide range of possible migration routes. Optimization, or identification of the "best" paths, was performed by ranking the paths according to how well they would have exposed birds to high levels of primary productivity. To generate primary productivity scores, the locations for each path were mapped upon a dynamic, real-world grid of values derived from the enhanced vegetation index (EVI) and the 28-day change in EVI (dEVI). We incorporated both of these data layers into the grid, but with rescaling to emphasize the importance of new plant growth expressed in dEVI. The resulting scaled EVI map data was updated every day during the fall migration period, which we designated as July 17 (day of year 198) to December 22 (day 356). As birds moved through this virtual landscape, they acquired a scaled EVI score for each day. Total scaled EVI scores for the entire migration provided a means of ranking migration paths.

Bridge et al. (2016) used several variants on the agent specifications described above as well as different optimization methods for finding optimal migration routes within the constraints of the model. For the sake of simplicity, here we only use what was arguably the best method in terms of processing speed and the resulting scaled EVI totals among the highest ranking routes. This method, referred to as the Constrained Random Optimized method, allowed birds to move randomly as described above, except that at the end of each time step, the virtual bird could search locally (within ±30 km along each of the x and y axes) for the map pixel with the highest scaled EVI score and then go to that pixel location and add its scaled EVI score to the list of daily values. In addition, we added constraints on the initiation of migration to reflect the distribution of departure dates in real birds, and we incorporated a stationary period of at least 30 days to simulate the molting period. For each of five selected origin locations, we simulated one million agents using the Constrained Random Optimized method. Optimal routes were designated as the top 100 routes ranked according to the total scaled EVI score from throughout the migration period.

We chose origin locations that represented the extent of the breeding range for Painted Buntings (Fig. 11.1). The choice of these particular locations was arbitrary although they correspond with current and potential study sites for our research group. Within the western population we chose four sites. The western most of these is Big Bend Texas (29.25°N, 103.29°W). Also among these sites is our original study site at the Wichita Mountains Wildlife Refuge in Southwestern Oklahoma (34.8°N, 98.8°W), for which we have tracking data from real birds. We also include sites in northeastern Arkansas (35.46°N, 94.15°W) and in southern Mississippi near its border with Louisiana (31.42, 91.46) which we know host robust populations of Painted Buntings. Within the eastern portion of the breeding range, we specified migration origins in coastal North Carolina (33.85°N, 78.0°W) and at an inland site in South Carolina (33.1°N, 80.9°W).

Upon examining the results from these simulations, we noted that it was common for birds from the east coast to move across the continent to northwestern Mexico (Fig. 11.2). Given that these migration routes are highly unlikely based on what we know about the species, we carried out a series of post hoc simulations in which we imposed a distance penalty in which presumed energetic gains relative to scaled EVI scores were reduced in proportion to the distances the agents moved. Hence, long transcontinental movements were less likely to result in net benefit, despite the resulting possibility of exposure to higher and rapidly increasing EVI scores. The penalty was assessed at every time step with one unit of scaled EVI subtracted for every 10 km traveled. This distance penalty could also be regarded as an energy penalty that affects the number of refueling days required in association with a long-distance movement. For example, the number of "refueling" days needed to offset a 500 km migration was, for example: 0.8 days at a theoretical maximum site (EVI = 1.0, dEVI = 1.0, scaled EVI = 125.0), 3.2 days at an average site (EVI = 0.5, dEVI = 0.5, scaled EVI = 15.625), or 25.6 days at a poor site (EVI = 0.25, dEVI = 0.25; scaled EVI = 1.95). To make a slightly more thorough search of the landscape, this second simulation was run with 20 million agents for each site of origin. Otherwise, all settings were identical to the previous simulation.

2.3   Results

As in Bridge et al. (2016), the model results emphasize the importance of the North American Monsoon region as a stopover site for molting birds (Fig. 11.2). Even some of the best virtual-bird tracks from the eastern population made the cross-continental journey to the monsoon region. We regard these routes as highly unlikely if not impossible due to the distances traveled, genetic signatures of Painted Buntings in the monsoon region (Contina, unpublished data), and the fact

Tracks with Highest Scaled EVI Totals

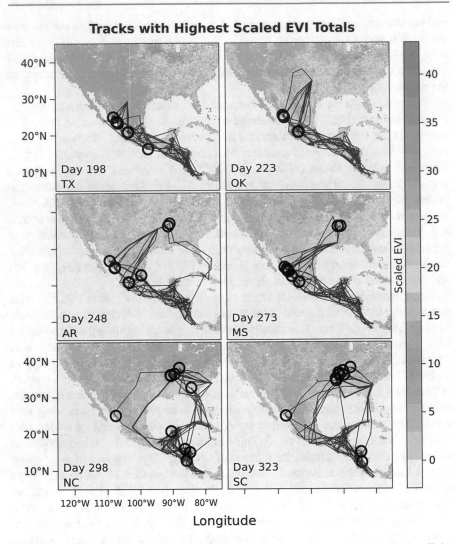

Fig. 11.2 Maps for each of the origin sites used for agent-based simulations without a flight-distance penalty. The 10 tracks with the highest total scaled EVI scores are shown with the highest scoring track of each set of 10 indicated by a red line. Black circles indicate molting sites for the simulated birds. The tracks are presented atop a series of background maps that illustrate how daily scaled EVI values vary in both space and time during the migration period

that the eastern birds molt on the breeding grounds before migrating (Thompson 1991). The conventional view of Painted Bunting migration is that birds in the eastern populations move south to winter in the tip of Florida, the Caribbean islands, and perhaps the Yucatan Peninsula (Lowther et al. 1999; Sykes et al. 2007). It is also apparent that the simulated birds are drawn to the Mississippi River valley, where high and perhaps increasing levels of primary productivity are maintained in the fall, which is largely due to agriculture (Hicke et al. 2004). We note that real Painted Buntings do not make use of this region as a migratory stopover location, which suggests a degree of oversimplification in our model. Clearly, current levels of primary productivity alone do not dictate migration routes, and it is likely that the migration routes of real Painted Buntings evolved prior to agricultural enhancement of primary productivity in the Mississippi River Valley.

Among the simulations originating from Oklahoma and Texas, which had the highest total accumulation of scaled EVI scores, all of the birds molted their feathers in the monsoon region. In contrast, there was a mix of strategies among the best simulations originating from Arkansas and Mississippi. This difference suggests that molting at or near the breeding grounds becomes a viable option as one moves east across the breeding range. This scenario is corroborated by museum specimens from the coastal plains region that were in molt when collected during the post-breeding period (see Thompson 1991). Also, eBird checklists from late fall and early winter corroborate the suggestion that many Painted Buntings molt near the north coast of the Gulf of Mexico (eBird 2012).

Our initial simulations (i.e., those without the distance penalty) of birds originating from North Carolina and South Carolina resulted in many high-ranking tracks where birds flew across North America to the monsoon region in Northwestern Mexico (Fig. 11.2). The behavior of these agents points to a potential weakness in the model, which is that birds can potentially fly great distances without constraints related to energetic costs. To address this problem, we carried out a post hoc set of simulations that entailed a distance penalty (described above) wherein extremely long flights would detract from total scaled EVI accumulations. These simulations yielded tracks that were more in line with expected migratory behavior than the simulations with no distance penalty. In particular, birds from the eastern populations did not fly to the monsoon region with the distance penalty in effect. Instead, these birds moved almost directly southward to areas within the known wintering range of Painted Buntings (Fig. 11.3). Interestingly, all these simulated birds reached wintering sites in Mexico and Central America. A large contingent of Painted Buntings are known to winter in Southern Florida and the Caribbean, but this strategy is not represented among the highest scoring simulations. Moreover, it is known that most if not all Painted Buntings in the eastern USA molt their feathers on or near their breeding grounds prior to fall migration. Again, the best of the simulated tracks did not hold to this strategy as molt locations (circles in Fig. 11.3) were more strongly associated with wintering locations than breeding locations, with considerable molt occurring in Cuba and Southern Mexico.

Although the distance penalty had little effect on optimal routes from Oklahoma and Texas, it did appear to influence migration from Arkansas and Mississippi. The top ten simulated birds from these areas were divided among migration routes that circumnavigated the Gulf of Mexico to the east and west (Fig. 11.3), although the majority of the birds went west. There was also some disagreement among these tracks with regard to molt strategies, as most birds molted after a substantial migratory movement, but a few molted near the breeding area. We note that the

Tracks with Highest Scaled EVI Totals

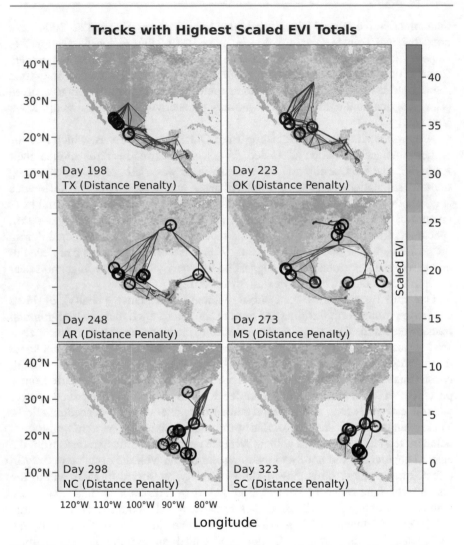

Fig. 11.3 Maps for each of the origin sites used for agent-based simulations with a flight-distance penalty. Colors and symbols are the same as in Fig. 11.2

highest scoring track from Mississippi did not leave North America when the distance penalty was in effect (Fig. 11.3), which is likely due to artificially high EVI scores associated with agriculture.

Although we enabled agents to cross the Gulf of Mexico, it rarely occurred in the highest ranking simulations either with or without the distance penalty. Reports from offshore drilling operations suggest that some Painted Buntings do cross the Gulf of Mexico, although we do not have sufficient data to speculate as to the length or path of this overwater flight. It is possible that gulf-crossing is performed

Table 11.1 Means and standard deviations for simulated fall migration track parameters associated with six different origin locations

Origin	Total scaled EVI	Departure date (day of year)	Molt initiation (day of year)
AR	2960 ± 34.6	201.9 ± 4.2	219.8 ± 11
MS	2949 ± 37.4	198 ± 0	222 ± 12
NC	2868 ± 29.7	198 ± 0	220.8 ± 13
OK	2973 ± 38.5	198 ± 0.41	219.6 ± 11
SC	2895 ± 32.7	198 ± 0	224 ± 13
TX	3073 ± 39.7	198 ± 0	217.4 ± 12
With distance penalty in effect			
AR	2517 ± 28.5	206.2 ± 6.8	224 ± 11
MS	2500 ± 40.2	198.3 ± 3.1	222.8 ± 13
NC	2464 ± 34	198 ± 0	223.5 ± 11
OK	2558 ± 43.1	198 ± 0	223.8 ± 10
SC	2537 ± 28.9	198 ± 0	225.3 ± 11
TX	2702 ± 39.4	198 ± 0	222.2 ± 13

primarily by birds that breed near the gulf coast. These birds would benefit most from an overwater shortcut in terms of flight time and distance. Our simulations did not include gulf-coast breeders, and examination of gulf-crossing per se is not the focus of this exercise. However, simulations that start at an array of locations along the gulf coast might make for a useful follow-up exercise.

Total scaled EVI scores for the top 100 tracks were remarkably similar across the different breeding origins (Table 11.1). Tracks originating from Southwestern Texas had slightly higher scaled EVI totals than the others, which is probably due to the site's proximity to the North American Monsoon Region. Molt dates were also slightly advanced for the best tracks from Texas, which is also likely due to the fact that these birds can reach the monsoon region sooner. Interestingly, departure dates (defined as the day that corresponds with a movement of at least 160 km from the origin) were uniform and early across the breeding sites. The same phenomenon was observed in Bridge et al. (2016), which implies that some selective force other than maximizing exposure to primary productivity is influencing the timing of migration. One possible explanation is that Painted Buntings are primarily granivorous and require plant growth that is mature enough to produce seeds. Hence, new growth might not be the best metric for evaluating favorable locations for Painted Buntings.

In summary, the simulations offer several testable hypotheses with regard to migratory connectivity among different populations of Painted Buntings. For example, birds from the eastern edge of the western population would most likely circumnavigate the Gulf of Mexico to the west, but an eastern movement is also viable. Also, it is predicted that many of these birds molt after leaving the breeding grounds. These hypotheses could be tested empirically through geolocator tracking or, in the case of molt initiation, simple field observations.

3 Further Expansion and Implementation of Agent-Based Modeling

The modeling effort presented above is one of several means by which agent-based modeling can be applied to aeroecology. Clearly, our approach could be fine-tuned more to introduce more species-specific movement parameters. Moreover, this approach could be modified to emulate the state-space samplers that underlie several animal-tracking analysis packages (Rakhimberdiev et al. 2016; Sumner et al. 2009; Wotherspoon and Sumner 2014).

A few of the potential applications for ABMs in aeroecology are mentioned above, but the list is expansive. ABMs are already used by epidemiologists, but Ross et al. (Chap. 17) offer an ABM customized to diseases spread by volant animals that not only simulates the disease agent but also the measures taken to detect and monitor the disease. Hence, this ABM lets us determine how much effort it will take to effectively track a disease. ABMs also provide a means of predicting responses to anticipated landscape changes (e.g., habitat degradation, fragmentation, corridor construction, etc.) as well as climate change. With regard to the latter, ABMs could be a valuable tool in predicting optimal timing of migratory movements under a variety of warming scenarios. Whether those optima fall within the range of phenotypic variation for a particular migratory species could provide a key means of quantifying extinction risk or range shifts as a result of climate change.

3.1 Validation

Of course, an agent-based model is ultimately a reasoned guess about how the mechanics of a particular system operate. A key element in any ABM-based study is validation. In the case of aerial and/or migratory animals, there are numerous avenues toward validation. In the example presented above, the ABM leads us to several hypotheses about widespread migration patterns in Painted Buntings that we can test in a number of ways. The most obvious means of validation might be to deploy tracking devices on birds at all of the breeding sites investigated in the ABM. However, this means of validation would have to overcome several major logistical hurdles, among them the fact that Painted Buntings, along with most other songbirds, are so small that they must be tracked with archival devices as opposed to transmitters (Bridge et al. 2011). Hence, the birds must be captured to deploy the devices and captured again, a year later, to recover them. Coordinated tracking efforts of this sort have been done (e.g., Fraser et al. 2013; Hallworth et al. 2015 #5124; Stanley et al. 2014), but they are not common.

Genetic markers offer an alternative means of validating ABM results, especially if one can identify geographical markers that distinguish among regional populations (e.g., Haig et al. 2011; Kelly et al. 2005; Ruegg et al. 2014; Rushing et al. 2014). For example, a set of genetic tools that distinguish among individuals from the six specified breeding origins would allow us to determine whether birds breeding in Arkansas and Mississippi actually travel to the North American Monsoon Region in large numbers. Similarly museum collections and observational data can serve to verify that a species does indeed occur when and where an ABM predicts it should. In the above example, we found both corroboration of ABM results from observational records and reason to dismiss some simulations as misleading. Finally, various types of radar can serve as a validation tool for ABMs. If output from an ABM is specific enough to render three-dimensional locations for flying individuals through time, it is possible to simulate how those agents would be represented by radar data (Stepanian 2015).

3.2   Tools for Agent-Based Modeling

There are numerous tools available for generating agent-based models (see Table 11.2). In choosing a modeling platform, one must generally balance ease of use with the scope and complexity of the envisioned model. The higher level modeling frameworks, such as NetLogo (Wilensky 1999) and Repast Simphony (North et al. 2013), are easy to implement, but they can be extremely slow to run if models are complex or require large numbers of agents and/or time steps (Railsback et al. 2006). Moreover, the specialized and feature-rich ABM platforms sometimes lack transparency with regard to how models actually work. For example, Romanowska (2014, #5304) complained that in NetLogo some operations are so obscure and that the user is sometimes unaware of the details that underlie a particular set of agent "decisions." If the behavior of each agent depends on several calculations, then it would be important to know the order of those calculations. Yet, such specifications are handled "behind the scenes" in some ABM platforms.

Our analysis of Painted Bunting migration was performed with Agent Analyst (ESRI 2010), which is an ABM package that operates within ArcGIS (ESRI 2011). This package greatly simplifies the process of using real-world GIS data in an ABM framework. However, our use of Agent Analyst for the exercise described above was complicated by an obscure programming language (Not-Quite Python) and a data-leak bug that consistently caused a system crash when accessing map data directly from ArcGIS while performing large (>100,000) numbers of simulations. Ultimately, we had to forego the use of Agent Analyst's GIS features to execute a sufficiently large series of simulations. Hence, we would only recommend this platform for models that involve static loading of map layers but do not require rapid and repeated switching of large map datasets (e.g., long sequences of daily EVI maps loaded during independent agent simulations).

Looking beyond the packages listed in Table 11.2, one can generally implement an agent-based model in any programming environment. Although ABM packages like NetLogo have appeal to the beginner and can be useful for many types of simulation, we think that most researchers who regularly build ABMs will eventually gravitate toward generalized programming tools like R, MatLab, and Python (Romanowska 2014; Zvoleff and Li 2014). A downside to this approach is that one must often start from scratch. Establishing agents, generating a virtual environment

Table 11.2 A brief list of some available agent-based modeling platforms with their relative advantages and disadvantages

Name	Description	Advantages	Disadvantages
Repast	Flexible and widely used agent-based modeling system	Capable of hosting a wide range of models and approaches Decent support among its user community	Requires considerable specialized learning Performance may be too slow for large simulations
Netlogo	A widely used, user-friendly, graphical modeling platform with implementations in several other platforms (see below)	Allows users to rapidly progress to an individual-based models by defining "turtles" that interact with a virtual environment and each other	Large simulations not possible Limited accessibility to lower level functions
ReLogo	A merger of the repast AMB engine and the netlogo graphical interface	Simpler to learn than Repast, with good support for beginners	Limited scope of models and too slow for extremely large simulations
HexSim	Stand-alone spatially explicit, individual-based modeling platform	User-friendly and intuitive graphic interface	Limited modeling scope that applies mainly to standard population models based on life-history parameters
PyABM	Library of agent-based scripts for python	Works in Python, which means virtually no restrictions on the nature of the ABM, and there is the potential for fast (i.e., large) simulations	Requires literacy in Python
Insight maker[a]	An online ABM platform	Runs in a web browser and allows for easy online sharing of models and results	Performance (speed) is sacrificed to increase accessibility
RNetlogo	A package for the R programming environment with functions for working with NetLogo models and reporters	Extends the capabilities of Netlogo to include R functionality[b]	Requires literacy in R and an understanding of the Netlogo model structure. May be too slow for large simulations
Agent analyst	An implementation of Repast as an ArcGIS package	Allows implementation of mapping functions and convenient displays of spatially explicit results	Uses an odd version of Python (NQPython) with limited support. Data leaks often derail large simulations, when mapping functions are employed

[a]The authors have no direct experience with this newly released platform. The information presented is based on Fortmann-Roe (2014)
[b]Note that it is also possible to embed R into NetLogo models using the Netlogo-R-Extension (http://r-ext.sourceforge.net)

for them, and implementing behaviors in a generalized computing language could involve reinventing several of the wheels already implemented in a higher level ABM package. Perhaps the future of agent-based modeling lies in shared packages and software libraries that work within generalized environments. For example, Zvoleff (2014) has recently offered PyABM, a toolkit for implementing ABMs in Python.

The emergence of new ABM platforms is accompanied by an increasing number of sources for real-world environmental data. With increased variety and resolution of environmental data, we can come ever closer to generating arenas for ABM that mirror the real world. One of the most exciting aspects of aeroecology is the rapid development of sensor data and data infrastructure that is relevant to life in the aerosphere. Although the Painted Buntings in this example use the aerosphere as a conduit for long-distance movement, this model does not make explicit use of data that describe the aerosphere, such as air temperatures, winds aloft, or precipitation patterns. Clearly these factors influence migration routes (see Chap. 8), and several agent-based studies have focused on phenomena in the aerosphere as drivers of migration patterns. The study of Golden Eagles by Dennhardt et al. (2015; described in Sect. 1) is one example of how localized aeroecology phenomena (updrafts) shape migration patterns. A more general example of how aerosphere phenomena can shape migration routes is the optimization study by Kranstauber et al. (2015), which sought optimal migration routes for songbirds given 21 years of empirical global wind data. Although the authors do not characterize their work as an agent-based model, they effectively simulated more than one million north-to-south migratory flights of bird agents moving between 2000 pairs of randomly generated departure and destination locations. They found that optimal migration routes (i.e., those that exploited winds to minimize migration time) were on average about 25% faster than the shortest distance routes. The optimization algorithm by Kranstauber et al. (2015) will likely be an important tool in the next generation of bird-migration ABMs directed at scaling up from individuals to populations.

As tracking technologies improve and more researchers are able to follow the movements of a limited number of individuals, we will likely make increasing use of data from the aerosphere as we attempt to scale up from individuals to populations. At the same time, improvements in agent-based modeling within the field of aeroecology will be linked to improvements in both computing infrastructure and the programming skills of future researchers. We predict that the use of ABMs will expand as ecologists become more concerned with what individuals are doing and as we attempt to provide wildlife managers with more detailed predictions about how species of conservation concern will respond to different management efforts. Aggregate phenomena such as migration, communal roosting, and swarming are key features of aeroecology (see Chaps. 7, 14 and 16). The basic premise behind emergent behavior, that the behavior of the group is determined by the actions of the independent agents within the group, demands that we understand the motivations of individuals if we want to fully describe aggregate phenomena in the field of aeroecology.

Acknowledgements This reasearch was aided by support from the National Science Foundation (awards 1340921, 1152356, and 0946685) and from the United States Department of Agriculture National Institute for Food and Agriculture (award 2013-67009-20369). All authors belong to the Applied Aeroecology Group, a University Sponsored Organization at the University of Oklahoma.

References

Botkin DB, Wallis JR, Janak JF (1972) Some ecological consequences of a computer model of forest growth. J Ecol 60:849–873

Bridge ES et al (2011) Technology on the move: recent and forthcoming innovations for tracking migratory birds. Bioscience 61:689–698

Bridge ES, Ross JD, Contina AJ, Kelly JF (2016) Do molt-migrant songbirds optimize migration routes based on primary productivity? Anim Behav 27:784–792

Contina A, Bridge ES, Seavy NE, Duckles J, Kelly JF (2013) Using geologgers to investigate bimodal isotope patterns in painted buntings. Auk 130:265–272

Coss-Custard JD, Stillman RA (2008) Individual-based models and the management of shorebird populations. Nat Resour Model 21:3–71

DeAngelis DL, Grimm V (2014) Individual-based models in ecology after four decades. F1000Prime Rep 6:39

Dennhardt AJ, Duerr AE, Brandes D, Katzner TE (2015) Modeling autumn migration of a rare soaring raptor identifies new movement corridors in central appalachia. Ecol Model 303:19–29

eBird (2012) eBird: an online database of bird distribution and abundance [web application]. eBird, Ithaca. http://www.ebird.org. Accessed 23 June 2015

Erni B, Liechti F, Bruderer B (2003) How does a first year passerine migrant find its way? Simulating migration mechanisms and behavioural adaptations. Oikos 103:333–340

ESRI (2010) Agent analyst: agent based modeling extension for arcgis users. Environmental Systems Research Institute, Redlands

ESRI (2011) ArcGIS Desktop: Release 10. Environmental Systems Research Institute, Redlands

Fortmann-Roe S (2014) Insight maker: a general-purpose tool for web-based modeling & simulation. Simul Model Pract Theory 47:28–45

Fraser KC et al (2013) Consistent range-wide pattern in fall migration strategy of purple martin (progne subis), despite different migration routes at the Gulf of Mexico. Auk 130:291–296

Haig SM et al (2011) Genetic applications in avian conservation. Auk 128:205–229

Hallworth MT, Sillett TS, Van Wilgenburg SL, Hobson KA, Marra PP (2015) Migratory connectivity of a Neotropical migratory songbird revealed using archival light-level geolocators. Ecol Appl 25:336–347

Hicke JA, Lobell DB, Asner GP (2004) Cropland area and net primary production computed from 30 years of usda agricultural harvest data. Earth Interact 8(10):1–20

Jahn AE et al (2013) Migration timing and wintering areas of three species of flycatchers (tyrannus) breeding in the great plains of North America. Auk 130:247–257

Kanarek AR, Lamberson RH, Black JM (2008) An individual-based model for traditional foraging behavior: investigating effects of environmental fluctuation. Nat Resour Model 21:93–116

Kaul H, Ventikos Y (2015) Investigating biocomplexity through the agent-based paradigm. Brief Bioinform 16:137–152

Kelly JF, Ruegg KC, Smith TB (2005) Combining isotopic and genetic markers to identify breeding origins of migrant birds. Ecol Appl 15:1487–1494

Kranstauber B, Weinzierl R, Wikelski M, Safi K (2015) Global aerial flyways allow efficient travelling. Ecol Lett 18:1338–1345

Leu M, Thompson CW (2002) The potential importance of migratory stopover sites as flight feather molt staging areas: a review for neotropical migrants. Biol Conserv 106:45–56

Lowther PE, Lanyon SM, Thompson CW (1999) Painted bunting (*Passerina ciris*). In: Poole A, Gill F (eds) The birds of North America no. 398. The Birds of North America, Philadelphia, pp 1–28

Marceau DJ (2008) What can be learned from multi-agent systems? In: Gimblett R (ed) Monitoring, simulation and management of visitor landscapes. University of Arizona Press, Tucson, pp 411–424

North MJ, Collier NT, Ozik J, Tatara ER, Macal CM, Bragen M, Sydelko P (2013) Complex adaptive systems modeling with Repast Simphony. Complex Adapt Syst Model 1:1–26

Parrish JK, Viscido SV, Grunbaum D (2002) Self-organized fish schools: an examination of emergent properties. Biol Bull 202:296–305

Railsback SF, Lytinen SL, Jackson SK (2006) Agent-based simulation platforms: review and development recommendations. Simul Trans Soc Model Simul Int 82:609–623

Rakhimberdiev E, Senner NR, Verhoeven MA, Winkler DW, Bouten W, Piersma T (2016) Comparing inferences of solar geolocation data against high-precision GPS data: annual movements of a double-tagged black-tailed godwit. J Avian Biol 47(4):589–596

Reynolds CW (1987) Flocks, herds, and schools: a distributed behavior model. Comput Graph 21:25–34

Rohwer S, Butler LK, Froehlich DR (2005) Ecology and demography of east-west differences in molt scheduling in neotropical migrant passerines. In: Greenberg R, Marra PP (eds) Birds of two worlds. Johns Hopkins University Press, Baltimore, pp 87–105

Romanowska I (2014) How the python ate the turtle. Accessed 29 June 2015

Ruegg KC, Anderson EC, Paxton KL, Apkenas V, Lao S, Siegel RB, Desante DF, Moore F, Smith TB (2014) Mapping migration in a songbird using high-resolution genetic markers. Mol Ecol 23:5726–5739

Rushing CS, Ryder TB, Saracco JF, Marra PP (2014) Assessing migratory connectivity for a long-distance migratory bird using multiple intrinsic markers. Ecol Appl 24:445–456

Stanley CQ et al (2014) Connectivity of wood thrush breeding, wintering, and migration sites based on range-wide tracking. Conserv Biol 29:164–174

Stepanian PM (2015) Radar polarimetry for biological applications university of oklahoma. Norman, Oklahoma

Sumner MD, Wotherspoon SJ, Hindell MA (2009) Bayesian estimation of animal movement from archival and satellite tags. PLoS One 4:7324

Sykes PW Jr, Holzman S, Inigo-Elias EE (2007) Current range of the eastern population of painted bunting (*Passerina ciris*) part II: winter range. North Am Birds 61:378–406

Tesfatsion L (2002) Agent-based computational economics: growing economies from the bottom up. Artif Life 8:55–82

Thompson CW (1991) The sequence of molts and plumages in painted buntings and implications for theories of delayed plumage maturation. Condor 93:209–235

Wang Z, Butner JD, Cristini V, Deisboeck TS (2015) Integrated PK-PD and agent-based modeling in oncology. J Pharmacokinet Pharmacodyn 42:179–189

Wilensky U (1999) NetLogo. Center for connected learning and computer-based modeling, Northwestern University. Evanston. http://ccl.northwestern.edu/netlogo/

Wotherspoon S, Sumner M (2014) SGAT: Solar/satellite geolocation for animal tracking. https://github.com/SWotherspoon/SGAT

Yorke JA, Anderson WN (1973) Predator-prey patterns. Proc Natl Acad Sci USA 70:2069–2071

Zvoleff A (2014) PyABM: an open source agent-based modeling toolkit. http://azvoleff.com/pyabm.html. Accessed 29 June 2015

Zvoleff A, Li A (2014) Analyzing human-landscape interactions: tools that integrate. Environ Manag 53:94–111

Radar Aeroecology

12

Phillip B. Chilson, Phillip M. Stepanian, and Jeffrey F. Kelly

Abstract

Aeroecology takes an integrative approach across several scientific disciplines to help further our understanding of biological patterns and processes. The use of radar systems to observe and monitor the airborne animals individually, as small groups, or as large-scale collective ensembles provides one example of this modality. Radar systems scanning the atmosphere are primarily used to monitor weather conditions and track the location and movements of aircraft. However, radar echoes regularly contain signals from other sources, such as airborne birds, bats, and arthropods. We briefly discuss how radar observations can be and have been used to study a variety of airborne organisms and examine some of the many potential benefits likely to arise from radar aeroecology for meteorological and biological research over a wide range of spatial and temporal scales. We also provide a brief background covering the fundamentals of radar operations and signal processing followed by a summary of the various radar systems commonly used for aeroecology. Throughout the chapter, we provide examples of biological scatter as detected by radar and describe how these observations can be used to provide meaningful biological information. Radar systems are becoming increasingly sophisticated with the advent of innovative signal processing and polarimetric capabilities. These capabilities are being harnessed to both promote

P.B. Chilson (✉) · P.M. Stepanian
School of Meteorology, Advanced Radar Research Center, and Center for Autonomous Sensing and Sampling, Norman, OK, USA
e-mail: chilson@ou.edu; step@ou.edu

J.F. Kelly
Oklahoma Biological Survey and Department of Biology, University of Oklahoma, Norman, OK, USA
e-mail: jkelly@ou.edu

© Springer International Publishing AG, part of Springer Nature 2017
P.B. Chilson et al. (eds.), *Aeroecology*,
https://doi.org/10.1007/978-3-319-68576-2_12

meteorological and aeroecological research and explore the interface between these two broad disciplines.

1   Introduction

The primary driving force behind the development of radar came with the pressing need to be able to detect and track the movement of airplanes and ships during the period surrounding World War II. Initially, the concept was developed under the vague name of radio direction finding (RDF) in an attempt to cloak the true nature of the technology from enemy spies. The term RADAR as an acronym for Radio Detection And Ranging was not adopted until 1940 by the USA and 1943 by the British. A fascinating account of the technological development of radar along with a collection of fortuitous serendipities in connection with the efforts can be found in such books as those of Buderi (1996) and Conant (2003). An unintended result of probing the atmosphere with radio waves was the detection of non-aircraft signals, which later were identified as either precipitation or so-called angels (Atlas 1959). The value of observations of weather systems motivated the adoption of radar technology in meteorological studies and has contributed to major operational and research applications. The source of this latter class of "angels" was a subject of controversy, with equally many believing in meteorological, inanimate particulate, and volant biological causes (Plank 1956; Tolbert et al. 1958; Kocurek and LaGrone 1967; Chadwick and Gossard 1983).

In the early days of radar development, there was unquestionable proof that airborne organisms could be—and were indeed—detected in radar measurements (Lack and Varley 1945; Crawford 1949). However, the extent to which this occurred was often subject to speculation in the absence of routine validation (Chadwick and Gossard 1983). This uncertainty is likely the reason that radar technology was not readily adopted into the biological sciences, as it was in meteorology. This uncertainty not withstanding, we can acknowledge that it has been known for over 70 years that radar can be used to study the behavior of volant animals in the planetary boundary layer and lower free atmosphere (i.e., the aerosphere).

Subsequent radar studies on biological subjects were primarily performed as a novelty topic in electrical engineering. Laboratory measurements performed by Edwards and Houghton (1959) investigated the dependence of radar cross section on polar look angle for a pigeon (*Columba livia*), starling (*Sturnus vulgaris*), and sparrow (*Passer domesticus*). Additionally, the scattering contribution of feathers was investigated by measuring a pigeon before and after plucking, and effects of wing position were investigated for a rook (*Corvus frugilegus*). Blacksmith and Mack (1965) performed crude lab measurements of ducks and chickens, but as with several similar studies of the time, emphasis was on novelty of the topic rather than serious analysis or biological application. Several studies attempted radar cross-section measurements by releasing live insects and birds from aircraft within a radar sampling volume (Glover et al. 1966; Konrad et al. 1968), others performed measurements on dead specimens tethered to lofted balloons (Richter and Jensen 1973), and some attempted such characterizations on wild birds flying within the airspace (Eastwood and Rider 1966).

Many of the other studies on biological radar scatter during this period were performed by meteorologists within the context of weather research. Atmospheric scientists used small insects as passive tracers of clear-air atmospheric motions (Lhermite 1966; Richter et al. 1973), while radar meteorologists performed studies to characterize and censor biological clutter. Through the latter half of this period, several biologists began applying radar measurements to pure biological studies, effectively founding the modern science of radar aeroecology (Eastwood 1967; Bruderer 1969; Gauthreaux 1970; Able 1970; Alerstam 1972; Williams et al. 1972; Alerstam and Bauer 1973).

The following decades generated a growth in biological radar studies, primarily consisting of an overlap among three distinct areas: (1) the mechanistic exploration of biological radio wave scatter by electrical engineers and radio physicists, (2) the application of radar measurements to biological studies by ecologists, and (3) the study of biological radar echoes by meteorologists for clear-air studies or clutter suppression. Some highlights of this period included the use of tracking-radar baseband signals for wing-beat characterization (Schaefer 1968; Bruderer and Steidinger 1972; Vaughn 1974), multi-radar networks for larger scale migratory studies (Gauthreaux 1971), Doppler observations of flying animals (Martinson 1973), and polarimetric measurements (Mueller and Larkin 1985; Riley 1985). It was also during this period that the concept of radar-based taxonomic classification began to gain traction as the future of biological radar (Martinson 1973; Williams and Williams 1980; Larkin 1980).

Developments through the 1990s established the contemporary landscape of radar aeroecology. These areas included the use of the newly upgraded weather surveillance radar network in the USA (Gauthreaux and Belser 1998; Russell and Gauthreaux 1998), the ornithological tracking and identification efforts by the Swiss (Bruderer 1994; Liechti et al. 1995; Bruderer 1997a, b), and Swedes (Alerstam and Gudmundsson 1999), British entomological studies (Riley and Reynolds 1990), deployments of portable marine radars for ecological studies (Riley and Reynolds 1990; Cooper et al. 1991), and radar-based aircraft birdstrike mitigation (Haykin et al. 1991). Additionally, studies focusing on meteorological data quality and echo classification continued to gain importance (Wilczak et al. 1995; Russell and Wilson 1997; Zrnić and Ryzhkov 1998).

Radar continues to be used to track aircraft and ships, but the applications now go far beyond these initial applications. For example, radar can be used to monitor the movement of land-based vehicles, image and map planetary terrains, detect the presence of objects under ground or under water, observe and characterize weather phenomena, and so forth. Although a continually evolving area, many of these applications have reached a reasonable level of maturity. This cannot be said, however, for the application of radar to the study of flying organisms.

2   Principles of Radar Operation

At its most basic level, the principle of operation of radar is quite simple. Radio waves are generated and transmitted into the atmosphere after being focused by an antenna. The radio waves scatter off an object or collection of objects, and the scattered radio waves are then received by the radar. Typically the range of the scatterers from the radar is determined by noting the time interval between when the radio wave was transmitted and the returned signal was received. The magnitude of the returned signal can be used to glean information about the scatterer or scatterers. A simple block diagram showing the key components of a standard radar system is provided in Fig. 12.1. In the diagram, the same antenna is used for both transmission and reception (monostatic); therefore, a switch is required to control whether the antenna is actively connected to the transmitter or receiver at a given time. The operations of the radar components are regulated by the controller. For Doppler radar systems, the radial velocity of the objects being detected is determined by evaluating the phase of the received radio wave with respect to the phase of the transmitted pulse. After receiving the backscattered signal, data are recorded, processed, and then presented in a form that is meaningful for the radar operator.

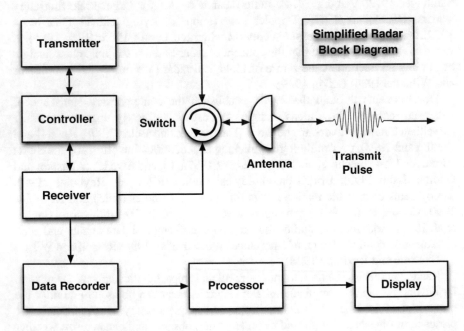

Fig. 12.1 A simplified block diagram of a typical radar system

2.1   Radar Equation: Discrete and Distributed Scatterers

To develop an understanding of how radar can be meaningfully used to advance aeroecology, we must first consider the physics governing the propagation of radio waves through the atmosphere and how these waves are scattered off airborne objects, such as birds, bats, and insects. We can begin by investigating the so-called radar equation, which is used to calculate the amount of backscattered electromagnetic radiation received by a radar in terms of power. The electromagnetic radiation transmitted via the antenna interacts with an object or objects and the scattered energy is typically received by the same antenna used for transmission (monostatic). The expected power of the received radiation is calculated using the basic radar equation. For the case of scatter from a single object located at a distance r [m] from a monostatic radar and angular position given by it's elevation angle θ [rad] and azimuth angle ϕ [rad], the received power P_r [W] is given by a simplified form of the radar equation:

$$P_r = P_t \frac{G^2 \lambda^2}{64\pi^3 r^4} f^4(\theta, \phi, \theta_o, \phi_o) \mid W^2(r, r_o) \mid \sigma_b, \qquad (12.1)$$

where P_t [W] is the transmit power of the radar, G is the gain of the antenna, and λ [m] is the wavelength of the radio waves (Doviak and Zrnić 1993). The pointing direction of the radar antenna is given by θ_o and ϕ_o and the center of the range weighting function by r_o. The monostatic or backscattered radar cross section (RCS) of the scatterer is designated as σ_b [m^2]. A discussion of the RCS, which is a measure of how reflective an object is for a given radar wavelength, is presented below. In general, the RCS of an object (σ) is a function of both the angle of incidence and the scattering angle of the radiation relative to the object. For the case of backscatter observed by a monostatic radar (σ_b), only the angle of the incident radiation need be considered. The weighting functions W and f represent the position of the scatterer within the beam and account for variations in sensitivity with the range sample (W) and beam width (f), with r_o [m] denoting the center of the radar sampling volume. The "size" of the sampling volume is described by W and f as discussed below. Analytical expressions for W and f can be found in several texts such as those of Probert-Jones (1962), Battan (1973), Doviak and Zrnić (1993), and Chilson et al. (2012b).

   The general approach when interpreting radio wave scatter depends on the underlying density and distribution of the scatterers sampled by the radar. In the case of *discrete scatterers*, it is assumed that the received power P_r can be attributed to scatter from individual animals. Another common case is known as *distributed bioscatter*. Throughout this chapter, we will use the term *bioscatter* when referring to radio-wave scatter off volant animals, as opposed to that from precipitation, aircraft, chaff, and such. In considering distributed bioscatter, we assume that the sampled animals are uniformly distributed in space and that they are sufficiently abundant to be treated as a "continuum" of scatterers. These cases are not unique to bioscatter. It is common to consider discrete and distributed scatter when using radar to study geophysical phenomena such as precipitation and ionized media (plasma) (Doviak and Zrnić 1993). A thorough understanding of the differences between discrete and distributed bioscatter is needed when attempting to quantify animal densities in the aerosphere using radar. Although the focus of the following discussion is predominantly on weather radar, the treatment applies to most radars used for biological studies.

The radar equation given by Eq. (12.1) only considers scatter from a single object. When multiple scatterers are present within a single sample volume, e.g., for the case of several discrete bioscatterers, their collective contributions to the backscattered power can be expressed as

$$P_r = P_t \frac{G^2 \lambda^2}{64\pi^3} \sum_i f^4(\theta_i, \phi_i, \theta_o, \phi_o) \mid W^2(r_i, r_o) \mid \frac{\sigma_{b,i}}{r_i^4}, \tag{12.2}$$

where the summation includes all scatterers. Each individual scatterer is indexed by the subscript i. In principle, the summation can be taken over all scatterers, but only those contained within the sampling volume will contribute significantly on account of the range and beam weighting functions.

If the scatterers are uniformly distributed in space within the region being probed by the radar and their number is sufficiently large (*volume-filling assumption*), then we can simplify Eq. (12.2) by introducing the concept of a radar sampling volume.

$$P_r = P_t \frac{G^2 \lambda^2}{64\pi^3} \left[\frac{1}{\Delta V} \sum_{\text{vol}} \frac{\sigma_i}{r_o^4} \right] V_{\text{rad}} = P_t \frac{G^2 \lambda^2}{64\pi^3} \frac{1}{r_o^4} \left[\frac{1}{\Delta V} \sum_{\text{vol}} \sigma_i \right] V_{\text{rad}}, \tag{12.3}$$

where Σ_{vol} represents the summation over all scatterers within a unit volume ΔV and we have used $r_i \approx r_o$. That is, the total number of scatterers being considered is the number of scatterers contained within a unit volume times the radar sampling volume given by Eq. (12.4). The effect of the weighting functions is accounted for by our definition of the radar sampling volume:

$$V_{\text{rad}} = r_o^2 \int_r W^2(r_o, r) dr \int_\Omega f^4(\theta, \phi, \theta_o, \phi_o) d\Omega, \tag{12.4}$$

where Ω is the solid angle subtended by the center of the sampling volume. Using common expressions for W and f, we find that

$$V_{\text{rad}} = \frac{0.35\sqrt{2\pi}}{2\ln 2} \left(\frac{\pi r_o^2 \theta_1 \phi_1 \Delta r}{4} \right), \tag{12.5}$$

where the quantity in parenthesis is equivalent to a truncated oval-based cone having diameters along the horizontal and vertical axes of $r_o\theta_1$ and $r_o\phi_1$, respectively, and a length of Δr (Chilson et al. 2012b).

2.2 Radar Cross Section

Here, we consider in more detail the backscattered radar cross section, which we will simply refer to as the RCS, since we are only considering monostatic radars in this chapter. The RCS of an object is a measure of how reflective it is to radio waves of a given wavelength and has units of area. The RCS depends on both the angle of incidence of the radio wave relative to the object's orientation and the object's dielectric properties. Analytic descriptions of the scattering properties of objects only exist for a small number of geometric shapes. More typically one must rely on laboratory measurements or numerical models to predict how radio waves are scattered off of selected objects. An example of the former approach for biological scatter can be found in the classic paper by Edwards and Houghton (1959) for birds and the comprehensive study by Riley (1985) for insects. A discussion of the latter approach can be found in Mirkovic et al. (2016). These authors applied a method-of-moments model to study radio-wave scatter at X-band in two orthogonal and linear polarizations off a particular species of bat. It must be said that both the laboratory and modeling approaches can be very labor intensive.

Returning our focus to analytic solutions, in the beginning of the twentieth century, Gustav Mie developed a comprehensive set of equations to describe electromagnetic scatter off dielectric spheres of any size (Mie 1908). Solutions to the equations for the case of backscatter from liquid water at different temperatures and radar wavelengths are presented in Fig. 12.2. The upper panel shows calculated RCS values corresponding to a temperature of $0°$ C. Values of RCS are plotted using a convention in which they have been normalized by the physical cross section of the water spheres. Along the abscissa, a normalized metric is used to represent the size of the sphere given by $\alpha = 2\pi a/\lambda$ (circumference divided by radar wavelength), where a is the radius of the sphere. This convention is commonly used in the literature when discussing the RCS values of dielectric spheres. As we see, the results for X-band ($\lambda = 3$ cm), C-band ($\lambda = 5$ cm), and S-band ($\lambda = 10$ cm), which are common weather radar wavelengths, are all similar using the normalized coordinates.

The plot of normalized RCS versus α shown in the upper panel of Fig. 12.2 can be grouped into three regimes depending on the size of the sphere. For $\alpha \lesssim 0.2$ ($D \lesssim 1/16\lambda$), values of the normalized RCS increase according to a power law (linearly when plotted on a logarithmic scale as shown) with α and is known as the Rayleigh regime. Here, the RCS is given as

$$\sigma_b = \frac{\pi^5}{\lambda^4} |K_m|^2 D^6, \tag{12.6}$$

where D is the sphere's diameter and $|K_m|^2$ is a quantity related to the properties of the dielectric material of the sphere, here liquid water. The quantity $K_m = (m^2 - 1)/(m^2 + 2)$ and $m = n + i\kappa$ is the complex refractive index of the material, with n being the conventional refractive index related to the phase speed of waves in a dielectric medium compared to its speed in a vacuum and κ being the absorption coefficient

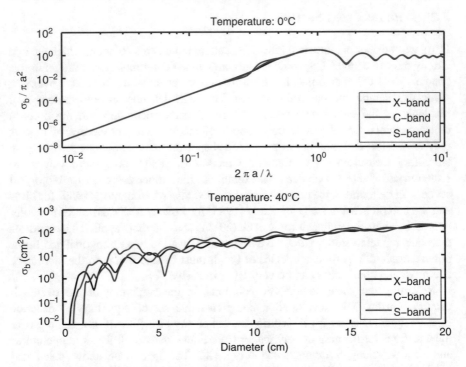

Fig. 12.2 Plots showing calculated values of RCS for spheres of liquid water as a function of their size. Calculations have been made for three common meteorological radar wavelengths: X-band ($\lambda = 3$ cm), C-band ($\lambda = 5$ cm), and S-band ($\lambda = 10$ cm)

(Battan 1973; Doviak and Zrnić 1993). This formulation is referred to as the Rayleigh approximation. For larger values of α, the radar wavelength becomes comparable with the circumference of the sphere and resonant oscillations in the electromagnetic field are created. This is known as the resonant, or sometimes Mie, regime. As the sphere size, and correspondingly α, continues to increase, the resonant oscillations dampen out and the normalized RCS approaches an asymptotic limit determined by the dielectric properties of the sphere. This is the geometric optical regime. For the case of precipitation observed by meteorological radar, the hydrometeors generally fall within the Rayleigh regime. Below we show how this assumption is built into the calculation of the radar reflectivity factor.

The lower panel of Fig. 12.2 depicts calculated RCS values (not normalized) as a function of sphere diameter for water spheres at a temperature of 40° C. This temperature was selected as being representative of the body temperature of flying vertebrates that may be observed by radar. The Mie solution has been invoked in past studies and discussions as a simplifying means of approximating the RCS of birds (e.g., Eastwood 1967; Alerstam 1990; Martin and Shapiro 2007). Clearly, the approximation has limitations but offers a starting point for analysis of bioscatter.

In such cases, the body mass of the animal may be used to calculate the diameter of an equivalent sphere of water. The plots demonstrate the complex dependence of RCS on both size and radar wavelength. In certain cases, larger objects can be less reflective than smaller ones, which can make quantitative interpretation of bioscatter difficult. For example, RCS values corresponding to a 10-cm wavelength radar as reported in Alerstam (1990) for several species of birds assuming mass equivalent spheres of water are willow warbler (9 g and $\sigma_b = 8$ cm^2), chaffinch (20 g and $\sigma_b = 20$ cm^2), song thrush (70 g and $\sigma_b = 2$ cm^2), lapwing (200 g and $\sigma_b = 60$ cm^2), woodpigeon (500 g and $\sigma_b = 40$ cm^2), buzzard (1000 g and $\sigma_b = 70$ cm^2), eider (2000 g and $\sigma_b = 110$ cm^2), and whooper swan (10,000 g and $\sigma_b = 300$ cm^2). RCS values for these species at other radar wavelengths ($\lambda = 3$ cm and $\lambda = 23$ cm) are also given in Alerstam (1990).

Being generally smaller in size, RCS values of insects do tend to increase monotonically at X-band and longer wavelengths and in a sense can be thought of as Rayleigh scatterers. For example, Riley (1985) shows that the RCS for a wide variety of insects exhibits a power-law relationship with the animal's body mass. However, it should be noted that the body geometry of many insects is such that the actual RCS is dependent on the radio wave's angle of incidence. That is, the scatter is aspect sensitive (Mueller and Larkin 1985; Melnikov et al. 2015). More information on radar observations of insects can be found in Chap. 9. Additionally, the reader is referred to Drake and Reynolds (2012) for an in-depth treatment of radar entomology.

3   Scanning Strategies

The scanning strategy used for a particular radar system depends upon its design, which in turn is largely driven by the type of observations the radar is intended to collect. As discussed earlier, radar samples the atmosphere by directing radiation through an antenna to the region of the airspace that is of interest and recording the backscattered signal (Doviak and Zrnić 1993). These resulting signals are transformed into data products, sometimes generally referred to as measurables or observables, which correspond to specific sampling volumes located radially along the antenna beam direction. That is, the retrieved information is a function of antenna pointing direction in azimuth and elevation, as well as range from the radar. Depending on the position and motion of the antenna, different spatial and temporal information may be extracted from the radar data. This discussion will focus on radar systems that operate in surveillance modes, as opposed to those that actively move the antenna to follow a defined object, that is, tracking radar. More information about tracking radar as used in aeroecology can be found in such resources as Bruderer (1994, 1997a, b) Liechti et al. (1995), and Alerstam and Gudmundsson (1999). A number of different sampling techniques have been developed and optimized for surveillance radar systems to extract specific information in time and space based on

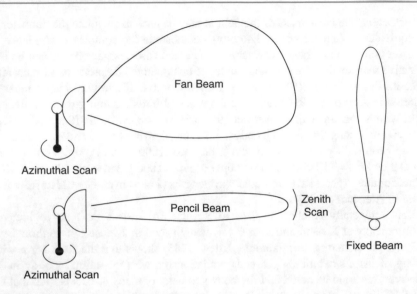

Fig. 12.3 Examples of three different types of radar antenna beam patterns used in aeroecological studies. The fan beam antenna is typically fixed in elevation and steered azimuthally. The pencil beam antenna can be steered in both azimuth and elevation. The fixed-beam pattern remains stationary in time and typically consists of a narrow beam

the application needs. In particular, three sampling strategies commonly utilized on radar platforms include fixed-beam sampling, azimuth scanning, and elevation scanning.

Radar antennas are designed to focus radio-wave signals into a particular region of the atmosphere. This offers the benefit of improving both sensitivity and spatial resolution of the radar. However, it also limits the amount of the atmosphere that can be sampled at a given moment. To systematically scan larger swaths of the atmosphere, one must implement a scanning strategy based on the geometry of the radar beam produced by the antenna, the region of the atmosphere to be sampled, and a desired update time of the observations. Shown in Fig. 12.3 are three typical beam geometries produced by the types of radar systems commonly used for aeroecological studies. We will discuss each of these in more detail below. A more complete discussion can be found in Larkin and Diehl (2012).

Fixed-beam, or spotlight, sampling is the most basic technique, requiring no antenna motion. In this case, the radar is pointed in a fixed direction, often vertically or near vertically, and repeatedly samples the airspace (Figs. 12.3 and 12.4a, left). As organisms pass through the sampled volumes, each pulse returns a profile of the radar measurables as a function of range along the beam (Fig. 12.4a, center). While each sample returns a one-dimensional profile of measurables, consecutive samples are typically displayed in series as a function of time to form a range-time indicator, or RTI display (Fig. 12.4a, right). In addition, it is common that the results of several

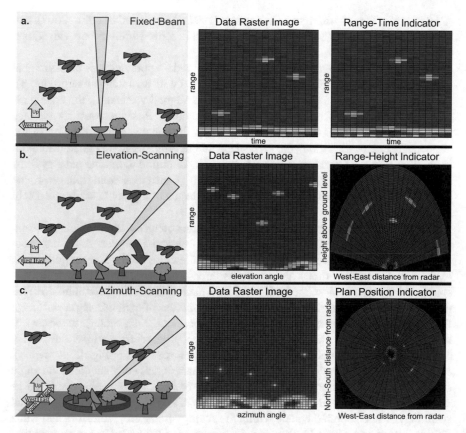

Fig. 12.4 Physical sampling technique (*left column*), data raster image (*center column*), and physical coordinate image (*right column*) for (**a**) fixed-beam sampling, (**b**) elevational scanning, and (**c**) azimuthal scanning. Reproduced from Stepanian et al. (2014)

consecutive pulses are averaged together to reduce the contamination by random noise, thereby improving the data product quality. It is important to note that the RTI display does not directly contain any two-dimensional spatial information but only the temporal evolution of the sampled radial. Examples of fixed-beam sampling in ecological applications have been demonstrated by, e.g., Chapman et al. (2003), Martin and Shapiro (2007), Schmaljohan et al. (2008), and Dokter et al. (2013a, b).

In the second technique, the antenna scans in elevation at a fixed horizontal azimuth angle (Fig. 12.3). In some cases, the antenna will scan from one horizon, through the vertical, to the opposite horizon (e.g., Fig. 12.4b, left), while other times, only a subset of elevation angles are covered. These resulting radar data products are a function of elevation angle and range from the radar (sometimes referred to as an E-Scope, Fig. 12.4b, center). Unlike fixed-beam sampling, elevation scanning contains explicit spatial information with the surveyed airspace encompassing a vertical cross section and is typically displayed as such in the range-height indicator

(RHI) format (Fig. 12.4b, right). Gauthreaux (1991), Martin and Shapiro (2007), and van Gasteren et al. (2008) provide ecological demonstrations of elevational scanning.

Perhaps the most common sampling method in radar applications is the azimuthal rotation of the antenna (Fig. 12.4c, left). If equipped with an antenna that produces a fan beam (with the broadest portion of the beam aligned vertically), the antenna is typically rotated at a fixed elevation angle, providing instantaneous surveillance across a range of heights. That is, the fan beam illuminates a broad region of the atmosphere in elevation along the vertical extent but offers relatively good resolution in azimuth along the horizontal extent (Fig. 12.3). Common to marine radar systems and airport surveillance radars, this scanning strategy provides rapid frame-to-frame updates but cannot provide the altitude of the detected objects (Larkin and Diehl 2012).

In the case of weather radars, a parabolic antenna creates a highly focused beam, which is swept azimuthally through a number of elevation angles. A typical beam width for weather radars is about one degree. While these volume coverage patterns lead to slower update times, they provide three-dimensional coverage of the airspace. In both cases, these methods yield information about the horizontal distribution of organisms, with radar products being a function of azimuth angle and range from the radar (sometimes referred to as a B-Scope, Fig. 12.4c, center). In the case of weather radar, the three-dimensional data products, often referred to as volume data, can be thought of as a collection of two-dimensional scans (as in Fig. 12.4c, center), each at a different elevation angle. These two-dimensional, constant elevation angle data products often referred to as sweeps, tilts, or cuts are typically displayed as viewed from above in the plan-position indicator (PPI) (Fig. 12.4c, right). Azimuthal scanning has been applied widely within the ecology literature, with Horn and Kunz (2008), Buler and Diehl (2009), and Dokter et al. (2011, 2013b) demonstrating several applications.

The examples of radar displays for particular scanning strategies shown in Fig. 12.4 have focused on bioscatter from discrete animals or groupings of animals. However, under many circumstances, a radar will observe collections of biological scatterers within a sampling volume. For example, PPI scans corresponding to migrating birds passing over radars operating in different scanning strategies and with different beam types are depicted in Fig. 12.5. The upper panel shows the assumed distribution of birds in height. Shown in the middle panels are representations of PPI plots for the case in which the migration is widespread with no variations in the height distribution of the birds across the sampled domain. The plots do not take into account azimuthal dependences in the RCS of the birds. For the pencil beam, it is possible to estimate the height distribution of the birds since range from the radar for a single elevation angle corresponds to height. The RHI plot shown is for a narrow radar beam. The height distribution of the birds can be discerned as a band of enhanced radar backscatter in range. Lastly, we show an example of how the migrating birds would be displayed for the case of fixed-beam

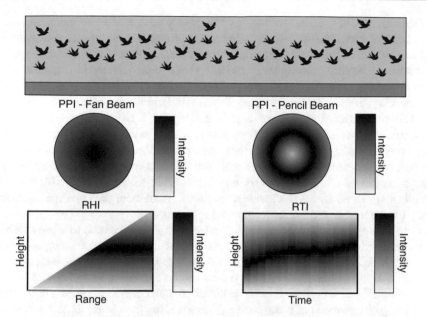

Fig. 12.5 Depiction of how a collection of migrating birds (*upper panel*) may be visualized using different types of radar beams and scanning strategies. In each case, we are assuming that the migration is uniformly distributed across the radar scanning domain. The PPI and RHI scans are representative of "snapshots" of the migration. In the RTI scan, the time evolution of the height distribution of birds (increasing in height with time) can be seen

radar with the antenna pointing vertically. Here, again the height distribution can be obtained by noting the backscattered power with height. In this case, the height distribution is changing over time.

4   Weather Radar in Aeroecology

There is a long tradition of incorporating radar technology into biological studies as a means of observing movements of airborne animals and quantifying their numbers in the lower atmosphere (Gauthreaux and Livingston 2006). Much of this research has been conducted using radar systems that have been specifically adapted for observations of volant species (e.g., Bruderer 1999; Harmata et al. 2003; Chapman et al. 2011; Alerstam et al. 2011). However, some investigators have chosen to incorporate data from operational radar systems (those designed for routine operation) into their biological studies. One example is the use of weather radar, such as those deployed by national weather services, to study birds, bats, and insects (Gauthreaux et al. 2008; van Gasteren et al. 2008; Dokter et al. 2011; Chilson et al. 2012a; Kelly et al. 2012). Indeed, Gauthreaux (1970) began using weather radars in the USA for biological research not long after these facilities were established in 1959. Radar technology has greatly advanced since that time, and it has become increasingly easier to access and process weather radar data with the advent of faster processing and data storage capabilities. Consequently, we are witnessing a rapidly growing interest in integrating radar data into biological studies. Moreover, the expanding availability of polarimetric radar information (see below) is providing even more opportunities to biologists to utilize weather radar data.

The conventional outputs of a pulsed, Doppler weather radar are the radar reflectivity factor Z, radial velocity v_r, and spectrum width σ_v, which are evaluated for every sampling volume. From these fundamental measurements, other derived meteorological products such as rainfall rate, vertical integrated liquid, echo tops, and such can be extracted (Doviak and Zrnić 1993; Rinehart 2010). With this in mind, we proceed with a discussion of meteorological radar data products and how they can be interpreted for aeroecological applications.

We have already considered how the radar equation can be used to relate received backscattered power to the number and position of sampled objects along with their respective RCS values. For the case of uniformly distributed scatterers, which fill the radar sampling volume, the radar equation can be expressed as in Eq. (12.3). The volume-filling assumption is well justified in most meteorological applications involving observations of precipitation. Therefore, the term in Eq. (12.3) describing the cumulative backscattering cross section per unit volume has been defined as the radar reflectivity or sometimes simply the reflectivity:

$$\eta \equiv \frac{1}{\Delta V} \sum_{\text{vol}} \sigma_{b,i} \tag{12.7}$$

Doviak and Zrnić (1993) and Rinehart (2010). Using Eqs. (12.3), (12.5), and (12.7), we can see how the radar reflectivity is related to known radar parameters and radar measurables and thus can be readily calculated:

$$\eta = \frac{64\pi^3 r_o^4}{G^2 \lambda^2 V_{\text{rad}}} \frac{P_r}{P_t} = \frac{512(\ln 2)\pi^2 r_o^2}{0.35\sqrt{2\pi}G^2\lambda^2\theta_1\phi_1\Delta r} \frac{P_r}{P_t}. \tag{12.8}$$

We next consider how the radar reflectivity can be related to biologically meaningful parameter. If the bioscatter can be described through a number density N_{bio} with each animal having an actual or "representative" RCS of σ_b, then the radar reflectivity simply becomes

$$\eta = \frac{1}{\Delta V} \sum_{\text{vol}} \sigma_b = N_{\text{bio}}\sigma_b. \tag{12.9}$$

Therefore, under these conditions, it is possible to calculate N_{bio} directly from the radar estimate of η. Moreover, the *effective* number of bioscatterers contributing to the receiver power in the radar equation is simply given as $V_{\text{rad}} \cdot N_{\text{bio}}$. Note that the radar sampling volume increases with range as shown in Eq. (12.5). A discussion of how a nonuniform distribution of bioscatterers can be treated can be found in Chilson et al. (2012b).

The magnitude of the backscattered signal for weather radars is not reported directly as radar reflectivity but rather as a radar reflectivity factor. Whereas radar reflectivity naturally lends itself to applications involving the quantitative analysis of bioscatter, this is not the case for the radar reflectivity factor. The radar reflectivity factor has been specifically developed for the study of precipitation and as such contains many assumptions that are customized for this application but could inhibit biological interpretability of radar signals.

$$\eta = \frac{\pi^5}{\lambda^4} |K_m|^2 z, \tag{12.10}$$

where the radar reflectivity factor z is defined as

$$z \equiv \frac{1}{\Delta V} \sum_{vol} D_i^6, \tag{12.11}$$

where the units are often given as $mm^6\ m^{-3}$. We are using a lowercase z here to distinguish this quantity from its logarithmic form denoted by Z as discussed below.

For the case of precipitation, we can further assume that the dielectric material of the scatterers is water; however, the actual value that we assign to m depends on the water's phase and temperature and the wavelength of the radar. Since these properties of the water are not in general known, values of the radar reflectivity factor reported by weather radars are given as the *equivalent* radar reflectivity factor z_e defined as

$$z_e = \frac{\lambda^4}{\pi^5 |K_m|^2} \eta, \tag{12.12}$$

where η is obtained from Eq. (12.8). Here, we not only assume that the Rayleigh approximation holds but also that all scatterers consist of liquid water, resulting in $|K_m|^2 = 0.93$, which is valid for wavelengths used by most weather radars. That is, the equivalent radar reflectivity value represents the value that would be associated with the observed backscatter if it were attributed to volume-filling liquid precipitation in the Rayleigh regime.

Because the radar reflectivity factor z (or equivalent radar reflectivity z_e) can have such a wide dynamic range, values are commonly reported and discussed using a logarithmic scale

$$Z[dBZ] = 10 \log_{10}\left(\frac{z}{1\ mm^6\ m^{-3}}\right). \tag{12.13}$$

Values of Z_e for meteorological phenomena might range from about 10–20 dBZ for light rain to 60–70 dBZ for severe weather events involving hail. As we can see, the assumptions used to derive values of radar reflectivity factor have little relevance for quantitative studies of bioscatter.

We next consider velocity data from meteorological radars. Doppler weather radar contain a coherent receiver, which is capable of detecting changes in phase of the received signal produced by motion of the detected scatterers (Doviak and Zrnić 1993). However, the radar can only detect motion oriented along the viewing angle of the radar beam, that is, motions directed either toward or away from the radar. As signals are received at the radar, the contributions of backscattered power are inherently tied to the motion of the scattering objects. That is, the total power can be decomposed by its frequency content into the contributions from scatterers of different velocities along the pointing direction of the radar beam. Consequently, when sorted by radial velocity, Z becomes $Z(v_r)$. That is, there exists a spectrum of radar reflectivity factors distributed by radial velocity, which is called the Doppler spectrum.

In Doppler weather radar applications, the Doppler spectrum expressed in terms of radar reflectivity factor per radial velocity is typically assumed to follow a Gaussian distribution and is characterized by the moments of the fitted distribution. The zeroth moment, or integrated area under the curve, is simply Z. The first moment corresponds to the peak of the Gaussian fit and designated v_r. The second central moment, or width of Gaussian fit, is designated as the spectrum width σ_v. Note that the moments can be calculated without assuming an underlying functional form of the spectrum.

The Doppler spectrum is the convolution of the frequency modulating effects from several sources. The first source of frequency content is from the scatterers themselves; the velocity components along the radial vector determine the mean frequency shift, while the variation in velocity broadens the peak. Other spectral variations are caused by sampling effects such as scanning motion and width of the beam, both of which broaden the overall spectrum. Finally, the spectrum is also affected by computational artifacts like the choice of windowing functions and discrete Fourier transform length.

In weather radar applications, the Doppler spectrum is assumed to follow a Gaussian distribution and is characterized by the moments of the fitted distribution. The zeroth moment, or integrated area under the curve, is the signal power and is related to Z and η. The first moment is the peak of the Gaussian fit and indicated the mean radial velocity of scatterers (v_r). A typical convention is that negative values denote motions toward the radar. The second central moment of the distribution is the variance of the Gaussian fit and is reported as the spectrum width of the mean Doppler radial velocity (σ_v). In general, spectrum width indicates the diversity of scatterer velocities within a sampling volume.

5   Radar Polarimetry

Many weather radar systems have been or are in the process of being converted to allow dual-polarization operation, which although motivated by improvements in meteorological applications offers many opportunities for aeroecology. Conventional

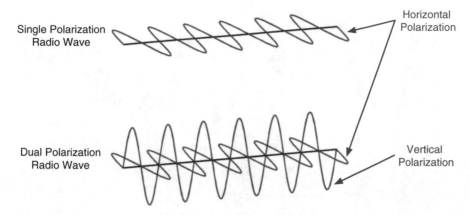

Fig. 12.6 Illustration of single- and dual-polarization radio waves transmitted by radar

radar systems transmit and receive radiation in a single horizontal polarization. There are several schemes for making polarimetric measurements, but weather radar rely on orthogonal dual polarizations. That is, waves are transmitted in the horizontal polarization plane with oscillations parallel to the horizon, as well as the orthogonal vertical plane. See Fig. 12.6. Two distinct modes of orthogonal dual polarization are used: Alternating Transmit and Receive (ATAR) and Simultaneous Transmit and Receive (STAR). In ATAR, a horizontally polarized pulse is transmitted, scattered, and received. Subsequently, a vertical pulse is transmitted, scattered, and received, and the process repeats. Radars operating in STAR configuration transmit a single pulse containing both horizontal and vertical components, which is scattered and received as a mixed-polarization packet. These mixed-polarization signals are finally sequestered upon reception.

Radar polarimetry has proven to be an important tool for meteorologist because it allows researchers to improve precipitation estimation, better discriminate between classes of hydrometeors, and discriminate between meteorological and non-meteorological scatter. The polarimetric data are also used to better interpret signals from complex shaped objects like hail stones. Not surprisingly the availability of polarimetric radar data is beneficial for the field aeroecology as well. For the case of bioscatter, not only are the shapes of the entities that produce radar backscatter complex but also evolve over time as wing beats and body orientations. An illustration depicting a case of bats and moths collated within a radar sampling volume can be seen in Fig. 12.7. In such a case, the polarimetric radar signature would be primarily driven by whichever species contributes most significantly to the overall radar return. If hydrometeors were present, then the polarimetric radar data could be used to discriminate between weather and biological signals.

The simplest of the polarimetric radar products to calculate and understand is differential reflectivity, which is defined as the linear ratio (or logarithmic difference) between radar reflectivity factors of orthogonal polarizations. For the case of weather radar,

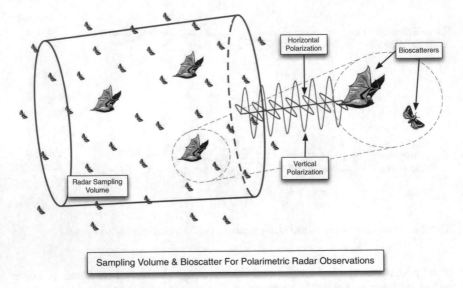

Fig. 12.7 Illustration of how bats and moths might be distributed within a single radar sampling volume, which is being probed by a polarimetric radar. The differences in size and shape of the bioscatterers would result in different radar signature

$$Z_{DR} = 10 \log_{10}\left(\frac{z_H}{z_V}\right) = Z_H - Z_V \qquad (12.14)$$

where the subscripts indicate the received polarization. Note that the differential reflectivity is given in dB. In meteorological applications, scatterers are generally assumed to follow the Rayleigh conditions, thereby linking Z_{DR} to the bulk aspect ratio of the scatterers (Doviak and Zrnić 1993). For larger scatterers, the conditions of the Rayleigh assumption are not valid, and Z_{DR} has a more complicated interpretation as the amplitude from orthogonal polarizations oscillates independently. Additional complications arise when scatterers are nonspherical and aligned (Ryzhkov and Zrnić 2007). In some cases, Z_{DR} can be used to identify insects in bioscatter data from polarimetric radar observations, if the particular insects have elongated body shapes.

When examining received signals from the horizontal and vertical channels of a polarimetric radar, it is useful to explore to what extent the data streams are correlated. The correlation coefficient is computed in an ensemble sense from backscattered signals recorded from the two polarizations and provides a measure of how similar the two are (Doviak and Zrnić 1993). The quantity is commonly denoted as ρ_{HV}. Values of ρ_{HV} are close to one for echoes from rain and snow. The values are typically significantly reduced for bioscatter.

As radio wave propagates through a medium, the phase is continuously modified along the path. One way of conceptualizing these phase shifts is by considering many forward-scattering events within the medium that each introduce a slight phase shift upon scattering (Bohren and Clothiaux 2006). When the medium is stratified, these forward scattering phase shifts are described in terms of ray refraction (Bean and Dutton 1966). When a medium is homogenous along the dimension perpendicular to wave propagation, these phase shifts are a continuous modification to the original phase. The magnitude of these phase shifts is dependent on the refractive index, size, shape, and orientation of the scattering medium at the polarization of the propagating wave. The quantity is often referred to as differential phase ϕ_{DP}, which depicts changes if phase shifts along a radial.

6   Brief Survey of Radar Used in Biology

Current advances in radar aeroecology are progressing across several research fronts. Radio-wave scatter by animals is still being approached as a measurement and modeling topic in electromagnetics (Lang and Rutledge 2004; Bachmann and Zrnić 2007; Melnikov et al. 2012, 2014, 2015). Some studies are focused on development of practical radar-based retrieval methods and method development (Larkin et al. 2002; Martin and Shapiro 2007; Schmaljohan et al. 2008; Nebuloni et al. 2008; Horn and Kunz 2008; Zaugg et al. 2008; Buler and Diehl 2009; Taylor et al. 2010; Cabrera-Cruz et al. 2013). Others are finding ways to exploit larger radar networks for ecological applications (Dokter et al. 2011; Chilson et al. 2012a). Currently, weather radar networks across Europe and the USA are being used for studies in biological sciences. A large volume of work is applying radar to basic ecology questions in migration (Alerstam et al. 2001; Dinevich et al. 2003; Hedenström et al. 2009; Dokter et al. 2013a; O'Neal et al. 2015), animal behavior (Shamoun-Baranes et al. 2010; Dokter et al. 2013a; Diehl 2013; Doren et al. 2015), and phenology (Kelly et al. 2012). Additionally, radar is playing a significant role in animal conservation, including land management (Bonter et al. 2009; Buler and Dawson 2014) and wind farm siting (Hüppop et al. 2006; Plonczkier and Simms 2012). Efforts directed toward human health and well-being include agricultural crop pest monitoring (Westbrook 2008; Bell et al. 2013) and aircraft birdstrike prevention (Zakrajsek and Bissonette 2001; Nohara et al. 2011). Radar studies of bioscatter are still an active topic in meteorology, with applications related to clear air studies (Contreras and Frasier 2008), hydrometeor classification (Park et al. 2009; Chandrasekar et al. 2013; Al-Sakka et al. 2013), and data quality (Lakshmanan et al. 2010; Martner and Moran 2001; Zhang et al. 2005).

Much of the increase in radar ecological studies can be attributed to an increase in radar technology and data availability. Several aeroecology labs have developed specialized biological radar units, such as the mobile marine radar system of Cooper et al. (2001), the vertically looking entomological radar of Chapman et al. (2003), and the Swiss Superfledermaus (Bruderer et al. 2010). The applications related to conservation and air safety have motivated several commercially available dedicated biological radar systems (Nohara et al. 2007). While specialized biological radar systems are becoming more common, much of the research is still conducted using atmospheric radars, including the polarimetric NEXRAD network in the USA

(Chilson et al. 2012a), the OPERA network in Europe (Shamoun-Baranes et al. 2014), and other serendipitous sites.

For some radar systems, particularly those dedicated to biological studies, it is possible to deduce the identity of scatterers in the airspace. Techniques directly enabling this capability include the analysis of wing flapping patterns (Williams and Williams 1980; Bruderer et al. 2010), airspeed (Cabrera-Cruz et al. 2013), or polarization diversity (Zrnić and Ryzhkov 1998). In other cases, the large-scale shape of radar patterns can yield information on scatterer identity (Horn and Kunz 2008). Additional taxonomic information can be gleaned from geographical, environmental, or phenological constraints that limit the range of potential taxa in the airspace. For example, temperature is often used as a constraining parameter for insect flight (Luke et al. 2008), and season can discriminate between some roosting bird species (Bridge et al. 2016). Taken together, these sources of information can often guide interpretation of scatterer identity in radar data.

In some cases, especially when the radar sampling volume is large, as is the case for operational weather radars, several animals may contribute to the received backscatter. The number of animals can be estimated if (i) the radar sampling volume is predominantly filled with a specific type of animal with a known RCS, (ii) the proportion of different types of animals and their respective RCS values can be assumed, or (iii) a representative RCS value for all animals can be used, then the radar reflectivity η can be used to calculate the number density of the animals (Diehl et al. 2003; Schmaljohan et al. 2008; Buler and Diehl 2009; Dokter et al. 2011; Shamoun-Baranes et al. 2011). As discussed in Chilson et al. (2012b), for meteorological radars operating at a particular wavelength the radar reflectivity factor Z can be readily mapped into radar reflectivity η in dB using

$$\eta[\text{dB}] = Z[\text{dBZ}] + \beta, \tag{12.15}$$

where $\beta = 10 \log_{10}(10^3 \pi^5 |K_m|^2/\lambda^4)$ with λ expressed in cm and $|K_m|^2 = 0.93$ and the factor of 10^3 is needed to account for unit conversion. A reference value of 1 cm^2 km^{-3} is used when converting η to dB. Just as Z is reported in dBZ with a standard reference value of 1 mm^6 m^{-3}, we could use 1 cm^2 km^{-3} as a standard reference and report η in logarithmic forms as dBη. That is, an η of 500 cm^2 km^{-3}, for example, would be 27 dBη [calculated using $10 \log_{10}(\eta/\eta_0)$], where $\eta_0 = 1$ cm^2 km^{-3}. Having calculated the value for η, Eq. (12.9) can be used to calculate N_{bio}, provided one of the assumptions provided above regarding RCS is fulfilled. For convenience, we report some values of β used in Eq. (12.15) for common weather radar wavelengths: $\beta= 14.54$ for $\lambda = 10$ cm (S-band), $\beta= 26.58$ for $\lambda = 5$ cm (C-band), and $\beta= 35.46$ for $\lambda = 3$ cm (X-band). Ideally one should use the exact wavelength of the radar, since the calculation of β is sensitive to the value of λ.

Radar is also widely used in aeroecology for the study of insects. For an extensive treatment of this topic, the interested reader is encouraged to consult Drake and Reynolds (2012) and the extensive collection of references contained within. Additionally, several chapters in this book contain material on this topic, especially Chaps. 7 and 9.

7   Specific Examples of Radar Aeroecology

7.1   Radar Observation of Bat Emergences

In some cases, large aggregations of single-species communities form identifiable signatures on radar. Such is the case for the Brazilian free-tailed bats (*Tadarida brasiliensis*) that inhabit caves and bridges in the south central plains. These bats roost communally and, immediately following sunset, exit the roost in a mass exodus that can be observed routinely on radar (Frick et al. 2012). To decrease competition for food among individuals, the bats disperse throughout the region to forage for insects. The result of this behavior with respect to radar sensing is twofold. First, the emergence signatures at these known roost sites are predominantly of a single known species. Second, because of the divergence behavior, the radar views ensembles of bats from most or all azimuth look angles. That is, some bats are flying toward the radar, yielding a head-on view; some are flying away, giving a tail-on view; and others are flying in each perpendicular direction.

In Fig. 12.8, we provide example of how emerging Brazilian free-tailed bats appear in radar images resulting from observations with a NEXRAD system. The NEXRAD station is KEWX located between Austin and San Antonio, Texas. Observations are for July 4th, 2014, at 01:43 UTC (July 3rd, 2014, at 20:43 local time), which is about 5 min after sunset. Values of Z (upper panel), v_r (middle panel), and σ_v (lower panel) taken from the lowest elevation angle ($0.5°$) displayed in a plan-position indicator format are shown. At this time and location, the most prominent feature in the radar images is not the bat signatures but rather a precipitation producing weather system in the western portion of the displays. Bioscatter signatures from the bats are visible as several ring-shaped regions having enhanced values of Z located throughout the image. The locations of three known bat roosting sites are shown for reference. Two of these sites are caves and one is associated with a cluster of bridges and overpasses along the interstate joining Austin and San Antonio. Of the three, the greatest population of Brazilian free-tailed bats can be found in Bracken Cave. In fact, during the summer months, Bracken Cave hosts the largest known bat colony in the world. Finally, we note that a line of enhanced radar reflectivity factor values oriented roughly east-west (just south of KEWX) is visible in the middle of the image. This feature is caused by radar scatter from insects caught in the wind outflow boundary from the storm. For this example, we have intentionally selected a period involving several varieties of radar echoes to illustrate some of the complexities, which can confront radar aeroecologists when attempting to discern bioscatter from weather returns.

The velocity information shown as v_r and σ_v provided further insight into the event shown in Fig. 12.8. A common convention used for weather radar displays is to indicate motions away from the radar in shades of red and motions toward the radar in shades of green. A divergence signature can be identified on radial velocity displays as a transition from inward (green) to outward (red) velocities along radials relative to the position of the radar. The signature can be clearly seen in the velocity data for Round Rock. The divergence signature for Bracken Cave is partially masked

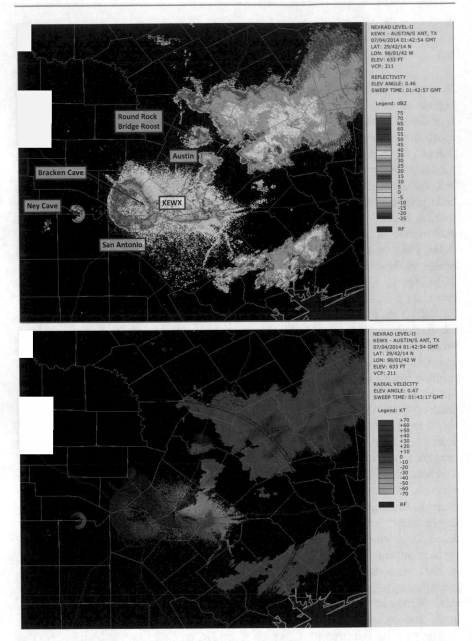

Fig. 12.8 Radar images from the NEXRAD station KEWX showing PPI plots of Z, v_r, and σ_v resulting from both meteorological and biological backscatter. The bioscatter is caused by emerging Brazilian free-tailed bats and insects. See text for more details

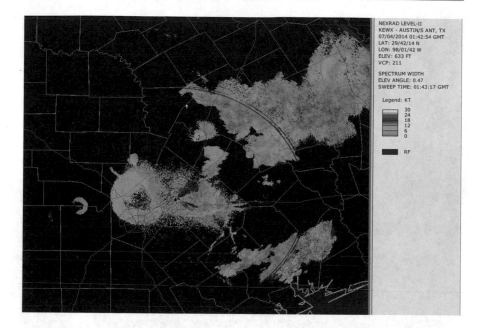

Fig. 12.8 (continued)

by the presence of insect bioscatter caught in the storm outflow, which is moving away from the radar while the bats are ostensibly moving toward the radar. However, some of the emerging bats may have slowed their outward motion to take advantage of the food source provided by the insects in the outflow. Moreover, winds near the surface for this time appear to be from the east, which would reduce the westward flight speeds of the bats (toward the radar in this case). For the cases of bat bioscatter, spectrum width data show high velocity variance in the head-on and tail-on measurements with low values at cross-beam flight directions. The physical interpretation is that the divergent velocity (i.e., the component directed radially away from the cave) is highly variable in magnitude but quite consistent in direction. For example, if the individual bat motions were predominantly isotropic in the horizontal with a slight divergent component, the spectrum width would be approximately equal around the circumference of the divergent ring. The lower values at cross-beam flight angles mean that the bats are actually flying perpendicular to the beam with little variance in the radial direction. In total, the behavior must be that the bats are flying strictly away from the cave at variable speeds. More research is needed regarding such signatures and how they can be interpreted.

We now examine the polarimetric radar values of Z_{DR}, ρ_{HV}, and ϕ_{DP} corresponding to the images shown in Fig. 12.8. These are shown in Fig. 12.9. For the case of rain, we expect values of Z_{DR} to be close to 0 dB (spherical drops) or slightly greater (oblate drops). The situation is more difficult for bioscatter. At S-Band, values of Z_{DR} for insects are usually expected to be >0 dB, on account of

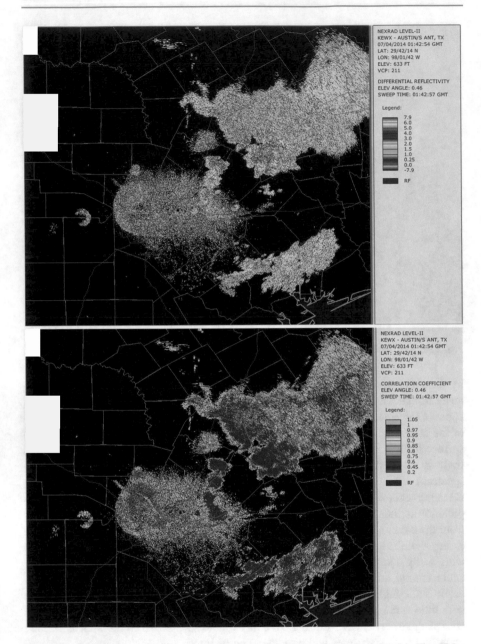

Fig. 12.9 Radar images of Z_{DR}, ρ_{HV}, and ϕ_{DP} corresponding to the data presented in Fig. 12.8

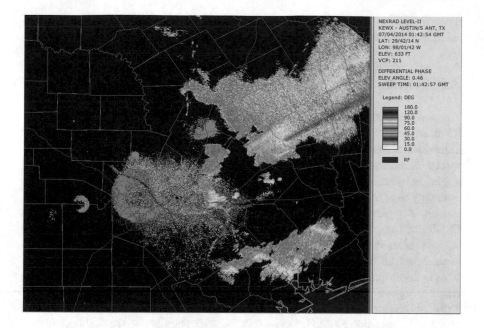

Fig. 12.9 (continued)

their body shape and orientation relative to the radar when flying. For larger animals, which fall within the resonant regime of the radar, such as birds and bats, Z_{DR} could be either positive or negative. Variability in the values is expected based on body orientation. As we see in the upper panel of Fig. 12.9, values of Z_{DR} for the regions of precipitation are close to 0 dB. For the emerging bats, the values seem to vary as a function of the heading of the bats. Note that the color scale has been optimized for meteorological echoes, which typically have positive Z_{DR}. Negative values are assigned to one gray scale bin. The insects in the outflow have consistently large values of Z_{DR}. From this single parameter, it becomes possible for this case to discriminate between the origins of the observed backscatter: rain, bats, and insects.

The correlation coefficient is another useful parameter when making distinctions between meteorological and non-meteorological scatter (Park et al. 2009; Van Den Broeke 2013). Radar scatter at horizontal and vertical polarizations from precipitation is typically highly correlated and results in values of ρ_{HV} being very close to one. The exception is irregularly shaped hail. Having irregularly shaped bodies, radar return at the two polarizations for bioscatter is less correlated in general. In the middle panel of Fig. 12.9, the regions of precipitation are clearly indicated by their values of ρ_{HV}. One could use the correlation coefficient as a means of filtering out meteorological signal. A more reliable filter, however, would likely need to incorporate several radar observables. For example, we see that differential phase also presents a stark contrast between regions of precipitation and regions of bioscatter. We hasten to note that much more work is needed both through observation and modeling, before polarimetric radar data can be fully harnessed in aeroecological studies.

7.2   Radar Observation of Bird Orientation

The phenomenon of avian migration is marked by the widespread, directional, purposeful movements of millions of individuals. This shared behavior results in aligned individuals across large regional expanses. As a seasonal phenomenon, not all widespread bioscatter is migration; however, polarimetry can indicate the presence of migration by azimuthal signatures resulting from mass alignment (Horton et al. 2016). For many birds, migration is a nocturnal activity, initiating at sunset and continuing until sunrise. A typical case of nocturnal bird migration is shown in the Spring case at Oklahoma City, Oklahoma (KLTX, Fig. 12.10) in which the radar sweeps are the 0.5° elevation angle immediately following midnight local time. This spring migration is characterized by high reflectivity, exceeding 25 dBZ, indicating a high number of birds in the airspace. Radial velocity magnitudes reach 30 m s^{-1}, indicating either strong winds or active, directional flight. Differential reflectivity ranges from −4 through 4 dBZ with strong azimuthal variability, forming apparent quadrants. Both differential phase and correlation coefficient display strong symmetry across the flight orientation vector (solid arrow), and correlation coefficient shows consistent antisymmetry across the perpendicular axis (i.e., head–tail axis, dashed line).

As spring transitions to summer, migrating birds leave Oklahoma for their northern breeding grounds. The summer resident birds of Oklahoma have no need to make high altitude flights at night, and as a result, there is an absence of birds in the nocturnal airspace during the summer. Nonetheless, nighttime bioscatter is still quite high in the summer, reaching 35 dBZ (Fig. 12.10). These signals are made up primarily of insects, most of which have been pumped aloft by thermals in the convective boundary layer, and remain suspended in the nocturnal residual layer. Additionally, the airspace in some cases may contain bats that are foraging on these insects. These processes result in low velocities, likely representative of the background winds. Differential phase and correlation coefficient have less azimuthal structure and are comparatively unorganized.

The onset of fall migration yields nearly identical patterns as the spring, only oriented southward (Fig. 12.10). Birds from Canada and the Northern USA again fill the airspace in nocturnal migration over central Oklahoma, resulting in high reflectivity and velocity magnitudes. Again, differential phase and correlation coefficient are both highly symmetric along the flight orientation of the organisms. Differential reflectivity is similar to the spring case but has developed a secondary lobe of negative values that is symmetric with the original lobe. Through the winter, bioscatter is limited. The migratory birds have moved south to their wintering grounds, and resident birds have no reason to fly at altitude, especially at night. The result is negligible reflectivity and velocity and unorganized polarimetric variables.

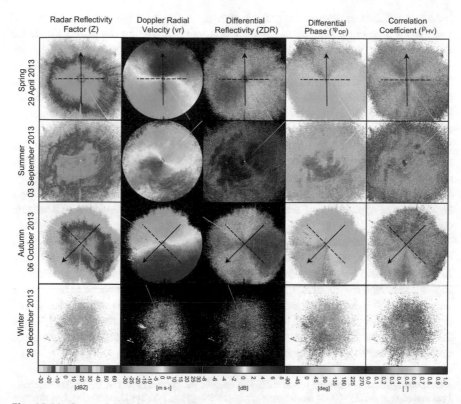

Fig. 12.10 Seasonal variations in the polarimetric fields of the 0.5° PPI from KTLX (Oklahoma City, Oklahoma), reproduced from Stepanian and Horton (2015). Tiles cover a 300 km by 300 km square domain, centered on the radar site, and are the first volume scan following midnight local time (i.e., 05 UTC in spring, summer, and autumn, and 06 UTC in winter). Cases showing widespread migratory alignment (i.e., spring and autumn) are annotated with an approximate orientation vector

8   Summary

Over the last 70 years, radar has proven itself to be a valuable tool for monitoring the behavior and movement of airborne animals. Often the location and movement of birds, bats, and arthropods are unknown to researchers because the animals are too small or too distant to be easily seen or there is inadequate visible light. For example, high flying raptors, many species of insects, and most of all nocturnal migrations are difficult to observe. But these animals can be monitored with radar under the appropriate conditions. Another advantage of radar is its ability to operate continuously over extended periods of time. In the case of networks of operational radars, such as those run by the national weather services, data are also available over large spatial extents with update times on the order of minutes. The downside of using radar for aeroecology is a lack of ability to identify individual species. Moreover, it becomes difficult to observe animals, which are flying near the ground, within the canopy, or near large structures. Nevertheless, we believe that the benefits of radar aeroecology far outweigh the limitations.

In this chapter, we have tried to cast some light on the fundamental principles of how radar works, some recent developments in the technology, and how it can be used to advance aeroecology with biologists being the target audience. Clearly, it has not been possible to touch upon the many types of radar and sampling strategies used to detect and observe volant animals, nor has it been possible to consider the wide range of biological studies, which have been aided through radar data. Many features of radar remain relatively unexplored or underutilized for biological studies. One example is radar polarimetry, which has the potential to significantly advance radar aeroecology. We briefly described the utility of such polarimetric parameters as Z_{DR}, ρ_{HV}, and ϕ_{DP}. However, the application of this technology for biological applications will require considerably more development before it can be seen as mature.

There is still much to learn about how radar can be optimally used to better promote the field of aeroecology. Going forward a more integrative approach involving ground-based measurements, including radar, trained observers, meteorological data, land use models, behavioral models, and so forth are needed to advance aeroecology. There is also a need for improved methods of characterizing and displaying radar imagery as it relates to aeroecology (Shamoun-Baranes et al. 2016). We hope that by introducing the reader to radar aeroecology more research and development will be triggered.

Acknowledgements Research presented in this chapter was funded in part by the National Science Foundation (Award # 1340921 from Macrosystems Biology) and the US Department of Agriculture National Institute of Food and Agriculture (NIFA-AFRI-003536). PBC is grateful to the support and accommodations from the Swiss Ornithological Institute during his sabbatical leave, which facilitated work on the chapter. Moreover, PBC acknowledges travel support from the European Network for the Radar surveillance of Animal Movement (ENRAM) program.

References

Able KP (1970) A radar study of the altitude of nocturnal passerine migration. Bird Band 41 (4):282–290. https://doi.org/10.2307/4511688. http://www.jstor.org/stable/4511688

Alerstam T (1972) Nocturnal bird migration in Skane, Sweden, as recorded by radar in autumn 1971. Ornis Scand 3(2):141–151. https://doi.org/10.2307/3676221

Alerstam T (1990) Bird migration. Cambridge University Press, Cambridge

Alerstam T, Bauer CA (1973) a radar study of the spring migration of the crane (*Grus grus*) over the southern Baltic area. Vogelwarte 27:1–16

Alerstam T, Gudmundsson GA (1999) Migration patters of tundra birds: tracking radar observations along the northeast passage. Arctic 52(4):346–371. https://doi.org/10.14430/arctic941

Alerstam T, Gudmundsson GA, Green M, Hedenström A (2001) Migration along orthoromic sun compass routes by arctic birds. Science 291(5502):300–303. https://doi.org/10.1126/science.291.5502.300

Alerstam T, Chapman JW, Bäckman J, Smith AD, Karlsson H, Nilsson C, Reynolds DR, Klaassen RHG, Hill JK (2011) Convergent patterns of long-distance nocturnal migration in noctuid moths and passerine birds. Proc R Soc B. https://doi.org/10.1098/rspb.2011.0058

Al-Sakka H, Boumahmoud AA, Fradon B, Frasier SJ, Tabary P (2013) A new fuzzy logic hydrometeor classification scheme applied to the French X-, C-, and S-band polarimetric radars. J Appl Meteorol Climatol 52(10):2328–2344. https://doi.org/10.1175/JAMC-D-12-0236.1

Atlas D (1959) Radar studies of meteorological "angel" echoes. J Atmos Sol Terr Phys 15:262–287

Bachmann S, Zrnić DS (2007) Spectral density of polarimetric variables separating biological scatterers in the VAD display. J Atmos Oceanic Tech 24:1186–1198

Battan LJ (1973) Radar observations of the atmosphere. University of Chicago Presss, Chicago, IL

Bean BR, Dutton EJ (1966) Radar meteorology. U.S. Govt Printing Office, Washington, DC, p 431

Bell JR, Aralimarad P, Lim KS, Chapman JW (2013) Predicting insect migration density and speed in the daytime convective boundary layer. PLoS One 8(1):1–9. https://doi.org/10.1371/journal.pone.0054202

Blacksmith P Jr, Mack RB (1965) On measuring the radar cross sections of ducks and chickens. Proc IEEE 53(8):1125

Bohren CF, Clothiaux EE (2006) Fundamentals of atmospheric radiation. Wiley, Weinheim

Bonter DN, Gauthreaux SA Jr, Donovan TM (2009) Characteristics of important stopover locations for migrating birds: remote sensing with radar in the Great Lakes basin. Conserv Biol 23:440–448

Bridge ES, Pletschet SM, Fagin T, Chilson PB, Horton KG, Broadfoot KR, Kelly JF (2016) Persistence and habitat associations of Purple martin roosts quantified via weather surveillance radar. Landsc Ecol 31(1):43–53

Bruderer B (1969) Zur registrierung und interpretation von echosignaturen an einem 3-cm-zielverfolgungsradar. Der Ornithologologische Beobachter 66:70–88

Bruderer B (1994) Nocturnal bird migration in the Negev (Israel) – a tracking radar study. Ostrich 65(2):204–212

Bruderer B (1997a) The study of bird migration by radar Part 1: the technical basis. Naturwissenschaften 84:1–8

Bruderer B (1997b) The study of bird migration by radar Part 2: major achievements. Naturwissenschaften 84:45–54

Bruderer B (1999) Three decades of tracking radar studies on bird migration in Europe and the Middle East. In: Leshem Y, Mandelik Y, Shamoun-Baranes J (eds) Proceedings international seminar on birds and flight safety in the Middle East, pp 107–142

Bruderer B, Steidinger P (1972) Methods of quantitative and qualitative analysis of bird migration with a tracking radar. In: Galler SR, Schmidt-Koenig K, Slotow R (eds) Animal orientation and navigation. National Aeronautic and Space Administration, Washington, DC, pp 151–167

Bruderer B, Peter D, Boldt A, Liechti F (2010) Wing-beat characteristics of birds recorded with tracking radar and cine camera. Ibis 152:272–291

Buderi R (1996) The invention that changed the world: how a small group of radar pioneers won the second world war and launched a technological revolution, Sloan technology series. Simon and Schuster, New York

Buler JJ, Dawson DK (2014) Radar analysis of fall bird migration stopover sites in the northeastern U.S. Condor 116(3):357–370. https://doi.org/10.1650/CONDOR-13-162.1

Buler JJ, Diehl RH (2009) Quantifying bird density during migratory stopover using weather surveillance radar. IEEE Trans Geosci Remote Sens 47(8):2741–2751

Cabrera-Cruz SA, Mabee TJ, Patraca RV (2013) Using theoretical flight speeds to discriminate birds from insects in radar studies. Condor 115(2):263–272. https://doi.org/10.1525/cond.2013.110181. http://www.jstor.org/stable/10.1525/cond.2013.110181

Chadwick RB, Gossard EE (1983) Radar remote sensing of the clear atmosphere – review and applications. Proc IEEE 71(6):738–753. https://doi.org/10.1109/PROC.1983.12661

Chandrasekar V, Keränen R, Lim S, Moisseev D (2013) Recent advances in classification of observations from dual polarization weather radars. Atmos Res 119:97–111. https://doi.org/10.1016/j.atmosres.2011.08.014

Chapman JW, Reynolds DR, Smith AD (2003) Vertical-looking radar: a new tool for monitoring high-altitude insect migration. Bioscience 53(5):503–511. https://doi.org/10.1641/0006-3568 (2003)053[0503:VRANTF]2.0.CO;2. http://bioscience.oxfordjournals.org/content/53/5/503.full

Chapman JW, Drake VA, Reynolds DR (2011) Recent insights from radar studies of insect flight. Annu Rev Entomol 56:337–356

Chilson PB, Frick WF, Kelly JF, Howard KW, Larkin RP, Diehl RH, Westrook JK, Kelly TA, Kunz TH (2012a) Partly cloudy with a chance of migration: weather, radars, and aeroecology. Bull Am Meteorol Soc 93(5):669–686. https://doi.org/10.1175/BAMS-D-11-00099.1

Chilson PB, Frick WF, Stepanian PM, Shipley JR, Kunz TH, Kelly JF (2012b) Estimating animal densities in the aerosphere using weather radar: to Z or not to Z? Ecosphere 3(8). https://doi.org/10.1890/ES12-00027.1

Conant J (2003) Tuxedo Park: a Wall Street Tycoon and The Secret Palace of science that changed the course of world war II. Simon and Schuster, New York

Contreras RF, Frasier SJ (2008) High-resolution observations of insects in the atmospheric boundary layer. J Atmos Oceanic Tech 25(12):2176–2187. https://doi.org/10.1175/2008JTECHA1059.1

Cooper BA, Day RH, Ritchie RJ, Cranor CL (1991) An improved marine radar system for studies of bird migration. J Field Ornithol 62:367–377

Cooper BA, Raphael MG, Mack DE (2001) Radar-based monitoring of marbled murrelets. Condor 103(2):219–229

Crawford AB (1949) Radar reflections in the lower atmosphere. Proc IRE 37:404–405

Diehl RH (2013) The airspace is habitat. Trends Ecolol Evol 28(7):377–379. https://doi.org/10.1016/j.tree.2013.02.015

Diehl RH, Larkin RP, Black JE (2003) Radar observations of bird migration over the Great Lakes. Auk 120(2):278–290

Dinevich L, Matsyura A, Leshem Y (2003) Temoporal characteristics of night bird migration above central Irael – a radar study. Acta Ornithol 38(2):103–110. https://doi.org/10.3161/068.038.0206

Dokter AM, Liechti F, Stark H, Delobbe L, Tabary P, Holleman I (2011) Bird migration flight altitudes studied by a network of operational weather radars. J R Soc Interface 8(54):30–43. https://doi.org/10.1098/rsif.2010.0116

Dokter AM, Åkesson S, Beekhuis H, Bouten W, Buurma L, van Gasteren H, Holleman I (2013a) Twilight ascents of common swits, *Apus apus*, at dwan and dusk: acquisition of orientation cues. Anim Behav 85(3):545–552. https://doi.org/10.1016/j.anbehav.2012.12.006

Dokter AM, Shamoun-Baranes J, Kemp MU, Tijm S, Holleman I (2013b) High altitude bird migration at temperate latitudes: a synoptic perspective on wind assistance. PLoS One 8 (1):1–8. https://doi.org/10.1371/journal.pone.0052300

Doren BMV, Sheldon D, Geevarghese J, Hochachka WM, Farnsworth A (2015) Autumn morning flights of migrant songbirds in the northeastern United States are linked to nocturnal migration and winds aloft. Auk 132(1):105–118. https://doi.org/10.1642/AUK-13-260.1

Doviak RJ, Zrnić DS (1993) Doppler radar and weather observations, 2nd edn. Dover Publications, New York

Drake VA, Reynolds DR (2012) Radar entomology. Centre for Agriculture and Biosciences International

Eastwood E (1967) Radar ornithology. Methuen & Co. Ltd, London

Eastwood E, Rider GC (1966) Grouping of nocturnal migrants. Nature 211:1143–1146

Edwards J, Houghton EW (1959) Radar echoing area polar diagram of birds. Nature 184:1059

Frick WF, Stepanian PM, Kelly JF, Howard KW, Kuster CK, Kunz TH, Chilson PB (2012) Climate and weather impact timing of emergence in bats. PLoS One 7(8). https://doi.org/10.1371/journal.pone.0042737

Gauthreaux SA Jr (1970) Weather radar quantification of bird migration. Bioscience 20:17–20

Gauthreaux SA Jr (1971) A radar and direct visual study fo passerine spring migration in southern Louisiana. Auk 88:343–365

Gauthreaux SA Jr (1991) The flight behavior of migrating birds in changing wind fields: radar and visual analysis. Am Zool 31(1):187–204. https://doi.org/10.1093/icb/31.1.187

Gauthreaux SA Jr, Belser CG (1998) Displays of bird movements on the WSR-88D: patterns and quantification. Weather Forecast 13:453–464

Gauthreaux SA Jr, Livingston JW (2006) Monitoring bird migration with a fixed-beam radar and a thermal imaging camera. J Field Ornithol 77(3):319–328

Gauthreaux SA Jr, Livingston JW, Belser CG (2008) Detection and discrimination of fauna in the aerosphere using Doppler weather surveillance radar. Int Comp Biol 48(1):12–23

Glover KM, Hardy KR, Sullivan TGKWN, Michaels AS (1966) Radar observations of insects in free flight. Science 154:967–972

Harmata AR, Leighty GR, O'Neil EL (2003) A vehicle-mounted radar for dual-purpose monitoring of birds. Wildl Soc Bul 31(3):882–886

Haykin S, Stehwien W, Deng C, Weber P, Mann R (1991) Classification of radar clutter in an air traffic control environment. Proc IEEE 79(6):742–772

Hedenström A, Alerstam T, Bächman J, Gudmundsson GA, Henningsson S, Kalrsson H, Rosen M, Strandberg R (2009) Radar observations of arctic bird migration in the Beringia region. Arctic 62(1):25–37. http://www.jstor.org/stable/40513262

Horn JW, Kunz TH (2008) Analyzing NEXRAD Doppler radar images to assess nightly dispersal patterns and population trends in Brazilian free-tailed bats (*Tadarida brasiliensis*). Int Comp Biol 48:24–39

Horton KG, Doren BMV, Stepanian PM, Hochachka WM, Farnsworth A, Kelly JF (2016) Nocturnally migrating songbirds drift when they can and compensate when they must. Sci Rep 6:21249. https://doi.org/10.1038/srcp21249

Hüppop O, Dierschke J, Exo KM, Fredrich E, Hill R (2006) Bird migration studies and potential collision risk with offshore wind turbines. Ibis 148:90–109. https://doi.org/10.1111/j.1474-919X.2006.00536.x

Kelly JF, Shipley JR, Chilson PB, Howard KW, Frick WF, Kunz TH (2012) Quantifying animal phenology in the continental scale using NEXRAD weather radars. Ecosphere 3(32). https://doi.org/10.1890/ES11–00,257.1

Kocurek W, LaGrone A (1967) Radar cross-section of a meteorological model of a coherent-dot radar angel. J Atmos Sol Terr Phys 29(8):975–985. https://doi.org/10.1016/0021-9169(67)90246-2. http://www.sciencedirect.com/science/article/pii/0021916967902462

Konrad TG, Hicks JJ, Dobson EB (1968) Radar characteristics of birds in flight. Science 159:274–280

Lack D, Varley GC (1945) Detection of birds by radar. Nature 156:446–446

Lakshmanan V, Zhang J, Howard K (2010) A technique to censor biological echoes in radar reflectivity data. J Appl Meteorol Climatol 49(3):435–462

Lang TJ, Rutledge SA (2004) Observations of quasi-symmetric echo patterns in clear air with the CSU-CHILL polarimetric radar. J Atmos Oceanic Tech 21(8):1182–1189. https://doi.org/10.1175/1520-0426(2004)021<1182:OOQEPI>2.0.CO;2

Larkin RP (1980) Transoceanic bird migration: evidence for detection of wind direction. Behav Ecol Sociobiol 6(3):229–232. http://www.jstor.org/stable/4599283

Larkin RP, Diehl RH (2012) Radar techniques for wildlife biology. In: Silvy N (ed) The wildlife techniques manual: research, vol 1, 7th edn. The Wildlife Society, Baltimore, MD, pp 320–335

Larkin RP, Evans WR, Diehl RH (2002) Nocturnal flight calls of dickcissels and Doppler radar echoes over south Texas in spring. J Field Ornithol 73(1):2–8

Lhermite RM (1966) Probing air motion by Doppler analysis of radar clear air returns. J Atmos Sci 23:575–591

Liechti F, Bruderer B, Paproth H (1995) Quantification of nocturnal bird migration by moonwatching – comparison with radar and infrared observations. J Field Ornithol 66(4):457–652

Luke EP, Kollias P, Johnson KL, Clothiaux EE (2008) A technique for the automatic detection of insect clutter in cloud returns. J Atmos Oceanic Tech 25(9):1498–1513

Martin WJ, Shapiro A (2007) Discrimination of bird and insect radar echoes in clear air using high-resolution radars. J Atmos Oceanic Tech 24:1215–1230

Martinson LW (1973) A preliminary investigation of bird classification by doppler radar. Technical report NASA-CR-137457, National Aeronautic and Space Adminsitration, USA

Martner BE, Moran KP (2001) Using cloud radar polarization measurements to evaluate stratus cloud and insect echoes. J Geophys Res Atmos 106(D5):4891–4897. https://doi.org/10.1029/2000JD900623

Melnikov VM, Lee RR, Langlieb NJ (2012) Resonance effects within S-band in echoes from birds. IEEE Geosci Remote Sens Lett 9(3):413–416. https://doi.org/10.1109/LGRS.2011.2169933

Melnikov V, Leskinen M, Koistinen J (2014) Doppler velocities at orthogonal polarizations in radar echoes from insects and birds. IEEE Geosci Remote Sens Lett 11(3):592–596. https://doi.org/10.1109/LGRS.2013.2272011

Melnikov VM, Istok MJ, Westbrook JK (2015) Asymmetric radar echo patterns from insects. J Atmos Oceanic Tech 32(4):659–674. https://doi.org/10.1175/JTECH-D-13-00247.1. http://journals.ametsoc.org/doi/abs/10.1175/JTECH-D-13-00247.1

Mie G (1908) Beiträge zur Optik trüber Medien, speziell kolloidaler Metallösungen. Ann Geophys 25(3):377–445

Mirkovic D, Stepainan PM, Kelly JF, Chilson PB (2016) Electromagnetic model reliably predicts radar scattering characteristics of airborne organisms. Nat Sci Rep 6:1–11. https://doi.org/10.1038/srep35637

Mueller EA, Larkin RP (1985) Insects observed using dual-polarization radar. J Atmos Oceanic Tech 2:49–54

Nebuloni R, Capsoni C, Vigorita V (2008) Quantifying bird migration by a high-resolution weather radar. IEEE Trans 46(6):1867–1875

Nohara TJ, Eng B, Eng M, Weber P, Ukrainec A, Premji A, Jones G (2007) An overview of avian radar development – past, present and future. In: 2007 Bird Strike Committee – USA/Canada, 9th annual meeting, Kingston, Ontario, pp 1–8

Nohara TJ, Beason RC, Weber P (2011) Using radar cross-section to enhance situational awareness tools for airport avian radars. Hum Wildl Interact 5(2):210–217

O'Neal BJ, Stafford JD, Larkin RP (2015) Migrating ducks in inland North America ignore rivers as leading lines. Ibis 57(1):154–161. https://doi.org/10.1111/ibi.12193

Park HS, Ryzhkov AV, Zrnić DS, Kim KE (2009) The hydrometeor classification algorithm for the polarimetric WSR-88D: description and application to an MCS. Weather Forecast 24(3):730–748. https://doi.org/10.1175/2008WAF2222205.1

Plank VG (1956) A meteorological study of radar angels. Geophysical research papers, U.S. Department of Commerce, Office of Technical Services

Plonczkier P, Simms I (2012) Radar monitoring of migrating pink-footed geese: behavioural responses to offshore wind farm development. J Appl Ecol 49(5):1187–1194. https://doi.org/10.1111/j.1365-2664.2012.02181.x

Probert-Jones JR (1962) The radar equation in meteorlogy. Q J R Meteorol Soc 88(378):485–495

Richter JH, Jensen DR (1973) Radar cross-section measurements of insects. Proc IEEE 61:1176–1178

Richter JH, Jesnen DR, Noonkester VR, Breasky JB, Stimmann MW, Wolf WW (1973) Remote radar sensing: atmospheric structure and insects. Science 180:1176–1178. https://doi.org/10.1126/science.180.4091.1176

Riley JR (1985) Radar cross sections of insects. Proc IEEE 73(2):228–232

Riley JR, Reynolds DR (1990) Nocturnal grasshopper migration in West Africa. Philos Trans R Soc B 328:655–672

Rinehart RE (2010) Radar for meteorologists, 5th edn. Rinehart Publications, Columbia, MO

Russell KR, Gauthreaux SA Jr (1998) Use of weather radar to characterize movements of roosting purple martins. Wildl Soc Bul 26(1):5–16

Russell RW, Wilson JW (1997) Radar-observed "fine lines" in the optically clear boundary layer: reflectivity contributions from aerial plankton and its predators. Bound Lay Meteorol 83:235–262

Ryzhkov AV, Zrnić DS (2007) Depolarization in ice crystals and its effect on radar polarimetric measurements. J Atmos Oceanic Tech 24(7):1256–1267. https://doi.org/10.1175/JTECH2034.1

Schaefer GW (1968) Bird recognition by radar: a study in quantitative radar ornithology. In: Murton RK, Wright EN (eds) The problems of birds as pests. Academic Press, London, New York, pp 53–86

Schmaljohan H, Liechti F, Bächler E, Steuri T, Bruderer B (2008) Quantification of bird migration by radar – a detection probability problem. Ibis 150:342–355

Shamoun-Baranes J, Bouten W, van Loon EE (2010) Integrating meteorology into research on migration. Int Comp Biol. https://doi.org/10.1093/icb/icq011:1–13

Shamoun-Baranes J, Dokter AM, van Gasteren H, van Loon EE, Leijnse H, Bouten W (2011) Birds flee en mass from New Year's Eve fireworks. Behav Ecol. https://doi.org/10.1093/beheco/arr102:1–5

Shamoun-Baranes J, Alves JA, Bauer S, Dokter AM, Hüppop O, Koistinen J, Leijnse H, Liechti F, van Gasteren H, Chapman JW (2014) Continental-scale radar monitoring of the aerial movements of animals. Movement Ecol 2:9

Shamoun-Baranes J, Farnsworth A, Aelterman B, Alves JA, Azijn K, Bernstein G, Branco S, Desmet P, Dokter AM, Horton K, Kelling S, Kelly JF, Leijnse H, Rong J, Sheldon D, den Broeck WV, Meersche JKVD, Doren BMV, van Gasteren H (2016) Innovative visualizations shed light on avian nocturnal migration. PLoS One 11(8):e0160106. https://doi.org/10.1371/journal.pone.0160106

Stepanian PM, Horton KG (2015) Extracting migrant flight orientation profiles using polarimetric radar. IEEE Tran Geosci Remote Sens 53(12):6518–6528. https://doi.org/10.1109/TGRS.2015.2443131

Stepanian PM, Chilson PB, Kelly JF (2014) An introduction to radar image processing in ecology. Methods Ecol Evol 5:730–738. https://doi.org/10.1111/2041-210X.12214

Taylor PD, Brzustowski JM, Matkovich C, Peckford ML, Wilson D (2010) radR: an open-source platform for acquiring and analysing data on biological targets observed by surveillance radar. BMC Ecol 10(22):1–8. https://doi.org/10.1186/1472-6785-10-22

Tolbert C, Straiton A, Britt C (1958) Phantom radar targets at millimeter radio wavelengths. IRE Trans Antennas Propag 6(4):380–384. https://doi.org/10.1109/TAP.1958.1144609

Van Den Broeke MS (2013) Polarimetric radar observations of biological scatterers in Hurricane Irene (2011) and Sandy (2012). J Atmos Oceanic Tech 30(12):2754–2767. https://doi.org/10.1175/JTECH-D-13-00056.1

van Gasteren H, Holleman I, Bouten W, van Loon E, Shamoun-Baranes J (2008) Extracting bird migration information from C-band Doppler weather radars. Ibis 150:674–686

Vaughn CR (1974) Intraspecific wingbeat rate variability and species identification using tracking radar. In: Gauthreaux SA Jr (ed) Proceedings of a conference on the biological aspects of the bird/aircraft collision problem, Department of Zoology, Clemson University, Clemson, SC, pp 443–476

Westbrook JK (2008) Noctuid migration in Texas within the nocturnal aeroecological boundary layer. Int Comp Biol 48(1):99–106

Wilczak JM, Strauch RG, Martin FM, Weber BL, Meritt DA, Jordan JR, Wolfe DE, Lewis LK, Wuertz DB, Gaynor JE, McLaughlin SA, Rogers RR, Riddle AC, Dye TS (1995) Contamination of wind profiler data by migrating birds: characteristics of corrupted data and potential solutions. J Atmos Oceanic Tech 12(3):449–467

Williams TC, Williams JM (1980) A Pertson's guide to radar ornithology? Am Birds 34:738–739

Williams TC, Settel J, O'Mahoncy P, Williams JM (1972) An ornithological radar. Am Birds 26:555–557

Zakrajsek EJ, Bissonette JA (2001) Nocturnal bird-avoidance modeling with mobile-marine radar. In: Bird Strike Committee – USA/Canada, third joint annual meeting, Calgary, AB, pp 185–194

Zaugg S, Saporta G, van Loon E, Schmaljohann H, Liechti F (2008) Automatic identification of bird targets with radar via patterns produced by wing flapping. J R Soc Interface 5(26):1041–1053

Zhang P, Liu S, Xu Q (2005) Identifying Doppler velocity contamination caused by migraging birds. Part I: feature extraction and quantification. J Atmos Oceanic Tech 22(8):1105–1113. https://doi.org/10.1175/JTECH1757.1

Zrnić DS, Ryzhkov AV (1998) Observations of insects and birds with polarimetric radar. IEEE Trans Geosci Remote Sens 36(2):661–668

Inferring the State of the Aerosphere from Weather Radar

13

Eric Jacobsen and Valliappa Lakshmanan

Abstract

Weather surveillance radars (see Chap. 12) are indispensable tools for characterizing the state of the aerosphere—the lower layer of the atmosphere used by flying animals—owing to their ability to remotely sense in-flight insects, birds, and bats at unprecedented temporal and spatial scales. The increasingly worldwide distribution of weather radars, their improving hardware and software, decades-long datasets, and capacity for revealing vertical and horizontal distributions make them well suited to mapping populations and movement from local to continental scales and for correlation with global datasets in other specializations like meteorology and geography. However, doing so requires care in processing the data and removing radar echoes that are unlikely to be biological. Moreover, radar design and the geometrically unique nature of their observations necessitate corrective methods for producing comparable data. Nevertheless, a parallel progression of meteorological and aerecological studies of how best to analyze radar data in spite of these observation inhomogeneities continues to yield fruitful research and new algorithms, continuously enhancing the ability to gain new insight into animal behavior in the aerosphere.

E. Jacobsen (✉)
University of Oklahoma School of Meteorology, Norman, OK, USA

Cooperative Institute for Mesoscale Meteorological Studies, The University of Oklahoma, Norman, OK, USA
e-mail: ericpj@ou.edu

V. Lakshmanan
Google Inc., Seattle, WA, USA
e-mail: vlakshmanan@google.com

© Springer International Publishing AG, part of Springer Nature 2017
P.B. Chilson et al. (eds.), *Aeroecology*,
https://doi.org/10.1007/978-3-319-68576-2_13

1   Introduction

Weather surveillance radars, described in Chap. 12, offer a four-dimensional perspective of the contents and motion of the atmosphere. Our interpretations of these data are continually enhanced by systematic hardware and algorithm development. Originally built and calibrated with the intention of monitoring both local and large-scale patterns of precipitation, many of the same capabilities can be employed for observations of birds, insects, and other biological objects in the aerosphere (Diehl and Larkin 2005; Chilson et al. 2012a). These radars are especially invaluable as a network, with composite datasets being able to depict regional- and even continental-scale spatial and temporal responses of biota to seasons, weather, and other factors (Kelly and Horton 2016; Kelly et al. 2016).

Observations from weather radars have increasingly proven to be a cost-effective, indispensable, and complementary tool to supplement the methods already in use by aeroecologists internationally (Gauthreaux and Belser 2003). For example, their data have been used for local and large-scale surveys of the roosting spots of birds and bats (Russell et al. 1998; Horn and Kunz 2008; Dokter et al. 2013; Bridge et al. 2015). Insects, too, are readily detected, with implications for agricultural management (Leskinen et al. 2011; Westbrook et al. 2014; Westbrook and Eyster 2017). Radars can observe the response of volant populations to natural hazards like earthquakes and storms and even to urbanization or man-made hazards such as fireworks and buildings (Buler and Dawson 2014; Shamoun-Baranes et al. 2011). Migration stopover sites and flyways can be mapped using radar networks to help understand and conserve habitats (Buler and Diehl 2009; Cohen et al. 2017), as well as to better grasp the strategies of long-distance flight, in particular using the ability to observe the preferred altitudes, orientations, and speed of migrants (Gauthreaux et al. 2003; Diehl et al. 2003; Dokter et al. 2011; Farnsworth et al. 2016; Horton et al. 2016b, c, d; Shamoun-Baranes et al. 2017; Van Doren et al. 2016). In many cases, radars are cited as providing the missing link to understanding migrant behavior, while bolstering the value of other datasets as they are integrated into a more complete picture of the aerosphere.

And yet as powerful a tool as radars are for observing the populations of the aerosphere, a discussion of their application would be incomplete without consideration of some important caveats which affect the interpretation of data.

1. Radars exist in finite locations and thus do not homogeneously observe the atmosphere in all places and particularly at all heights. Thus, data gaps in between radar placements require some generous extrapolation from the end user to intuit the true regional coverage of organisms.
2. Even where radar coverage exists, the heights at which observations are collected vary due to the tilt geometry with which beams penetrate the atmosphere. These geometric aspects of radar observations necessitate care with quantitative and qualitative interpretation. This is discussed in Sect. 4 of this chapter.
3. A radar beam is large in comparison with individual organisms (Nebuloni et al. 2008). At present, it remains a considerable challenge to determine exactly how

many organisms might be represented by an observed signal, let alone what types (Chilson et al. 2012b). As radar products and signal processing techniques improve, there is some growing hope that simple, high-level distinctions might be made within observations, such as between echoes due to birds and those due to insects (Gauthreaux et al. 2008; Zrnic and Ryzhkov 1998; Melnikov et al. 2013).

4. Though perhaps obvious, it bears reminding that by design and nature, radars also observe signals due to weather, dust, smoke, debris, certain ground features (termed ground clutter), and also various forms of electronic interference such as from the sun and other signal-emitting sources (Lakshmanan et al. 2010). These require methodical filtering, but fortunately much of this work has been accomplished by the meteorology and engineering communities. Some of these techniques are explained in this chapter in Sect. 3.

Having listed these caveats, it should be said that many of the challenges enumerated above are also faced by the meteorological community, which uses these radars every day. And, through a combination of algorithm development and trained interpretation, these would-be limitations are prevailed over by countless actionable insights which routinely save life and property. Though the use of weather radar by the aeroecology community is still in its infancy, it is poised to grow rapidly as synergies between it and meteorology are leveraged.

This chapter will first present examples of biological activity as observed by weather radar, particularly of data collected by a network of radars showing large-scale patterns. These examples will be coupled with explanations of radar beam characteristics which help to explain the appearance of the data. It then shifts to an introduction of algorithms which are even now being developed by the aeroecological community, to assist in the interpretation of radar for the study of airborne biology. These algorithms include techniques for isolating biological signals from other radar-detected echoes, as well as for normalizing observations to a common reference value to account for systematic biases in radar observations, the latter arising from variations in the height and width of observation volumes.

2   Interpreting Radar Observations of the Aerosphere

Two seemingly inseparable themes occurring in radar aeroecology literature include, first, the indisputable value of weather radar networks as a developing tool for near real-time monitoring of the aerosphere, and, second, at the same time the technical and conceptual hurdles to understanding what radar data depict (Gauthreaux and Belser 2003; Diehl and Larkin 2005; Kunz et al. 2008; Stepanian et al. 2016). The latter difficulty is certainly to be expected when adopting a new tool, especially one which is arguably being re-purposed outside of its original design intentions. Nevertheless, a growing body of aeroecology literature is at once proving the usefulness of two-, three-, and even four-dimensional radar observations of the aerosphere (when including time) and building momentum behind interpretative methods. In this section, we aim to present a few such studies and provide qualitative insight into the appearance of radar data, building on the discussions of Chap. 12.

2.1   Basic Geometric Considerations in Observations

Weather radar observations have a distinct geometry of data collection which must be understood to be interpreted. First of all, weather surveillance radars almost always exist in a permanently fixed ground position. These radars tend to be strategically positioned near major population or commercial centers, as well as near critical infrastructure elements such as airports, thus offering detailed local coverage and weather insight near these centers with some degradation in coverage quality as range from these positions is increased. Often, a mosaic of observations across multiple radar sites is used in meteorology for deducing large-scale patterns and flows, and the same can be applied to observations of birds, bats, and insects which are detected by these instruments (see Sect. 5 for more on mosaicing radars). Figure 13.1 shows a mosaiced image of reflectivity for several radars on an early spring evening in Texas and surrounding areas. At each location, the maximum reflectivity from all the observed heights is shown (this is often termed the "composite reflectivity"; see Sect. 2.2). This figure provides a sense of the spatial distribution common to weather radars, although dependent on the specific network (in this case the set of operational, long-range S-band radars in the USA), as well as a fairly typical method of visualizing data from multiple radar sites. The reader will better understand the appearance of this data and the aforementioned aspect of range degradation after the coming paragraphs.

We recall from Chap. 12 that weather radars scan radially outward; thus, the data are encoded in spherical geometry. That is, each observation is encoded with the beam orientation valid when the data were collected, comprised of an elevation (or zenith) angle ϕ with respect to the horizon that the antenna is tilted, the azimuthal orientation (compass direction) θ at which the antenna is pointed, and finally the distance r from the radar along the beam at which the observation occurred.

It is crucial to appreciate that a radar beam is tilted at some elevation angle with respect to the horizon. Because of this, while the beam radiates outward starting at the height of the antenna, its height systematically increases with range, as seen in the left-panel depiction in Fig. 13.2. To reiterate, observations at increased ranges are understood (in standard atmospheric conditions) to depict higher altitudes. In a more typical, top-down visualization of radar data showing a full sweep or rotation over many azimuths (such as the right panel in Fig. 13.2), this range-height relation means that nearby rings about the antenna (e.g. h1) show lower altitudes while the farthest (e.g. h3) are increasingly high in the atmosphere. In the case of biological organisms which inhabit a lower portion of the atmosphere, one impact of the upward tilting beam geometry is that observations intersecting dense biological activity—which occur at low altitudes—are restricted to nearby ranges before the effect of the beam's upward progression is too large (Drake and Farrow 1988; Mueller and Larkin 1985; Diehl and Larkin 2005; Kunz et al. 2008). The farther and higher the radar beams

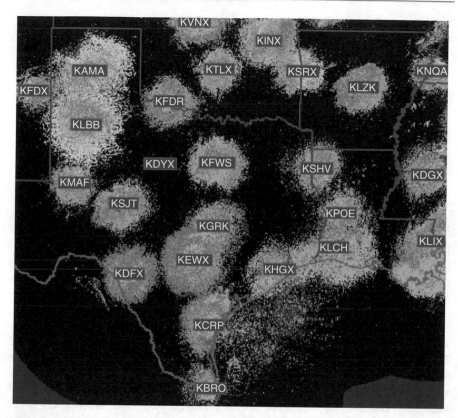

Fig. 13.1 A multi-radar mosaiced observation of reflectivity, shown without quality control, on April 2, 2015 at 2128 UTC. At each location, the maximum reflectivity from all heights is depicted. Radar IDs are also superimposed. Omitting quality control ensures that non-meteorological objects, which are useful for studying airborne biology, have not been filtered out. Non-meteorological objects are likely the dominant source of reflectivity near to the radars, where the beams still penetrate low altitudes. MRMS UnQC-CREF data and product viewer screenshots used with permission of NSSL, see https://mrms.nssl.noaa.gov/

travel, the less the low-altitude envelope of biological activity is directly observed, and the more some manner of inference must be relied upon for determining the near-surface values. In comparison, a similar problem exists in the meteorological use of weather radars. Weather phenomena such as storms are of such a scale that they often reach high enough into the atmosphere to be detected at great distances from the radar, even when the beam is sampling tens of thousands of feet above ground. However, the actual effect of the weather near to the ground is of paramount importance to forecasters and their customers, and this critical information—as it is not directly observed—can only be inferred via well-trained judgment or assisting algorithms.

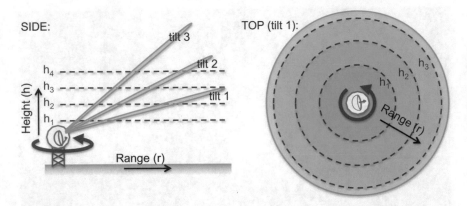

Fig. 13.2 A schematic depiction showing the variation of height with range in a radar scan. From a side-on view (*left image*), the upward tilt of multiple elevation angles can each be seen passing through greater heights with range. Viewing one such tilt from the top (*right image*), as is commonly done over maps, the relation of range and height leads to the farther rings of radar data sampling successively higher altitudes. This figure (*left*) also illustrates the limitations of direct sampling in terms of coverage gaps in between, below, and even above the available tilt angles

Because the weather radar is capable of observing low-altitude targets—such as the majority of airborne organisms—primarily near to its position, we observe a hotspot of activity clustered within some range around the radar when we look at maps of radar data. This is the case in Fig. 13.1, where clusters of reflectivity collocated with radar sites are clearly visible. The lack of observations outside this near-range cluster is not indicative of a zone with no biological presence, but rather that the radar beams are already too high in the atmosphere to observe biota at the low altitudes to which they are confined. The data in Fig. 13.1 represent a typical snapshot of nighttime biological activity, which will evolve over the evening as this activity fluctuates. For example, as insects and birds take flight and increase their altitude, they climb high enough to be visible by radar volumes which were previously positioned above them, and which then become gradually filled. Though not presented here, this is evident in progressive scans of the radar as an expansion of the cluster outward from the radar site, as the more distant and correspondingly higher-altitude observation volumes begin to be occupied by biota.

Weather surveillance radars of course conduct a complete range of azimuthal scans at more than one tilt angle with respect to the horizon (Rhinehart 1991). For example, a common volume coverage pattern (VCP) used by US WSR-88Ds to scan clear skies in fair weather (referred to as VCP32) will collect data at tilts of $0.5°$, $1.5°$, $2.5°$, $3.5°$, and $4.5°$ (Warning Decision Training Division 2016). Collectively, these scans produce a three-dimensional snapshot of the atmosphere, with multiple altitudes sampled over a given point on the ground. Aeroecology studies can directly leverage the whole of this data to study three-dimensional patterns of activity within a radar volume. Figure 13.3 from Diehl et al. (2003) presents two cross sections of radar data from Buffalo, New York, in May 1999 during northward migration of birds. Each cross section is generated by sampling the multiple, vertically stacked

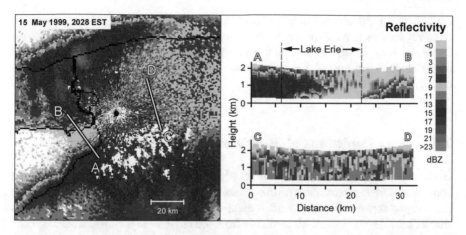

Fig. 13.3 A figure from Diehl et al. (2003) showing the start of evening migration near Buffalo, NY, in the spring, as observed by a local weather radar. Two distinct migration patterns are depicted by the vertical cross sections of reflectivity, both over the lake (*top-right*) and over land (*bottom-right*). A steep ascent from the south shore over lake Erie is apparent, while the land route lacks such structure. Used with permission of the American Ornithological Society

scans which a typical surveillance radar collects. In the top-right cross section, sampled across Lake Erie, the steep ascent pattern adopted by departing, lake-crossing birds is well detected, particularly in contrast to the over-land migration pattern to its east (shown by the lower-right cross section). Both cross sections are from a single radar volume and yet portray meaningful and significant variation in the structure of bird activity. Given that weather radars collect complete, volumetric datasets on the order of every 5–10 min, the potential for characterizing the detailed, evolving structure of biological activity is apparent.

One further element of spherical geometry which is critical to understand is that the radar beam has a fixed angular width, usually of about 1° in the operational WSR-88Ds. In spherical geometry, a solid angle of 1° width emanating out from a point is inherently going to have a much larger absolute width far from its origin at the radar than near to it. Thus, for a given beam width and a fixed increment of distance along the beam by which observations are delimited, a far-range observation will encompass a much larger volume and potentially more airborne objects than a near-range one. The absolute width of the beam for radars in the US network of WSR-88Ds is over a kilometer at 60 km in range and twice as wide at 120 km (Rhinehart 1991). Meanwhile, the increment of distance along the beam which is used to demarcate one observation in weather radars is generally of a scale far larger than an organism which might occupy that airspace (Nebuloni et al. 2008). For the WSR-88D network, the range increment in the direction of beam propagation is typically 250 m. For a more specific discussion relating radar variables to animal density, we refer the reader back to Chap. 12. Other factors such as the angle of incidence of the beam relative to the organism's orientation are also crucial for understanding the signals received from airborne biology, and this is an area of ongoing and promising research (Melnikov et al. 2015; Mirkovic et al. 2016; Stepanian et al. 2016) but beyond the scope of this chapter. In summary, geometric considerations are crucial for interpreting the sample volume and ultimately the density of organisms contained within it.

2.2   Map Depictions of Radar Data

Given that weather radar data contain multiple tilts, one challenge of radar use is the comprehensive representation of these data on a single display. In aeroecology, the most important layer of study is often relatively close to the ground. Thus, numerous studies consider the lowest tilt available from the radar to be the most relevant, even though it can often be the most sensitive to contamination by ground clutter (Doviak and Zrnic 1993). Alternatively, for a given coordinate on the map, one may be interested in the maximum value of organism density, regardless of the tilt elevation. This is easily obtained as the maximum value among the set of observations overlying that point. This is illustrated in Fig. 13.4 and is commonly referred to as the composite reflectivity. It is important to note for such depictions that the reflectivity value for a given pixel can be pulled from any elevation where the maximum happens to occur, potentially resulting in a diverse mix of values being portrayed on the map without any indication of their source altitude.

No doubt for many needs it is essential to consider the full, three-dimensional dataset, and so we do not mean to imply that a two-dimensional representation is

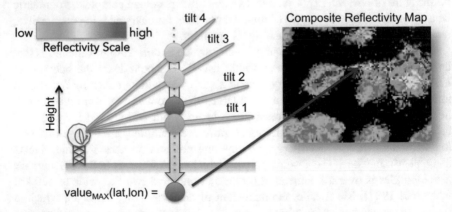

Fig. 13.4 A schematic depiction of how composite reflectivity is determined over a given surface coordinate. Multiple elevation scans are depicted by the blue beams, each observing some value of reflectivity (represented by the *filled circles*) at their respective altitudes above the ground point under consideration. The maximum value from this set of scans is taken as the representative value for the coordinate and plotted for that location on a map of composite reflectivity. MRMS UnQC-CREF (inset map) data and product viewer screenshots used with permission of NSSL, see https://mrms.nssl.noaa.gov/

always sufficient. Moreover, simple composites do not convey the inherent variation in altitude among observations on the map. In this case, techniques to normalize observations to a common reference value can be used (such as an inferred surface density or total column value). A discussion of possible corrections is provided in Sect. 4. Nevertheless, it is common in aeroecology applications to simply use either the lowest good-quality tilt or some form of composite value when plotting reflectivity. In addition, when plotting multiple overlapping radar images, the mosaic method (Sect. 5) can be used to combine observations.

2.3   Discerning Vertical Structure

We have already seen in Fig. 13.3 how the multiple elevations scanned by radars can be used to identify the structure or distribution of populations in the aerosphere. In fact, since even a single scan at one tilt probes the aerosphere over a range of altitudes, researchers have routinely extracted information from radar data about the different altitudes populated by organisms for different activities, such as by migrating birds (Lack and Varley 1945; Able 1970; Gauthreaux et al. 2003; Kemp et al. 2013). This can be done, for example, by segmenting the total horizontal domain observed by the radar into discrete range groups, as done by Gauthreaux et al. (2003) (recall the correspondence of height and range in radar scans, shown in Fig. 13.2). This creates a set of annulus shapes whose data can be spatially averaged into representative, range-specific values. By nature of radar geometry, these ranges will also correspond to altitude layers, thereby allowing us to characterize the observations as a function of height. In this way, variables such as the density, ground speed, heading, and track direction of embedded organisms can be described for a given layer or even as they vary with height. When specifically computed for reflectivity, this segmentation produces a rudimentary version of what is known as a vertical profile of reflectivity, or VPR. The level of discretization determines the resolution of the VPR. There are indeed many techniques for obtaining VPRs, which play a crucial role in representing or correcting the data collected by radars, and this will be further discussed in Sect. 4.

Many studies have extended the consideration of such vertical profiles to a regional or even continental scale (Gauthreaux et al. 2003; Diehl et al. 2003; Shamoun-Baranes et al. 2014; Dokter et al. 2011; La Sorte et al. 2015; Horton et al. 2016c). Indeed, it is in this scope that large-scale patterns of movement are best characterized. Gauthreaux et al. (2003), using the operational WSR-88D weather radar network, observed flight corridors across the central USA and diverging flyways in the northern states and also described their temporal trends and maximum migration density. Dokter et al. (2011) likewise analyzed vertical profiles of bird density, speed, and direction with a network of C-band radars in Europe, leveraging automated algorithms, weather data, and uniquely parameterized interpretations of the vertical profiles collected by the radars. Through this, multilayer structure in the preferred altitudes of migrating birds was identified, with possible relation to the birds' origins and meteorological factors. More recent leverage of dual-polarization radar variables, as in Horton et al. (2016b, c, d), has even enabled a discrimination of migrant heading from track, particularly in response to seasonal or topographical fluctuations. In addition to migration, the capability of observing vertical distributions can be used to study the heights attained by bird or insect populations for other purposes such as for feeding, exploration, and more (Dokter et al. 2013; Leskinen et al. 2011; Contreras and Frasier 2008; Drake 1984; Achtemeier 1991; Geerts and Miao 2005). Certainly, the availability of information in the vertical dimension may be as intriguing to aeroecologists as the vast horizontal coverage of the radar networks.

2.4   Spatial Mapping of Populations

Whichever means of analysis are employed by the researcher, weather radar can be used for cost-effective study of populations and the factors that influence them over a large domain. In addition to monitoring flight, important inferences can be made about stopover locations along migration paths, with implications for preservation of those habitats (Buler and Diehl 2009; Buler and Dawson 2014; Horton et al. 2016a; Cohen et al. 2017). Another example of the radar's capability to detect biological signatures is shown in Fig. 13.5, where rings of reflectivity seen to emanate outward with time are an example of birds or bats departing roosts. In this case, the outward expansion of flying organisms from a roost registers as a growing ring shape in radar reflectivity data. It is possible to track the daily emergence of birds or bats, as well as their sensitivity to weather and climate, by studying these unmistakable signals in radar data (Russell and Gauthreaux 1998; Horn and Kunz 2008; Frick et al. 2012; Dokter et al. 2013; Bridge et al. 2015). While these patterns are representative of birds in the morning, nighttime roost emergence is also used to study bats (Horn and Kunz 2008; Frick et al. 2012). Such data, when cataloged, allow inference about the spatial distributions, size, and temporal permanence of said roosts. Modern weather radars in particular, with their higher resolution and ease of comparison to other geospatial datasets, have been especially welcomed by aeroecologists for this purpose (Russell and Gauthreaux 1998). Also, with the maturation of computer vision and digital pattern recognition techniques, it is becoming possible to conceive of automated monitoring of these roosting sites and other movements (Horn and Kunz 2008; Dokter et al. 2011; Farnsworth et al. 2016).

Weather surveillance radars are uniquely suited to the study of airborne organisms in terms of their interaction with large- and small-scale weather patterns. Active shifts of warm and cold air masses, part of a weather phenomenon known as an extratropical cyclone (referenced in Chap. 2 of this book), occur particularly often in the USA during spring and fall. Such systems involve predictable wind regimes, with northward moving warm air to the south/east of a cold front and southward moving air behind the front to the north/west. Figure 13.6 shows a composite radar map of such a pattern, with the cold front identifiable as a southwest-to-northeast line

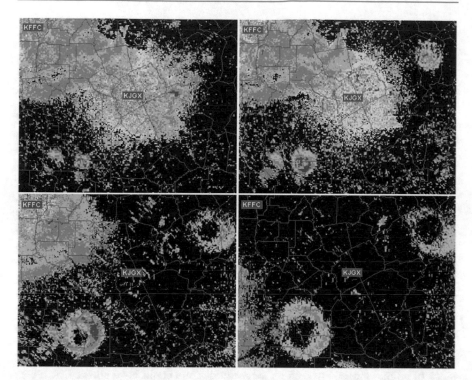

Fig. 13.5 A 4-panel time series of two bird roosts emerging in the morning hours of July 7, 2014, in central Georgia (follow *left-to-right, top-to-bottom*). Each image is approximately 15 min apart and shows the growth and eventual dissipation of the donut pattern in reflectivity as birds leave their roosts for daily activities. MRMS UnQC-CREF data and product viewer screenshots used with permission of NSSL, see https://mrms.nssl.noaa.gov/

of stronger radar echoes representing frontal thunderstorms and rainfall. Since northward winds favor spring migrants, it would seem reasonable for these organisms to leverage the prefrontal wind regime (Gauthreaux 1991; Kemp et al. 2013; Drake and Farrow 1988; Shamoun-Baranes et al. 2017), and this indeed appears to be corroborated by an abundance of radar-detected echoes in the portion of the southeastern states which lie ahead of the storms. By contrast, the often strong northerly winds occurring behind such a front would be unfavorable to spring migrants, and the radars northwest of the front indeed detect fewer airborne organisms.

As with the earlier reference to Diehl et al. (2003), where the radar's complete volumetric dataset was used to analyze migratory response to geographic obstacles, the interaction of airborne organisms with weather phenomena can also be studied in three spatial dimensions. While Fig. 13.6 depicts a biological response in two dimensions, Gauthreaux (1991) observed the three-dimensional structure of migrating songbird response to the passage of cold fronts of various strengths. As weak, relatively shallow fronts propagated past the radar, Gauthreaux (1991) was able to surmise that the birds maintained migration but simply at higher altitudes, above the newly present layer of unfavorable wind. However, if the front was strong and vertically deep—implying a scarcity of favorable winds in the low-altitude layers accessible to migrants—migration appeared to cease altogether. One considerable advantage of weather radar in such studies lies in the multidimensionality of its data, for example enabling simultaneous analysis of the vertical and horizontal consistency of these shapes and behaviors. Moreover, particularly in cases of nocturnal migration or when conditions are unfavorable for other observation methods, the ability of weather radar to continuously detect biological movement wherever they have coverage makes them a considerable asset to aeroecologists.

Population distribution in response to meteorological conditions can certainly also be observed with other instruments, such as vertically pointed radars. Wainwright et al. (2016) demonstrated for example, using a vertically pointed radar and lidar in Oklahoma, the altitude preference of migrants in relation to strong winds in the low level jet (usually below the jet). In comparison to other remote sensing instruments, however, the range, distribution, and increasingly the polarimetric capabilities of weather surveillance radars make them indispensable for sampling over areas where other instruments cannot reach (e.g., over oceans), or in conditions that others cannot operate. Weather radars helped to resolve a long debate over the route of migrants near the Gulf of Mexico, simply by their ability to detect organisms far out over the water (Cohen et al. 2017). Moreover, by pairing weather radar observations with traditional techniques, researchers are able to tell a more complete story of animal movement than either method alone could have allowed. A recent study by Van Doren et al. (2016) paired coastal, polarimetric WSR data at higher altitudes with visual counts of morning flight at altitudes below the radar's observing capability. In doing so, they strengthened a hypothesis that morning flights in a direction inconsistent with the overall migration vector were corrective measures for wind drift, by directly estimating the drift with polarimetric radar techniques.

The vertical layering of different wind and temperature regimes in weather patterns, often referred to as stratification, is a key influence in migratory behavior. Radar studies may be invaluable in the further examination of this behavior. For example, several studies have provided support for the theory that favorable tailwinds are one factor in altitude selection by migrant birds (Gauthreaux 1991; Dokter et al. 2011; Kemp et al. 2013; Horton et al. 2016c; Shamoun-Baranes et al. 2017). But wind speeds alone are insufficient to fully characterize migrant flight. Networks of weather surveillance radars have also been used to study flight response to seasonal land cover changes, geographical features, and terrain level, to name just a few of the other relevant factors (La Sorte et al. 2015; Kelly et al. 2016; Horton et al. 2016b, d). Section 2.5 reviews some of the means by which aeroecologists are increasingly incorporating meteorological data into their research for insight into the spatial distribution of airborne organisms (Fig. 13.6).

While radar networks provide the opportunity to study mass movement and identify habitats, they also are providing a view of unique events which impact airborne organisms. A pronounced jump in the amount of airborne organisms has

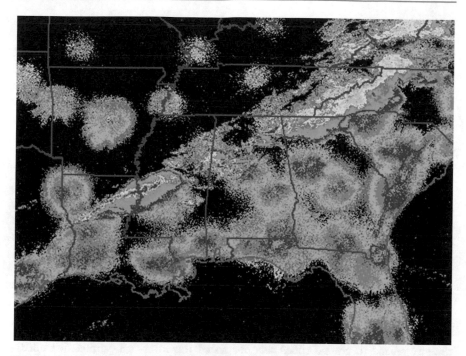

Fig. 13.6 A regional radar reflectivity mosaic of a frontal passage on April 4, 2015, depicts biological response to synoptic scale wind shifts. The frontal position is evident as a southwest-to-northeast line of stronger reflectivity (yellows and reds) where there are storms, which moves eastward. Meanwhile, winds from the north on the backside are unfavorable to spring migration and thus fewer airborne objects are detected on radars behind the front. Ahead of the front, winds from the south are supportive of migration. MRMS UnQC-CREF data and product viewer screenshots used with permission of NSSL, see https://mrms.nssl.noaa.gov/

been detected in conjunction with earthquakes, suggesting that radar may even be used in a study of population impacts by geologic hazards. A further part of the conservation effort which is enabled by weather radars is a consideration of human factors on the flight behavior of organisms. For example, the impact of new man-made structures on migration flyways can be studied with radars by considering long-term trends in density and orientation. Even temporary disturbances to populations such as by fireworks displays can be considered, which have been shown to cause abrupt flight of birds (Shamoun-Baranes et al. 2011).

Clearly, with the insight they offer into population locations and fluctuation, in addition to the regular monitoring which provides a rich dataset of susceptibility to human and meteorological factors, weather radars should be regarded as an invaluable tool for the conservation of many volant species (Gauthreaux and Belser 2003). Recognizing the importance of continental-scale monitoring, efforts are ongoing in Europe to cooperate in observing biology with the OPERA radar network (Shamoun-Baranes et al. 2014), and research groups in the USA are beginning to integrate with meteorological software developers on multi-radar products, with the aim of generating nation-wide datasets of bird density (Buler et al. 2012).

Presenting all the insights obtained from recent studies of the aerosphere would fill a much larger chapter than this. Good references discussing a range of methods and considerations in studying the aerosphere include Gauthreaux and Belser (2003), Diehl and Larkin (2005), Kunz et al. (2008), Shamoun-Baranes et al. (2010), and Chilson et al. (2012a).

2.5   Accounting for Meteorological Effects

As has been alluded to already in this book, a grasp of meteorology is indispensable for the radar aeroecologist. Understanding the dispersion, orientation, and speed of airborne organisms, for example, is highly dependent on the consideration of winds, temperature, precipitation, and other meteorological factors. Indeed, the parallel progression of meteorological and aeroecological analysis techniques for weather radar causes it to be in the radar aeroecologist's best interests to have their finger on the pulse of radar meteorology.

Deducing the true flight speed of organisms requires a knowledge of the wind in which they are embedded. Doppler weather radar observations of velocity are more directly related to ground speed than air speed [however, we caution that understanding how weather radars depict velocity requires some study, for which we refer the reader to Rhinehart (1991) or other weather radar handbooks]. To help correct for winds, radar-observed speeds can be compared against other observations which measure the meteorological background conditions, such as balloon-launched atmospheric sounding data (Gauthreaux et al. 2003; Gauthreaux 1991; Buler and Dawson 2014) or, as is becoming more common, against comparatively more frequent and geographically high-resolution weather models (Dokter et al. 2011, 2013; Kemp et al. 2013; Farnsworth et al. 2016; Horton et al. 2016d). For passive or slow-flying organisms like insects and spores, the wind itself may be the primary influence in transport (Westbrook et al. 2014; Westbrook and Isard 1999; Leskinen et al. 2011). Even if one's aim in research is more focused on organism position than velocity, it may still be necessary to incorporate meteorological data in a quality control stage, since accurately determined flight speed has been repeatedly demonstrated to be a helpful filtering criteria in sorting echoes into biological classes (Gauthreaux et al. 2003; Dokter et al. 2011; Kemp et al. 2013; Buler and Dawson 2014).

Use of numerical weather model data in particular will certainly become more important as radar aeroecology matures. These models provide a host of meteorological data pertaining to temperature, moisture, wind, visibility, precipitation, and even turbulence, all of which may be related to biological flight. The temperature and humidity at different altitudes can pose risks to flying organisms, such as ice aggregation and possibly moisture loss in very dry air (Schmaljohann et al. 2007; Kemp et al. 2013). These and other factors, such as precipitation, play a role in supporting, inhibiting, or modifying flight behavior (Eastwood 1967; Shamoun-

Baranes et al. 2010). The increasingly high temporal and spatial resolution available from these models makes them favorable for spatial correlation with radar versus comparatively sparse observations like balloon-launched measurements. In addition, precedents in the meteorological community for comparing models to radar data are available to aid aeroecologists in their own research. Therefore, weather models will play a growing role in considering the relationship of meteorology with airborne organisms in increasingly greater detail (Shamoun-Baranes et al. 2010; Kunz et al. 2008; Kemp et al. 2013).

Having presented several examples of aeroecological applications for weather radars, we move to an overview of some more advanced techniques and concepts. First, a technique for isolating radar echoes due to biology is presented which employs dual-polarization variables. Following that is a discussion of how VPRs are used to normalize the range-height bias of radar data discussed already in this chapter. One VPR technique is presented in special detail near the conclusion of the chapter.

3 Removing Nonbiological Echoes

Although the primary purpose of weather radar is to observe radio scatter from hydrometeors aloft, some of the radio wave scatter in weather radars can be attributed to the presence of airborne animals, such as birds, bats, and arthropods in the planetary boundary layer and lower free atmosphere (Chilson et al. 2012a). In addition to weather echoes and bioscatter, weather radar data is susceptible to several artifacts due to anomalous propagation, ground clutter, electronic interference, sun angle, and second-trip echoes (Lakshmanan et al. 2010). Lakshmanan et al. (2007a), Kessinger et al. (2003), Steiner and Smith (2002), and Lakshmanan et al. (2014) describe the causes, effects, and characteristics of such contamination in weather radar data and lay out algorithms based on texture features to remove non-weather echoes from weather radar data. Unfortunately, these techniques will also remove bioscatter.

To remove weather echoes and non-weather artifacts, while retaining bioscatter, we suggest the following approach, with details tailored for the US network of WSR-88Ds, sometimes referred to as the next-generation radars (or NEXRAD):

1. Near the time of sunrise and sunset, when the radar beam points directly at the sun, the resulting weather radar data has a number of thin, fine lines of enhanced reflectivity (called sun spikes) that point at the sun (see Fig. 13.7a, the East Coast of Florida). It is at sunrise and sunset that biological activity peaks, so removal of these sun spikes is important. One way to remove sun spikes is to calculate the sun angle based on the time and location of the radar and remove any echo-filled radials in that direction. Because the height of the radar beam at the limit of the range of a NEXRAD is around 3.6 km, and because the width of the beam at that range is also on that order, it is unlikely that a completely filled radial can

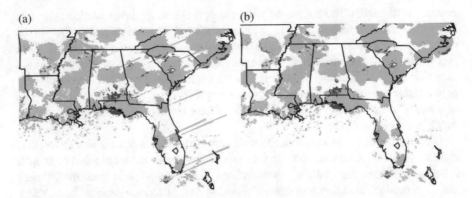

Fig. 13.7 *Left*: a multi-radar mosaic of raw radar reflectivity data shows a number of thin, fine lines of enhanced reflectivity called sun spikes (see, for example, Florida). *Right*: after quality control is carried out on the radar data, these sun spikes have been removed

correspond to biological echoes. There are weather situations (such as hurricanes) where the entire radar coverage area in both the vertical dimension (around 3.6 km) and horizontal dimension (around 460 km) can be filled. If we need to retain weather echoes while removing sun spikes, it would be necessary to differentiate between these two situations. However, since we are interested in bioscatter only, it is safe to remove mostly filled radials in directions that correspond to the sun angle at times of sunrise and sunset. The removed data could be filled in by interpolating nearby, unaffected radials.

2. When the air near the ground is colder than the air aloft, the radar beam gets bent downward toward the ground. Refraction that bends the beam toward the ground (called ducting) is likely in the morning since tropospheric inversions tend to be stronger in the morning (colder air under warmer air). In extreme cases ("super-refraction"), the resulting radar data may correspond to ground targets (buildings and trees) rather than to the aerosphere. The presence of ground targets in radar data is termed ground clutter. Besides being caused by super-refraction, ground clutter may also be caused when the sidelobes of low-angle radar beams reflect ground targets, or if there is beam blockage because of mountainous terrain (Steiner and Smith 2002). The expected height of the radar beam should be calculated assuming standard refraction, and data at and beyond terrain features that would block more than 50% of the beam should be removed. In addition, depending on the type of bioscatter one is interested in, echoes over water could be removed. However, this sort of deterministic calculation of radar beam height does not address the problem of anomalous propagation, when the beam is bent downward from its expected path due to tropospheric inversion. To handle this situation, while retaining shallow bioscatter, methods from Grecu and Krajewski (2000) may be adapted as follows. Compare the radar reflectivity at the lowest two tilts, and if the difference in reflectivity is more than 20 dBZ (this threshold comes from trial and error on NEXRAD, and may need to be adapted elsewhere),

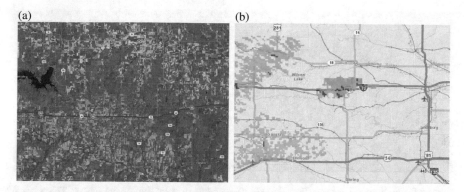

Fig. 13.8 Wind farms, such as those west of Salina, KS (positions shown in the *left* image from USGS dataset), impact weather data (*right* image), appearing as unwanted reflectivity for most research applications

then treat that echo as being suspect. However, if the echo is connected to non-suspect echoes closer into the radar (i.e., to echoes with vertical extent in areas where such vertical extent is observable, and the beam is at a high enough angle to not suffer ducting), then retain it.

3. Wind farms can cause turbulence aloft and may, therefore, not be removed by the vertical continuity check suggested in the previous step. It is suggested that echoes in the vicinity of wind farms[1] be censored. As wind energy increases in importance, this is a step that cannot be ignored (see Fig. 13.8). Telecommunication towers occasionally interfere with the radio waves from the radar and result in very high reflectivity returns. Unlike wind farms, however, the location of such electronic interference cannot be known in advance.

4. Depending on the application (the type of bioscatter of interest), we suggest speckle filtering the weather radar data and retaining only echoes <20 dBZ.

5. From the point of view of applications that wish to infer the state of the aerosphere, the final type of echoes that need to be removed from weather radar data are echoes that are meteorological in nature. The US network of weather radars was upgraded to dual polarization starting in 2013, and if the weather radar data are polarimetric, hydrometeor classification algorithms (such as Park et al. 2009; Liu and Chandrasekar 2000) can be employed to identify and remove weather pixels while retaining echoes due to aerosphere inhabitants. For example, insects have positive differential reflectivity (Z_{DR}) values in the range of 6–7 dB, while the range of values for birds is much wider (anywhere from −2 to 7 dB) because it is greatly affected by viewing angle and Mie scattering effects. While these are general tendencies, using polarimetric measures to discriminate between insects and birds has met with little success (Stepanian and Horton 2015).

---

[1]In the USA, you can obtain the locations of wind farms from the US Geographic Survey at http://eerscmap.usgs.gov/windfarm/

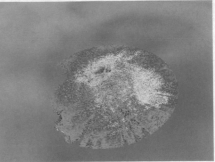

Fig. 13.9 An illustration of weather filtering of radar data to isolate biological echoes. The original PPI of reflectivity (*left*) shows a line of higher reflectivity storms impinging on the region from the lower left. Through weather masking (discussed in Sect. 3), it is possible to retrieve only the reflectivity data representative of airborne organisms (*right*). Note that the final product (*right*) also includes application of the Buler–Diehl VPR technique, discussed in Sect. 4, and that the color scales are different in these images

Weather echoes tend to have Z_{DR} in the $(-2.3, 2.3)$ dB range. The correlation coefficient (ρ_{HV}) for weather echoes tends to be higher than 0.95, whereas for bioscatter, it tends to be in the $(0.2–0.65)$ range (Lakshmanan et al. 2014). Thus, removing pixels with low absolute values of Z_{DR} and high values of ρ_{HV} will censor weather echoes. If the weather radar data at hand is not polarimetric, then values of Z_{DR} and ρ_{HV} will not be available, and one would have to determine weather echoes in some other manner. We suggest the calculation of vertical integrated liquid [VIL; Greene and Clark (1972)] from the lowest 5 tilts of weather radar data and removing all pixels connected to VIL values >20 kg/m^3 (Fig. 13.9).

4   Correcting for Uneven Observations

One of the challenges faced by aeroecologists and meteorologists alike is the need to extrapolate radar data to a reference value, such as to the expected surface intensity or to a column-weighted value. This need exists because radars do not sample heights in a consistent manner across their domain, and thus their geometry does not inherently lend itself to comparison on a horizontal plane, despite this being an important condition for geospatial analysis. We recall the image of nighttime biological density spanning multiple radar sites presented in Fig. 13.1. In this image, in the vicinity of each radar site, a general decay in intensity corresponding with range can be easily mistaken as suggesting that the greatest abundance of birds and insects occurs specifically at radar sites. Any biologist knows of course that radar sites have no relation with populations of airborne organisms. With that said, it might not be as evident how to correct radar data for this bias.

Techniques addressing this problem have been researched by meteorologists studying precipitation rates but are still relatively novel in the field of aeroecology; thus, we make a special effort to demystify the concept here. This section opens with a very basic introduction to the usefulness of VPRs, followed by a more specific discussion of techniques used by Buler and Diehl (2009), as it is a reasonably sophisticated technique accounting for multiple aspects of radar beam geometry and thus is useful for discussion here. Some additional techniques for VPR computation will be referenced at the conclusion of this section for the reader interested in pursuing this subject more.

4.1 The Basic Use of VPRs

Fundamentally, vertical profiles of reflectivity are a one-dimensional representation of how reflectivity varies in the vertical. As discussed in Sect. 2.3, these simple depictions are useful for studying the abundance of organisms at different heights. They can be easily acquired from dedicated vertically oriented radars, or, alternatively, computed from the multidimensional datasets collected by weather surveillance radars. When computing a one-dimensional VPR from an initial dataset with multiple spatial dimensions, the primary objective of computation is to collapse the data to a single dimension. The mathematics in this transformation can be crude or complex depending on the degree of precision required, but all methods account to some degree for the radar beam's geometry, such as for its upward tilt and in some cases even for beam broadening.

But why compute a VPR and reduce the dimensionality when studies like Diehl et al.'s (2003) have specifically leveraged radar data's three-dimensional nature to better understand the structure of the atmosphere and organisms in it? One reason is that it may be useful to statistically summarize radar observations for long time series analyses. However, a more pertinent issue to this section is related to the fact that a surveillance radar does not always sample where and how you wish it to sample.

We recall that radars do not scan in flat disks but at an angle to the ground and over a limited number of tilts. Depending on the beam width and the set of tilts used, gaps in coverage between samples may be numerous, especially at farther ranges where beams have diverged. In addition, in standard atmospheric conditions, the lowest tilt will climb increasingly far above the ground as it propagates outward, leaving an ever-larger slice of the lower aerosphere unobserved. We refer the reader back to the schematic in Fig. 13.2 to understand where gaps between observations, especially at comparable heights, may exist. We can see that it is not always possible to find a suitable observation in one of the provided tilts at the desired combination of height and range.

A more realistic portrayal is given in Fig. 13.10 in the left-side panel. Here, using an actual dataset, we have computed the true beam elevations (accounting for atmospheric refraction effects) and the width of the beam, including its broadening with range. Colors are used to distinguish beams of one tilt from beams of a different tilt, and the sizes of the line segments correspond to the beam width at that location. The

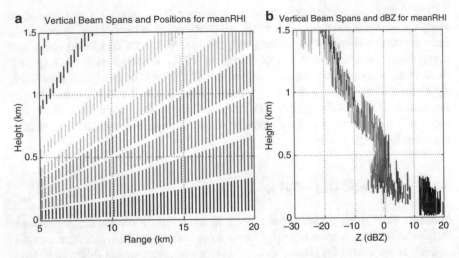

Fig. 13.10 A depiction (*left*) of the horizontal and vertical positions of multiple radar beams (as collected in a standard VCP 21 from KVAX on 2014/04/07 at 1500Z), as well as the height spanned by the beam width at each collection point. A plot of observed reflectivity by height for each of these bins is also shown (*right*), where the vertical span of each beam is again indicated. Colors are used to differentiate the tilts to which the radar bins belong and are consistent between both panels. (Note that in the *left-hand panel*, an additional scan at the each of the lowest two tilts is obscured by the foreground color, while in the *right-hand panel* some additional colors are apparent which represent these additional low-level scans)

horizontal axis shows range from the radar and the vertical axis shows height, both in units of km. This figure again emphasizes the challenge of heterogeneous observations in terms of both bin size and inter-beam gaps. Moreover, the increment of altitude spanned by each observation can pose a further challenge to analysis, such as when attempting to extract fine-scale variation in height. Although the beam width appears to enhance coverage, a radar volume especially at longer ranges can overlap multiple height intervals which we may wish to differentiate from each other. Recall the earlier statistic that for a one-degree beam width, as is characteristic of the US network of S-band weather radars, the beam will already have a kilometer-wide cross section at 60 km in range (Rhinehart 1991). Note that this figure shows only the nearest 20 km to the radar; beam widths and inter-beam gaps will be more exaggerated at farther ranges.

And so, even where coverage gaps are small, radar observation volumes at different ranges will frequently differ in size and are rarely centered at the same height. It would therefore be imprudent to conduct a study of horizontal variation in the aerosphere on any plane without first correcting for these inconsistencies, accounting for lack of coverage, different sample heights, and even the sample volumes represented by the data.

However, let us temporarily consider the data in Fig. 13.10 from another perspective. Since we are interested in the vertical structure of reflectivity observed by a weather radar (as an analogue for biological distribution with height), we might simply plot this value for each of the radar bins to see what they show. The right-hand plot in

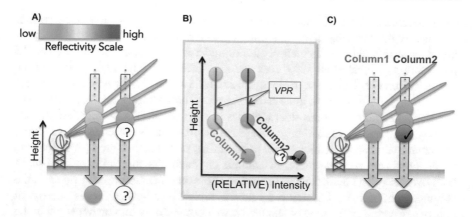

Fig. 13.11 A schematic depiction of how vertical profiles of reflectivity (VPR) can be used to infer non-observed values. Comparing observations at constant height can be difficult due to gaps in coverage (panel **a**). But, if a pattern (VPR) is known to be valid over the radar domain which dictates relative intensity as a function of height (panel **b**), then the missing value can be inferred (panel **c**)

Fig. 13.10 retains the same height, beam width, and color information as the left-hand plot but changes the value on the horizontal axis to reflectivity instead of range along the surface. This plot now indicates how reflectivity varies with the altitude of the samples. Despite the varied bin geometries and sample locations, a trend is clear. From this plot, we wish the reader to observe that, with reasonable agreement, reflectivity appears to be a function of height.

The notion that reflectivity is predominately a function of height and that there can thus be said to be a common vertical profile of reflectivity (VPR) across some domain is a useful one for extrapolating observations at uneven heights to a common reference. Figure 13.11 is a simplistic portrayal which attempts to illustrate this point. A set of radar observations at three comparable heights is shown for two ranges in Panel a, with the farther range clearly lacking any direct sampling at the lowest desired height. In the near-range column, we can discern a pattern of relative intensity with height consisting of an increase in value for the lowest observation, relative to the upper levels (Panel b, Column 1). Now suppose that enough observations are available to confirm that this pattern of relative intensity with height, which for convenience we refer to as a VPR, is consistent across the radar domain. If so, we need no longer be concerned about gaps in our observations so long as we have at least some observations at any level to fit our pattern to. Specifically, we may match the VPR pattern to our partial set of observations at the farther range (Panel b, Column 2) and thereby infer the value of the lowest observation even when it is not directly observed (Panel c).

In fact, with even one value of reflectivity at some height, it is possible to use the known, relative vertical profile of reflectivity to determine the expected reflectivity at any other height within that profile. A sizable portion of original work in applying this technique has been performed in the field of radar hydrology, where extrapolation of elevated observations of water droplets and ice crystals to near-surface rainfall amounts is the key objective. Any curious reader is referred to the wide body of literature in that field for theoretical studies underpinning the application of vertical profiles (e.g., Vignal et al. 1999; Joss and Lee 1995; Koistinen and Pohjola 2014), which lend credence to its use in biological studies. Essentially, the notion that vertical structure in the atmosphere is quite consistent across broad horizontal extents enables a great deal of inference about other altitudes in spite of varying observation heights. While overall intensity of airborne objects in one area may be different from another, the relative vertical distribution is said to be similar, such that they might each have the same normalized vertical profile, an example of which was shown in Fig. 13.10b. In both hydrology and aeroecology, validation and experimentation with respect to the assumption of horizontal homogeneity for these relative vertical profiles is a persistent area of research. For the sake of simplicity, we consider only one such method in the following section which implicitly assumes the technique to be appropriate.

4.2   Computing a Vertical Profile of Reflectivity

Having explained the relevance of VPRs to correcting and studying radar data, how can one be computed? Many techniques are possible including those referenced later in Sect. 4.3. Here, we consider the method outlined by Buler and Diehl (2009), which accounts for both the position and beam width of each radar bin. Ultimately, the Buler–Diehl technique is a computation of the expected mean reflectivity throughout the atmospheric column. As mentioned previously, a VPR can be used to compute many quantities such as the adjusted value at a reference altitude, or as in this case a more holistic quantity like the column mean. The final quantity is somewhat secondary to understanding the VPR itself, which we proceed with next.

The mathematical expressions for computing a VPR involve a transformation of reflectivity Z from its native radar coordinates to reflectivity at the set of heights which construct the vertical profile. For example, just one sample radar observation might initially be logged as occurring at an azimuth (θ) of 15° clockwise from north, at an elevation angle (ϕ) of 1.5° above the horizon, and at a range (r) along this beam of 50 km. Instead of a massive dataset with thousands of these spherical coordinates, we simply wish to know how reflectivity fluctuates at each height in a layer and resolution of interest, such as the lowest 3 km of the atmosphere in 20 m height increments. So, we may translate this spatial information in spherical coordinates to a height value and begin to build a model of reflectivity values observed at various altitudes. To actually accomplish this, we must not only account for this one sample but all the other samples which were collected at a similar height, as well as repeating the process for different heights by aggregating those sets of samples too. We refer the reader again to Fig. 13.10b to see that many different samples are available for producing a weighted estimate of the final vertical profile.

To construct mathematical notation for this process, it is helpful to begin by symbolically representing each possible pair of original coordinates in range r and tilt ϕ with an index c. Likewise, we use the index i to represent each of the possible height intervals in the final VPR. Thus, in terms of notation we ultimately seek to transform from Z_c to Z_i. This is done with the help of a transformation coefficient β_{ci}. Because many radar bins c overlap with and thus will contribute to the estimate of Z_i at a single level, we shall see that Z_i is expressed as an aggregate of many Z_c values, each multiplied by the factor β_{ci} which determines their specific contribution to that height interval i.

Before any such computation, one must first determine in what region to sample reflectivity data Z_c. It is advantageous in terms of ensuring dense, low-level sampling to restrict the domain to several tens of kilometers from the radar (Vignal et al. 1999). Buler and Diehl (2009) use 5–20 km in range as their selection, where ranges under 5 km are excluded due to undesired artifacts very close to the radar. Note that Fig. 13.10a shows these ranges for one azimuth, suggesting the density of radar data available for this computation.

We first collapse our radar data along the azimuthal dimension. This is done using a simple trimmed mean of values from all azimuths for each range r and elevation angle ϕ. This is written as $\overline{Z_c}$,

$$\overline{Z_c}(r, \phi) = \frac{\Sigma_{n=1}^N Z(r, \phi, \theta)|_{r,\phi}}{N} \tag{13.1}$$

where Z is the reflectivity at each coordinate (r, ϕ, θ) in the domain, θ is the azimuthal dimension that will be averaged over, and where each tail of the data is trimmed by 25% before computing the mean in order to reduce sensitivity to outliers. The azimuth coordinate is not a factor in determining height, unlike tilt and range, which is why it is easily averaged. The result, $\overline{Z_c}$, is a set of reflectivity statistics on a single two-dimensional plane to which the three-dimensional observations have been collapsed. For those familiar with radar scan strategies (see Chap. 12), the new two-dimensional form is now analogous to a typical range-height-intensity (RHI) scan, although it was not directly collected as such but rather computed from the mean of volumetric information. Note that it is not definitively superior to use the trimmed mean of the azimuthally varying values in computing this RHI, as done here. In some cases, azimuthal homogeneity about the radar is not a valid assumption and alternative statistics or VPR methods may be considered.

Next the two-dimensional data must be collapsed to a single spatial dimension, height. At this stage, we face the challenge of how to aggregate multiple values which co-occur at certain heights. Figure 13.12 is presented to illustrate this challenge. This figure shows the sampling of multiple beam volumes in order to determine a representative value for one height interval i, with width Δh, shown as Δh_i in the illustration. Some single representative statistic of reflectivity for a height increment should be obtained by combining all of these observations. Here, the mean is again employed but with the caveat that not all observations are weighted equally.

This is due to the fact that rarely will multiple observations, each of which actually corresponds to a rather large volume, be geometrically centered at identical heights. The circular cross section with which a radar beam penetrates the airspace yields a volume estimate of reflectivity which should be weighted more toward its central height and less so toward the upper and lower fringes where at best a small sliver of the cross section might intersect a height interval.

Now, for a height increment $\left[h_i^-, h_i^+\right]$ we can provide a function determining the weight of each individual bin estimate to be applied to this height interval. This is written as β_{ci} where c denotes a bin at some (r, ϕ) combination and i is a specific height interval. One bin at radar coordinate c will, as mentioned, overlap with multiple height intervals i and should be weighted differently in its contribution at each height, hence the need for this dual-index notation for β. The value of β can be determined in a multitude of ways. One method is to determine the proportion of the circular beam cross section lying between chords positioned at $\left[h_i^-, h_i^+\right]$, as indicated by the red shaded segment within the circle in Fig. 13.12. One may opt for a more realistic but complex method of weighting using a function to describe the energy falloff from the center of a radar beam, as described in detail in Chap. 3 of Doviak and Zrnic (1993) and as used in an implementation of the Buler–Diehl algorithm in the WDSS-II software package, presented in Buler et al. (2012). By whatever method it is obtained, β_{ci} is some value between 0 and 1 describing the percentage of the beam sample which is bracketed by the layer of interest. With this we arrive at

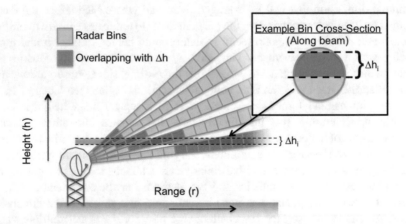

Fig. 13.12 A schematic depicting the vertical cross section through five radar beams as they originate from a radar and propagate upward to the right. Each radar bin along a beam is represented by a *box*, and for a given altitude shown by the horizontal line (with thickness Δh), multiple bins (red shaded boxes) from multiple beams may to some degree sample that altitude. The inset image shows the circular cross section of an example bin in the direction of beam propagation, with a red shaded segment showing how a height interval like Δh might occupy some proportion of the cross section. Based on the methods of Buler and Diehl (2009)

the following equation for a weighted average of reflectivity Z_i at height interval i and over domain D_i:

$$\overline{Z_i}(D_i) = \frac{\sum_{c=1}^{n_c} \overline{Z_c} \beta_{ci}}{\sum_{c=1}^{n_c} \beta_{ci}} \qquad (13.2)$$

where n_c is the number of coordinate positions c. Evaluated over each height interval i, this provides the beam-weighted mean reflectivity as a function of height. The reference to D_i is simply to remind us that this reflectivity profile was computed over a specific domain near to the radar.

Equation (13.2) is the formula for the vertical profile of reflectivity, obtained from a statistical, geometric integration of reflectivity values within a 20 km radius of the radar. In computing this, we have considered the beam heights and widths and weighted their contributions to each height interval i. Therefore, this function would make a suitable summary of the abundance of organisms at various heights near the radar, which can be studied over time, compared among different radars, and so on.

However, for the purposes outlined in Sect. 4.1 of correcting data at other ranges within a radar volume, a few more steps are required. Equation (13.2) gives the magnitude of reflectivity itself at each height in our local domain D_i, but these values may not apply to the broader radar domain. Within the entire domain sampled by the radar, cumulative animal density—and reflectivity in turn—will likely vary over different locations. For example, over more heavily vegetated areas such as parks or rivers, more biological organisms will be present in the airspace, raising the column mean Z. This would manifest in a shift of the entire VPR to a higher or lower value for more or less airborne animals, respectively, though critically we make the assumption still that its shape remains constant. In this case, the utility of a VPR is not in assuming that the absolute profiles are constant everywhere but that the relative ones are. Thus, a relative profile is required to characterize the vertical distribution of scatterers.

Using column-mean reflectivity as a normalizer yields a unitless, generic function which can be thought of as depicting object distribution with height. To compute this, the previous expression for height-specific reflectivity $\overline{Z_i}(D_i)$ is divided by the column mean reflectivity. The following formulation accomplishes that:

$$z(i) = \frac{Z_i(D_i) n_z}{\sum_{i=1}^{n_z} Z_i(D_i)} \qquad (13.3)$$

where n_z is the number of height increments.

Equation (13.3) is a normalized version of the vertical profile of reflectivity. Though unitless, it represents the structure of reflectivity with height. Values of $z(i)$

for some height i represent the ratio of reflectivity at that height with respect to the mean reflectivity of the entire column. For example while the column may have a mean 20 dBZ reflectivity, a sample at one height registering 30 dBZ would represent 1.5 times the column-mean reflectivity, while another registering 15 dBZ would represent 0.75 times the column mean. Integration of the full set of ratios over all the finite height increments must yield 1, since a consolidation of values over the entire VPR by definition results in just the mean itself, or a ratio of 1. A profile of ratios like this makes it a trivial matter to take any observed reflectivity value at some vertical point and translate it into the column mean.

In the correction stage, let us briefly consider a single radar observation at some height which we wish to convert to the column mean. As repeatedly emphasized, a single observation volume typically spans multiple height increments over which the vertical profile of reflectivity was computed. Therefore, any single interval i will be insufficient to characterize the full range of heights which the observed reflectivity sample reflects, and we should not expect one ratio value from $z(i)$ to accurately reflect the entire radar bin in question. Rather, we must determine the weighted average of the VPR values to use based on the spatial overlap of the radar bin with each of the height intervals in the VPR. That is, each $z(i)$ value for height increment i contributes to a weighted ratio estimate based on the proportion of the beam cross section which overlaps that height interval. This again involves the transformation term β_{ci}. Perhaps the reader correctly recognizes that this is in some sense a reverse of the geometric weighting process used to compute the VPR but now applied to a single bin. For a full description of how to apply the VPR to correct the domain values, the reader is referred to equations in Buler and Diehl (2009).

In summary, a vertical profile of reflectivity, once obtained, can be used to correct values over the broader domain of radar observations to the column mean reference value. The effect of this correction is to remove systematic changes in reflectivity which are due merely to varying measurement heights and not reflective of horizontal variability, which is the desired measure. By applying a VPR correction, we correct to a reference value which can more be effectively used to identify legitimate horizontal variations in organism quantities.

4.3   Some Additional Notes on VPRs

The approach by Buler and Diehl (2009) described here is just one of the possible implementations of a range-correction technique for radar data. As previously referenced, this challenge is shared by those in the meteorological community studying quantitative precipitation estimation (QPE), and a wide body of literature precedes the adoption of VPRs by the aeroecology community. For example, rather than computing a VPR in real time from nearby observations, some meteorological applications use a fixed climatological VPR appropriate for the region, with demonstrated improvement over doing nothing at all (Joss and Lee 1995; Vignal et al. 2000). In fact, in the event that limited or poor data is available within the near ranges typically used for on-the-fly VPR computation, a predetermined VPR may be the only way to correct other observations in the radar domain. One improvement on the fixed VPR is to define a general profile but allow it to vary using parameterization (Kitchen 1997; Koistinen and Pohjola 2014). In some respects, Dokter et al. (2013) used this approach in their study of the ascent behaviors of swifts. In this, they fit a parameterized Gaussian distribution to vertical reflectivity observations to identify birds, while simultaneously fitting an exponential curve to low-level reflectivity as a means of filtering out the effect of insects and other echoes.

Whether using observed, fixed, or parameterized VPRs, one further decision involves whether to apply a single VPR to the entire radar domain or to apply different profiles to different regions according to some rationale. For example, topographic or land-cover variations may call into question the relative homogeneity of airspace usage across the domain. The parameterized technique in fact allows a generally similar but slightly varying profile to fit the observations in each locale as best as possible. But even more novel approaches may involve using different parameterizations depending on some knowledge of the echoes, such as in QPE when regions of convective and stratiform rain are handled with different criteria, as suits their distinct structures (Smyth and Illingworth 1998). As proficiency with the VPR increases among radar aeroecologists, it seems assured that techniques will advance, such as by relaxing the assumption that vertical profiles of reflectivity are homogeneous over the entire radar domain. In fact, asymmetries are known to appear in radar variables even in a relatively homogeneous layer of birds or insects aloft. This is due not to variation in the organisms but rather to the fact the radar observations are a function of viewing angle and thus may vary systematically by azimuth (Mueller and Larkin 1985; Lang et al. 2004; Melnikov et al. 2015; Mirkovic et al. 2016; Stepanian et al. 2016). The appearance of complex biological shapes to weather radar is somewhat unique to aeroecology and will require more study, but as understanding in this area grows, it will also inform more modern correction methods including VPR computation techniques.

5   Multi-Radar Mosaics

The areal coverage of a single weather radar is on the order of a few hundreds of kilometers, while the area of interest for studies of migratory inhabitants of the aerosphere may be on the order of thousands of kilometers. Because a weather radar scans the aerosphere from a fixed point on the ground, the data from a weather radar are collected in a spherical coordinate system. Finally, it is possible that multiple weather radars may scan a given geographic area, and it might be best to combine their estimates. For these three reasons, it is preferable to deal with mosaics of radar data where the data from multiple radars are mapped onto the surface of the earth and interpolated onto a standard map projection.

Another advantage of creating such a mosaic is to bring the data into a manageable temporal resolution. An individual NEXRAD collects an elevation scan in under a minute and moves on the next elevation scan. There are 150 radars that cover the continental USA. It is far more manageable to carry out analysis on mosaiced radar grids that combine data collected over, say, 5-min periods.

Spatial interpolation techniques to create 3D multiple-radar grids have been examined by Trapp and Doswell (2000), Askelson et al. (2000), Lakshmanan et al. (2006), and Dean and Ghemawat (2008). Due to radar beam geometry, each range gate from each radar contributes to the final grid value at multiple points within a 3D dynamic grid. Additionally, in areas of overlap between multiple radars, the contributions from the separate radars should be weighted by the distance of the grid point from the radar because beams from the closer radar suffer less beam spreading.

For a grid point in the mosaiced 3D grid at latitude α_g, longitude β_g, and at a height h_g above mean sea level, the range gate that fills it, under standard atmospheric conditions, is the range gate that is a distance r from the radar [located at (α_r, β_r, h_r) in 3D space] on a radial at an angle a from due north and on a scan tilted e to the earth's surface where a is given by (Lakshmanan et al. 2006):

$$a = \sin^{-1}\left(\sin\left(\pi/2 - \alpha_g\right)\sin\left(\beta_g - \beta_r\right)/\sin\left(s/R\right)\right) \tag{13.4}$$

where R is the radius of the earth and s is the great-circle distance, given by (Beyer 1987):

$$s = R\cos^{-1}\left(\cos\left(\pi/2 - \alpha_r\right)\cos\left(\pi/2 - \alpha_g\right) + \sin\left(\pi/2 - \alpha_r\right)\sin\left(\pi/2 - \alpha_g\right)\cos\left(\beta_g - \beta_r\right)\right) \tag{13.5}$$

the elevation angle e is given by (Doviak and Zrnic 1993):

$$e = \tan^{-1}\frac{\cos\left(s/IR\right) - \frac{IR}{IR+h_g-h_r}}{\sin\left(s/IR\right)} \tag{13.6}$$

where I is the index of refraction, which under the same standard atmospheric conditions may be assumed to be 4/3 (Doviak and Zrnic 1993), and the range r is given by

$$r = \sin\left(s/IR\right)\left(IR + h_g - h_r\right)/\cos\left(e\right) \tag{13.7}$$

Since the $\sin^{-1}$ function has a range of $[-\pi, \pi]$, the azimuth, a, is mapped to the correct quadrant ($[0, 2\pi]$) by considering the signs of $\alpha_g - \alpha_r$ and $\beta_g - \beta_r$. The voxel at α_g, β_g, h_g can be affected by any range gate that includes (a, r, e), and the final result at α_g, β_g, h_g is given by a distance-weighted blend of all the available values.

The above coordinate system transformation to go from each voxel α_g, β_g, h_g to the spherical coordinate system (a, r, e) can be precomputed. If the radar operates only at a limited number of elevation angles (NEXRADs do), the voxels impacted by data from an elevation scan can be computed beforehand and employed whenever that elevation angle from that radar is received. Dean and Ghemawat (2008) describe a MapReduce approach to compute radar mosaics on a distributed cluster of compute nodes.

6   Summary

Weather surveillance radars offer an extraordinarily useful means of remote sensing biological presence in the atmosphere, both in a local sense with single radars and on a regional or even continental scale when observations from multiple units are merged onto a single map. Although the practice of studying radar data explicitly to quantify airborne organisms may be said to be in its infancy, it dovetails from a long history of algorithm development for radar data. Among these are techniques to categorize radar-detected objects, which are greatly empowered by the recent upgrade of many radars to dual polarization, making this an exciting era for additional classification schemes which stand to benefit all users of radar data. These algorithms can be effectively employed, as discussed in this chapter, to filter out unwanted signals and enable more accurate study of biological objects in the aerosphere. Moreover, techniques borrowed from radar hydrology help reduce the burden of complex radar geometry by providing a reference value, such as column-mean reflectivity, with which to meaningfully interpret actual horizontal distributions of biological activity. Although these techniques, applied to aeroecology, still require further verification, they are already aiding in the study of how birds and other organisms are influenced by spatial and temporal trends which might previously have been difficult to observe, such as related to land cover, man-made hazards, and the three-dimensional nature of meteorological and climate systems.

Acknowledgements Several of the algorithms described in this chapter have been implemented within the Warning Decision Support System Integrated Information [WDSSII; Lakshmanan et al. (2007b)] as the w2merger and w2birddensity algorithms. WDSS-II is available for download at www.wdssii.org.

References

Able KP (1970) A radar study of the altitude of nocturnal passerine migration. Bird Band 41(4):282. https://doi.org/10.2307/4511688

Achtemeier GL (1991) The use of insects as tracers for "clear-air" boundary-layer studies by Doppler radar. J Atmos Ocean Technol 8(6):746–765. https://doi.org/10.1175/1520-0426 (1991)008<0746:TUOIAT>2.0.CO;2

Askelson M, Aubagnac J, Straka J (2000) An adaptation of the Barnes filter applied to the objective analysis of radar data. Mon Weather Rev 128(9):3050–3082

Beyer W (ed) (1987) CRC standard math tables, 18th edn. CRC Press, Boca Raton

Bridge ES, Pletschet SM, Fagin T, Chilson PB, Horton KG, Broadfoot KR, Kelly JF (2015) Persistence and habitat associations of Purple Martin roosts quantified via weather surveillance radar. Landsc Ecol 31(1):43–53. https://doi.org/10.1007/s10980-015-0279-0

Buler JJ, Dawson DK (2014) Radar analysis of fall bird migration stopover sites in the northeastern U.S. Condor 116(3):357–370. https://doi.org/10.1650/CONDOR-13-162.1

Buler J, Diehl R (2009) Quantifying bird density during migratory stopover using weather surveillance radar. IEEE Trans Geosci Remote Sens 47(8):2741–2751

Buler J, Lakshmanan V, La Puma D (2012) Improving weather radar data processing for biological research applications: final report. Technical report G11AC20489, Patuxent Wildlife Research Center, USGS, Laurel, MD

Chilson PB, Frick WF, Kelly JF, Howard KW, Larkin RP, Diehl RH, Westbrook JK, Kelly TA, Kunz TH (2012a) Partly cloudy with a chance of migration: weather, radars, and aeroecology. Bull Am Meteorol Soc 93(5):669–686. https://doi.org/10.1175/BAMS-D-11-00099.1

Chilson PB, Frick WF, Stepanian PM, Shipley JR, Kunz TH, Kelly JF (2012b) Estimating animal densities in the aerosphere using weather radar: to Z or not to Z? Ecosphere 3(8):1–19. https://doi.org/10.1890/ES12-00027.1

Cohen EB, Barrow WC, Buler JJ, Deppe JL, Farnsworth A, Marra PP, McWilliams SR, Mehlman DW, Wilson RR, Woodrey MS, Moore FR (2017) How do en route events around the Gulf of Mexico influence migratory landbird populations? Condor 119(2):327–343. https://doi.org/10.1650/CONDOR-17-20.1

Contreras RF, Frasier SJ (2008) High-resolution observations of insects in the atmospheric boundary layer. J Atmos Ocean Technol 25(12):2176–2187. https://doi.org/10.1175/2008JTECHA1059.1

Dean J, Ghemawat S (2008) Mapreduce: simplified data processing on large clusters. Commun ACM 51(1):107–113

Diehl RH, Larkin RP (2005) Introduction to the WSR-88D (NEXRAD) for ornithological research. In: Ralph CJ, Rich TD (eds) Bird conservation implementation and integration in the Americas: Proceedings of 3rd international partners in flight conference, Asilomar, CA. 20 24 March 2002. Gen. Tech. Rep. PSW-GTR-191, vol 2, U.S. Department of Agriculture, Forest Service, Pacific Southwest Research Station, Albany, NY, pp 876–888

Diehl R, Larkin R, Black J, Moore F (2003) Radar observations of bird migration over the Great Lakes. Auk 120(2):278–290

Dokter AM, Liechti F, Stark H, Delobbe L, Tabary P, Holleman I (2011) Bird migration flight altitudes studied by a network of operational weather radars. J R Soc Interface 8:30–43

Dokter AM, Åkesson S, Beekhuis H, Bouten W, Buurma L, van Gasteren H, Holleman I (2013) Twilight ascents by common swifts, Apus apus, at dawn and dusk: acquisition of orientation cues? Anim Behav 85(3):545–552. https://doi.org/10.1016/j.anbehav.2012.12.006

Doviak R, Zrnic D (1993) Doppler radar and weather observations, 2nd edn. Academic Press, Cambridge, MA

Drake VA (1984) The vertical distribution of macro-insects migrating in the nocturnal boundary layer: a radar study. Bound Lay Meteorol 28(3–4):353–374. https://doi.org/10.1007/BF00121314

Drake VA, Farrow RA (1988) The influence of atmospheric structure and motions on insect migration. Annu Rev Entomol 33(1):183–210. https://doi.org/10.1146/annurev.en.33.010188.001151

Eastwood E (1967) Radar ornithology. Methuen, London

Farnsworth A, Van Doren BM, Hochachka WM, Sheldon D, Winner K, Irvine J, Geevarghese J, Kelling S (2016) A characterization of autumn nocturnal migration detected by weather surveillance radars in the northeastern USA. Ecol Appl 26(3):752–770

Frick WF, Stepanian PM, Kelly JF, Howard KW, Kuster CM, Kunz TH, Chilson PB (2012) Climate and weather impact timing of emergence of bats. PLoS One 7(8):e42737. https://doi.org/10.1371/journal.pone.0042737

Gauthreaux SA (1991) The flight behavior of migrating birds in changing wind fields. Am Zool 31 (1):187–204

Gauthreaux SA, Belser C (2003) Radar ornithology and biological conservation. Auk 120(2):266–277

Gauthreaux SA, Belser CG, van Blaricon D (2003) Avian migration. Springer, Berlin. https://doi.org/10.1007/978-3-662-05957-9

Gauthreaux SA, Livingston JW, Belser CG (2008) Detection and discrimination of fauna in the aerosphere using Doppler weather surveillance radar. Integr Comp Biol 48(1):12–23

Geerts B, Miao Q (2005) Airborne radar observations of the flight behavior of small insects in the atmospheric convective boundary layer. Environ Entomol 34(2):361–377. https://doi.org/10.1603/0046-225X-34.2.361

Grecu M, Krajewski W (2000) An efficient methodology for detection of anamalous propagation echoes in radar reflectivity data using neural networks. J Atmos Oceanic Tech 17:121–129

Greene DR, Clark RA (1972) Vertically integrated liquid water – a new analysis tool. Mon Weather Rev 100:548–552

Horn JW, Kunz TH (2008) Analyzing NEXRAD doppler radar images to assess nightly dispersal patterns and population trends in Brazilian freetailed bats (*Tadarida brasiliensis*). Integr Comp Biol 48(1):24–39. https://doi.org/10.1093/icb/icn051

Horton KG, Shriver WG, Buler JJ (2016a) An assessment of spatio-temporal relationships between nocturnal bird migration traffic rates and diurnal bird stopover density. Mov Ecol 4:1. https://doi.org/10.1186/s40462-015-0066-1

Horton KG, Van Doren BM, Stepanian PM, Farnsworth A, Kelly JF (2016b) Seasonal differences in landbird migration strategies. Auk 133(4):761–769. https://doi.org/10.1642/AUK-16-105.1

Horton KG, Van Doren BM, Stepanian PM, Farnsworth A, Kelly JF (2016c) Where in the air? Aerial habitat use of nocturnally migrating birds. Biol Lett 12(11). https://doi.org/10.1098/rsbl.2016.0591

Horton KG, Van Doren BM, Stepanian PM, Hochachka WM, Farnsworth A, Kelly JF (2016d) Nocturnally migrating songbirds drift when they can and compensate when they must. Sci Rep 6:21249. https://doi.org/10.1038/srep21249

Joss J, Lee R (1995) The application of radar–gauge comparisons to operational precipitation profile corrections. J Appl Meteorol 34(12):2612–2630. https://doi.org/10.1175/1520-0450(1995)034<2612:TAORCT>2.0.CO;2

Kelly JF, Horton KG (2016) Toward a predictive macrosystems framework for migration ecology. Glob Ecol Biogeogr 25(10):1159–1165. https://doi.org/10.1111/geb.12473

Kelly JF, Horton KG, Stepanian PM, de Beurs KM, Fagin T, Bridge ES, Chilson PB (2016) Novel measures of continental-scale avian migration phenology related to proximate environmental cues. Ecosphere 7(8):e01434. https://doi.org/10.1002/ecs2.1434

Kemp M, ShamounBaranes J, Dokter A (2013) The influence of weather on the flight altitude of nocturnal migrants in mid-latitudes. Ibis 155:734–749

Kessinger C, Ellis S, Van Andel J (2003) The radar echo classifier: a fuzzy logic algorithm for the WSR-88D. In: 3rd conference on artificial applications to the environmental sciences, American Meteor Society, Long Beach, CA, P1.6

Kitchen M (1997) Towards improved radar estimates of surface precipitation rate at long range. Q J R Meteorol Soc 123(537):145–163. https://doi.org/10.1002/qj.49712353706

Koistinen J, Pohjola H (2014) Estimation of ground-level reflectivity factor in operational weather radar networks using VPR-based correction ensembles. J Appl Meteorol Climatol 53(10):2394–2411. https://doi.org/10.1175/JAMC-D-13-0343.1

Kunz TH, Gauthreaux SA, Hristov NI, Horn JW, Jones G, Kalko EKV, Larkin RP, McCracken GF, Swartz SM, Srygley RB, Dudley R, Westbrook JK, Wikelski M (2008) Aeroecology: probing and modeling the aerosphere. Integr Comp Biol 48(1):1–11. https://doi.org/10.1093/icb/icn037

La Sorte FA, Hochachka WM, Farnsworth A, Sheldon D, Van Doren BM, Fink D, Kelling S (2015) Seasonal changes in the altitudinal distribution of nocturnally migrating birds during autumn migration. R Soc Open Sci 2(12):150347. https://doi.org/10.1098/rsos.150347

Lack D, Varley G (1945) Detection of birds by radar. Nature 156(3963):446

Lakshmanan V, Smith T, Hondl K, Stumpf GJ, Witt A (2006) A real-time, three dimensional, rapidly updating, heterogeneous radar merger technique for reflectivity, velocity and derived products. Weather Forecast 21(5):802–823

Lakshmanan V, Fritz A, Smith T, Hondl K, Stumpf GJ (2007a) An automated technique to quality control radar reflectivity data. J Appl Meteorol 46(3):288–305

Lakshmanan V, Smith T, Stumpf GJ, Hondl K (2007b) The warning decision support system – integrated information. Weather Forecast 22(3):596–612

Lakshmanan V, Zhang J, Howard K (2010) A technique to censor biological echoes in radar reflectivity data. J Appl Meteorol 49(3):435–462

Lakshmanan V, Karstens C, Krause J, Tang L (2014) Quality control of weather radar data using polarimetric variables. J Atmos Oceanic Tech 31:1234–1249

Lang TJ, Rutledge SA, Stith JL (2004) Observations of quasi-symmetric echo patterns in clear air with the CSUCHILL polarimetric radar. J Atmos Ocean Technol 21(8):1182–1189. https://doi.org/10.1175/1520-0426(2004)021<1182:OOQEPI>2.0.CO;2

Leskinen M, Markkula I, Koistinen J, Pylkkö P, Ooperi S, Siljamo P, Ojanen H, Raiskio S, Tiilikkala K (2011) Pest insect immigration warning by an atmospheric dispersion model, weather radars and traps. J Appl Entomol 135:55–67. https://doi.org/10.1111/j.1439-0418.2009.01480.x

Liu H, Chandrasekar V (2000) Classification of hydrometeors based on polarimetric radar measurements: development of fuzzy logic and neuro-fuzzy systems, and in situ verification. J Atmos Oceanic Tech 17(2):140–164

Melnikov V, Leskinen M, Koistinen J (2013) Doppler velocities at orthogonal polarizations in radar echoes from insects and birds. IEEE Geosci Remote Sens Lett 11(3):592–596. https://doi.org/10.1109/LGRS.2013.2272011

Melnikov VM, Istok MJ, Westbrook JK (2015) Asymmetric radar echo patterns from insects. J Atmos Ocean Technol 32(4):659–674. https://doi.org/10.1175/JTECH-D-13-00247.1

Mirkovic D, Stepanian PM, Kelly JF, Chilson PB (2016) Electromagnetic model reliably predicts radar scattering characteristics of airborne organisms. Sci Rep 6:35637. https://doi.org/10.1038/srep35637

Mueller EA, Larkin RP (1985) Insects observed using dual-polarization radar. J Atmos Ocean Technol 2(1):49–54. https://doi.org/10.1175/1520-0426(1985)002<0049:IOUDPR>2.0.CO;2

Nebuloni R, Capsoni C, Vigorita V (2008) Quantifying bird migration by a high-resolution weather radar. IEEE Trans Geosci Remote Sens 46(6):1867–1875

Park H, Ryzhkov A, Zrnic D, Kim K (2009) The hydrometeor classification algorithm for the polarimetric WSR-88D: description and application to a mcs. Weather Forecast 24(3):730–748

Rinehart R (1991) Radar for meteorologists, or, You, too, can be a radar meteorologist, 2nd edn. Rinehart Pubilcations, Grand Forks, ND

Russell KR, Gauthreaux SA (1998) Use of weather radar to characterize movements of roosting Purple martins. Wildl Soc Bull 26(1):5–16

Russell KR, Mizrahi DS, Gauthreaux SA (1998) Large-scale mapping of Purple martin pre-migratory roosts using WSR-88D weather surveillance radar. J Field Ornithol 69(2):316–325

Schmaljohann H, Liechti F, Bruderer B (2007) Songbird migration across the Sahara: the non-stop hypothesis rejected! Proc Biol Sci 274(1610):735–739. https://doi.org/10.1098/rspb.2006.0011

Shamoun-Baranes J, Bouten W, van Loon EE (2010) Integrating meteorology into research on migration. Integr Comp Biol 50(3):280–292. https://doi.org/10.1093/icb/icq011

Shamoun-Baranes J, Dokter AM, van Gasteren H, van Loon EE, Leijnse H, Bouten W (2011) Birds flee en mass from New Year's Eve fireworks. Behav Ecol 22(6):1173–1177. https://doi.org/10.1093/beheco/arr102

Shamoun-Baranes J, Alves JA, Bauer S, Dokter AM, Hüppop O, Koistinen J, Leijnse H, Liechti F, van Gasteren H, Chapman JW (2014) Continental-scale radar monitoring of the aerial movements of animals. Mov Ecol 2(1):9. https://doi.org/10.1186/2051-3933-2-9

Shamoun-Baranes J, Liechti F, Vansteelant WMG (2017) Atmospheric conditions create freeways, detours and tailbacks for migrating birds. J Comp Physiol A 203(6–7):509–529. https://doi.org/10.1007/s00359-017-1181-9

Smyth TJ, Illingworth AJ (1998) Radar estimates of rainfall rates at the ground in bright band and non-bright band events. Q J R Meteorol Soc 124(551):2417–2434. https://doi.org/10.1002/qj.49712455112

Steiner M, Smith J (2002) Use of three-dimensional reflectivity structure for automated detection and removal of non-precipitating echoes in radar data. J Atmos Oceanic Tech 19:673–686

Stepanian PM, Horton KG (2015) Extracting migrant flight orientation profiles using polarimetric radar. IEEE Trans Geosci Remote Sens 53(12):6518–6528. https://doi.org/10.1109/TGRS.2015.2443131

Stepanian PM, Horton KG, Melnikov VM, Zrnić DS, Gauthreaux SA (2016) Dual-polarization radar products for biological applications. Ecosphere 7(11):e01539. https://doi.org/10.1002/ecs2.1539

Trapp RJ, Doswell CA (2000) Radar data objective analysis. J Atmos Ocean Tech 17:105–120

Van Doren BM, Horton KG, Stepanian PM, Mizrahi DS, Farnsworth A (2016) Wind drift explains the reoriented morning flights of songbirds. Behav Ecol 27(4):1122–1131. https://doi.org/10.1093/beheco/arw021

Vignal B, Andrieu H, Creutin JD (1999) Identification of vertical profiles of reflectivity from volume scan radar data. J Appl Meteorol 38:1214–1228

Vignal B, Galli G, Joss J, Germann U (2000) Three methods to determine profiles of reflectivity from volumetric radar data to correct precipitation estimates. J Appl Meteorol 39(10):1715–1726. https://doi.org/10.1175/1520-0450-39.10.1715

Wainwright CE, Stepanian PM, Horton KG (2016) The role of the US great plains low-level jet in nocturnal migrant behavior. Int J Biometeorol 60(10):1531–1542. https://doi.org/10.1007/s00484-016-1144-9

Warning Decision Training Division (2016) VCP training. http://wdtb.noaa.gov/modules/vcpTraining/index.html

Westbrook J, Eyster R (2017) Doppler weather radar detects emigratory flights of noctuids during a major pest outbreak. Remote Sens Appl Soc Environ 8:64–70. https://doi.org/10.1016/j.rsase.2017.07.009

Westbrook J, Isard S (1999) Atmospheric scales of biotic dispersal. Agric For Meteorol 97(4):263–274. https://doi.org/10.1016/S0168-1923(99)00071-4

Westbrook JK, Eyster RS, Wolf WW (2014) WSR-88D Doppler radar detection of corn earworm moth migration. Int J Biometeorol 58(5):931–940. https://doi.org/10.1007/s00484-013-0676-5

Zrnic D, Ryzhkov A (1998) Observations of insects and birds with a polarimetric radar. IEEE Trans Geosci Remote Sens 36:661–668

Part III

Aeroecological Applications

Linking Animals Aloft with the Terrestrial Landscape

14

Jeffrey J. Buler, Wylie C. Barrow, Jr, Matthew E. Boone,
Deanna K. Dawson, Robert H. Diehl, Frank R. Moore,
Lori A. Randall, Timothy D. Schreckengost,
and Jaclyn A. Smolinsky

Abstract

Despite using the aerosphere for many facets of their life, most flying animals (i.e., birds, bats, some insects) are still bound to terrestrial habitats for resting, feeding, and reproduction. Comprehensive broad-scale observations by weather surveillance radars of animals as they leave terrestrial habitats for migration or feeding flights can be used to map their terrestrial distributions either as point locations (e.g., communal roosts) or as continuous surface layers (e.g., animal densities in habitats across a landscape). We discuss some of the technical challenges to reducing measurement biases related to how radars sample the aerosphere and the flight behavior of animals. We highlight a recently developed methodological approach that precisely and quantitatively links the horizontal spatial structure of birds aloft to their terrestrial distributions and provides novel insights into avian ecology and conservation across broad landscapes. Specifically, we present case studies that (1) elucidate how migrating birds contend with crossing ecological

J.J. Buler (✉) • M.E. Boone • T.D. Schreckengost • J.A. Smolinsky
Department of Entomology and Wildlife Ecology, University of Delaware, Newark, DE, USA
e-mail: jbuler@udel.edu; boone@udel.edu; tschreck@udel.edu; jsmo@udel.edu

W.C. Barrow, Jr • L.A. Randall
U.S. Geological Survey, Wetland and Aquatic Research Center, Lafayette, LA, USA
e-mail: barroww@usgs.gov; randalll@usgs.gov

D.K. Dawson
U.S. Geological Survey, Patuxent Wildlife Research Center, Laurel, MD, USA
e-mail: ddawson@usgs.gov

R.H. Diehl
U.S. Geological Survey, Northern Rocky Mountain Science Center, Bozeman, MT, USA
e-mail: rhdiehl@usgs.gov

F.R. Moore
Department of Biological Sciences, University of Southern Mississippi, Hattiesburg, MS, USA
e-mail: frank.moore@usm.edu

© Springer International Publishing AG, part of Springer Nature 2017
P.B. Chilson et al. (eds.), *Aeroecology*,
https://doi.org/10.1007/978-3-319-68576-2_14

barriers and extreme weather events, (2) identify important stopover areas and habitat use patterns of birds along their migration routes, and (3) assess waterfowl response to wetland habitat management and restoration. These studies aid our understanding of how anthropogenic modification of the terrestrial landscape (e.g., urbanization, habitat management), natural geographic features, and weather (e.g., hurricanes) can affect the terrestrial distributions of flying animals.

1   Using Weather Radar to Link Flying Animals Aloft with Terrestrial Habitats

Despite using the aerosphere for many facets of their life, flying animals are still bound to terrestrial habitats for resting, feeding, and/or breeding. The alpine swift (*Tachymarptis melba*) exhibits perhaps the most extreme use of the aerosphere of any vertebrate by being able to remain airborne for 200 days while performing all vital physiological processes including sleeping (Liechti et al. 2013). However, it still returns to cliffs or caves to breed. Accordingly, the ecologies of animals in the air and on the ground coevolve and are inextricably linked. Terrestrial habitat quality, abundance, and distribution are of vital importance for maintaining species of flying animals. Additionally, terrestrial habitat plays an important role in shaping morphological, physiological, behavioral, and life history traits that impact the aerial ecology of flying animals. For example, habitat structure and selection are closely related to the wing morphology and the foraging and flight activity of birds (Robinson and Holmes 1982; Janes 1985; Winkler and Leisler 1985), bats (Brigham et al. 1997; Hodgkison et al. 2004), and insects (Hassell and Southwood 1978; Vandewoestijne and Van Dyck 2011). In turn, the aeroecology of flying animals shapes their terrestrial life histories and habitat use. For example, the improved dispersal capabilities of flying animals are shaped by global wind patterns and influence the evolution of migratory syndromes and pathways that expose animals to a changing suite of terrestrial habitats throughout the annual cycle (Able 1972; Gauthreaux 1980; Alerstam 1993; Moore et al. 1995; Drake and Gatehouse 1996; Gauthreaux et al. 2005; Dingle and Drake 2007; La Sorte et al. 2014a; Kranstauber et al. 2015).

Among available research tools for measuring the abundance of animals aloft, weather surveillance radars (WSR) offer the distinct advantage of being able to detect animal movements across broad spatial domains (~100 km radius) at both a comparatively fine spatial resolution (~1–150 ha) and at frequent intervals (~4–10 min). Large networks of radars in the United States (Gauthreaux et al. 2003), Canada (Gagnon et al. 2011), and Europe (Shamoun-Baranes et al. 2014) provide opportunities for continental-scale monitoring of animal movements. Routine local movements of individual animals between aerial and terrestrial habitats are challenging to measure with WSR due to their closeness to the ground (typically within the atmospheric surface layer, i.e., <100 m above ground level) and potential contamination from ground clutter. However, some animals initiate periodic and well-synchronized flights *en masse* that are comprehensively sampled by WSR in a way that can be used for rigorous scientific research. Recent radar-observed movements include seasonal migratory flights of insects (e.g., Westbrook et al. 2014) and birds (e.g., Buler and Dawson 2014), foraging flights of birds (e.g., Randall et al. 2011; Sieges et al. 2014), dispersal flights from communal roosts of birds (e.g., Laughlin et al. 2014) and bats (e.g., Frick et al. 2012), and twilight ascending flights by swifts (e.g., Dokter et al. 2013). WSR has provided novel information about population dynamics (see Chap. 15) and the phenology of annual life-cycle events (see Chap. 16) of flying animals that is not easily attainable by other means. WSR operating at low tilt angles can also provide information about where animals are concentrated or distributed in the terrestrial landscape by capturing patterns of flying animals as they enter the airspace *en masse* (Diehl and Larkin 2005, also see Chap. 12).

The nature of how various animals move and congregate gives rise to two main approaches for mapping their terrestrial distributions with WSR: (1) point mapping of roosts and other concentrations of animals and (2) continuous-surface mapping of animal distributions over broad geographic areas when initiating high-altitude synchronized flights. The two approaches differ in their consideration of when to sample animals aloft to accurately determine their on-ground locations and abundance and how to validate radar data with ground observations. We focus this chapter on discussing the application of these two approaches to mapping terrestrial distributions of animals with WSR. We highlight a recently developed methodological approach that precisely and quantitatively links the horizontal spatial structure of birds aloft to their ground distributions. We then present several case studies that improve our understanding of how anthropogenic modification of the terrestrial landscape (e.g., urbanization, habitat management), natural geographic features, and weather (e.g., hurricanes) can affect the terrestrial distributions of flying animals. We close by identifying some issues that present challenges for the use of WSR for mapping terrestrial distributions of flying animals and suggest areas for future research.

2   Point Mapping of Terrestrial Concentrations of Flying Animals

2.1   The Mapping Process

Roosts and some other animal concentrations are relatively easy to identify and locate in radar data. Animals taking flight from roosts often appear on radar as expanding rings or arcs emanating from a point source (Fig. 14.1) (Lack and Varley 1945; Elder 1957; Eastwood 1967). Roosting swallows, in particular, tend to leave the roost *en masse*, flying upward to high altitudes before turning out in nearly all directions, thus giving rise to these characteristic rings (Winkler 2006). Other manifestations of animal roosts can also be seen on WSR. Bats exiting roosts can take on a funnel shape, expanding outward as distance from the source increases

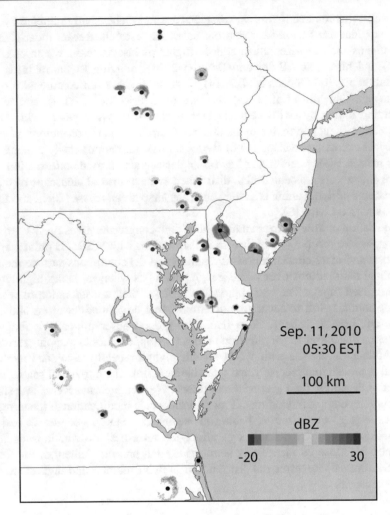

Fig. 14.1 Composite of radar reflectivity scans from 6 WSR-88D sites depicting the morning exodus flight of birds from 34 communal roosts on September 11, 2010, in the mid-Atlantic region of the United States. Black dots denote center of roost rings and presumed location of roosts. Reflectivity not attributed to bird roosts was removed for clarity

(Horn and Kunz 2008). Congregations of waterfowl leaving a discrete wetland site for migration appear as an irregularly shaped polygon growing laterally in size as they move away from the source (O'Neal et al. 2010).

These patterns allow roost locations to be easily pinpointed and mapped. Early radar studies typically involved visually analyzing "analog" photographs of radar scans to pinpoint locations of roosts. Modern approaches to pinpointing roost locations rely on georeferenced digital radar data and involve manually fitting a circle around a roost ring or arc within a geographic information system and calculating the geographic position of the circle center (Laughlin et al. 2013), or more-sophisticated automated image analysis algorithms such as patch identification (Horn and Kunz 2008) or a Hough transform (Larkin 2006) to recognize roosts and calculate their centers.

Pinpointing the location of roosts by observing animals as they enter roosts is sometimes difficult because of the disorganization and low altitude of aggregating flights. Large-bodied martins (*Progne* spp.) often fly low to the ground and under the radar beam or enter roosts as individuals or in small flocks over tens of minutes as sunset approaches (Russell and Gauthreaux 1998, 1999). By contrast, smaller swallows (*Tachycineta bicolor, Stelgidopteryx* spp.), as described by Winkler (2006), "congregate a few hundred meters above the roost site, milling around the site in an increasingly large and dense cloud of birds. Finally, as the last daylight fades, a few courageous birds make the plunge downward into the reeds of the roost site, followed immediately by a swirling stream of birds pouring into the vegetation, with hundreds of thousands of birds settling in only a few minutes' time." Other bird species may take as long as an hour or more to assemble at a location away from the ultimate roost location. For example, Eastwood et al. (1962) observed that roosting European starlings (*Sturnus vulgaris*) vary their pre-assembly location (up to a mile from the roost location) and time on a daily basis. The flock then moves from the pre-assembly location to the roost, occasionally circling the roost once or twice before entering and finally settling. This activity appears on the radar as a form of random variation in reflectivity or "effervescence" and can complicate pinpointing of the ultimate roost location.

Sampling expanding radar rings of animals aloft as they exit roosts does not require precise timing to accurately pinpoint roost locations. This is because expanding rings will typically appear on several radar scans in succession as animals expand out from their roost locations. This allows researchers either to select an individual scan from the series of scans depicting roost rings or to integrate information across the scans to pinpoint roost locations. For example, Laughlin et al. (2013, 2014) used the center of the circle in the first scan in which the roost was detected. Using a similar approach, Bridge et al. (2016) found that estimated locations of purple martin (*Progne subis*) roosts from radar images occurred within 10 km of actual ground-truthed locations. Moreover, some animals leave roosts in a succession of organized waves producing a sequence of expanding rings that appear much like the ripples in water after a tossed stone breaks the surface. Successive exodus waves increase the window of time for determining roost locations. For example, emerging waves of European starlings produce upward of 18 rings from the roost center at consistent average intervals of 3 min, with complete evacuation of the roost taking at least 50 min (Eastwood et al. 1962). Brazilian free-tailed bats (*Tadarida brasiliensis*) at peak colony size typically exit roosts in 2–4 waves separated by 15–30 min intervals (Horn and Kunz 2008).

The temporal flexibility in sampling roost rings is important since the exact timing of roost exodus, especially for bats, can be quite variable. The evening exodus of bat colonies is one of the most easily and frequently studied aspects of bat behavior (reviewed by Jones and Rydell 1994). The timing of bat emergence appears to be primarily a function of diet and foraging strategy that balances predation risk and foraging time. Insectivorous bats that feed primarily on dipterans (i.e., the majority of temperate bat species) emerge relatively early during the peak flight activity of small insects about 20–40 min after sunset. However, emerging before dark exposes insectivorous bats to greater predation risk from raptors and possible competition from insectivorous birds. Bats that eat moths, flightless prey, or plants typically leave later at around 40–70 min after sunset, thus reducing predation risk (Jones and Rydell 1994). Roost exodus timing of bats also fluctuates daily and seasonally with weather and food availability. In dry seasons in central Texas when fewer insects likely exist, cave-dwelling bats emerged as early as 1.5 h before sunset, risking predation to increase foraging time (Frick et al. 2012). In contrast, during moist seasons bats emerged 30 min after sunset.

Variability in roost exodus timing is lower in birds than in bats. Russell and Gauthreaux (1999) found that the onset of roost exodus for purple martins across 32 days ranged from 31 to 48 min before sunrise, with flights lasting 68 ± 12 min in duration. The fine-scale variability in the timing of exodus was related to atmospheric pressure and cloud cover, indicating that martins likely cued on daily fluctuations in ambient light intensity when departing roosts (Harper 1959; Russell and Gauthreaux 1999).

Complete rings are usually observed only when animals disperse away in all directions at a consistent speed under no or light winds (<3 m/s). Under these conditions, the center of a ring remains stationary over the roost location until exodus flight is completed (Ligda 1958; Eastwood 1967). However, displacement of complete rings downwind is sometimes observed under stronger winds (3–6 m/s) without drift correction by animals (Eastwood 1967; Russell and Gauthreaux 1999). With yet further increase in wind speed, radars observe arcs of animals that move downwind, as opposed to complete rings. In these cases, arc formations are not caused by an absence of animals in the upwind direction but rather from a tendency for animals to fly into the wind at a lower altitude and thus remain undetected by the radar. However, arcs may also be seen when animals show directionality in their movement. For example, arcs are observed for some bird roosts located near water bodies, since landbirds may avoid dispersing over water (Eastwood 1967). Arcs are also observed when birds leave roosts to embark on directed migratory flights (Harper 1959; Russell and Gauthreaux 1999). Horn and Kunz (2008) found that surface winds do not appear to affect the initial direction or the speed at which Brazilian free-tailed bats set out to forage. Rather, bats show directed flight toward masses of emerging, dispersing, or migrating insects (Cleveland et al. 2006). In situations where drifting rings or arcs are seen, a correction for wind velocity can be applied (e.g., Eastwood et al. 1962) or the trajectory of the arc can be traced back to its origin with an object-tracking algorithm (e.g., Horn and Kunz 2008). In other studies, no correction is made for potential drifting of roost rings or arcs (e.g., Russell et al. 1998; Laughlin et al. 2013).

Identifying the species composition of animal roosts that appear on radar is easily confirmed on the ground by placing observers at the center of the expanding rings or arcs as animals leave roosts (e.g., Eastwood et al. 1962) or by matching the center location of radar rings with already known roosts (e.g., Kelly et al. 2012).

Alternatively, in one of the first studies to document observations of bat roosts on radar, Williams et al. (1973) confirmed by helicopter that the reflectivity on airport surveillance radars in Texas was caused by groups of Brazilian free-tailed bats dispersing from caves to forage. O'Neal et al. (2010) used other remote-sensing tools to identify the coarse taxonomic identity of individual waterfowl emanating from a discrete wetland complex in Illinois for nocturnal migratory flight—a thermal infrared camera to record animal shape and a portable stationary-beam X-band radar to measure wing beat frequencies. The time of day or season when roost rings appear on radar can also be used as a clue to the identity of animals producing them. For example, purple martins roost communally from the end of their breeding season until they depart for tropical wintering areas in late July and August, while roosts of species that winter in temperate regions (e.g., tree swallow, red-winged blackbird) are used for much longer (Winkler 2006). Foraging bats and foraging or migrating waterfowl leave diurnal roosts in the evenings after sunset, while diurnally active birds generally leave nocturnal roosts shortly before sunrise.

2.2   Ecological Insights

Communal roosting is often a regional phenomenon that involves broad spatial and temporal relationships among roosts (Caccamise et al. 1983). However, early radar studies of spatial and temporal dynamics of roosts were limited in scope to providing "local"-scale information (e.g., Ligda 1958; Harper 1959; Eastwood et al. 1962). For example, Eastwood et al. (1962) used a single radar in England to study the dynamics of European starling roost locations over the course of 2 years. The lack of any distinctive topographic features in the landscape associated with persistent ring centers led the researchers to suspect starlings as the source of the rings, which was confirmed by observers on the ground. By monitoring roost activity over time, they were able to determine there were upward of 12 persistent roosts of resident starlings and several temporary roosts of migrating starlings. Of particular note was a large persistent roost that was detectable within the center of Trafalgar Square in London despite clutter radar echoes produced from surrounding tall buildings. Eastwood et al. (1962) documented changes in the locations of starling roosts in response to anthropogenic disturbance. For example, one roost location changed after local farmers shot many individual starlings.

Most roost mapping studies in the United States have been conducted since the establishment of the network of more than 150 WSR-88D stations and digital archiving of radar data in the mid-1990s. The archived observations from these networked radars have allowed for examination of roost location dynamics at a sufficiently large scale to detect regional and seasonal patterns over many years, which wasn't possible in earlier radar studies.

2.2.1   Bat Roosting Ecology

The first observations of bat colonies on WSR-88D were made in Oklahoma and Texas soon after the installation of the network (Ruthi 1994; McCracken and

Westbrook 2002). By studying the dynamics and locations of maternity colonies of Brazilian free-tailed bats for 11 years around a WSR-88D station in Texas, Horn and Kunz (2008) provided new insight into bat roost dynamics. They found that bats occupy bridge roosts more frequently than cave roosts early in the spring and late in the fall despite the potential negative aspects of noise, air pollution, and disturbance from humans. Concrete bridges with expansion joints may offer a significant thermal advantage over caves during the cooler periods of the roosting season by having warmer and more stable temperatures. Additionally, they hypothesized that occupancy of bridges may relieve overcrowding in the limited number of suitable roosting spaces in caves.

2.2.2 Bird Roosting Ecology

The first large-scale mapping of purple martin roosts in the United States, using 45 networked WSR-88D stations across 19 states, was conducted by Russell et al. (1998). Roost sites were consistently detected if they were within 100 km of a radar and occasionally up to 240 km from a radar, likely when the radar beam was being strongly refracted. More recent studies use 175 km as a threshold distance for detecting large swallow (Hirundinidae) roosts (Kelly et al. 2012; Laughlin et al. 2013). The detection range is limited in part by "range" bias (sensu Diehl and Larkin 2005), which is caused by the increasing height of the radar beam above the ground as it travels away from the radar. Orographic terrain relief can also impact the range coverage of a radar and, thus, detecting roosts. Radar detection of animals leaving roosts is also limited by their flight heights. For example, Russell and Gauthreaux (1998) found that WSR-88D detected 80% of roost departures observed during ground surveys at a purple martin roost located 28 km from the radar. Roost exodus went undetected when rain obscured radar detection of birds (2 days), birds flew just above the treetops because of fog or low cloud ceiling (4 days), or the radar beam experienced extreme sub-refraction and passed above the birds (2 days).

The Russell et al. (1998) and Russell and Gauthreaux (1998) studies provided a proof of concept that bird roosts over a large geographic area could be mapped easily with minimal resources using the radar network. Most of the 33 identified roosts were associated with areas of open water. Subsequently, more extensive identification and ground validation of radar-observed roost locations, conducted by volunteers of the Purple Martin Conservation Association, have documented 358 suspected martin roost sites within eastern North America (www.purplemartin. org/research/19/project-martinroost/). Furthermore, the comprehensive study by Bridge et al. (2016) evaluated habitat associations and persistence of 234 purple martin roosts over most of their range. They found that martins actually use a diverse array of roosting habitats including forest, cropland, and urban development in addition to areas adjacent to open water. Moreover, martins appear to prefer urban sites, and urban roosts were associated with the high year-to-year persistence.

Kelly et al. (2012) developed an approach that extends the use of WSR beyond simple mapping of roost locations to automated monitoring of the phenology of roost activity and relative sizes of the 358 roosts identified by the Purple Martin Conservation Association. They used the daily maximum radar reflectivity value recorded within the hour before sunrise within a 9-km^2 grid centered at each roost location as an index of the number of individuals within a roost (i.e., roost size). They found that roosts farther away from a radar station do not appear as consistently as those closer in and that roost size was negatively related to distance from the radar. Furthermore, Laughlin et al. (2014) found that variability in precision of swallow roost locations was positively related to distance from the radar. These studies highlight that range bias in detection, location, and size estimation of roosts remains a challenge that needs to be addressed for future quantitative studies, especially for those that seek to directly compare point locations across the radar domain.

Phenology of roosts has provided insight into the stopover and overwintering habitat use of tree swallows (*Tachycineta bicolor*) too. Laughlin et al. (2013) integrated WSR-88D observations with telemetry of individually marked birds to study stopover habitat use in southeastern Louisiana by several breeding populations of tree swallows from across North America. Numbers of tree swallows in sugarcane roosts decreased as harvest of sugarcane commenced. Large roosts of >1 million birds decreased to the tens of thousands near the end of harvest. It appears that the sugarcane fields are used as autumn stopover habitat for swallows, and wetland areas are used as overwintering habitat for the few birds that remain. More recently, Laughlin et al. (2016) found strong correlations between radar-estimated and eBird-estimated occupancy dynamics of tree swallow roosts. The long-term occupancy dynamics based on radar data provided evidence that Louisiana acts as a combined stopover and overwintering region, whereas Florida occupancy dynamics were akin to a traditional winter region.

Radar observations have revealed that over much of their migratory and wintering ranges, swallow roost sites in general appear to have fairly consistent spacing (Winkler 2006). Roost sites in eastern North America tend to be about 100–150 km apart. Based on these observations, Winkler (2006) proposed a series of testable questions about swallow migration strategies and roost-site dynamics. For example, given that roost spacing is within the range of a day's flight, he hypothesized that birds migrating through the day could easily reach the next roost site in the successive chain of nocturnal roosts. He predicted the mechanism for individuals to find roosts along the chain could be for experienced birds to recall a navigational map of previously used sites (i.e., roost fidelity) and/or for individuals to simply forage in the preferred migratory direction and then be recruited to the next roost by aggregating near the end of the day with other birds that used that roost the previous night (i.e., conspecific attraction). Understanding the mechanisms of roost selection could illuminate why certain sites are consistently used from year to year while others are more ephemeral in nature. These mechanisms may also inform the causes of the intra- and interannual changes in roost sizes.

In a related study, Laughlin et al. (2014) examined tree swallow roost dynamics during fall migration. They observed roosts forming in the same places each night, which indicated a fairly high level of individual roost fidelity. However, radio-marked individuals switched between roosts at a rate of at least 22% each night and showed some attraction to conspecifics going toward other roosts. Thus, the pattern of tree swallow roosting dynamics seems largely explained by individuals exhibiting a combination of moderately high roost-site fidelity coupled with moderate conspecific attraction.

2.2.3 Stopover Ecology of Waterfowl Congregations

O'Neal et al. (2010) developed a rigorous approach to use WSR to identify and enumerate the ducks emigrating from a 12,000 ha wetland stopover complex in central Illinois over the course of a fall migration season. This approach entailed determining an average radar cross section for an individual duck and summing the total reflectivity of the discrete patch of reflectivity (i.e., flock of birds) emanating from the wetland complex on a nightly basis. The system was unique in that the reflectivity of the emanating flock of birds did not mix with reflectivity from other sources and could be confidently attributed to the "point" ground source.

In a subsequent study, O'Neal et al. (2012) used WSR data in combination with data from weekly aerial censuses to estimate an average seasonal stopover duration for fall-migrating dabbling ducks across eight seasons. Their rather elegantly simple approach entailed dividing the total numbers of emigrants leaving the complex over the course of a migration season by the total duck-use days for the season determined via the aerial surveys. The quotient of the two measures provides an estimate of stopover duration in units of days. The mean seasonal stopover duration was 28 days (range 11–48 days), which is identical to a 28-day estimate for mallards (Bellrose and Crompton 1970) and consistent with a 21-day estimate for all dabbling duck species (Bellrose et al. 1979). Furthermore, O'Neal et al. (2012) found a positive relationship between seasonal stopover duration with a seasonal index of foraging habitat quality. Thus, migrating ducks appear to assess local stopover site conditions and adjust the amount of time they stop over accordingly. The wide flexibility in stopover duration that they observed is consistent with the hypothesis that ducks allocate their time adaptively among individual stopovers (Harper 1982), although the extent to which this flexibility reflects interindividual adaptation (e.g., Harper 1982) or adaptive stopover (e.g., Weber et al. 1999; Beekman et al. 2002; McLaren et al. 2013) is untested.

3   Continuous-Surface Mapping of Terrestrial Distributions of Broadly Dispersed Animals

The advent of digital measurement and recording of radar reflectivity by WSR networks in the United States (Crum et al. 1993) and Europe (Holleman et al. 2008) allows for more quantitative treatment of reflectivity and, combined with modern geographic information systems, the ability to summarize and map continuous surfaces of bioscatter over space and time within radar domains (Gauthreaux et al. 2003). Surface mapping of the terrestrial distributions of flying animals with WSR is a technically challenging and novel application that requires animals to be sampled immediately upon entering the radar-swept airspace before they have dispersed far from their ground sources (Diehl and Larkin 2005). Because animals of interest typically are much closer to the ground than the precipitation that the WSR networks were designed to sample, there are often large gaps in the network where flying animals cannot be detected (Fig. 14.2). Thus, unlike precipitation data, WSR networks don't provide comprehensive continuous coverage of bioscatter across the entire network. In fact, Buler and Dawson (2014) estimated that only about one-third of the land area in the northeastern United States is robustly sampled by WSR for surface mapping of animal distributions. To date, continuous surface mapping has only been conducted using the United States' WSR-88D network to sample bird distributions at the onset of nocturnal migratory flights (e.g., Bonter et al. 2009; Buler and Moore 2011; Ruth et al. 2012; Buler and Dawson 2014) and nocturnal feeding

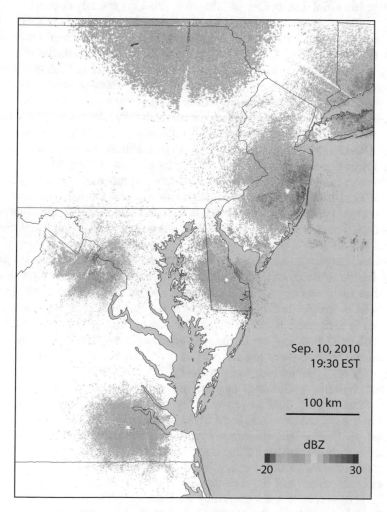

Fig. 14.2 Composite of radar reflectivity scans from 6 WSR-88D sites depicting the evening onset of migratory flights of birds on September 10, 2010, in the mid-Atlantic region of the United States. Areas of greater reflectivity (dBZ) coincide with areas of greater bird density

flights of waterfowl (e.g., Buler et al. 2012b; Sieges et al. 2014). The remainder of this chapter will therefore focus on continuous-surface mapping of bird distributions. However, most aspects of mapping bird distributions are applicable to mapping terrestrial distributions of other flying animals. Mapping ground distributions of emerging migrating insects or bats may be possible when either taxa is dominant within the airspace, but such attempts have not yet been made.

3.1 The Mapping Process

Sampling terrestrial distributions of birds with WSR is generally done with a single near-instantaneous radar scan for a given night at the lowest elevation sweep as birds leave their terrestrial habitats near the initial onset of well-synchronized *en masse* flights (Buler and Diehl 2009). This approach helps to preserve the geographic fidelity and structure in animal distributions (Buler et al. 2012b). Widely distributed birds appear as sudden "blooms" of reflectivity centered around radars as they take flight and enter radar-sampled airspace (Fig. 14.2). At an early stage of the onset of flight, bird reflectivity measures throughout the radar domain can vary by orders of magnitude, with greater values coinciding with habitats that contained more birds. Within minutes, as the initial birds continue to gain altitude and disperse away from their point of departure, the bloom of reflectivity expands and reflectivity measures become more homogeneous. Migrating landbirds and foraging flights of waterfowl continue to emerge from terrestrial habitats for at least 30 min (Hebrard 1971; Åkesson et al. 1996; Buler et al. 2012b; Wingo and Knupp 2014). Gauthreaux and Belser (2003) used multiple radar scans over a time period of 20–50 min during the onset of bird migration on a given night to map generalized stopover areas of high relative bird density. In contrast, continuous-surface mapping requires more precise timing for the selection of a single radar scan. How the accuracy and precision of radar data for mapping ground distributions decline through time has not been empirically evaluated for migrating landbirds, but for feeding flights of waterfowl, it appears to peak about 10 min after the initiation of flight (Buler et al. 2012b).

As with point mapping, continuous-surface mapping by sampling birds aloft as they descend into terrestrial habitats may be nearly impossible because the descent is generally not synchronized. Gauthreaux (1971) described the behavior of small flocks of migrating landbirds as they arrived on the northern coast of the Gulf of Mexico following an 18–24 h over-water flight. He wrote that "as a flock high aloft moved over a coastal woodland some of the individuals hesitated, hovered, or flew in broad, shallow spirals while the remaining flock members continued farther inland. The individuals that left the flock then closed their wings and dove nearly straight down." From a radar perspective, birds that remain aloft would continue to produce radar echoes and largely obscure the activity of individual birds making landfall. The radar echoes disappear only when the last birds descend into terrestrial habitats. Subsequently, in the only published study to examine landfall patterns, Gauthreaux (1975) observed that migrating birds disappeared from radar scans at forests 46–140 km inland from the coastline even when flying into adverse weather.

His study helped explain the scarcity of migratory birds within the coastal plain of Louisiana (i.e., coastal hiatus; Lowery 1945). Additionally, flocks of wintering waterfowl have been anecdotally observed to terminate their nocturnal feeding flights at flooded rice fields in the Central Valley of California (Buler et al. 2012b). However, only limited qualitative inferences can be made about terrestrial habitat use based on monitoring the termination of bird flights on radar.

More generally, there are several sources of measurement bias related to how radars sample the air and the flight behavior of animals that should be addressed to maximize accuracy in mapping animal density on the ground (Diehl and Larkin 2005). We highlight recently developed and increasingly sophisticated methodological approaches that minimize several of these biases and other data quality control issues to produce continuous-surface maps (Fig. 14.3) that precisely and quantitatively link the horizontal spatial structure of birds aloft to their terrestrial distributions.

3.1.1   Range Bias

Perhaps the biggest source of measurement bias is caused by the increase in altitude of the radar beam above the earth's surface with increasing range from the radar (Fig. 14.4). This leads to a systematic decline in reflectivity values with increasing range when sampling animals close to the ground. This range bias precludes direct comparison of raw reflectivity measures across ranges. Furthermore, the bottom of the radar beam eventually passes completely over animals in the airspace (i.e., beam overshoot) and will record no reflectivity even when animals are present in the airspace. Although dependent on the heights of animals in the air, the refractive conditions of the atmosphere, and the elevation angle of the radar beam, beam overshoot from WSR-88D occurs on average at ~80 km from the radar when sampling birds at exodus in relatively flat terrain at the lowest elevation angle (Buler and Diehl 2009; Buler et al. 2012b). The very narrow beam at ranges closest to the radar combined with relatively sparse densities of animals aloft creates greater uncertainty in reflectivity measures that justify filtering data close to the radar (Chilson et al. 2012).

In the absence of range bias correction, researchers generally limit comparisons of terrestrial habitat use to a subset of data at similar ranges. For example, Bonter et al. (2009) compared a few selected locations of high bird density paired with areas of low density at the same distance from the radar to characterize land covers and landscape contexts corresponding to high-use areas by migrating landbirds in the Great Lakes region. Thus, they effectively reduced continuous-surface data into coarse point location data (i.e., 5-km radius areas). However, they acknowledged they were unable to analyze radar data at a finer scale, which is key for informing conservation efforts on a site-by-site basis. As an alternative approach to limit the effect of range bias, Ruth et al. (2012) maintained continuous-surface data but restricted their analysis of land cover and migrant stopover density in the southwestern United States to a 15-km band ranging from 35 to 50 km away from a radar. While this restriction was severe, they were still able to document differences in bird densities among a variety of land cover types by analyzing data from several radars.

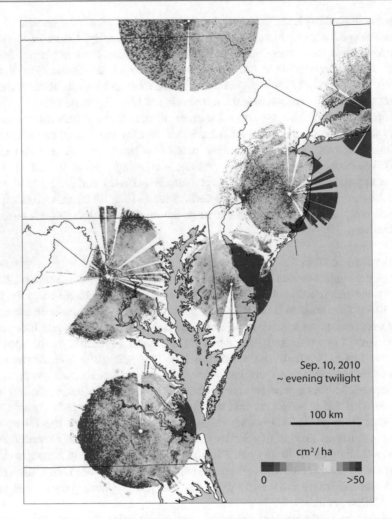

Fig. 14.3 Composite of bias-adjusted vertically integrated reflectivity scans from 6 WSR-88D sites depicting the evening onset of migratory flights of birds on September 10, 2010, in the mid-Atlantic region of the United States following Buler and Dawson (2014). Areas of greater reflectivity (cm²/ha) coincide with areas of greater bird density

Range bias can be minimized by adjusting radar reflectivity measures from individual radar sample volumes (the basic radar sampling unit, measuring 250 m in range by 0.5° in diameter) into equivalent reflectivity measures with respect to a common height reference across all sample volumes [i.e., vertically integrated reflectivity (VIR)]. Essentially two pieces of information are needed for this: the height limits of the radar beam for every sample volume and the mean vertical profile of reflectivity (VPR) of animals aloft at the time of sampling. Established methods for determining radar beam limits include estimating where the beam is

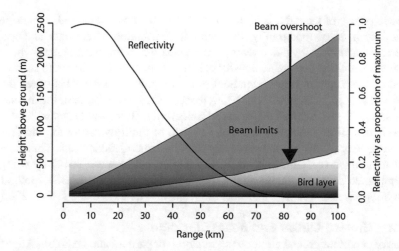

Fig. 14.4 Example illustration of declining radar reflectivity with increasing range measured within the limits of a 3-dB radar beam propagating in a standard atmosphere at 0.5° elevation (gray region bounded by narrow black lines) for a homogeneous layer of birds across ranges. The bird layer extends from the ground to 500 m above ground and is presented as the gradient-filled box to depict increasing bird density closer to the ground. The radar beam completely overshoots the bird layer starting at around 82 km range, which coincides with measured reflectivity of zero despite birds in the air

blocked by topography using digital elevation models (Bech et al. 2003) and modeling beam propagation paths assuming a standard atmosphere (i.e., simple) or using a piecewise linear model of the refractive-index gradient constructed from ancillary radiosonde observations (i.e., complex) (Doviak and Zrnic 1993). The VPR is a function that describes the ratio of the reflectivity at a given height interval with respect to a reference, generally the average reflectivity from the ground to the highest beam height considered. Buler and Diehl (2009) developed an algorithm to derive a high-resolution (i.e., at 10-m height intervals) mean apparent VPR by integrating radar data from the five lowest elevation angle sweeps near the radar. The quotient of the observed reflectivity and the beam-area-weighted mean VPR ratio sampled within a given sample volume gives the estimated VIR that can be compared directly across the radar domain. VIR in volumetric units of cm^2 per km^3 can be "flattened" to the ground by multiplying it by the reference height range measured in kilometers to derive a surface bioscatter estimate in units of cm^2 per km^2. The accuracy of VIR estimates relies on developing fine resolution beam propagation models and VPRs since animals occupy only a few hundred meters of airspace when sampled near the onset of flight. For example, using a high-resolution refractive index from radiosonde observations to model beam propagation can help improve the correlation between the VIR of birds aloft and their ground densities (Buler and Diehl 2009).

Range bias also gives rise to the need for considering effective detection range and deciding how to censor unreliable radar data. Unfortunately, the effective detection range of the radar is highly dynamic within and among radar scans due to variability in local topography, the vertical distribution of animals aloft, atmospheric refraction of the radar beam, and attenuation of the radar signal, which can impact the minimal detectable density of birds. Thresholds can be set to censor data based on the extent of beam overshoot or the magnitude of the VPR-based adjustment factor (e.g., where the radar beam passes over 90% of birds in the airspace or there is a >20-fold increase in observed reflectivity; Buler and Dawson 2014). The variable detection limits also create a scenario of multiply censored data that can introduce error when deriving summary statistics among radar scans. This error has been minimized (e.g., Buler et al. 2012b; Buler and Dawson 2014) by using the semiparametric robust linear regression on order statistics (ROS) method for estimating summary statistics of multiply censored data (Lee and Helsel 2005).

3.1.2 Ground Clutter and Beam Blockage

Areas of persistent ground clutter contamination or partial radar beam blockage from human infrastructure or topography need to be identified to avoid mistaking strong clutter echoes as flying animals or blocked areas as being devoid of animals. These masking maps are an important component of data quality control and are easily created by summarizing the detection probabilities of reflectivity from thousands of scans over long time periods (Kucera et al. 2004). Masking maps help augment the inherent dynamic clutter suppression algorithms of WSR as well. Partial beam blockage from topographic relief can also be modeled (Bech et al. 2003). Radars in areas of wide relief may have extensive beam blockage, which can reduce the efficacy of some radars for mapping terrestrial distributions of flying animals. Moreover, these radars may also be prone to violating assumptions of homogeneous VPRs throughout the radar domain and, consequently, "overcorrection" of range bias algorithms and thus may warrant special considerations (e.g., Ruth et al. 2012).

3.1.3 Sun Angle Bias and Exodus Timing

En masse initiations of animal flight are often closely synchronized to the elevation of the sun. These include migratory flights of landbirds (Gauthreaux 1971; Hebrard 1971; Åkesson et al. 1996), waterfowl (O'Neal et al. 2010), and insects (Westbrook 2008; Westbrook et al. 2014) and feeding flights of wintering waterfowl (Raveling et al. 1972; Baldassarre and Bolen 1984; Ely 1992; Cox and Afton 1996; Randall et al. 2011; Buler et al. 2012b). Many of these flights occur under the cover of darkness and begin shortly after sunset near the end of civil twilight when atmospheric conditions are stable, multiple environmental cues are optimally available for flying animals, and visual predators are thwarted. For example, the magnitude and variability in horizontal and vertical winds reach their minimum about 20 min after sunset (Wingo and Knupp 2014). Additionally, important navigational cues from the sun's position on the horizon, skylight polarization, and astronomical cues are only all available at the end of civil twilight (Kerlinger and Moore 1989; Åkesson et al. 1996). In particular, the skylight polarization pattern during twilight is stable, intense, and closely aligned along the north–south axis (Cronin et al. 2006), which provides important directional information for migrating birds

(Moore and Phillips 1988; Able 1989; Helbig and Wiltschko 1989; Helbig 1990; Muheim et al. 2007; Muheim 2011).

Mapping terrestrial distributions of birds among multiple radars typically involves careful selection of individual radar scans at similar sun angles or timing in relation to the onset of migration (Bonter et al. 2009; Buler and Moore 2011; Ruth et al. 2012). However, small differences among radars in the exact sampling timing of radar scans with respect to relative sun angle can introduce bias in reflectivity measures (Buler and Diehl 2009). These small differences in timing can arise due to the lack of synchronization of radars to the onset of bird flights and the relatively coarse sampling rate of WSR-88D (i.e., one scan per 4–10 min) during the sudden onset of bird flight when the number of birds aloft can double every few minutes (Hebrard 1971; Åkesson et al. 1996). Furthermore, there is about an 8-min time differential in sun elevation along an east–west axis across the radar domain at mid-latitudes. Thus, there is potential for significant sun angle bias even within a single radar domain.

As an example, we measured the onset of autumn nocturnal migration flight of birds at the KLIX radar in Slidell, Louisiana, USA, during 13 nights. We divided the radar domain within 50–60 km of the radar into 18 7-km-wide longitudinal bands. For each night, we fit a logistic growth curve within each band through the mean reflectivities from a series of radar scans during the onset of migration (Fig. 14.5) and extracted the time of the inflection point of the "exodus" curve. We then used the slope of the linear regression between inflection point time (dependent variable) and longitude (independent variable) to derive the speed at which the inflection point of exodus curves moves to the west for each night. The mean observed speed of inflection points of flight exodus curves (21.2 ± 4.7 km/ min) was not different ($T = -1.76$, $df = 12$, $P = 0.9$) from that of the predicted speed of the sunset terminator (23.6 km/min).

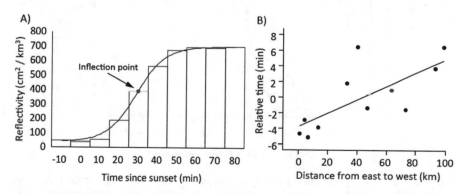

Fig. 14.5 (a) Example "exodus curve" at the onset of bird migration within a single 7-km wide longitude band to derive the inflection point of peak change in mean reflectivity (red dot) among radar scans (bars). (b) This inflection point was combined with similarly derived inflection points from other longitude bins within the same radar domain (black dots) in a linear regression model to determine the speed of the onset of migration for a given night

Approaches have been developed to reduce sun angle bias both within and among radars. These approaches involve interpolating reflectivity measures to a static relative time point with respect to sun elevation at every sample volume across all sampling nights and radars (Buler et al. 2012b; Buler and Dawson 2014). However, using a static sun angle for all nights and radars may still introduce bias because there is variability in the onset of flights between nights within and among radars. For example, Gauthreaux (1971) found a 15-min range in timing of the onset of nocturnal bird migration across nights relative to sunset at a single radar. Thus, a new approach further reduces sun angle bias; it dynamically samples at the sun angle at the peak rate of change in reflectivity during exodus (i.e., when the greatest numbers of birds depart) for each night and radar (McLaren et al. 2018).

Visual observations and radio telemetry of individual birds show that peak exodus times occur close to those observed by nearby radar (Gauthreaux 1971; Hebrard 1971; Åkesson et al. 1996). However, mean peak exodus was earlier on radar compared to that of radio-marked Swainson's thrushes (*Catharus ustulatus*) using data from two separate studies with overlapping study areas in coastal Alabama, USA (Fig. 14.6). While exhibiting similar duration of flight exodus, results from Smolinsky et al. (2013) indicate that the peak exodus time of migrating thrushes in autumn (sun angle = −9.0°) was about 9 min after the mean peak exodus time of bird-dominated flights on the nearby Mobile, Alabama radar (sun angle = −7.0°; Buler and Moore 2011). Others have found that departures of the earliest cohort of Swainson's thrush are also largely restricted to the period of

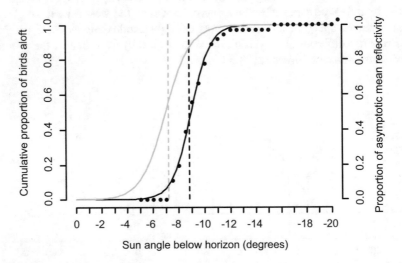

Fig. 14.6 Modeled logistic growth curves relative to sun angle depicting (1) the cumulative portion of 37 radio-marked Swainson's thrushes (black dots) aloft at the onset of nocturnal flights during autumn migration at Fort Morgan, Alabama 2008 and 2009 (black line), and (2) the increase in mean reflectivity among 12 autumn nights during 2002 and 2003 around the KMOB (Mobile, Alabama) radar (gray line). Vertical dashed lines indicate where the inflection point of fitted curve lines occurs. Sun angle is −6° at the end of civil twilight. Data from Smolinsky et al. (2013) and Buler and Moore (2011)

nautical twilight (sun angle between $-6°$ and $-12°$) (Cochran 1987; Mills et al. 2011). Additionally, mean peak radar departure in spring from Gauthreaux's (1971) study in coastal Louisiana, USA, was at $-9°$, matching that of thrushes from Smolinsky et al. (2013). Admittedly, the data from this two-study comparison were from different seasons and years. This may highlight the possibility of annual variability in mean departure times. It also highlights the fact that individual species are known to vary in their departure timing; other species exhibit earlier departure than Swainson's thrush (Åkesson et al. 1996, 2001). It is also possible that other biota (e.g., insects) may contribute to the early exodus reflectivity around the end of civil twilight (Westbrook 2008).

3.1.4   Displacement and Dispersion

Dispersion of birds throughout the radar domain is evident over time and can affect the spatial accuracy and precision of the georeferenced data. Wintering waterfowl tend to disperse heterogeneously in multiple directions when initiating feeding flights. An effective correction for this spreading behavior of waterfowl during feeding flights is currently elusive. Estimates of median dispersal distances of birds from their ground sources over time closely match the optimal bandwidth for kernel smoothing of ground waterfowl densities and the scale at which radar data are spatially autocorrelated (Buler et al. 2012b). Consequently, dispersal leads to increased loss of spatial structure in reflectivity scans and to autocorrelation of reflectivity measures at greater distances. It also leads to a slight weakening of the association between waterfowl density aloft and observed waterfowl density on the ground. Despite autocorrelation in reflectivity measures out to about 4 km at the optimal sampling time, finer-scale associations of birds with their ground sources can still be discerned when summarizing data across multiple sampling nights, particularly where discrete patches of suitable habitat within an unsuitable habitat matrix are sampled by multiple radar sample volumes.

In contrast, migrating landbirds leave their stopover sites in a relatively uniform mean speed and direction, so adjusting for their displacement is possible although difficult to implement (Buler and Diehl 2009). Even after adjusting for displacement, the strength of associations between the reflectivity of birds aloft and the dominant habitats of their ground sources does not always improve, likely due to uncertainty in details of estimating the extent of displacement. However, adjusting for displacement may improve associations with coastal habitats and with small habitat patches or narrow, linear habitats like riparian corridors. This is because birds may take off from one habitat type and be detected above another. Associations of birds from habitat patches or sites that are smaller than the physical dimensions of a single radar sample volume are likely not possible due to mixing from birds emanating from other locations within the sample volume (Buler et al. 2012b). Moreover, determining terrestrial habitat associations at far ranges from the radar may be less reliable because the radar beam passes farther above the earth's surface, increasing the displacement of animals from their ground source in the time it takes for them to fly up into the beam (Diehl and Larkin 2005). This possibility served as a rationale for trimming data at far ranges for Ruth et al. (2012), yet remains to be empirically tested.

3.1.5 Taxonomic Identification

Identification of flying animals observed by WSR remains a key challenge, even for coarse taxonomic discrimination of birds from insects and bats. This is true for all radar studies of animals aloft, not just when mapping their terrestrial distributions. During spring and autumn migratory seasons, birds, insects, and bats likely overlap in their flight activity (Cryan 2003; Fleming and Eby 2003; Gauthreaux and Livingston 2006; Alerstam et al. 2011; Larkin and Diehl 2012). Thus, WSR sample volumes can contain multiple taxa since they are large enough to contain tens to thousands of individual animals. Accordingly, species discrimination on WSR is not absolute but focuses instead on determining the classes of animals that dominate the airspace within the radar domain. Insects are typically discriminated from vertebrates by their slower horizontal airspeeds (Larkin 1991). Radar measures of the radial velocity of animals during the peak of flight activity (typically about 3 h after sunset) are used to characterize dominant taxa aloft for a given sampling day (e.g., Gauthreaux and Belser 1998; Ruth et al. 2012; Buler and Dawson 2014).

3.2 Ecological Insights

3.2.1 Understanding How Migrating Birds Contend with Crossing Ecological Barriers and Extreme Weather Events

Long-distance, intercontinental migratory birds must negotiate ecological barriers during their biannual journeys, and movement in relation to the Gulf of Mexico (GOM) is a highly conspicuous feature of the Nearctic–Neotropical bird migration system. Habitats along the northern coast of the GOM provide the last possible stopover before migrants make a nonstop flight of greater than 1000 km in fall, and the first possible landfall for birds returning north in spring, so migrant–habitat relations are critical when birds negotiate this geographical barrier (e.g., Kerlinger and Moore 1989; Moore et al. 1990; Deppe et al. 2015).

Radar mapping of migratory stopover distributions of birds along the northern GOM has provided valuable insight into how birds negotiate ecological barriers. In particular, the broad scale of radar observations has helped to reveal the interplay between extrinsic factors (e.g., weather and energetic condition) and finer-scale intrinsic factors (e.g., food) on stopover habitat selection, which may be especially acute along large ecological barriers (Gauthreaux 1975; Gauthreaux and Belser 1999; Gauthreaux et al. 2005; Buler and Moore 2011; LaFleur et al. 2016).

The most extensive continuous-surface mapping effort of spring bird stopover densities to date spans the northern GOM from Texas to Florida within 100 km of the coastline and incorporates 4 years of radar data (LaFleur et al. 2016). Most of the spatial variability in bird density is related to longitude, proximity to the coast, and the amount of hardwood forest cover in the landscape. At the broadest scale, winds and weather over the GOM have likely shaped the evolution of migration routes in this region (Able 1972; Moore and Kerlinger 1989; Moore et al. 1990; Rappole and Ramos 1994; Gauthreaux et al. 2005; Kranstauber et al. 2015). Accordingly, LaFleur et al. (2016) found exceptionally high bird densities in western Louisiana consistent with the existence of a trans-Gulf route from Mexico's Yucatan Peninsula that was identified from radar analysis of flight activity (Gauthreaux et al. 2005) and observed bird movements from telemetry studies (Callo et al. 2013; Stanley et al. 2015). However, annual variability revealed that this region does not always support the highest densities of migrants. In some years, the eastern panhandle of Florida had the greatest densities of migrants, suggesting a largely overlooked eastern trans-Gulf route also consistent with telemetry studies (Fraser et al. 2013; Stanley et al. 2015) and the influence of winds in shaping broad-scale distributions of migrants crossing the GOM (Gauthreaux et al. 2005; Russell 2005).

Trans-Gulf migrants typically overfly coastal marshes to stop over in forested landscapes (Gauthreaux 1971, 1975; Gauthreaux and Belser 1998) or stop in forested landscapes at the immediate coastline when wide coastal marshes are absent (Buler and Moore 2011). However, high migrant densities at the immediate coast at some longitudes regardless of land cover type (LaFleur et al. 2016) and weakening of the relationship between migrant density and forest cover closer to the coastline (Buler and Moore 2011) provide evidence that some migrants are limited in their ability to select among habitats of different quality or are forced to land at the first dry ground they encounter. These constraints are likely the result of physiological stress (Moore et al. 1990; Kuenzi and Moore 1991; Spengler et al. 1995) or adverse weather experienced while aloft (Lowery 1945; Gauthreaux 1971).

The GOM is also a region that experiences some of the most severe weather events in the form of hurricanes that often coincide with the autumn migration season of landbirds and can impact their stopover distributions. For example, Hurricane Katrina caused extensive damage to bottomland hardwood forests within the Lower Pearl River Basin (Chambers et al. 2007; Chapman et al. 2008; Wang and Xu 2009) that migrants preferentially use at high densities during stopover (Buler and Moore 2011). Barrow et al. (2007) compared maps of bird stopover density around a single radar immediately after the storm to composite maps from previous years near the path of the hurricane. They found migrating birds stopping over in the region for several weeks after Katrina's passage increased their use of less-disturbed, upland pine forests near the damaged bottomlands. About 5 weeks after the hurricane struck, much of the surviving forest canopy in the Pearl River bottomlands began to grow new foliage and migrant use of these forested wetlands mostly returned to pre-Katrina levels. Thus, radar observations can reveal the flexibility of migrating birds for responding to acute, short-term, and broad-scale disturbance to stopover habitats. This unique ability of networked radars is possible given the comprehensive, broad-scale extent of radar coverage and their constant surveillance operation. These radar data can improve our understanding of the effects of hurricane events on the population dynamics of migrating birds and a macrosystem-scale understanding of the potential scope and importance of eco-evolutionary processes in the face of climate change (sensu Heffernan et al. 2014).

3.2.2 Identifying Important Stopover Areas and Habitat Use Patterns of Birds Along Their Migration Routes

In an analysis of autumn stopover densities of migrants in the northeastern United States, Buler and Dawson (2014) summarized reflectivity measures for each sample volume from 16 WSR-88Ds across nights and years (2008, 2009) to estimate the seasonal mean and coefficient of variation of daily bird stopover density. Using these metrics, they classified areas with the highest mean and lowest variability in reflectivity as the most important bird stopover use areas within the sampling ranges of individual radars (locally) and across all radars (regionally). Locally important areas generally were associated with deciduous forests within landscapes dominated by developed or agricultural lands, or near the shores of major water bodies. These results are consistent with those of Bonter et al. (2009), who analyzed data collected during spring migration by six WSR-88Ds in the Great Lakes basin. Large regionally important stopover areas were located along the north shore of Long Island Sound (Connecticut, New York), on the Delmarva Peninsula (Delaware, Maryland), near Baltimore and Washington, DC (Maryland, Virginia), along the western edge of the Adirondack Mountains (New York), and within the Appalachian Mountains (southwestern Virginia, West Virginia) (Buler and Dawson 2014).

Concentrations of migrants near coastlines are consistently found where stopover distributions have been mapped with radar (Bonter et al. 2009; Buler and Moore 2011; Buler and Dawson 2014; LaFleur et al. 2016). However, the availability of high-quality stopover habitats for birds is at odds with rapid human population growth in coastal regions (Crossett et al. 2004; Buler and Moore 2011). Anthropogenic disturbances and habitat loss likely limit birds' use of coastal habitats and exacerbate other constraints on habitat use by migrants in coastal areas (Buler and Moore 2011). Thus, the conservation of coastal stopover sites is of high priority for protection of migratory bird populations (Mehlman et al. 2005).

Radar mapping studies have also revealed high-density use of forests in human-dominated landscapes, particularly urban parks within large cities (Bonter et al. 2009; Buler and Dawson 2014). Migrant stopover use of human-dominated landscapes has also been documented by large-scale citizen-science survey efforts in the United States (La Sorte et al. 2014b). Urban forests can provide resources for migrants to refuel during their stopovers (Seewagen and Slayton 2008). Forest patches may be particularly sought after by migrants who find themselves in human-dominated landscapes; the strongest positive relationships between bird stopover density and forest cover occurred in areas, like cities, with low amounts of forest cover in the landscape (Buler and Dawson 2014). Moreover, bird density was positively related to the amount of human development in the landscape near large cities. Thus, large cities actually had relatively greater bird densities within them after accounting for the amount of forest cover and other factors shaping migrant distributions. This supports the hypothesis that migrating birds may actually be attracted to large urban areas at a landscape scale. A possible mechanism could be phototaxis of migrants to the artificial light glow of big cities (Gauthreaux and Belser 2006), which has also been cited as a possible explanation for altitudinal shifts in migrants during flight (Bowlin et al. 2015).

Ruth et al. (2012) analyzed data from seven WSR-88Ds spanning the US–Mexico border to examine stopover habitat use patterns of migratory birds moving through the arid southwestern United States. Migrating birds are known to use riparian habitats in the Southwest (Carlisle et al. 2009), but the intervening arid habitats have received little study. Ruth et al. (2012) documented relatively low migrant densities in arid scrub/shrub habitats across the region. However, the dominance of this habitat type suggests that, collectively, scrub/shrub supports a large number of migratory birds during stopover and that the importance of a given habitat is more than a function of density alone.

Although radar-based maps of important stopover areas do not characterize their intrinsic qualities or ecological function, they can focus conservation actions on areas and habitats where they likely will benefit the largest numbers of migrants, regardless of how or why birds use them, and ensure that a network of suitable stopover sites are protected along migration routes (Hutto 2000; Mehlman et al. 2005; Faaborg et al. 2010). As such, they can provide direct input to local or regional conservation plans.

3.2.3 Assess Waterfowl Response to Wetland Habitat Management and Restoration

Radar observations of waterfowl distributions at landscape and regional scales have had important implications for waterfowl conservation and management planning. Buler et al. (2012a) used 13 years of radar data to quantify wintering waterfowl response to restored wetlands in California's Central Valley by sampling birds at the onset of evening feeding flights. Two radars provided data for approximately 57% of the land area enrolled in the United States Department of Agriculture's (USDA) Wetland Reserve Program (WRP). They found that daytime (i.e., roosting) use of WRP wetlands by wintering waterfowl increased dramatically after wetland restoration and was sustained for up to 8 years post-restoration. Variability in the magnitude of waterfowl densities after restoration was greater with greater density of birds in the local area before restoration, lower amount of surrounding wetland habitat within a 1.5 km radius, greater increase in site soil wetness (i.e., flooding) after restoration, and closer proximity to flooded rice fields that serve as important feeding grounds. The study corroborated the understanding that the juxtaposition of diurnal roosts and nocturnal feeding habitats is critical to effectively conserve and manage landscapes for wintering waterfowl (Haig et al. 1998; Stafford et al. 2010).

In addition to providing evidence of the value of restored wetlands, Buler et al. (2012a) used WSR observations to document a long-term shift in habitat use by waterfowl. Within the Central Valley, diurnal use of flooded rice fields by waterfowl nearly tripled over time from 1995 to 2007 relative to use of natural wetland habitats. Waterfowl use of flooded rice fields relative to use of natural wetlands was also greater during wetter winters. This corroborated a long-term radio-telemetry study that found similar shifts in habitat use by waterfowl (Fleskes et al. 2005).

In response to the Deepwater Horizon oil spill in summer 2010, the USDA implemented the Migratory Bird Habitat Initiative (MBHI) to provide temporary wetland habitat via managed flooding of agricultural lands for migrating and wintering waterfowl, shorebirds, and other birds along the northern GOM. Sieges et al. (2014) used WSR observations to show that birds responded positively to MBHI management by exhibiting greater relative bird densities within sites relative to prior years when no management was implemented and also concurrently relative to non-flooded agricultural lands. Like waterfowl in California, the magnitude of bird densities in managed wetlands was related to the surrounding landscape context. WSR observations provided strong evidence that MBHI sites offered birds wetland habitat inland from coastal wetlands impacted by the oil spill.

3.3 The Future

We highlight many of the issues of mapping terrestrial distributions of animals with radar that continue to pose future challenges and opportunities for improving aeroecological studies. These issues are either subjects of ongoing research efforts or are of great importance for advancing quantification of animal distributions using radar. We are hopeful that improved data quality control and processing techniques will continue to unleash the full research potential of WSR to advance our understanding of aeroecology in the future.

The natural history of some insects presents the opportunity to map the distributions of insects initiating flights. Many migratory insects initiate migratory flights around dusk in well-synchronized flights similar to birds (Reynolds and Riley 1979; Riley and Reynolds 1979; Riley et al. 1983). In fact, insect activity is often a source of contamination for efforts to map migrating bird distributions. Some aquatic insects, like many short-lived adult mayflies, emerge in high concentrations from aquatic habitats in well-synchronized crepuscular flights (Brittain 1982), which are readily detected by WSR. For example, in 1999, emerging mayflies of the genus *Hexagenia* were observed with WSR along the shores of Lake Erie after a 30-year absence due to hypoxia resulting from cultural eutrophication (Masteller and Obert 2000). There is interest in monitoring mayfly distributions and population dynamics in the face of anthropogenic alterations of water quality and other habitat threats (e.g., Fremling 1973; Corkum 2010). However, despite the fact that emergence flights are very distinct in nature and finite in duration, we could find no studies that have developed techniques to map distributions of aquatic insect emergence flights or terrestrial insect migratory flights. Thus, future studies are needed that merge radar observations with traditional insect survey techniques to map and monitor population distribution and dynamics of insects.

Further study is needed to comprehensively describe broad-scale flight initiation timing patterns among and within radars and to determine the factors that explain variability in exodus timing such as geographic position, weather, or species composition (Åkesson et al. 2001). For example, the impact of cloud cover on flight initiation is not consistent. While Cochran et al. (1967) found that several North American thrush species depart later at night under cloudy skies, Gauthreaux (1971) and Hebrard (1971) did not find that cloud cover delayed exodus among the collection of all migrating landbirds.

Another important assumption that needs future consideration is whether the terrestrial distribution derived from the first cohort of animals to initiate nightly flights is representative of the broader population of animals of interest. Automated tracking networks have documented that the first-departing migrants tend to move long distances in the direction of their ultimate migratory destination, while later-departing individuals may engage in short-distance relocation flights, sometimes in directions away from their ultimate migratory destination and may be in poorer energetic condition (Mills et al. 2011; Taylor et al. 2011; Schmaljohann and Naef-Daenzer 2011; Smolinsky et al. 2013). If first flyers are biased toward birds in good energetic condition prepared for making a long duration flight, they may bias the densities of birds emanating from high-quality habitats or along main flyways. Future field work can test this assumption, and improved telemetry technology should continue to provide new insights into the exact flight timing and behavior of individual animals (Bridge et al. 2011).

Flying animals reflect radio energy in extremely complex ways (Edwards and Houghton 1959; P. Chilson, unpub. data) that continue to challenge efforts to quantify migrant density using radar. No progress has been made in confirming the existence of or correcting for bias in reflectivity measures due to the aspect of animals (i.e., their shape and exposure with respect to the radar) when quantifying animal densities aloft since Diehl and Larkin (2005) first broached the subject. Considerably more research in this area is needed to better understand both the effects of aspect on radar reflectivity and the impact of this inherent uncertainty on quantification.

The expansion and integration of terrestrial distribution maps of migrating birds from WSR data with large-scale field mapping efforts from eBird data (Fink et al. 2010) may help link seasonal changes in stopover habitat distributions with known compositional changes in migratory bird taxa (La Sorte et al. 2014a, b). Such efforts to merge eBird and radar data have not yet been attempted in earnest but hold promise to disentangle the aggregate density information provided by radar. Broad-scale efforts to integrate ground surveys of bird-use days with radar-derived estimates of emigrant densities could be a cost-effective approach to get an index of stopover duration across multiple sites (sensu O'Neal et al. 2012) and broaden our understanding of the stopover ecology of migrating birds.

References

Able KP (1972) Fall migration in coastal Louisiana and the evolution of migration patterns in the Gulf region. Wilson Bull 84:231–242

Able KP (1989) Skylight polarization patterns and the orientation of migratory birds. J Exp Biol 141:241–256

Åkesson S, Alerstam T, Hedenström A (1996) Flight initiation of nocturnal passerine migrants in relation to celestial orientation conditions at twilight. J Avian Biol 27:95–102

Åkesson S, Walinder G, Karlsson L, Ehnbom S (2001) Reed warbler orientation: initiation of nocturnal migratory flights in relation to visibility of celestial cues at dusk. Anim Behav 61:181–189

Alerstam T (1993) Bird migration. Cambridge University Press, New York

Alerstam T, Chapman JW, Bäckman J, Smith AD, Karlsson H, Nilsson C, Reynolds DR, Klaassen RHG, Hill JK (2011) Convergent patterns of long-distance nocturnal migration in noctuid moths and passerine birds. Proc R Soc B Biol Sci 278:3074–3080

Baldassarre GA, Bolen EG (1984) Field-feeding ecology of waterfowl wintering on the southern high plains of Texas. J Wildl Manag 48:63–71

Barrow WCJ, Buler JJ, Couvillion B, Diehl RH, Faulkner S, Moore FR, Randall L (2007) Broad-scale response of landbird migration to the immediate effects of hurricane Katrina. In: Farris GS, Smith GJ, Crane MP, Demas CR, Robbins LL, Lavoire DL (eds) Science and the storms – the USGS response to the hurricanes of 2005. Circular 1306. U.S. Geological Survey, Reston, pp 131–136

Bech J, Codina B, Lorente J, Bebbington D (2003) The sensitivity of single polarization weather radar beam blockage correction to variability in the vertical refractivity gradient. J Atmos Ocean Technol 20:845–855

Beekman JH, Nolet BA, Klaassen M (2002) Skipping swans: fuelling rates and wind conditions determine differential use of migratory stopover sites of Bewick's Swans *Cygnus bewickii*. Ardea 90:437–460

Bellrose FC, Crompton RD (1970) Migrational behavior of mallards and black ducks as determined from banding. Ill Nat Hist Surv Bull 30(3):167–234

Bellrose FC, Paveglio F, Steffeck DW (1979) Waterfowl populations and the changing environment of the Illinois River valley. Ill Nat Hist Surv Bull 32(1):1–54

Bonter DN, Gauthreaux SA, Donovan TM (2009) Characteristics of important stopover locations for migrating birds: remote sensing with radar in the Great Lakes Basin. Conserv Biol 23:440–448

Bowlin MS, Enstrom DA, Murphy BJ, Plaza E, Jurich P, Cochran J (2015) Unexplained altitude changes in a migrating thrush: long-flight altitude data from radio-telemetry. Auk 132:808–816

Bridge ES, Thorup K, Bowlin MS, Chilson PB, Diehl RH, Fleron RW, Hartl P, Roland K, Kelly JF, Robinson WD, Wikelski M (2011) Technology on the move: recent and forthcoming innovations for tracking migratory birds. Bioscience 61:689–698

Bridge ES, Pletschet SM, Fagin T, Chilson PB, Horton KG, Broadfoot KR, Kelly JF (2016) Persistence and habitat associations of Purple Martin roosts quantified via weather surveillance radar. Landsc Ecol 31(1):43–53

Brigham RM, Grindal SD, Firman MC, Morissette JL (1997) The influence of structural clutter on activity patterns of insectivorous bats. Can J Zool 75:131–136

Brittain JE (1982) Biology of mayflies. Annu Rev Entomol 27:119–147

Buler JJ, Dawson DK (2014) Radar analysis of fall bird migration stopover sites in the northeastern U.S. Condor 116:357–370

Buler JJ, Diehl RH (2009) Quantifying bird density during migratory stopover using weather surveillance radar. IEEE Trans Geosci Remote Sens 47:2741–2751

Buler JJ, Moore FR (2011) Migrant–habitat relationships during stopover along an ecological barrier: extrinsic constraints and conservation implications. J Ornithol 152:S101–S112

Buler JJ, Barrow WC, Randall LA (2012a) Wintering waterfowl respond to wetlands reserve program lands in California's Central Valley. USDA NRCS CEAP Conservation Insight

Buler JJ, Randall LA, Fleskes JP, Barrow WC, Bogart T, Kluver D (2012b) Mapping wintering waterfowl distributions using weather surveillance radar. PLoS One 7:e41571

Caccamise DF, Lyon LA, Fischl J (1983) Seasonal patterns in roosting flocks of starlings and common grackles. Condor 85:474–481

Callo PA, Morton ES, Stutchbury BJ (2013) Prolonged spring migration in the Red-eyed Vireo (Vireo olivaceus). Auk 130:240–246

Carlisle JD, Skagen SK, Kus BE, Riper CV III, Paxtons KL, Kelly JF (2009) Landbird migration in the American West: recent progress and future research directions. Condor 111:211–225

Chambers JQ, Fisher JI, Zeng H, Chapman EL, Baker DB, Hurtt GC (2007) Hurricane Katrina's carbon footprint on US Gulf coast forests. Science 318:1107–1107

Chapman EL, Chambers JQ, Ribbeck KF, Baker DB, Tobler MA, Zeng H, White DA (2008) Hurricane Katrina impacts on forest trees of Louisiana's Pearl River basin. For Ecol Manag 256:883–889

Chilson PB, Frick WF, Stepanian PM, Shipley JR, Kunz TH, Kelly JF (2012) Estimating animal densities in the aerosphere using weather radar: to Z or not to Z? Ecosphere 3:art72

Cleveland CJ, Betke M, Federico P, Frank JD, Hallam TG, Horn J, López JD, McCracken GF, Medellín RA, Moreno-Valdez A, Sansone CG, Westbrook JK, Kunz TH (2006) Economic value of the pest control service provided by Brazilian free-tailed bats in south-central Texas. Front Ecol Environ 4:238–243

Cochran WW (1987) Orientation and other migratory behaviours of a Swainson's thrush followed for 1500 km. Anim Behav 35:927–929

Cochran WW, Montgomery GG, Graber RR (1967) Migratory flights of Hylocichla thrushes in spring: a radiotelemetry study. Living Bird 6:213–225

Corkum LD (2010) Spatial-temporal patterns of recolonizing adult mayflies in Lake Erie after a major disturbance. J Great Lakes Res 36:338–344

Cox RR, Afton AD (1996) Evening flights of female northern pintails from a major roost site. Condor 98:810–819

Cronin TW, Warrant EJ, Greiner B (2006) Celestial polarization patterns during twilight. Appl Opt 45:5582–5589

Crossett KM, Culliton TJ, Wiley PC, Goodspeed TR (2004) Population trends along the coastal United States: 1980–2008. National Oceanic and Atmospheric Administration, Silver Spring, MD

Crum TD, Alberty RL, Burgess DW (1993) Recording, archiving, and using WSR-88D data. Bull Am Meteorol Soc 74:645–653

Cryan PM (2003) Seasonal distribution of migratory tree bats (Lasiurus and Lasionycteris) in North America. J Mammal 84:579–593

Deppe JL, Ward MP, Bolus RT, Diehl RH, Celis-Murillo A, Zenzal TJ, Moore FR, Benson TJ, Smolinsky JA, Schofield LN, Enstrom DA, Paxton EH, Bohrer G, Beveroth TA, Raim A, Obringer RL, Delaney D, Cochran WW (2015) Fat, weather, and date affect migratory songbirds' departure decisions, routes, and time it takes to cross the Gulf of Mexico. Proc Natl Acad Sci 112:E6331–E6338

Diehl RH, Larkin RP (2005) Introduction to the WSR-88D (NEXRAD) for ornithological research. In: Ralph CJ, Rich TD (eds) Bird conservation implementation and integration in the Americas: proceedings of the third international partners in flight conference. USDA Forest Service, General Technical Report PSW-GTR-191, pp 876–888

Dingle H, Drake VA (2007) What is migration? Bioscience 57:113–121

Dokter AM, Åkesson S, Beekhuis H, Bouten W, Buurma L, van Gasteren H, Holleman I (2013) Twilight ascents by common swifts, Apus apus, at dawn and dusk: acquisition of orientation cues? Anim Behav 85:545–552

Doviak RJ, Zrnic DS (1993) Doppler radar and weather observations. Academic, San Diego

Drake V, Gatehouse A (1996) Population trajectories through space and time: a holistic approach to insect migration. In: Floyd RB, Sheppard AW, De Barro PI (eds) Frontiers of population ecology. Collingwood, CSIRO, pp 399–408

Eastwood E (1967) Radar ornithology. Methuen, London

Eastwood E, Isted GA, Rider GC (1962) Radar ring angels and the roosting behaviour of starlings. Proc R Soc Lond B Biol Sci 156:242–267

Edwards J, Houghton EW (1959) Radar echoing area polar diagrams of birds. Nature 184:1059–1059

Elder FC (1957) Some persistent ring echoes on high powered radar. In: Proceedings of the sixth weather radar conference. Cambridge, pp 281–286

Ely CR (1992) Time allocation by greater white-fronted geese: influence of diet, energy reserves and predation. Condor 94:857–870

Faaborg J, Holmes RT, Anders AD, Bildstein KL, Dugger KM, Gauthreaux SA, Heglund P, Hobson KA, Jahn AE, Johnson DH, Latta SC, Levey DJ, Marra PP, Merkord CL, Nol E, Rothstein SI, Sherry TW, Sillett TS, Thompson FR, Warnock N (2010) Recent advances in understanding migration systems of New World land birds. Ecol Monogr 80:3–48

Fink D, Hochachka WM, Zuckerberg B, Winkler DW, Shaby B, Munson MA, Hooker G, Riedewald M, Sheldon D, Kelling S (2010) Spatiotemporal exploratory models for broad-scale survey data. Ecol Appl 20:2131–2147

Fleming TH, Eby P (2003) Ecology of bat migration. In: Kunz TH, Fenton MB (eds) Bat ecology. The University of Chicago Press, Chicago, pp 156–208

Fleskes JP, Yee JL, Casazza ML, Miller MR, Takekawa JY, Orthmeyer DL (2005) Waterfowl distribution, movements, and habitat use relative to recent habitat changes in the Central Valley of California: a cooperative project to investigate impacts of the Central Valley joint venture and changing agricultural practices on the ecology of wintering waterfowl. Final report. U.S. Geological Survey – Western Ecological Research Center, Dixon Field Station, Dixon, CA

Fraser KC, Silverio C, Kramer P, Mickle N, Aeppli R, Stutchbury BJ (2013) A trans-hemispheric migratory songbird does not advance spring schedules or increase migration rate in response to record-setting temperatures at breeding sites. PLoS One 8:e64587

Fremling CR (1973) Factors influencing the distribution of burrowing mayflies along the Mississippi River. In: Proceedings of the first international conference on Ephemeroptera. EJ Brill, Leiden, pp 12–15

Frick WF, Stepanian PM, Kelly JF, Howard KW, Kuster CM, Kunz TH, Chilson PB (2012) Climate and weather impact timing of emergence of bats. PLoS One 7:e42737

Gagnon F, Ibarzabal J, Savard J-PL, Bélisle M, Vaillancourt P (2011) Autumnal patterns of nocturnal passerine migration in the St. Lawrence estuary region, Quebec, Canada: a weather radar study. Can J Zool 89:31–46

Gauthreaux SA (1971) A radar and direct visual study of passerine spring migration in southern Louisiana. Auk 88:343–365

Gauthreaux SA (1975) Coastal hiatus of spring trans-Gulf bird migration. A rationale for determining Louisiana's coastal zone: Baton Rouge, Louisiana State University, Center for Wetland Resources, Coastal Zone Management Series report no 1, pp 85–91

Gauthreaux SA (1980) The influences of long-term and short-term climatic changes on the dispersal and migration of organisms. In: Gauthereaux SA (ed) Animal migration, orientation, and navigation. Academic, New York, pp 103–174

Gauthreaux SA, Belser CG (1998) Displays of bird movements on the WSR-88D: patterns and quantification. Weather Forecast 13:453–464

Gauthreaux SA, Belser CG (1999) Bird migration in the region of the Gulf of Mexico. In: Adams NJ, Slotow RH (eds) Proceedings of the 22nd international ornithological congress. Birdlife South Africa, Durban, pp 1931–1947

Gauthreaux SA, Belser CG (2003) Radar ornithology and biological conservation. Auk 120:266–277

Gauthreaux SA, Belser CG (2006) Effects of artificial night lighting on migrating birds. In: Rich C, Longcore T (eds) Ecological consequences of artificial night lighting. Island Press, Washington, DC, pp 67–93

Gauthreaux SA, Livingston JW (2006) Monitoring bird migration with a fixed-beam radar and a thermal-imaging camera. J Field Ornithol 77:319–328

Gauthreaux SA, Belser CG, Blaricom DV (2003) Using a network of WSR-88D weather surveillance radars to define patterns of bird migration at large spatial scales. In: Berthold P, Gwinner E, Sonnenschein E (eds) Avian migration. Springer, Berlin, pp 335–346

Gauthreaux SA, Michi JE, Belser CG (2005) The temporal and spatial structure of the atmosphere and its influence on bird migration strategies. In: Greenberg R, Marra PP (eds) Birds of two worlds. Smithsonian Institution, Washington, DC, pp 182–196

Haig SM, Mehlman DW, Oring LW (1998) Avian movements and wetland connectivity in landscape conservation. Conserv Biol 12:749–758

Harper WG (1959) Roosting movements of birds and migration departures from roosts as seen by radar. Ibis 101:201–208

Harper DGC (1982) Competitive foraging in mallards: "Ideal free" ducks. Anim Behav 30:575–584

Hassell MP, Southwood TRE (1978) Foraging strategies of insects. Annu Rev Ecol Syst 9:75–98

Hebrard JJ (1971) The nightly initiation of passerine migration in spring: a direct visual study. Ibis 113:8–18

Heffernan JB, Soranno PA, Angilletta MJ Jr, Buckley LB, Gruner DS, Keitt TH, Kellner JR, Kominoski JS, Rocha AV, Xiao J (2014) Macrosystems ecology: understanding ecological patterns and processes at continental scales. Front Ecol Environ 12:5–14

Helbig AJ (1990) Depolarization of natural skylight disrupts orientation of an avian nocturnal migrant. Experientia 46:755–758

Helbig AJ, Wiltschko W (1989) The skylight polarization patterns at dusk affect the orientation behavior of Blackcaps, Sylvia atricapilla. Naturwissenschaften 76:227–229

Hodgkison R, Balding ST, Zubaid A, Kunz TH (2004) Habitat structure, wing morphology, and the vertical stratification of Malaysian fruit bats (Megachiroptera: Pteropodidae). J Trop Ecol 20:667–673

Holleman I, Delobbe L, Zgonc A (2008) Update on the European weather radar network (OPERA). In: Proceedings of the 5th European conference on radar in meteorology and hydrology, Helsinki, 30 June–4 July 2008

Horn JW, Kunz TH (2008) Analyzing NEXRAD doppler radar images to assess nightly dispersal patterns and population trends in Brazilian free-tailed bats (Tadarida brasiliensis). Integr Comp Biol 48:24–30

Hutto RL (2000) On the importance of en route periods to the conservation of migratory landbirds. Stud Avian Biol 20:109–114

Janes SW (1985) Habitat selection in raptorial birds. In: Cody ML (ed) Habitat selection in birds. Academic, New York, pp 159–188

Jones G, Rydell J (1994) Foraging strategy and predation risk as factors influencing emergence time in echolocating bats. Philos Trans R Soc B Biol Sci 346:445–455

Kelly JF, Shipley JR, Chilson PB, Howard KW, Frick WF, Kunz TH (2012) Quantifying animal phenology in the aerosphere at a continental scale using NEXRAD weather radars. Ecosphere 3:art16

Kerlinger P, Moore F (1989) Atmospheric structure and avian migration. Curr Ornithol 6:109–142

Kranstauber B, Weinzierl R, Wikelski M, Safi K (2015) Global aerial flyways allow efficient travelling. Ecol Lett 18:1338–1345

Kucera PA, Krajewski WF, Young CB (2004) Radar beam occultation studies using GIS and DEM technology: an example study of Guam. J Atmos Ocean Technol 21:995–1006

Kuenzi AJ, Moore FR (1991) Stopover of Neotropical landbird migrants on East Ship Island following trans-Gulf migration. Condor 93:869–883

La Sorte FA, Fink D, Hochachka WM, Farnsworth A, Rodewald AD, Rosenberg KV, Sullivan BL, Winkler DW, Wood C, Kelling S (2014a) The role of atmospheric conditions in the seasonal dynamics of North American migration flyways. J Biogeogr 41:1685–1696

La Sorte FA, Tingley MW, Hurlbert AH (2014b) The role of urban and agricultural areas during avian migration: an assessment of within-year temporal turnover. Glob Ecol Biogeogr 23:1225–1234

Lack D, Varley GC (1945) Detection of birds by radar. Nature 156:446–446

LaFleur JM, Buler JJ, Moore FR (2016) Geographic position and landscape composition explain regional patterns of migrating landbird distributions during spring stopover along the northern coast of the Gulf of Mexico. Landsc Ecol 31:1–13

Larkin RP (1991) Flight speeds observed with radar, a correction: slow "birds" are insects. Behav Ecol Sociobiol 29:221–224

Larkin RP (2006) Locating bird roosts with Doppler radar. In: Timm R, O'Brien J (eds) Proceedings of the 22nd vertebrate pest conference. University of California, Davis, pp 244–249

Larkin RP, Diehl RH (2012) Radar techniques for wildlife biology. In: Silvy NJ (ed) The wildlife techniques manual. Research, vol 1, 7th edn. Johns Hopkins University Press, Baltimore, pp 319–335

Laughlin AJ, Taylor CM, Bradley DW, Leclair D, Clark RC, Dawson RD, Dunn PO, Horn A, Leonard M, Sheldon DR (2013) Integrating information from geolocators, weather radar, and citizen science to uncover a key stopover area of an aerial insectivore. Auk 130:230–239

Laughlin AJ, Sheldon DR, Winkler DW, Taylor CM (2014) Behavioral drivers of communal roosting in a songbird: a combined theoretical and empirical approach. Behav Ecol 25:734–743

Laughlin AJ, Sheldon DR, Winkler DW, Taylor CM (2016) Quantifying non-breeding season occupancy patterns and the timing and drivers of autumn migration for a migratory songbird using Doppler radar. Ecography 39(10):1017–1024

Lee L, Helsel D (2005) Statistical analysis of water-quality data containing multiple detection limits: S-language software for regression on order statistics. Comput Geosci 31:1241–1248

Liechti F, Witvliet W, Weber R, Bachler E (2013) First evidence of a 200-day non-stop flight in a bird. Nat Commun 4:art2554

Ligda MG (1958) Radar observations of blackbird flights. Tex J Sci 10:255–265

Lowery GH (1945) Trans-Gulf migration of birds and the coastal hiatus. Wilson Bull 57:92–121

Masteller EC, Obert EC (2000) Excitement along the shores of Lake Erie – Hexagenia – Echoes from the past. Great Lakes Res Rev 5:25

McCracken GF, Westbrook JK (2002) Bat patrol. Natl Geogr 201:14–23

McLaren JD, Buler JJ, Schreckengost T, Smolinsky JA, Boone M, Dawson DK, Walters EL (2018) Artificial light confounds broad-scale habitat use by migrating birds. Ecol Lett (in press)

McLaren JD, Shamoun-Baranes J, Bouten W (2013) Stop early to travel fast: modelling risk-averse scheduling among nocturnally migrating birds. J Theor Biol 316:90–98

Mehlman DW, Mabey SE, Ewert DN, Duncan C, Abel B, Cimprich D, Sutter RD, Woodrey MS (2005) Conserving stopover sites for forest-dwelling migratory landbirds. Auk 122:1281–1290

Mills AM, Thurber BG, Mackenzie SA, Taylor PD (2011) Passerines use nocturnal flights for landscape-scale movements during migration stopover. Condor 113:597–607

Moore FR, Kerlinger P (1989) Atmospheric structure and avian migration. Curr Ornithol 6:109–142

Moore FR, Phillips JB (1988) Sunset, skylight polarization and the migratory orientation of yellow-rumped warblers, Dendroica coronata. Anim Behav 36:1770–1778

Moore FR, Kerlinger P, Simons TR (1990) Stopover on a Gulf Coast barrier island by spring trans-gulf migrants. Wilson Bull 102:487–500

Moore FR, Gauthreaux SA, Kerlinger P, Simons TR (1995) Habitat requirements during migration: important link in conservation. In: Martin TE, Finch DM (eds) Ecology and management of neotropical migratory birds. Oxford University Press, New York, pp 121–144

Muheim R (2011) Behavioural and physiological mechanisms of polarized light sensitivity in birds. Philos Trans R Soc Lond B Biol Sci 366:763–771

Muheim R, Åkesson S, Phillips J (2007) Magnetic compass of migratory Savannah sparrows is calibrated by skylight polarization at sunrise and sunset. J Ornithol 148:485–494

O'Neal BJ, Stafford JD, Larkin RP (2010) Waterfowl on weather radar: applying ground-truth to classify and quantify bird movements. J Field Ornithol 81:71–82

O'Neal BJ, Stafford J, Larkin RP (2012) Stopover duration of fall-migrating dabbling ducks. J Wildl Manag 76:285–293

Randall LA, Diehl RH, Wilson BC, Barrow WC, Jeske CW (2011) Potential use of weather radar to study movements of wintering waterfowl. J Wildl Manag 75:1324–1329

Rappole JH, Ramos MA (1994) Factors affecting migratory bird routes over the Gulf of Mexico. Bird Conserv Int 4:251–262

Raveling DG, Crews WE, Klimstra WD (1972) Activity patterns of Canada geese during winter. Wilson Bull 84:278–295

Reynolds DR, Riley JR (1979) Radar observations of concentrations of insects above a river in Mali, West Africa. Ecol Entomol 4:161–174

Riley JR, Reynolds DR (1979) Radar-based studies of the migratory flight of grasshoppers in the middle Niger area of Mali. Proc R Soc Lond B Biol Sci 204:67–82

Riley JR, Reynolds DR, Farmery MJ (1983) Observations of the flight behaviour of the army worm moth, Spodoptera exempta, at an emergence site using radar and infra-red optical techniques. Ecol Entomol 8:395–418

Robinson SK, Holmes RT (1982) Foraging behavior of forest birds: the relationships among search tactics, diet, and habitat structure. Ecology 63:1918–1931

Russell RW (2005) Interactions between migrating birds and offshore oil and gas platforms in the northern Gulf of Mexico: final report. U.S. Department of the Interior, Minerals Management Service, Gulf of Mexico OCS Region, New Orleans

Russell KR, Gauthreaux SA (1998) Use of weather radar to characterize movements of roosting purple martins. Wildl Soc Bull 26:5–16

Russell KR, Gauthreaux SA Jr (1999) Spatial and temporal dynamics of a purple martin pre-migratory roost. Wilson Bull 111:354–362

Russell KR, Mizrahi DS, Gauthreaux SA (1998) Large-scale mapping of purple martin pre-migratory roosts using WSR-88D weather surveillance radar. J Field Ornithol 69:316–325

Ruth JM, Diehl RH, Felix RK Jr (2012) Migrating birds' use of stopover habitat in the Southwestern United States. Condor 114:698–710

Ruthi L (1994) Observation of bat emergence from Reed Bat Cave (HR-004) with the WSR-88D radar. Okla Undergr 17:54–56

Schmaljohann H, Naef-Daenzer B (2011) Body condition and wind support initiate the shift of migratory direction and timing of nocturnal departure in a songbird. J Anim Ecol 80:1115–1122

Seewagen CL, Slayton EJ (2008) Mass changes of migratory landbirds during stopovers in a New York City park. Wilson J Ornithol 120:296–303

Shamoun-Baranes J, Alves J, Bauer S, Dokter A, Huppop O, Koistinen J, Leijnse H, Liechti F, van Gasteren H, Chapman J (2014) Continental-scale radar monitoring of the aerial movements of animals. Mov Ecol 2:9

Sieges ML, Smolinsky JA, Baldwin MJ, Barrow WC, Randall LA, Buler JJ (2014) Assessment of bird response to the migratory bird habitat initiative using weather-surveillance radar. Southeast Nat 13:G36–G65

Smolinsky J, Diehl R, Radzio T, Delaney D, Moore F (2013) Factors influencing the movement biology of migrant songbirds confronted with an ecological barrier. Behav Ecol Sociobiol 67:2041–2051

Spengler TJ, Leberg PL, Barrow WC (1995) Comparison of condition indices in migratory passerines at a stopover site in coastal Louisiana. Condor 97:438–444

Stafford J, Horath M, Yetter A, Smith R, Hine C (2010) Historical and contemporary characteristics and waterfowl use of Illinois River valley wetlands. Wetlands 30:565–576

Stanley CQ, McKinnon EA, Fraser KC, Macpherson MP, Casbourn G, Friesen L, Marra PP, Studds C, Ryder TB, Diggs NE, Stutchbury BJM (2015) Connectivity of wood thrush breeding, wintering, and migration sites based on range-wide tracking. Conserv Biol 29:164–174

Taylor PD, Mackenzie SA, Thurber BG, Calvert AM, Mills AM, McGuire LP, Guglielmo CG (2011) Landscape movements of migratory birds and bats reveal an expanded scale of stopover. PLoS One 6:e27054

Vandewoestijne S, Van Dyck H (2011) Flight morphology along a latitudinal gradient in a butterfly: do geographic clines differ between agricultural and woodland landscapes? Ecography 34:876–886

Wang F, Xu YJ (2009) Hurricane Katrina-induced forest damage in relation to ecological factors at landscape scale. Environ Monit Assess 156:491–507

Weber TP, Fransson T, Houston AI (1999) Should I stay or should I go? Testing optimality models of stopover decisions in migrating birds. Behav Ecol Sociobiol 46:280–286

Westbrook JK (2008) Noctuid migration in Texas within the nocturnal aeroecological boundary layer. Integr Comp Biol 48:99–106

Westbrook JK, Eyster RS, Wolf WW (2014) WSR-88D doppler radar detection of corn earworm moth migration. Int J Biometeorol 58:931–940

Williams TC, Ireland LC, Williams JM (1973) High altitude flights of the free-tailed bat, Tadarida brasiliensis, observed with radar. J Mammal 54:807–821

Wingo S, Knupp K (2014) Multi-platform observations characterizing the afternoon-to-evening transition of the planetary boundary layer in northern Alabama, USA. Bound-Layer Meteorol 155:29–53

Winkler DW (2006) Roosts and migrations of swallows. Hornero 21:85–97

Winkler H, Leisler B (1985) Morphological aspects of habitat selection in birds. In: Cody ML (ed) Habitat selection in birds. Academic, New York, pp 415–434

The Lofty Lives of Aerial Consumers: Linking Population Ecology and Aeroecology

15

Winifred F. Frick, Jennifer J. Krauel, Kyle R. Broadfoot, Jeffrey F. Kelly, and Phillip B. Chilson

Abstract

Integrating population ecology and aeroecology is important for conservation for species that depend on aerial habitats. Assessing population response to anthropogenic stressors is key to predicting species at risk of extirpation or extinction, yet can be particularly challenging for species that predominately use the aerosphere. In this chapter, we focus on two aerial vertebrate consumers (purple martins and Brazilian free-tailed bats) as well-studied model systems to explore interactions between environmental forcing, behaviors, and population ecology. We also explore new recent progress on using radar-based methods to quantify aerial populations and discuss the utility of these novel approaches for a more promising integration of aeroecology and population ecology.

W.F. Frick (✉)
Bat Conservation International, Austin, TX, USA

Department of Ecology and Evolutionary Biology, University of California, Santa Cruz, CA, USA
e-mail: wfrick@batcon.org

J.J. Krauel
Department of Ecology and Evolutionary Biology, University of Tennessee, Knoxville, TN, USA
e-mail: jennifer@krauel.com

K.R. Broadfoot · J.F. Kelly
Oklahoma Biological Survey and Department of Biology, University of Oklahoma, Norman, OK, USA
e-mail: krbroadfoot@ou.edu; jkelly@ou.edu

P.B. Chilson
School of Meteorology, Advanced Radar Research Center, and Center for Autonomous Sensing and Sampling, University of Oklahoma, Norman, OK, USA
e-mail: chilson@ou.edu

© Springer International Publishing AG, part of Springer Nature 2017
P.B. Chilson et al. (eds.), *Aeroecology*,
https://doi.org/10.1007/978-3-319-68576-2_15

1   Introduction

Population dynamics are the outcome of an organism's intrinsic life history traits interacting with environmental forcing. For an enormous abundance and diversity of animal life, conditions in the aerosphere are a key aspect of their environment (Root and Hughes 2005; Kunz et al. 2008), yet how biota respond to structural and functional changes in the aerosphere at any scale remains poorly understood (Root and Hughes 2005; Bowlin et al. 2010). More sophisticated investigations of how life in the aerosphere influences population dynamics of volant species will be required if aeroecology is going to fulfill its promise to contribute to conservation (Kunz et al. 2008). The original conception of aeroecology recognized that species that depend on the aerosphere face direct anthropogenic threats and may be particularly sensitive to climate change (Kunz et al. 2008). The aeroecological perspective provides a framework for focusing on species interactions in the dynamic lower atmosphere and emphasizes investigating how environmental forces associated with the aerosphere influence population dynamics of species that rely on aerial habitats.

Integrating population ecology and aeroecology involves studying both patterns of population trends of aerial species and uncovering processes that drive population fluctuations. Mechanistic understanding of factors responsible for regulating population sizes could be viewed as the holy grail of ecological science and yet remains a formidable challenge in many systems. However, effective conservation depends on science that connects population-level responses to anthropogenic changes to the landscape—whether that landscape be terrestrial, marine, or aerial. Moreover, establishing monitoring systems to track population trends over time can serve as a vital tool for discovering species or populations at risk, even when mechanisms behind population trends remain obscured (Lovett et al. 2007; Nebel et al. 2010; Smith et al. 2015).

In this chapter, we examine integration of population ecology and aeroecology by synthesizing research focused on populations of aerial consumers—vertebrate species that depend on the aerosphere to forage for aerial invertebrate prey. We first summarize case study research on aerial consumers that investigates how weather and climate influence aerial consumer foraging behaviors and then explore how new approaches in radar aeroecology can be used to establish long-term population monitoring opportunities and potentially lead to new understanding of space use by animals in the aerosphere in these systems. Our focus on radar methods stems from our own research experiences and interests, and we recognize other techniques and approaches, such as recent advances in miniaturized tracking systems (Bridge et al. 2011), are contributing greatly to the field and are covered elsewhere in this volume (Chap. 11). We also note that work in the case studies we explore here has not yet achieved that "holy grail" of integrating process and pattern for a thorough mechanistic understanding of population dynamics of these species. Instead, our aim here is to explore how future integration of population ecology and aeroecology in these systems promises to contribute to a broader understanding of ecology and conservation.

1.1   Aerial Consumers

Aerial consumers are useful models for integrating population ecology and aeroecology for several reasons. Firstly, their basic natural history is well studied in their terrestrial habitats, yet technological limitations have historically thwarted ability to study their behaviors in the aerosphere. Secondly, aerial insectivores are generally of conservation interest and some species have experienced wide-scale population declines (Nebel et al. 2010; Smith et al. 2015). They may be particularly vulnerable to climate change given that interactions and movements in the aerosphere are likely highly sensitive to weather (Frick et al. 2010; Krauel et al. 2015). High mobility and ability to potentially track resources over large areas makes their response to climate and land use change potentially different from animals constrained to terrestrial landscapes. Aerial consumers also provide ecosystem services in terms of consumption of aerial insects, including agricultural pest species (Cleveland et al. 2006; Kelly et al. 2013; Maine and Boyles 2015). Their sensitivity to climate and importance to ecosystems makes them useful as bioindicators (Furness and Greenwood 1993; Jones et al. 2009). Appropriately designed monitoring programs could serve to indicate ecosystems at risk from climate change and provide valuable conservation benefit beyond species-specific research goals (Williams et al. 2008; Jørgensen et al. 2015).

We focus in particular on two species of aerial insectivores: Brazilian free-tailed bats (*Tadarida brasiliensis*) and purple martins (*Progne subis*) as two well-studied model systems in aeroecology (Russell et al. 1998; Russell and Gauthreaux 1999; Horn and Kunz 2008; Kelly et al. 2012; Frick et al. 2012; Krauel et al. 2015). Brazilian free-tailed bats aggregate by the millions in cave roosts in Texas during the summer maternity period and emerge like twisting stream of smoke at sunset to forage both high (e.g., several kilometers above ground) (McCracken et al. 2008) and wide (e.g., as far as 50 km from their roosts) (Wilkins 1989). Aggregations of purple martins range from south-central Texas to the southeastern pine barrens, where they form enormous roosts that stop over for days to weeks to feed on local insect populations (Tarof and Brown 2013). Aggregations of Brazilian free-tailed bats and purple martins are easily identified and monitored with weather surveillance radars (WSR-88D) (Russell et al. 1998; Horn and Kunz 2008) (Fig. 15.1).

The impact of these consumers on the ecosystem is substantial. The Brazilian free-tailed bat (12 g) forms some of the largest aggregations of vertebrates known, wherein millions of females amass to give birth and raise young during summer months (Davis et al. 1962). Nightly and migratory movements of these animals are capable of redistributing upward of 48,000 kg of vertebrate biomass across landscapes. These bats are also significant predators of agricultural pests such as corn earworm and cotton bollworm moths (Lee and McCracken 2005; Westbrook 2008) and provide real economic value through ecological services in agroecosystems in Texas (Cleveland et al. 2006; Maine and Boyles 2015). Similarly, purple martins are abundant and widely distributed aerial insectivores that consume aerial insects. Kelly et al. (2013) developed a range-wide energetics simulation model of insect consumption by purple martins and showed that based

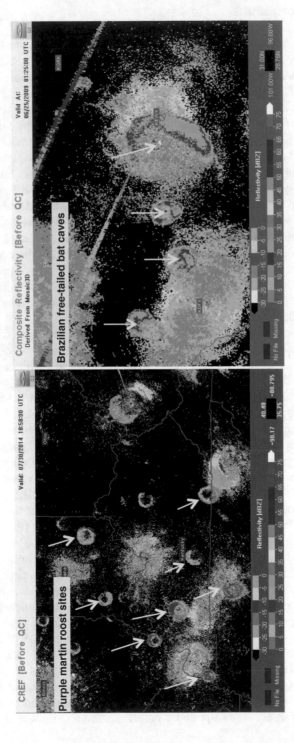

Fig. 15.1 Dawn emergence flights of purple martins (*Progne subis*) from known summer roost locations in Kentucky and surrounding states (left) and evening emergence flights of Brazilian free-tailed bats (*Tadarida brasiliensis*) in south-central Texas (right) visible on the national composite radar mosaic available from http://nmq.ou.edu

on a pre-breeding population of 10 million individuals (Rich et al. 2004), purple martins would consume a peak biomass of 600 tonnes of insects per day in July (Kelly et al. 2013). Consumption of such mass quantities of insects undoubtedly has pervasive ecosystem consequences in regions with large roost aggregations.

2   Weather, Climate, and Resource–Consumer Dynamics in the Aerosphere

Because aerial consumers forage in the air, where and when they forage is not necessarily tied to fluctuations in prey availability and terrestrial productivity at ground level. However, there is no primary productivity in the air and neither predators nor prey reproduce or spend their entire life cycle in the aerosphere. In this sense, the trophic system in the aerosphere is ultimately allochthonous and linked back to the terrestrial ecosystem and biological interactions in the aerosphere may be spatially separated from land but ultimately depend upon the productivity and integrity of the terrestrial landscape (Chaps. 3, 7, and 14). Both the quantity and quality of insect prey in the aerosphere vary hugely in space and time (Chapman et al. 2011a) as aerial insect prey consists mostly of migratory or dispersing animals and their movements are largely directed and determined by wind (Drake and Gatehouse 1995). For example, density of aerial insects can vary in altitude and be concentrated by turbulence into "eddies" coinciding with favorable air speeds and temperatures (Wood et al. 2010; Reynolds et al. 2010; Chapman et al. 2011b). Aerial consumers must then contend with the challenges of predicting and finding shifting aerial prey patches in the aerosphere (Krauel et al. 2015).

The primary mechanism driving variations in aerial insect availability appears to be weather and season (Drake and Gatehouse 1995). Syntopic (continental) patterns of shifting barometric pressure and winds can create fast-moving currents of air known as nocturnal jets (Dingle 2014). Nocturnal wind currents flow typically 200–600 m above ground level and can travel hundreds of kilometers over the landscape (Bonner 1968; Bonner and Paegle 1970) carrying insects across continents in different directions depending on seasonal fluctuations (Drake and Gatehouse 1995; Burt and Pedgley 1997). A set of studies on diets of Brazilian free-tailed bats in Texas highlight how aerial consumer diets are influenced by temporal variation in both quantity and composition of insect prey among nights and seasons (Lee and McCracken 2002; Lee and McCracken 2005; Krauel et al. 2015).

Phenology of when Brazilian free-tailed bats have peak energetic demands associated with reproduction and pre-migratory fat accumulation should match with seasonal periods of availability of high quantity and quality of aerial insect prey (Krauel et al. 2015). During late summer, when the bat pups have fledged and adult females are at their lowest weight (Pagels 1972; O'Shea 1976), the bats have the greatest need to add body fat that will fuel autumn migration. However, primary productivity in Texas is very low and local insect emergences are reduced to occasional post-rain termite flights. Although the bats are generalist predators and not strongly linked to any specific insect prey, the arrival of south-bound migratory noctuid moths in September provides a high-value prey when few other options exist (Krauel et al. 2015). Krauel et al. (2015) show how cold front passages in autumn in Texas correspond to increased insect prey availability, including crop pests, and increased body mass in Brazilian free-tailed bats. Pre-migratory fat accumulation in Brazilian free-tailed bats likely depends on the passage of the moths in the aerosphere and the ability of the bats to locate them in time and space. If the cycle of moth movement is interrupted or changed due to shifts in climate, populations of Brazilian free-tailed bats may be impacted by lower survival or changes in migratory phenology (Krauel et al. 2015).

Mass emergences of Brazilian free-tailed bats are easily detectable on the network of Doppler weather radars (WSR-88D) known as NEXRAD (Horn and Kunz 2008). Advances in radar processing of the networked 160 WSRs in the conterminous USA have created a mosaic of radar reflectivity mapped onto a Cartesian grid spanning the entire continental USA (Zhang et al. 2004, 2005; Langston et al. 2007; Vasiloff et al. 2007). These national scale radar mosaics were maintained publically available and viewable in near real time online at http://nmq.ou.edu by the National Severe Storms Laboratory at NOAA (Zhang et al. 2011). Use of the so-called un-quality controlled composite reflectivity (UnQC'd CREF) products were useful for answering a host of biological questions about aerial consumers (Kelly et al. 2012; Chilson et al. 2012a; Frick et al. 2012). One of the advantages to this system was the ability to visualize phenomena in the aerosphere at the temporal and spatial scales that are otherwise not visible. For example, the phenomena of Brazilian free-tailed bats foraging along a "buffet line" of insects being pushed along by a cold front can be seen in Fig. 15.2.

Similarly, Russell and Gauthreaux (1999) used earlier NEXRAD radar imagery to visualize and study the spatial and temporal dynamics of a purple martin pre-migratory roost. After young fledge, purple martins aggregate in huge numbers at nocturnal roosts prior to fall migration (Tarof and Brown 2013). Aggregations can reach upward of several hundreds of thousands of birds at a site and birds depart in the morning for aerial foraging and return again in the evening (Allen and Nice 1952). Russell and Gauthreaux (1999) used images of radar reflectivity to characterize daily timing and spatial patterns of martin flights and investigated the influence of weather on flight patterns. Timing of departures and movement distances of martins from the roost were influenced by weather conditions, supporting the hypothesis that weather influences aerial foraging behaviors (Finlay 1976; Russell and Gauthreaux 1999). Furthermore, by using NEXRAD imagery they were able to confirm that purple martins fly much longer distances than was previously thought at the time (80–100 km from the roost) because of the long-distance vantage radar imagery provides (Russell and Gauthreaux 1999).

More recently, Kelly et al. (2012) expanded on the earlier work of Russell and Gauthreaux (Russell et al. 1998; Russell and Gauthreaux 1999) and used the aforementioned national mosaic of radar reflectivity to identify and monitor seasonal phenology of over 350 summer roosts of purple martins across the species range in the USA. The ability to use remote-sensing methods such as NEXRAD to monitor and measure phenology of aerial consumers is timely given concerns about

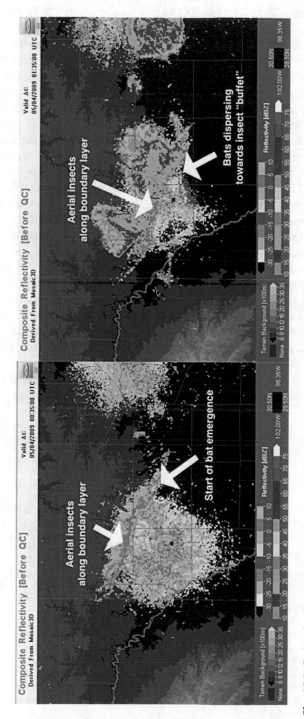

Fig. 15.2 Radar visualization of bats emerging and dispersing toward aerial insects pushed by a cold front in southern Texas on May 5, 2009

how increased warming correlates to changes in phenology (Ibáñez et al. 2010; Hodgson et al. 2011) and its potential for decoupling resource–consumer interactions (Both et al. 2006). In this volume, Kelly and colleagues explore further the connections between aeroecology and phenology of avian migration (Chap. 16).

Timing of roost departure to forage is an adaptive behavior that has important fitness consequences in terms of trade-offs between increased risk of predation and forfeiting foraging opportunities during peak prey availability (Duvergé et al. 2000). For bats, early departures from a roost may increase risk of predation but provide greater access to crepuscular blooms of aerial insects (Jones and Rydell 1994; Rydell et al. 1996; Duvergé et al. 2000). Several studies have demonstrated that reproductive condition of bats influences roost departure in ways that support this hypothesis (Duvergé et al. 2000; Reichard et al. 2009). In particular, lactating females are the most energetically stressed and therefore should emerge earlier if energetic demands outweigh costs of increased risk of predation (Reichard et al. 2009). The hypothesis that increased physiological stress leads to earlier emergence times leads to predictions about how climate variation may influence collective animal behavior. Frick et al. (2012) used archived radar data to test whether the onset of emergence of Brazilian free-tailed bats during the maternity season in Texas was associated with variation in summer drought conditions over an 11-year time series. Frick et al. (2012) showed that bats emerged significantly earlier in the evening in years with severe drought conditions than in normal or moist years and responded to daily temperatures in dry years, suggesting that bats behaviorally respond to both daily weather and seasonal climatic conditions in ways consistent with adaptive trade-offs of balancing reduced predation with increased probability of foraging success (Fig. 15.3).

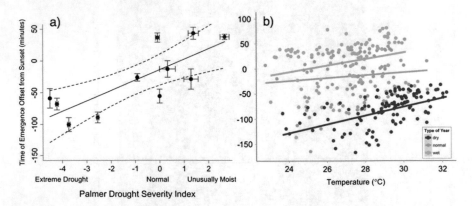

Fig. 15.3 Roost emergence behavior of Brazilian free-tailed bats (*Tadarida brasiliensis*) in response to annual climate and daily temperature adapted from Frick et al. (2012). (**a**) Mean time of colony emergence times at five maternity colonies in south-central Texas emerged earlier in the evening as drought severity increased. Error bars show standard errors around means for drought severity (*x*-bars) and emergence times (*y*-bars). (**b**) Bats emerged earlier on increasing daily temperatures in dry years, but daily temperature did not have a strong effect on emergence times in normal and wet years

Flexible consumer response to climate and weather may mediate impacts of shifting climates, if consumers can compensate for poor conditions by increasing foraging times (Frick et al. 2012). Alternatively, emerging to forage earlier may signal stressful conditions and correspond to lower survival or lower reproductive success. Understanding how flexibility in behaviors such as foraging departures relates to fitness is a necessary next step to predicting how climate conditions will influence population of aerial consumers. Frick et al. (2012) proposed linking annual variation in drought conditions and emergence behavior with changes in population sizes of bat colonies over time as a way to determine the long-term impacts of climate on bat populations, yet did not test that hypothesis due, in part, to lack of available population data over the same time interval.

3 Quantifying Aerial Populations

Methods to estimate aerial densities of biological scatterers from weather radar products have been progressing over the past few years (Bruderer and Steidinger 1972; Buler and Diehl 2009; Buler et al. 2012a; Chilson et al. 2012a, b) and should prove fruitful for integrating population ecology with aeroecology. There are a number of challenges to using radar products to estimate aerial densities—spanning from technical aspects of radar imagery and processing to ecological factors such as needing large aggregations of animals that take to the skies within detectable ranges of existing radar installations. Buler and Diehl (2009) developed ways to quantify bird densities at migratory stopover points after adjusting radar measurements for vertical variation in reflectivity, nonstandard beam refraction, and spatial displacement of scatterers. Buler and colleagues have made significant progress toward integrating population ecology and aeroecology using the network of weather surveillance radars in the USA to quantify and map migratory stopover and wintering waterfowl distributions (Buler et al. 2012b) (Chap. 14). They have also been at the forefront of developing new tools, such as the w2birddensity tool integrated into the WDSS-II software package used for analysis and processing of radar products (Buler et al. 2012a). These recent methods and tools allow a wider group of nontechnical users to benefit from the archive of radar data to investigate ecological questions and implement population monitoring. A thorough synthesis of the work by this group and others on linking animals aloft with terrestrial landscapes is covered in Chap. 14 which obviously relates to linking population ecology and aeroecology more broadly.

In contrast to the progress made to quantifying birds aloft, relatively little progress has been made to quantify bat populations. Earlier methods for bat monitoring used radar reflectivity factor (Z) as a relative index of bat colony size (Horn and Kunz 2008) and described seasonal shifts in bat occupancy in cave and bridge roosts throughout Texas. However, use of radar reflectivity factor is not particularly well suited for biological questions given it is based on assumptions developed for meteorological phenomena (Chilson et al. 2012b) and therefore has limited utility for modeling population dynamics and response to stressors. Other innovative approaches to estimating colony size of bats have used thermal imaging and computer imaging algorithms (Betke et al. 2008), but this approach requires investment in expensive thermal imaging hardware and extensive field deployment, which limits its utility for retrospective or time series analysis.

One of the advantages of using radar-based methods for estimating aerial densities is access to the 20-year archive of NEXRAD data, which would permit investigations into how aerial consumer aggregations (such as Brazilian free-tailed bats or purple martins) vary over time and in response to phenomena such as climate or land use change (Chilson et al. 2012a). Here, we describe recent efforts to implement methods developed by Chilson et al. (2012b) to quantify aerial densities of both purple martins and Brazilian free-tailed bats. Calculation of the radar reflectivity factor requires a set of assumptions that are reasonably met for precipitation (i.e., a uniform distribution of small spherical hydrometeors) but may be questionable for flying animals (Chilson et al. 2012b). As a solution, Chilson et al. (2012b) proposed that future biological radar products use units of radar reflectivity (η), which requires fewer assumptions and is more easily interpreted with respect to biology. For example, a value of $Z = 15$ dBZ as measured using NEXRAD (S-band) corresponds to $\eta = 690$ cm^2 km^3, which represents the collective cross-sectional area of scatterers (cm^2) per volume (km^3). Implementation of this approach is still relatively nascent (Dokter et al. 2011), but we demonstrate our early efforts to show "proof of concept" and hopefully inspire future applications for answering ecological questions.

Before exploring how animal density can be retrieved from weather radar, it is useful to first consider how radar reflectivity is used to estimate rates of precipitation. The degree to which an object reflects or scatters electromagnetic radiation of a particular wavelength is known as the radar cross section (RCS) and is expressed as an area. The larger the RCS, the "brighter" an object will appear to the radar. For the case of precipitation observed with meteorological radar, the hydrometeors are typically much smaller than the probing wavelength. In this case, the RCS increases monotonically with the particle diameter; in particular, the RCS is proportional to D^4, where D is the diameter (Doviak and Zrnić 1993). Using a probability distribution function of the hydrometeor sizes along with the measured backscatter power and assuming that the sampled volume of space is uniformly filled with hydrometeors, it is possible to calculate the collective RCS per unit volume (Chap. 12). This is the radar reflectivity η and from it we can calculate the quantity of water per unit volume using the D^4 relationship. Finally, knowing the fall speed of the hydrometeors as a function of size, we can find a precipitation rate.

In much the same way that precipitation rates are calculated, weather radar can be used to find number densities of animals (Chilson et al. 2012b). However, for the case of biological scatter the situation is much more complicated, because the aerosphere in general contains a rich variety of species of varying size, body shapes, and concentrations. Unlike hydrometeors, which are typically geometrically symmetric, the body shapes of animals are complex. Values of RCS for an animal obtained at different viewing angles can be dramatically different (Edwards and Houghton 1959; Bruderer and Joss 1969). Moreover, the body size of many vertebrate animals and some invertebrates is comparable to the wavelength of meteorological radars. As a consequence, we cannot assume that RCS increases monotonically with the size of the scatterer as is done for hydrometeors. Unfortunately, the radar scattering properties of volant animals are currently poorly understood. Many of the studies reported in the literature simply assume that the radio-wave scatter from animals is isotropic (Chap. 12).

Despite its complexity, the process of converting radar reflectivity values from radar products into a biological density product allows estimation of the number of individuals (or aerial density) of targeted species aloft (Chilson et al. 2012b). During the post-breeding, pre-migratory season, purple martins form dense aggregations at roost sites throughout the eastern USA, leaving roosts near dawn every morning to forage on insects in the lower atmosphere. The exoduses of these birds are regularly detected on weather radar as ring-shaped echoes (Figs. 15.1 and 15.4) (Russell et al. 1998; Kelly et al. 2012). Under these conditions, it is reasonable to assume that the preponderance of bioscatter is from a single species of animal. We used unfiltered Level II weather radar data available from the National Climatic Data Center's online portal (website here) and spatially clipped to the maximum extent of the roost echo (Fig. 15.4). Radar reflectivity values, reported in dBZ, were converted to the value of η cm^2 km^{-3} (Fig. 15.5) using Eq. (15) in Chilson et al. (2012b). Values of η were then multiplied by the size of the radar pixel volume and divided by the radar cross section (i.e., the reflected surface area of the organism as encountered by the radar beam), approximated for purple martins at 13 cm^2 after (Eastwood 1967) (Fig. 15.5). We estimated the number of purple martins in a scan by summing across the lowest two elevation sweeps (0.5° and 1.5°) and subtracting a corresponding value of "noise" after Stepanian et al. (2014) (Fig. 15.5) Using this method, we estimated an approximate maximum of 133,000 purple martins at a site near Garland, Texas, on the morning of July 13, 2014 (Fig. 15.5c). There are a number of caveats that come with an estimation derived from this technique; chief among them are the need for an accurate RCS value and the exclusion of nonbiological scatterers. While these potential concerns temper the generalizability of this method to other days, sites, and species, it has the potential to offer insights into the population trends of roosting organisms at unprecedented spatial and temporal scales.

A similar approach has been applied to investigate the emergence of Brazilian free-tailed bats in Texas. Free-tailed bats emerge at dusk from their roosting site to forage for insects. Upon departure, bats will typically rapidly ascend to 100s or 1000s of meters to avoid predation and then disperse as they begin to forage. The emergence can often be detected by NEXRAD as a pattern similar to that for purple martins—a localized region of enhanced radar reflectivity that expands radially outward as a full or partial ring shape. Radar reflectivity data from each volume scan are retrieved from an archive and converted into a measure of animal density by applying an appropriate RCS value. Since the sampling volumes within the full radar scan are known, the number densities are used to find the total number of bats

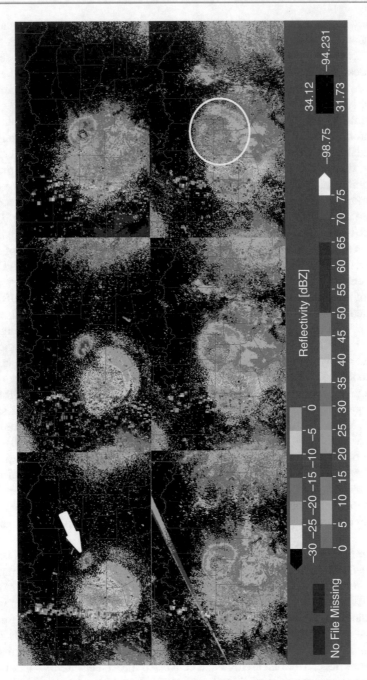

Fig. 15.4 Mosaicked radar images before quality control taken from http://nmw.ou.edu centered on Dallas, TX (KFWS). Time series (10:58–12:42 UTC) shows characteristic dawn emergence of purple martins from a roost in Garland, Texas. White ring in bottom-right panel corresponds to maximum roost domain used to calculate aerial densities of purple martins (see text for details)

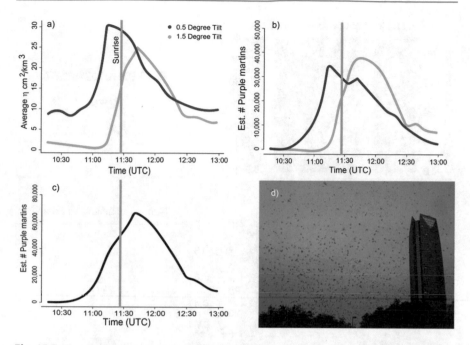

Fig. 15.5 Populations size of purple martins aloft based on NEXRAD radar products from KFWS station near Garland, Texas, on July 13, 2014. (**a**) Averaged values of η per sample volume for the two lowest elevation angles (blue = 0.5° and green = 1.5°) for each timestamp from 1 h before until 1.5 h after sunrise. Orange bar indicates time of local sunrise (11:29 UTC). (**b**) Number of purple martins aloft estimated by dividing cumulative η by a radar cross section of 13 cm^2 for each elevation angle separately. (**c**) Total population size of purple martins aloft in the sampling volume combining the two elevation angles. (**d**) Purple martins filling the sky as they return to a roost in Oklahoma City, OK (photo: Jeff Kelly)

observed at a given time. The application of these methods can be seen in Fig. 15.6, where the emergence of free-tailed bats from Frio Cave measured with NEXRAD is quantified in terms of number of bats observed based on measurements of reflectivity and assuming a bat has an isotropic RCS of 5 cm^2. This new method of quantifying bat emergences could aid conservation efforts by monitoring population changes over time at these roosts (Wiederholt et al. 2013, 2015). In particular, measuring changes in population size of roosts over time could help determine whether these populations are impacted by climate change (Frick et al. 2012) or other anthropogenic stressors, such as land use changes and siting of wind turbines (Kunz et al. 2007). Wind turbines kill large numbers of bats (Arnett et al. 2016), and understanding the behavior and density of bats in the air (Amorim et al. 2014) is necessary for identifying areas that are unsuitable for wind energy development.

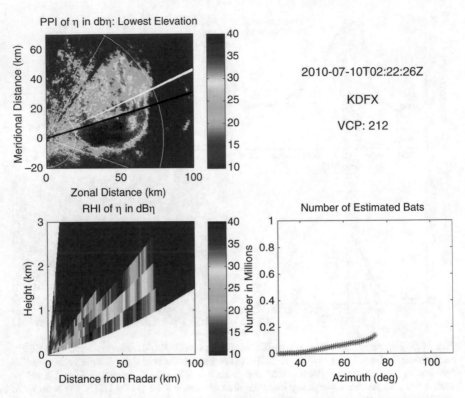

Fig. 15.6 Calculation of number of Brazilian free-tailed bats (*Tadarida brasiliensis*) emerging from Frio Cave using level II data from KDFX RADAR station on July, 7 2010, at 9:22 CDT. The summed values of η for each latitude and longitude (as viewed from above in upper left panel) are sampled across azimuths. The black line in the upper left plot indicates the current azimuth. The distribution of bats in height at this azimuth is displayed as a range height indicator plot (lower left). The summed number of bats across sampled azimuths is displayed in the lower right based on the methods of Chilson et al. (2012b)

3.1 Radar Simulator

Another approach to modeling aerial populations has been to develop a realistic radar simulator capable of scanning a virtual aerosphere populated with known numbers, distributions, and types of animals under different scenarios. The simulator outputs base-level radar data, which can be used to generate other products such as level II radar data. Our initial simulator [described by Chilson et al. (2009)] was adapted from modeling tools developed for weather studies (Palmer et al. 2008). However, we are currently in the process of developing a new radar simulator specifically designed for biological applications. Computer-based simulation allows us to compare radar data derived using mechanistic behavioral models to empirical radar observations across multiple spatial and temporal scales. Moreover, these simulations allow us to test new radar signal processing algorithms under controlled conditions. The radar simulator has been a critical analytic tool for linking radar visualizations of aerial biodiversity to expectations based on numbers of animals in the air. For example if we detect four purple martins using a thermal camera in an airspace being sampled by the RaXPol mobile radar, we can validate the echo returned to that radar against the signal returned to the nearest NEXRAD station for the same sampling time. We can then compare both of these reflectivity values to an expected value using the radar simulator based on the behavior and reflective properties of purple martins.

To demonstrate one application of the simulator, we consider the departure of purple martins from a roosting site in northeastern Oklahoma. One of the popular roosting sites for purple martins is in Tulsa, Oklahoma. Every year during the summer between 100,000 and 250,000 of these birds will take up temporary residence in the downtown area. Their early morning departure can be observed in the radar displays from the WSR-88D KINX (Tulsa, OK) as roost rings similar to those shown in Fig. 15.1. We have attempted to simulate such an emergence by allowing "virtual" purple martins to be observed in the simulator. The simulated roost was located 40 km from the virtual weather radar to match the spacing between the Tulsa martin roost and KINX. The number of martins considered in the simulation was limited to 10,000. At sunrise, the martins were allowed to leave the roost in a continuous stream over a fixed time interval. The flights of the birds were simulated for 310 computing steps with 1 s time interval between each step and the flight speed and directions of the birds were broken down into radial (horizontal) and vertical components. Each simulated bird departed in a random cardinal direction and was assigned a radial velocity taken from a Gaussian probability distribution with a mean of 10 m s^{-1} and a variance of 5 m s^{-1}. For each subsequent time step, the radial direction for every bird that had left the roost was allowed to slightly deviate from its original trajectory and a new radial velocity was assigned. The vertical trajectory of each bird was described by a rapid ascent followed by slower approach to a maximum height.

The virtual weather radar considered in the simulations was configured to match that of a NEXRAD unit. The radar had a 1° beam width and range resolution of 250 m. It scanned 19 elevation angles in steps of 1° with the lowest elevation angle being 0.5°. The wavelength was set to 10 cm. When considering the scattering properties of the martins, a Gaussian probability distribution was used. The backscattering characteristics of the birds were determined through assigned radar cross-section (RCS) values. Each bird was given an RCS according with a mean of 15 cm^2 and a variance of 1 cm^2. The geometric characteristic of the bird was modeled as oblate spheroids with an aspect ratio 0.5 and variance of 0.1.

The primary purpose of this simulation was to ascertain if the scattering physics and the crude model of the birds' flight trajectories would produce reasonable results. We realize that many refinements are needed in order to produce realistic roost ring echo patterns. But this initial attempt to present an example of the spatial distribution of the simulated purple martins during one time step appears in Fig. 15.7. Note that the scale for the z-axis (representing height above the surface) has been exaggerated with respect to the scales for the horizontal dimensions. The

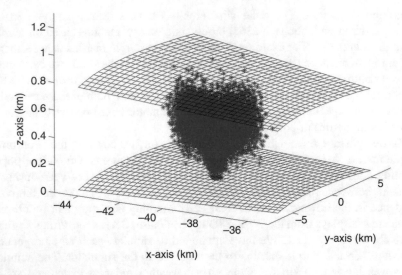

Fig. 15.7 A snapshot of the spatial distribution of purple martin locations during one time step of the simulation run after all birds have left the roost. Positions correspond to the 20th set of radar data or 190 s into the simulation

simulation was configured such that radar data were collected every 10 s. At this point in time, all of the martins will have left the roost. The corresponding simulated radar image is presented in Fig. 15.8. Data are for a sector scan with an elevation angle of 0.5°. Time series (I & Q) data are shown in the upper left-hand panel for a particular sampling volume. There are 50 points in the time series. The corresponding Doppler spectrum is shown in the upper right-hand panel. Also shown with the Doppler spectrum are fits to the data and an estimate of the radial velocity. The middle three panels show the reflectivity (dBZ), radial velocity (m s^{-1}), and the spectrum width (m s^{-1}) for a horizontally polarized radio wave. The data shown in the upper two panels were collected in the sampling volume denoted by the "x" mark. The lower three panels are the same as for the middle three, except that data from a vertically polarized radio wave. As the simulation progresses in time, the radar signatures grow in horizontal extent and become ring shaped.

Development of the simulator takes a mechanistic approach to interpreting radar data with specific applications to quantifying aerial densities or flows of animals aloft from radar imagery. The simulator can be used to better understand how animals distribute across aerial space which can be applied to questions in behavioral ecology, movement ecology, and population ecology. In some ways, this is similar to the advantages of agent-based modeling described in Chap. 11 by simulating emergent properties at a population level from specified model inputs at the individual (or agent) level. One of the key advantages of using radar methods is being able to show how animals are dispersed across the landscape in real terms. Many studies have focused on tracking individuals to make inference about how populations use a landscape (e.g., use of utilization distributions), which essentially

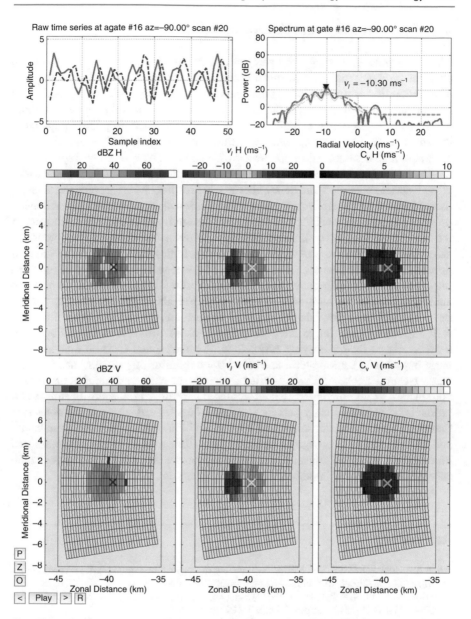

Fig. 15.8 Simulated radar data for the bird positions shown in Fig. 15.6. See text for further explanation of the format

takes data from individuals to scale to the process of populations. In contrast, radar imagery (once calibrated by the methods above to relate back to density of bioscatterers aloft) can directly measure utilization distributions at the true scale of the population.

4   Conclusions and Future Directions

Linking population ecology and aeroecology is still in many respects a "work in progress" and presents some formidable challenges within this emerging discipline. The difficulties of uncovering mechanisms that regulate population dynamics are not unique to aeroecology, yet the habits of species that largely depend on aerial habitats have often made traditional approaches to studying their population dynamics particularly vexing. Studies within aeroecology less likely focus directly on traditional demographic approaches of estimating survival and reproductive output, but rather focus on spatial and temporal patterns and trends in populations of both aerial consumers and their airborne food resources. Aeroecology provides a useful framework for studying drivers of population change for species that depend on the air as a primary habitat by focusing on the environmental forces that occur in the aerosphere (e.g., how climate and weather influence consumer behaviors in the aerosphere) as well as leveraging new technologies and methods to monitor populations that take to the skies en masse. Use of radar to quantify populations may not have broad applicability beyond a few focal taxa that form large aggregations and fly high enough to be detected above ground clutter, but for these species radar provides an unusual opportunity to monitor populations as well as allow retrospective analyses going back for more than 20 years. Radar tools available for quantifying aerial food abundance and their consumers offer great opportunity for long-term and broad-scale monitoring programs and in many cases existing networked radar systems already provide and archive data. New approaches to automatically extract ecological information from existing radar products promise to greatly benefit a variety of ecological research and conservation endeavors. Aerial consumers are often good bioindicators for ecosystem health and perform valuable ecosystem services. Therefore, future efforts to monitor and investigate how aerial consumers respond to climate and land use changes could serve as a particularly important tool for conservation.

References

Allen RW, Nice MM (1952) A study of the breeding biology of the Purple Martin (Progne subis). Am Midl Nat 47:606–665. https://doi.org/10.2307/2422034

Amorim F, Cryan PM, Rebelo H et al (2014) Behavior of bats at wind turbines. Proc Natl Acad Sci USA 111:15126–15131. https://doi.org/10.1073/pnas.1406672111

Arnett EB, Baerwald EF, Mathews F et al (2016) Impacts of wind energy development on bats: a global perspective. In: Voight CC, Kingston T (eds) Bats in the anthropocene: conservation of bats in a changing world. Springer, Cham, pp 295–323

Betke M, Hirsh DE, Makris NC et al (2008) Thermal imaging reveals significantly smaller Brazilian freetailed bat colonies than previously estimated. J Mammal 89:18–24. https://doi.org/10.1644/07-MAMM-A-011.1

Bonner WD (1968) Climatology of the low level jet. Mon Weather Rev 96:833–850

Bonner WD, Paegle J (1970) Diurnal variations in boundary layer winds over the south-central united states in summer. Mon Weather Rev 98:735–744

Both C, Bouwhuis S, Lessells CM, Visser ME (2006) Climate change and population declines in a long-distance migratory bird. Nature 441:81–83. https://doi.org/10.1038/nature04539

Bowlin MS, Bisson IA, Shamoun-Baranes J et al (2010) Grand challenges in migration biology. Integr Comp Biol 50:261

Bridge ES, Thorup K, Bowlin MS et al (2011) Technology on the move: recent and forthcoming innovations for tracking migratory birds. Bioscience 61:689–698

Bruderer B, Joss J (1969) Methoden und Probleme der Bestimmung von Radarquerschnitten frei fliegender Vögel

Bruderer B, Steidinger P (1972) Methods of quantitative and qualitative analysis of bird migration with a tracking radar

Buler JJ, Diehl RH (2009) Quantifying bird density during migratory stopover using weather surveillance radar. IEEE Trans Geosci Remote Sens 47:2741–2751. https://doi.org/10.1109/TGRS.2009.2014463

Buler JJ, Lakshmanan V, La Puma D (2012a) Improving weather radar data processing for biological research applications: Final report

Buler JJ, Randall LA, Fleskes JP et al (2012b) Mapping wintering waterfowl distributions using weather surveillance radar. PLoS One 7:e41571. https://doi.org/10.1371/journal.pone.0041571

Burt PJA, Pedgley DE (1997) Nocturnal insect migration: effects of local winds. In: Begon M, Fitter AII (eds) Advances in ecological research. Advances in ecological research. Academic, London, pp 61–92

Chapman JW, Drake VA, Reynolds DR (2011a) Recent insights from radar studies of insect flight. Annu Rev Entomol 56:337–356

Chapman JW, Klaassen RHG, Drake VA et al (2011b) Animal orientation strategies for movement in flows. Curr Biol 21:R861–R870. https://doi.org/10.1016/j.cub.2011.08.014

Chilson PB, Bolognini K, Cheong BL (2009) Using a dual-polarimetric weather radar simulator to investigate microwave backscatter from birds

Chilson PB, Frick WF, Kelly JF et al (2012a) Partly cloudy with a chance of migration: weather, radars, and aeroecology. Bull Am Meteorol Soc 93:669–686. https://doi.org/10.1175/BAMS-D-11-00099.1

Chilson PB, Frick WF, Stepanian PM et al (2012b) Estimating animal densities in the aerosphere using weather radar: to Z or not to Z? Ecosphere 3:art72. https://doi.org/10.1890/ES12-00027.1

Cleveland CJ, Betke M, Federico P et al (2006) Economic value of the pest control service provided by Brazilian free-tailed bats in south-central Texas. Front Ecol Environ 4:238–243

Davis RB, Herreid CF, Short HL (1962) Mexican free-tailed bats in Texas. Ecol Monogr 32:311. https://doi.org/10.2307/1942378

Dingle H (2014) Migration: the biology of life on the move, 2nd edn. Oxford University Press, Oxford

Dokter AM, Liechti F, Stark H et al (2011) Bird migration flight altitudes studied by a network of operational weather radars. J R Soc Interface 8:30–43. https://doi.org/10.1098/rsif.2010.0116

Doviak RJ, Zrnić DS (1993) Dopper radar and weather observations, 2nd edn. Academic, New York

Drake VA, Gatehouse AG (eds) (1995) Insect migration: tracking resources through space and time. Cambridge University Press, Cambridge

Duvergé PL, Jones G, Rydell J, Ransome RD (2000) Functional significance of emergence timing in bats. Ecography 23:32–40

Eastwood E (1967) Radar ornithology. Methuen, London

Edwards J, Houghton EW (1959) Radar echoing area polar diagrams of birds. Nature 184:1059

Finlay JC (1976) Some effects of weather on Purple Martin activity. Auk 93:231–244. https://doi.org/10.2307/4085041

Frick WF, Reynolds DS, Kunz TH (2010) Influence of climate and reproductive timing on demography of little brown myotis Myotis lucifugus. J Anim Ecol 79:128–136. https://doi.org/10.1111/j.1365-2656.2009.01615.x

Frick WF, Stepanian PM, Kelly JF et al (2012) Climate and weather impact timing of emergence of bats. PLoS One 7:e42737. https://doi.org/10.1371/journal.pone.0042737.t003

Furness RW, Greenwood JJD (eds) (1993) Birds as monitors of environmental change. Chapman & Hall, Beckenham

Hodgson JA, Thomas CD, Oliver TH et al (2011) Predicting insect phenology across space and time. Glob Chang Biol 17:1289–1300

Horn JW, Kunz TH (2008) Analyzing NEXRAD doppler radar images to assess nightly dispersal patterns and population trends in Brazilian free-tailed bats (Tadarida brasiliensis). Integr Comp Biol 48:24–39. https://doi.org/10.1093/icb/icn051

Ibáñez I, Primack RB, Miller-Rushing AJ et al (2010) Forecasting phenology under global warming. Philos Trans R Soc Lond Ser B Biol Sci 365:3247–3260

Jones G, Rydell J (1994) Foraging strategy and predation risk as factors influencing emergence time in echolocating bats. Philos Trans R Soc Lond Ser B Biol Sci 346:445–455

Jones G, Jacobs DS, Kunz TH, Willig MR (2009) Carpe noctem: the importance of bats as bioindicators. Endang Spec Res 8:93–115

Jørgensen PS, Böhning-Gaese K, Thorup K, et al (2015) Continent-scale global change attribution in European birds – combining annual and decadal time scales. Glob Chang Biol. https://doi.org/10.1111/gcb.13097

Kelly JF, Shipley JR, Chilson PB et al (2012) Quantifying animal phenology in the aerosphere at a continental scale using NEXRAD weather radars. Ecosphere 3:art16. https://doi.org/10.1890/ES11-00257.1

Kelly JF, Bridge ES, Frick WF, Chilson PB (2013) Ecological energetics of an abundant aerial insectivore, the Purple Martin. PLoS One 8:e76616. https://doi.org/10.1371/journal.pone.0076616

Krauel JJ, Westbrook JK, GF MC (2015) Weather-driven dynamics in a dual-migrant system: moths and bats. J Anim Ecol 84:604–614. https://doi.org/10.1111/1365-2656.12327

Kunz TH, Arnett EB, Erickson WP (2007) Ecological impacts of wind energy development on bats: questions, research needs, and hypotheses. Front Ecol Environ 5:315–324

Kunz TH, Gauthreaux SA, Hristov NI et al (2008) Aeroecology: probing and modeling the aerosphere. Integr Comp Biol 48:1–11. https://doi.org/10.1093/icb/icn037

Langston C, Zhang J, Howard K (2007) Four-dimensional dynamic radar mosaic. J Atmos Ocean Technol 24:776–790

Lee YF, McCracken GF (2002) Foraging activity and food resource use of Brazilian free-tailed bats, Tadarida brasiliensis (Molossidae) on JSTOR. Ecoscience 9:306–313. https://doi.org/10.2307/42901406

Lee YF, McCracken GF (2005) Dietary variation of Brazilian free-tailed bats links to migratory populations of pest insects. J Mammal 86:67–76

Lovett GM, Burns DA, Driscoll CT et al (2007) Who needs environmental monitoring? Front Ecol Environ 5:253–260. https://doi.org/10.1890/1540-9295(2007)5[253:WNEM]2.0.CO;2

Maine JJ, Boyles JG (2015) Bats initiate vital agroecological interactions in corn. Proc Natl Acad Sci USA 112:12438–12443. https://doi.org/10.1073/pnas.1505413112

McCracken GF, Gillam EH, Westbrook JK et al (2008) Brazilian free-tailed bats (Tadarida brasiliensis: Molossidae, Chiroptera) at high altitude: links to migratory insect populations. Integr Comp Biol 48:107–118. https://doi.org/10.1093/icb/icn033

Nebel S, Mills A, McCracken JD, Taylor PD (2010) Declines of aerial insectivores in North America follow a geographic gradient. Avian Conserv Ecol 5:1

O'Shea TJ (1976) Fat content in migratory central arizona brazilian free-tailed bats, Tadarida brasiliensis (Molossidae). Southwest Nat 21:321–326. https://doi.org/10.2307/3669717

Pagels JF (1972) The effects of short and prolonged cold exposure on arousal in the free-tailed bat, Tadarida brasiliensis cynocephala (Le Conte). Comp Biochem Physiol A Physiol 42:559–567. https://doi.org/10.1016/0300-9629(72)90134-X

Palmer RD, Xue M, Cheong BL, Xue M (2008) A time series weather radar simulator based on high-resolution atmospheric models. 25:230–243. https://doi.org/10.1175/2007JTECHA923.1

Reichard JD, Gonzalez LE, Casey CM et al (2009) Evening emergence behavior and seasonal dynamics in large colonies of Brazilian free-tailed bats. J Mammal 90:1478–1486. https://doi.org/10.1644/08-MAMM-A-266R1.1

Reynolds AM, Reynolds DR, Smith AD, Chapman JW (2010) A single wind-mediated mechanism explains high-altitude "non-goal oriented" headings and layering of nocturnally migrating insects. Proc Biol Sci 277:765–772. https://doi.org/10.1098/rspb.2009.1221

Rich TD, Beardmore C, Berlanga H et al (2004) Partners in Flight North American Landbird Conservation Plan. Cornell Lab of Ornithology Ithaca Partners in Flight, New York

Root TL, Hughes L (2005) Present and future phenological changes in wild plants and animals. In: Lovejoy TE, Hannah L (eds) Climate change and biodiversity. See Lovejoy & Hannah, New Haven, pp 61–69

Russell KR, Gauthreaux SA Jr (1999) Spatial and temporal dynamics of a Purple Martin pre-migratory roost on JSTOR. Wilson Bull. https://doi.org/10.2307/4164099

Russell KR, Mizrahi DS, Gauthreaux SA Jr (1998) Large-scale mapping of Purple Martin pre-migratory roosts using WSR-88D weather surveillance radar (Mapas a Larga Escala de los Dormideros Pre-Migratorios de Progne subis Utilizando Radares WSR-88D Para el Monitoreo del Clima) on JSTOR. J Field Ornithol. https://doi.org/10.2307/4514321

Rydell J, Entwistle A, Racey PA (1996) Timing of foraging flights of three species of bats in relation to insect activity and predation risk. Oikos 76:243–252

Smith AC, Hudson M-AR, Downes CM, Francis CM (2015) Change points in the population trends of aerial-insectivorous birds in North America: synchronized in time across species and regions. PLoS One 10:e0130768. https://doi.org/10.1371/journal.pone.0130768

Stepanian PM, Chilson PB, Kelly JF (2014) An introduction to radar image processing in ecology. Methods Ecol Evol 5:730–738. https://doi.org/10.1111/2041-210X.12214

Tarof SA, Brown CR (2013) Purple Martin (*Progne subis*). In: Poole A (ed) The Birds of North America Online. Cornell University, Ithaca

Vasiloff SV, Seo D, Howard KW et al (2007) Improving QPE and very short term QPF. Bull Am Meteorol Soc 88:1899–1911

Westbrook JK (2008) Noctuid migration in Texas within the nocturnal aeroecological boundary layer. Integr Comp Biol 48:99

Wiederholt R, López-Hoffman L, Cline J et al (2013) Moving across the border: modeling migratory bat populations. Ecosphere 4:article 114. https://doi.org/10.1890/ES13-00023.1

Wiederholt R, López-Hoffman L, Svancara C et al (2015) Optimizing conservation strategies for Mexican free-tailed bats: a population viability and ecosystem services approach. Biodivers Conserv 24:63–82. https://doi.org/10.1007/s10531-014-0790-7

Wilkins KT (1989) Tadarida brasiliensis. Mamm Species 331:1–10

Williams SE, Shoo LP, Isaac JL et al (2008) Towards an integrated framework for assessing the vulnerability of species to climate change. PLoS Biol 6:e325. https://doi.org/10.1371/journal.pbio.0060325

Wood CR, Clark SJ, Barlow JF, Chapman JW (2010) Layers of nocturnal insect migrants at high-altitude: the influence of atmospheric conditions on their formation. Agric For Entomol 12:113–121. https://doi.org/10.1111/j.1461-9563.2009.00459.x

Zhang J, Howard K, Langston C et al (2004) Three-and four-dimensional high-resolution national radar mosaic. ERAD Publ Ser 2:105–108

Zhang J, Howard K, Gourley JJ (2005) Constructing three-dimensional multiple-radar reflectivity mosaics: examples of convective storms and stratiform rain echoes. J Atmos Ocean Technol 22:30–42

Zhang J, Howard K, Langston C et al (2011) National mosaic and multi-sensor QPE (NMQ) system: description, results, and future plans. Bull Am Meteorol Soc 92:1321–1338

The Pulse of the Planet: Measuring and Interpreting Phenology of Avian Migration

16

Jeffrey F. Kelly, Kyle G. Horton, Phillip M. Stepanian, Kirsten de Beurs, Sandra Pletschet, Todd Fagin, Eli S. Bridge, and Phillip B. Chilson

Abstract

Changing phenology of bird migration has become a flagship example of the biological impacts of climate change. Bird migration phenology data come from a limited number of time series in idiosyncratic locations. Improved understanding of these relationships requires new data collected in a standardized method that spans spatial and temporal scales. We used weather surveillance radar data and eBird data to show that it is feasible to measure bird migration phenology

J.F. Kelly (✉)
Oklahoma Biological Survey and Department of Biology, University of Oklahoma, Norman, OK, USA
e-mail: jkelly@ou.edu

K.G. Horton
Oklahoma Biological Survey, Department of Biology, and Advanced Radar Research Center, University of Oklahoma, Norman, OK, USA
e-mail: hortonkg@ou.edu

P.M. Stepanian
School of Meteorology and Advanced Radar Research Center, University of Oklahoma, Norman, OK, USA
e-mail: step@ou.edu

K. de Beurs
Department of Geography and Environmental Sustainability, University of Oklahoma, Norman, OK, USA
e-mail: kdebeurs@ou.edu

S. Pletschet · T. Fagin · E.S. Bridge
Oklahoma Biological Survey, University of Oklahoma, Norman, OK, USA
e-mail: spletschet@ou.edu; tfagin@ou.edu; ebirdge@ou.edu

P.B. Chilson
School of Meteorology, Advanced Radar Research Center, and Center for Autonomous Sensing and Sampling, University of Oklahoma, Norman, OK, USA
e-mail: chilson@ou.edu

© Springer International Publishing AG, part of Springer Nature 2017
P.B. Chilson et al. (eds.), *Aeroecology*,
https://doi.org/10.1007/978-3-319-68576-2_16

from local to regional scales and that these data provide geographically informative and temporally consistent patterns of migration phenology. We examine
both a single species (Purple Martin) case and widespread nocturnal songbird
migration. We also analyzed patterns in potential environmental cues that
migrants might use to adjust their phenology *en route*. These analyses suggest
that temperature does form a thermal wave that is a plausible cue for adjusting
migration timing, but that the Normalized Difference Vegetation Index (NDVI)
as a measure of the start-of-spring does not produce a green wave that would be
informative to migrants for gauging seasonal phenology. This result calls into
question the use of vegetation indices to understand how migrants adjust their
timing *en route*.

1   Introduction

For centuries natural historians have used the timing of seasonal events to track
progress of Earth's annual cycle (Sparks and Carey 1995; Ahas 1999). Seasonal
phenology is reflected in both abiotic (snowmelt, first frost, degree days) and biotic
(budburst, flowering, nesting) components of ecosystems, and some of the most
informative phenology metrics have been long-term records of animal migration
systems (Lehikoinen et al. 2004). These systems have unique potential to provide
insights into ecological consequences of global-scale environmental change.
Nearly all animal taxa contain lineages that migrate, and understanding the unifying
characteristics of these diverse migratory life histories would inform us about the
biological impacts of global environmental change (Alerstam et al. 2003; Dingle
2014; Kelly and Horton 2016).

The potential of migrants as sensors of global change arises from each migrant's
need to arrive in a distant location in time to achieve sufficient survival and
reproduction (i.e., fitness) to replace itself in future generations. Natural selection
diminishes the fitness of mistimed migrants: those that arrive too early have reduced
survival owing to harsh climatic conditions, and those that arrive too late have
reduced fitness due to missed opportunities for reproduction. Timing arrival
correctly requires use of exogenous environmental cues that are dynamic over
space and time (e.g., temperature) and cues that vary in a fixed manner with time
of year (day length; Cohen et al. 2012). Improving our understanding of migratory
phenology depends on our ability to quantify both the dynamic spatiotemporal
distribution of migrants *en route* and the environmental cues that shape these
distributions.

Migratory phenology is an enormously diverse topic ranging from the autecol-
ogy of individual migrants to the synecology of migratory communities and the
dynamics of migratory macrosystems. Birds comprise the vast majority of vertebrate airborne migrants and have long been a focus of migration biology.
Migrations of many other aerial taxa, such as bats and insects, are equally important
but less well understood. This chapter builds on our understanding of bird migration

toward new ways of both quantifying the phenology (timing) of avian life-history phenomena (i.e., migration and post-breeding aggregation) and linking these phenologies to key environmental cues that might explain regional and year-to-year variation in the timing of seasonal behaviors.

There is accumulating evidence that avian migrants are changing their *en route* stopover dates, breeding-ground arrival times, and nesting dates, in response to climate warming (Lehikoinen et al. 2010). Our understanding of these changes has been evolving rapidly, with early concern over global trophic mismatches giving way to a more nuanced view of the roles of both phenotypic plasticity and directional selection in shifting migration phenology (Both and Visser 2001; Visser et al. 2015). Even with this progress in our thinking, the potentially disruptive effects mismatches between seasonal timing of productivity and energy needs of consumers remain a concern (Saino et al. 2011). A generalized understanding of how the behavior of migrants changes in response to environmental change is key for forecasting ecological impacts of climate change on seasonal migration systems.

Most of our inference about widespread impacts of climate change on phenology comes from accumulated local-scale case studies (Root et al. 2003; Parmesan 2006). The cumulative evidence from these studies has been influential in convincing broad groups of stakeholders of the ubiquitous ecological impacts of climate change (Solomon et al. 2007). However, most of the accumulated phenology data come from sedentary organisms that are not obligated to move long distances each annual cycle. Migrants, therefore, have the potential to provide a new perspective with regard to the biological impacts of climate change. Consider, for example, the vulnerability of migratory birds to climate change, which might differ systematically from that of more sedentary species, given that they rapidly traverse large spatial extents in anticipation of resources that may or may not be realized at the end of a journey (Robinson et al. 2009; Runge et al. 2014).

There are a substantial number of studies that document recent changes in migrant phenology associated with climate change (Cotton 2003; Tottrup et al. 2006; Van Buskirk et al. 2009), yet large gaps remain in the spatiotemporal coverage of existing phenological data, which limit our ability to understand the impact of environmental change on ecosystems (Ibáñez et al. 2010). As we increasingly alter ecosystems at regional and global scales, we need to increase the rigor with which we track phenology at these scales (Jetz et al. 2012). It would be desirable to have a measure of migration phenology that enables comparisons with existing phenology network data and individual migrant tracking data. Linking phenology across levels of biological organization from individuals to ecosystems through time and space would help us understand the biological impacts of climate change (Kelly and Horton 2016).

Here we compare relationships among primary productivity, air temperature, and bird phenology from a continental-scale perspective. We use the only two methods that we know of that are able to measure bird phenology at a continental spatial extent at a daily frequency: the NEXRAD weather surveillance radar (WSR) network and the eBird consortium of citizen scientists. These two methods present an interesting contrast because WSRs sample birds exclusively in flight, primarily nocturnally, and cannot distinguish species, whereas eBird observations are of birds on the ground, made primarily diurnally, with species identifications.

Several studies have attempted to combine ground-based observation with those made by WSRs to make migratory bird forecasting models, with mixed results (DiGaudio et al. 2008; Peckford and Taylor 2008; Fischer et al. 2012a, b; Horton et al. 2015). Our goal is somewhat different in that we want to demonstrate a continental-scale approach to monitoring seasonal flux in avian distributions that can address fundamental ecological questions. For example, how much biomass of migrant birds flows seasonally into and out of the temperate, boreal, and arctic regions of the northern hemisphere each year? What are the economic impacts of this flow of consumers? How is this flow of biomass changing over time and space and what is driving these changes? In pursuit of this goal, we describe two related studies: one focused on a single species, the Purple Martin (*Progne subis*), and one that applies to all nocturnal migrants in eastern North America. Our study of Purple Martins relies on the unusual ecology of this species, which makes it an excellent candidate for both radar-based observation and citizen science initiatives (Bridge et al. 2016). The second study applies these same resources more broadly to encompass 182 species of nocturnal migrants (Kelly et al. 2016).

Both studies derive avian phenologies from WSR and eBird data and compare these with potential environmental exogenous cues that might influence avian life-history schedules, specifically, air temperature and primary productivity. Although there is an enormous literature on environmental cues associated with the onset of migratory restlessness in the laboratory as well as studies of departure timing in the field (Ramenofsky and Wingfield 2007), a synthetic understanding of how changes in the global environment will impact migration timing is still beyond our horizon. A primary knowledge gap is how average-sized free-living birds respond to environmental variation (Bridge et al. 2010). While day length is clearly the primary fixed environmental cue that entrains the migratory syndrome, we are interested in the impacts of variable environmental cues that migrants use to adjust their timing *en route*. Therefore, we focus our attention on primary production and air temperature because they are arguably the two most frequently cited variable exogenous cues for the timing of seasonal events in birds. Expansion of our approach to examine other potential cues (e.g., insect abundance, soil conditions, precipitation, etc.) should be possible.

2   Phenology of Environmental Cues

2.1   Primary Production, the Green Wave, and Migration

Seasonal variation in photosynthesis is a primary driver of seasonal changes in trophic systems worldwide. People have long recognized the value of plant phenology in forecasting seasonal fluctuations. Several plant phenological time series go back centuries (Ge et al. 2015), and a few go back millennia (Sparks and Carey

1995). These invaluable sources of information tell us that spring phenology of plants has advanced worldwide in recent decades (Menzel et al. 2006; Parmesan 2007; Rosenzweig et al. 2008), with particularly rapid advances in plant phenology at northern latitudes. Consequently, broad-scale plant phenology has become a primary source of evidence that climate warming is having fundamental impacts on biological systems (Solomon et al. 2007). In addition, changing phenology of plants in northern temperate regions is of particular interest in understanding the potential for trophic mismatches for long-distance seasonal migrants (Marra et al. 2005; Both et al. 2006; Faaborg et al. 2010).

Although there has long been intense interest in the response of annual cycles of herbivores to changes in the phenology of primary productivity (Post and Forchhammer 2008), there are relatively few long time series of herbivore phenology. Most studies report phenology at a local scale over decadal time spans (Ahas 1999). In general, herbivore phenology advances as much or more than plant phenology in terms of days per decade (Parmesan 2007). Surprisingly few studies of any duration measure both local plant phenology and the phenology of herbivores that consume those plants (Miller-Rushing et al. 2010; Thompson and Gilbert 2014). A primary limitation on the inference from the few existing long-term phenology datasets on plants and herbivores is their small spatial extent (Ibáñez et al. 2010; Ge et al. 2015). This limitation has motivated establishment of phenology networks over the past few decades, such as the national phenology network (https://www.usanpn.org/) and project budburst (http://budburst.org/). These networks demonstrate the value of widespread and standardized phenology data (Schwarz et al. 2013).

The advent of satellite-based radiometry has enabled indices of vegetation greenness (e.g., Normalized Difference Vegetation Index, NDVI; Tucker 1979) that serve as a proxy for plant phenology across a broad range of spatial scales and at a high temporal frequency. The availability of these data has significantly advanced our understanding of the spatiotemporal dynamics of plant phenology (White et al. 2005) and has fueled interest in the spatial and temporal matching of primary productivity and herbivore life histories (Si et al. 2015). In general, studies of herbivorous migrants, both terrestrial (Post and Stenseth 1999) and aerial (Bauer et al. 2008), suggest advancing phenology in both satellite-derived vegetation indices and seasonal movements of primary consumers.

A fundamental question in aeroecology is what are the exogenous cues that drive the phenology of massive global annual aerial migrations? One widely held answer is that migrants track the phenology of spring green-up, an idea that has been termed the "green-wave hypothesis." This idea is at least 40 years old and can be traced to Drent et al. (1978) who suggested that ". . . .by making the northward shift the geese are riding the crest of the wave as concerns digestibility (early growth being characterized by lower cell wall content hence higher digestibility)." This seemingly innocuous statement made with no elaboration is the source of the "green-wave hypothesis," which predicts a correspondence between the greening of vegetation and the timing of migration in herbivores. This hypothesis received little attention until satellites made it possible to delineate the "green wave" of vegetation on a daily basis (Schwartz 1998). Recent studies testing the green-wave hypothesis have focused primarily on cursorial mammalian herbivores (Wang et al. 2010; Bischof et al. 2012) and migratory birds (LaSorte et al. 2014; Thorup et al. 2017; Socolar et al. 2017).

Two overlapping conceptualizations of the green-wave hypothesis have caused confusion in the literature. The first idea is that the predominant natural vegetation at a particular locality reliably greens-up during the spring as part of a localized temporal wave. There is strong support for the concept of a temporal green wave at a local scale that reflects local seasonal vegetation phenology (Schwartz 1998; Friedl et al. 2002; Emmenegger et al. 2014). The shortcoming of this version of the green wave is that local phenology may not relate to a broader spatial pattern in green-up. That is, the timing of local green-up among localities could be discordant. It is difficult to imagine how a migratory animal could use such disjunct spatial information to time long-distance movements. For this reason, we suggest that the conceptualization of the green-wave hypotheses that requires both a spatial and temporal pattern is more germane to migration biology.

The spatiotemporal green-wave hypothesis also requires a local green-up in the spring (as described above) and for that green-up to proceed along a directional spatial gradient. For north temperate migrants, a useful migration cue would be a wave of green-up that moves from south to north in a predicable fashion in the spring, but spatiotemporal phenological waves could proceed in other directions as well. Surprisingly, there is no evidence of a latitudinal gradient in green-up times, as measured by vegetation indices, over most of North America (White et al. 2009) and many other regions. An exception is the spatiotemporal green wave that is apparent when crossing from temperate to subarctic and arctic latitudes in North America (White et al. 2009). Recent studies have found some support for Drent et al.'s (1978) "green-wave hypothesis" among herbivorous migrants that cross latitudes from temperate to arctic regions (Bauer et al. 2008; Shariatinajafabadi et al. 2014; Si et al. 2015). However, the lack of latitudinal gradient in green-up within temperate latitudes suggests that the green wave may be an implausible cue for migration phenology for migrants heading to breeding locations within new world temperate zones.

There is a mixture of results among studies that examine changes in vegetation indices associated with avian migration phenology (Thorup et al. 2007; Tøttrup et al. 2012; Renfrew et al. 2013; Paxton et al. 2014; Cohen et al. 2015). This mixture could have many sources. It is possible that the green-wave hypothesis does not really apply to secondary consumers. As initially conceived, the hypothesis was about herbivores following the green-up of plants that they consumed. It is not clear that this framework makes any predictions about the timing of insectivore migration relative to the timing of primary productivity. Nonetheless, it is common for studies to include data on insectivores when testing predictions of the green-wave hypothesis. The logic applied is that green-up is accompanied by a concomitant increase in availability of prey populations (arthropods), and therefore, insectivorous birds will also be expected to follow the green wave. There are a couple of difficult assumptions with this expectation. First, that the lags between herbivorous and insectivorous migrant movements and the green wave are equivalent. Another is that the many significant life-history, phylogenetic, and body-size differences that exist between herbivorous and insectivorous birds do not impact migration timing (Bennett and Owens 2002). In summary, there are studies that report both support for the green wave as a migration cue and the absence of support. This is perhaps not surprising because these studies vary in many aspects including the definition of the green wave.

This chapter addresses several basic questions about the green wave (as we define it) as a cue potentially used by avian migrants in the eastern USA to adjust their spring arrival timing. First, is there evidence of a spring spatiotemporal green wave that could be used as (or correlated with) a latitudinal signal of the advancement of spring? If so, does the timing of this wave correlate with the timing of migration? Finally, are there other environmental cues that might also act to gauge the latitudinal advancement of spring, and would they correspond as well or better with the timing of migration?

2.2   Phenology of Air Temperature

Aside from day length, temperature is the most obvious environmental indicator of seasonal change and may serve as an alternative to primary production for assessing and predicting seasonal resource availability. Many analyses of plant phenology time series indicate that timing of spring life-history events is closely tied to spring temperatures (Sparks et al. 2000; Menzel et al. 2003). Various studies have found that degree days, soil temperature, the North Atlantic Oscillation, and other manifestations of temperature are the best predictors of plant phenology in different instances (Post and Stenseth 1999). Despite these variations, there is broad agreement that temperature is a primary driver of seasonal phenology of both primary productivity and consumers on an annual basis (Ahas 1999; Tøttrup et al. 2010). The consistency of phenological responses to temperature across trophic levels leads us to ask whether there might be a thermal wave that is as, or more, plausible as an avian migration cue than primary productivity (Dunn and Winkler 1999; Jonzen et al. 2006; Emmenegger et al. 2014). To compare these potential migration cues, we examine the relationships among spring green-up, spring temperatures, and the seasonal phenology of migrant birds.

3   Phenology of Avian Migrants

The three most common metrics used in avian phenology are timing of (1) passage through stopover locations, (2) arrival at a breeding location (e.g., first-of-season dates), and (3) nesting (or egg-laying). Each of these measures has different attributes and limitations (Lehikoinen et al. 2004). Some of the most impressive time series come from Marsham, UK, where the arrival of swallows (*Hirundo rustica*), cuckoos (*Cuculus canores*), nightingales (*Luscinia megarhynchos*), and nightjars (*Caprimulgus europaeus*) have been recorded for at least 110 years, non-continuously, between 1736 and 1947. There are also many other time series of this sort that span 40–50 years (Lehikoinen et al. 2004). Inferring general patterns of phenology often requires extrapolating from localized sampling, which can be misleading.

Standardized surveys, such as the Breeding Bird Survey in the USA, provide both long time series and large spatial coverage (Sauer et al. 2012). However, this survey and others like it are designed to estimate population abundance and not phenology. More recent development of citizen science networks, such as eBird (Sullivan et al. 2014), has potential for phenology measurements because of their broad spatial coverage and continuous recording through time. Limitations of these data are the relatively short time series available at present and that sampling effort is not standardized, which can create difficulties for interpreting spatial and temporal patterns.

Given the available data sources, we think that migration biologists interested in phenology face future challenges that include (1) increasing the spatial scope and temporal resolution with which we can quantify system-level migration phenology and (2) quantifying the spatiotemporal distributions of the primary environmental cues that determine migratory phenology. If we could accomplish these tasks, it would be possible to develop and test quantitative migration forecasts. To illustrate our evolving approaches to overcoming these challenges, we describe our work on Purple Martin phenology and phenology of nocturnal migration.

4   Integrating Weather Surveillance Radar and eBird to Measuring Migration Phenology

4.1   Purple Martin Phenology from eBird and WSR Data

We are interested in comparing the phenology of Purple Martins' life history at multiple points in their annual cycle. Martins arrive in the USA in the spring prior to nesting. We estimated the timing of this arrival using eBird observations. After arrival, the birds form pairs and nest, which is comprised of nest building, egg laying, incubation, nestling, and fledgling stages. Once nesting is finished, the birds form large communal roosts that are detected by Weather Surveillance Radars (WSR). Therefore, we are able to measure phenology of arrival using eBird data and phenology of roost initiation and roost dissipation using WSR data. Having multiple measures of different stages of the annual cycle enables comparisons of within- and between-year drivers of phenological variation.

4.1.1   Methods

First of Season (FOS) Index Based on eBird Data
To compare radar-derived phenologies with those based on direct observations, we compiled Purple Martin occurrence information from eBird checklists from 2011

for the contiguous USA (Sullivan et al. 2014). These data are observations from citizen scientists and are compiled by the Cornell Laboratory of Ornithology and the National Audubon Society. Each checklist contains information describing the geographic location sampled, time and date, species observed, protocol, and sampling effort. We used observations from complete checklists using stationary and traveling protocols at locations on land between 26° and 48°N latitude. We only used traveling counts with distances <8.1 km. We only used checklist where observations were made for <180 min. We only used but one set of observations at the same location on the same calendar date. From a total of approximately 3.17 million observations, a total of 56, 294 unique Purple Martin observations were recorded.

To examine latitudinal variation in spring migration arrival dates for Purple Martins, we categorized eBird observations into 1° intervals of latitude and calculated the mean occurrence for each Julian date (0–364). We included data from 9 January to 9 July in our analysis of spring patterns (La Sorte et al. 2013). To determine spring migration start date, we fit a generalized additive model (GAM) to occurrence observations for each 1° interval. Spring migration start dates were defined as the first Julian date when the fitted occurrence estimate exceeded the minimum upper 99% confidence interval. GAMs were computed using the mgcv package in R.

WSR Radar Data and Martin Roosts

Weather radar networks detect flying animals and thus provide a large-scale, remote-sensing system for monitoring animal migration phenologies (Gauthreaux and Belser 2003). The NEXRAD network of WSRs spans the continental USA, which gives them significant potential for sampling aerofauna (Gauthreaux 2006). Realization of this potential for monitoring has been limited by difficulties associated with accessing and processing large data volumes from the WSR network. An approach that overcomes some of the big data problem is to focus on specific aggregations of a particular species that have known locations or characteristic radar signatures (Russell et al. 1998; Frick et al. 2012; Kelly et al. 2012; Bridge et al. 2016, 2017). This approach reduces the amount of data needed and thereby increases the feasibility of monitoring.

In the late summer, Purple Martins roost in large communal aggregations at night and often fly at relatively high altitudes while departing from these aggregations in the hour prior to local sunrise. This characteristic flight behavior has enabled routine detection of Purple Martin roost sites with WSRs throughout the eastern USA, which has motivated several previous studies (Russell and Gauthreaux 1998, 1999; Russell et al. 1998). Russell et al. (1998) used WSR data to identify 33 martin roosts throughout 13 states in the southeastern USA. Since that time, it has become easier to use mosaics of WSR data to locate and monitor martin roosts (Zhang et al. 2011; Kelly et al. 2012; Bridge et al. 2016).

To illustrate the potential of radar-based phenologies as widespread seasonal indicators, we analyze initiation dates of Purple Martin roosts from the summer of 2011 (Fig. 16.1). Methods for finding and validating these roost locations and start

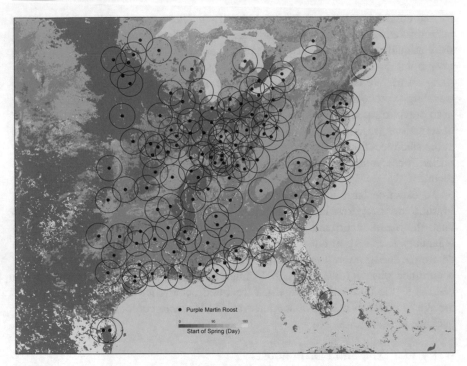

Fig. 16.1 Locations of Purple Martin roosts in 2011 (red circles) overlaid on start-of-spring dates (green scale) based on NDVI values for spring 2011. For each roost, the start-of-spring date used in analyses was the average of those pixels within 100 km of the roost (blue circle) that were not agricultural land (gray). Note the late start-of-spring dates throughout Florida and around the Gulf of Mexico

dates were the same as those reported previously (Kelly et al. 2012; Bridge et al. 2016; Kelly and Pletschet 2017). Briefly, we visually examined the mosaicked non-quality controlled composite reflectivity data from the eastern half of the USA (east of the 100th meridian) for each morning from June 1 through September of 2011 to locate Purple Martin roosts. We are relatively certain that no roosts occurred west of our search zone, but we did not inspect the entire continental mosaic on a daily basis and it is possible that a few roosts went undetected. Other species of birds also form roosts and some of those are detected by radar. Based on our ground validation studies, it is likely that 5 of the roosts included in our analysis from the northeastern USA were formed by Tree Swallows (*Tachycineta bicolor*) and/or Bank Swallows (*Riparia riparia*) rather than by Purple Martins (Kelly and Pletschet 2017).

We searched for radar-detected roost emergence signatures manually by systematically examining radar mosaic images. We recorded the locations of all roosts that showed daily emergences over a contiguous 7-day period. We determined the first date on which a signal of martin emergence could be discerned by a human observer at each roost and used this date as the initiation (start) date in our analyses.

Phenology of Migration Cues

To examine seasonal phenology of temperatures relative to that of vegetation indices and bird phenology, we extracted 2011 temperature data for roost locations from PRISM (http://www.prism.oregonstate.edu/). For each roost, we identified the latest spring date on which the average minimum nighttime temperature within 100 km of each roost site dipped below 5 °C. This particular threshold is arbitrary, but examination of other temperatures suggests that these isotherms are largely parallel. We compare the slopes of these patterns across spring dates to those of roost start dates and start-of-spring(SOS) based on NDVI (R. Core Team 2013).

To compare the radar-based and eBird-based phenologies of Purple Martin migration to potential environmental cues, we determined a SOS date based on NDVI data for the area around each roost following the midpoint pixel method of White et al. (2009). We used the 16-day composite MODIS product from within 100 km radius of each roost site (MCD43C4; http://glovis.usgs.gov/). To avoid phenology biases introduced by cropland, we excluded all areas (0.05° pixels) that were identified as crops in the MOD12C1 data product. For all other pixels, we determined a SOS date. These SOS dates were then averaged for each roost (Fig. 16.1). SOS date determined with this method has been referred to as the green-wave index (Shariatinajafabadi et al. 2014). To examine whether difference from the seasonal trend could be an alternative migration cue, we also examined the anomaly from the 30-year mean SOS date for each roost.

4.1.2 Results

Air temperature, Purple Martin arrival dates (first of season), and roost initiation dates were all strongly related to latitude with similar slopes (Fig. 16.2). In contrast, SOS had no relationship with latitude (Fig. 16.2), which is consistent with the ensemble pattern across these latitudes described by White et al. (2009). These patterns suggest that temperature, or the thermal wave, is a plausible cue for Purple Martins to use to gauge phenology across latitudes, but that green-up, or the green wave, is not. In particular, the SOS at southern latitudes does not forecast the timing of SOS further north. Start of spring anomaly was also not related to avian phenology or latitude. The possibility that a thermal wave might be used as a cue for timing of migration in an aerial insectivore is not particularly surprising. Dunn et al. (2011) showed strong correlations between temperature and the availability of airborne insect prey. There are also strong relationships between insect development and temperature (Emmenegger et al. 2014).

There were about 120 days between the arrival of martins and the time when post-breeding roosts were initiated at particular latitudes. For example at 30°N, martins arrived on about day 55 (25 February) and roosts were initiated on about day 171 (21 June). In contrast, at 45°N arrival was on day 95 (6 April) and roost initiation was on day 219 (8 August). The consistency of the time lag between arrival dates and roost initiation dates suggests the timing of these life-history stages is tightly constrained by seasonal effects that are strongly associated with

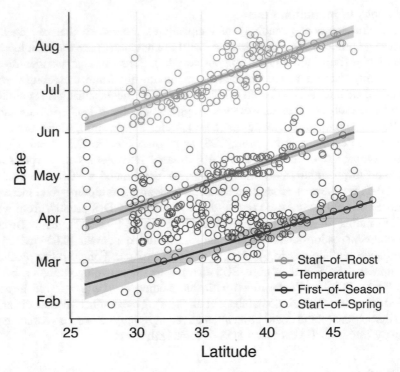

Fig. 16.2 Relationships between spring phenology and latitude for first-of-season observations of Purple Martins ($R^2 = 0.67$, $p < 0.001$), last date with a 5 °C minimum nighttime temperature ($R^2 = 0.60$, $p < 0.001$), and date of roost initiation by Purple Martins ($R^2 = 0.64$, $p < 0.001$). Slopes of the start-of-roost, first-of-season, and temperature lines do not differ from each other. Start-of-spring dates at Purple Martin roost locations are not related to latitude of the roost ($R^2 = 0.02$, $p = 0.095$)

latitude. Overall, the correspondence between eBird and WSR data suggests that both provide accurate phenological data for distinct life-history stages and that analysis of variation in these patterns across time and space might yield important clues about how migrants are responding to climate change.

4.2 Phenology of Nocturnal Bird Migration

Using traditional methods, nocturnal songbird migration is extremely difficult to perceive, much less quantify, over a large spatial extent. This difficulty is evinced by a massive research effort in the fall of 1951, wherein researchers recruited nearly 1400 observers at 235 locations to observe the moon through spotting scopes and count the silhouettes of what amounted to over 35,000 nocturnal avian migrants as they flew south over North America (Lowery and Newman 1966). The US national network of WSR offers a much more comprehensive and accessible index of bird migration. Almost from its inception, radar technology has been applied to the nocturnal-bird-migration problem (Nisbet 1963; Gauthreaux 1970; Williams et al. 1981; Liechti et al. 1995; Diehl et al. 2003; Schmaljohann et al. 2008; Dokter et al. 2010; Buler and Moore 2011; Shamoun-Baranes et al. 2014). Despite this history, WSR data are a relatively underused resource for understanding large-scale patterns in avian migration and conservation (Gauthreaux and Belser 2003). Because of the difficulty of obtaining and processing WSR data, most radar migration studies use data from only a few radars over a relatively narrow time window, with some recent exceptions (Kelly et al. 2012; Buler and Dawson 2014; Horton et al. 2016a, b; Farnsworth et al. 2016; Williams and Laird 2017). A primary constraint on broader-scale studies is the data processing required to convert radar data into biologically relevant measurements. Consequently, the potential for near-real-time migration monitoring at the continental scale has not materialized (but see Dokter et al. 2010).

Here we summarize a recent study in which we analyzed data from 31 radars in eastern North America across the spring migration season of 2013 (Kelly et al., 2016). Our objective was to test whether a large-scale, automated approach to radar data analysis has the potential to provide a useful index of migration phenology. In particular, we processed and filtered Level II (non-quality controlled) radar data into a measurement of nightly migration intensity in the region surrounding each radar. We compare this measure of migration intensity to species richness based on eBird observations in the area sampled by each radar.

4.2.1   Methods

Our methods are described in detail in Kelly et al. (2016), but we review them here briefly. Our analyses are based on level-II radar reflectivity data for each of 31 radars (Fig. 16.3). We used data from 0300UTC to 0800UTC (10 p.m. to 3 a.m. Eastern Standard Time). Our study period was January through June of 2013. Data were from the lowest elevation angle (0.5°) of each radar scan (360 or 720 azimuths with 250 m range gates—see Chap. 12). Data were available at 5–10 min intervals and we collected a total of 234,007 radar sweeps. We processed these data to filter out reflectivity from weather and keep reflectivity from migrating birds. To gauge the effectiveness of our filtering, we visually inspected thousands of images of the censored data; the vast majority of the analyzed reflectivity corresponded to easily recognizable patterns attributable only to bird migration. Our data filtering was based on the hydrometeor classification algorithm of Park et al. (2009). We also employed a static ground clutter mask. It is likely that we retained some data from other biological (not migrant birds) and nonbiological sources (ground clutter) of radar reflectivity.

We calculated the mean reflectivity each night for each radar. We interpret this mean as a representation of the intensity of bird migration. Because we were interested in characterizing the temporal distribution in seasonal migration, we fit a GAM to the data at each radar to smooth out among-day variation. We used these

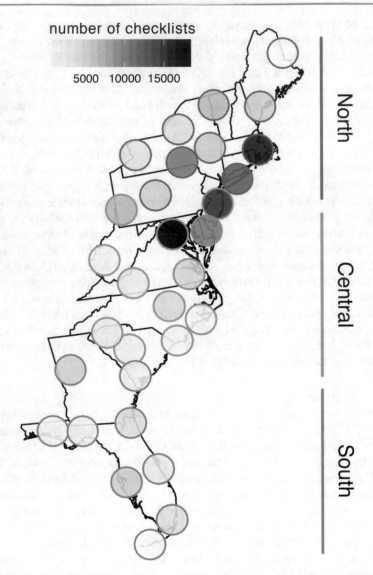

Fig. 16.3 Locations of 31 weather surveillance radars (WSRs) and density of eBird checklists (shading) within 100 km of those radars in the spring of 2013. WSR and eBird data were divided into north, central, and south regions. Redrawn from Kelly et al. (2016)

smoothed reflectivity values for each day and the date associated with the seasonal maximum in smoothed reflectivity at each radar. Smoothed reflectivity values were highly variable among radars. To facilitate comparison among radars, we normalized these data across radars using a *z*-score standardization. To examine

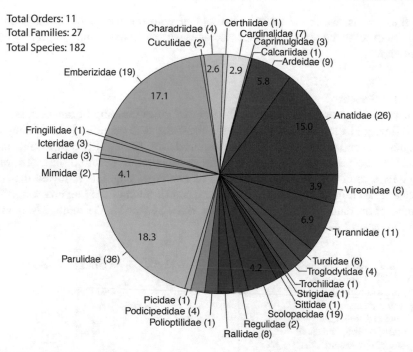

Fig. 16.4 Taxonomic distribution of eBird data used to estimate the phenology of nocturnal bird migration. The majority of 182 bird species encountered were warblers (Parulidae, 36), sparrows (Emberizidae, 19), and ducks (Anatidae, 26). Numbers inside the pie slices represent the percent of all species encountered that were in these families; e.g., 6.9% of all species encounters were Tyrannids

regional patterns in migration timing, we fit GAMs to these normalized data after grouping radars into south, central, and north regions (Fig. 16.3). Geographic regions were arbitrarily defined to contain about a third of the radars each.

We also obtained eBird checklists from within 80 km of the 31 radars for 2013. We extracted occurrence from these lists for 182 species known to be nocturnal migrants (Fig. 16.4). We used the same protocols to select checklists that we described previously for the Purple Martin analysis (Sect. 4.1.1). From these checklists, we determined the number of nocturnal migrant species encountered each day. Daily richness values were highly variable within and among radars. We normalized these data to a daily richness index using the same z-score standardization described above. We fit a GAM to data from each radar to determine the date associated with maximum richness. Then we fit a GAM for each geographic regions (i.e., south central, and north, Fig. 16.3).

Last, we examined the latitudinal trends in the date of maximum migration intensity as measured with radar, maximum species richness from eBird, temperature, and SOS. For temperature and SOS, we replicated the methods described in Sect. 4.1.1 for the area within 80 km of each radar for the spring of 2013. For each of these areas, we found the date associated with the maximum in the seasonal pattern of reflectivity and the date associated with the maximum in migrant species richness.

4.2.2 Results

Nocturnal migration produced a unimodal pattern in intensity of radar reflectivity. The timing of these seasonal maxima in the radar data occurred in the anticipated sequence; southern radar maximum occurred first (26 April), central maximum occurred second (12 May), and northern maximum occurred last (18 May; Fig. 16.5, left). The maximum richness of nocturnal migrants followed the same sequence as migration intensity (Fig. 16.5, right). Each of the maxima was within 3 days of the date estimated from the radar data: 24 April in the south, 7 May in the central, and 17 May in the north region.

Fig. 16.5 General Additive Model fits to Radar-based measure of nocturnal migration intensity (Normalized Reflectivity) in the north, central, and south regions of the eastern USA (top). Number of species on eBird checklists (Normalized Species Richness) in those same regions (bottom); both variables are plotted against date. Note the similarity in the maximum dates, labeled with day and month for each curve by region for both figures. Gray shading is the standard error around each fit

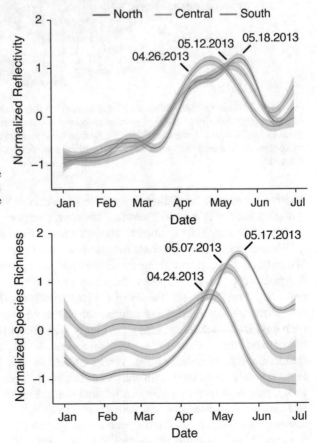

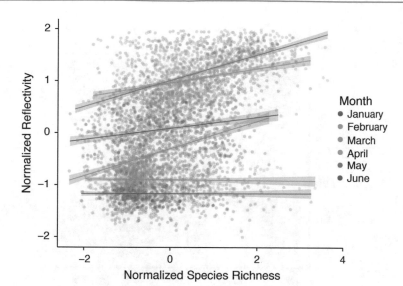

Fig. 16.6 Radar-based migratory intensity (Normalized Reflectivity) plotted against species richness for January through June of 2013. Lines are fit to daily samples taken at each of 31 radars in the eastern USA. Note the slopes of the correlations increase in each month from February through May. The relationships are significant for March through June (January slope = 0.00, R^2 = 0.00, p = 0.92; February slope = 0.00, R^2 = 0.00, p = 0.64; March slope = 0.26, R^2 = 0.14, $p < 0.01$; April slope = 0.14, R^2 = 0.07, $p < 0.01$; May slope = 0.25, R^2 = 0.30, $p < 0.01$; June slope = 0.11, R^2 = 0.03, $p < 0.01$). Lines are least-squares regressions and shaded regions are standard errors of the regression lines. Redrawn from Kelly et al. (2016)

Radar and eBird measures of migration intensity spanning the entire study area showed that they were strongly correlated during high-intensity migration months and less well correlated than outside of the peak migration period (Fig. 16.6). Radar and eBird maxima were highly correlated in May ($r = 0.55, p < 0.001$, slope = 0.25) and significantly correlated from March through June (all $r > 0.18$ and $p < 0.001$). Increase in migration intensity over the spring season is evident in both radar reflectivity and eBird richness indices as both the slopes and intercepts of the fit lines shift across the season (Fig. 16.6). The latitudinal patterns in timing of maximum migration intensity, peak species richness, and temperature across the spring all have similar positive slopes (Fig. 16.7). As was true with the Purple Martin phenology, the SOS dates from NDVI data for migrant birds also do not match the expected positive latitudinal patterns that would indicate a green wave. Rather the SOS values calculated within 80 km of WSRs in 2013 were significantly negatively related to latitude.

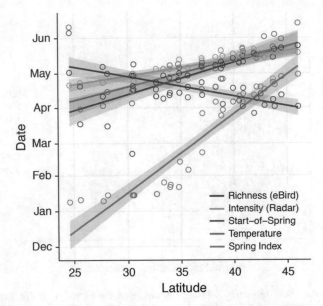

Fig. 16.7 Radar-based intensity of migration (slope $= 1.46$, $p < 0.001$), species richness of migrants (eBird-based, slope $= 2.73$, $p < 0.001$), start-of-spring (NDVI-based, slope -1.70, $p < 0.001$), and temperature (last day to dip below 5 °C, slope 2.37, $p < 0.001$) all plotted against latitude of the sampling location for 2013. Note that the phenology of migration intensity, migrant richness, and temperature are significantly positively related to latitude in the spring as expected whereas start-of-spring does not follow this pattern. Gray shading represents the standard error for each linear fit. Redrawn from Kelly et al. (2016)

5   Summary and Conclusions

Accumulated worldwide phenological data are a primary source of evidence that climate change is altering ecosystems and threatening biodiversity (Bellard et al. 2012). Most of these data come from local studies of sedentary species (Walther et al. 2002; Root et al. 2005; Rosenzweig et al. 2008). Migrants present a unique opportunity to link behaviors of individual animals to phenological changes associated with climate across large spatial extents (Faaborg et al. 2010). Using migration phenology to understand our impact on the Earth will require that we link these phenologies to environmental drivers, which could act through complex population, community, and ecosystem level pathways. Uncovering the scaling relationships and constraints on these systems will be key to understanding their complexity. Migration phenology indices derived from WSR and eBird data are particularly well suited to meet this particular research challenge.

We advocate a research program that would integrate the existing 25-year continental scale WSR data in North America with a decade of eBird data to produce a migration phenology index. If successful, this research program would create a standardized measure of continental scale migration phenology, which would advance our understanding of current and past spatial and temporal dynamics of climate and land use change and form a basis for a quantitative migration forecasting system with value for a wide diversity of land and resource managers (Fischer et al. 2012b).

We presented two possible research approaches to this problem. The more mechanistic example uses data for a single species, the Purple Martin, with an unusual life history. In this case, we demonstrate that the timing of arrival and roost formation follow parallel latitudinal phenologies and that these patterns closely match that of spring temperature thresholds. These results are consistent with the findings of Dunn and Winkler (1999) and Dunn et al. (2011) who show that spring phenology of Tree Swallows and their insect prey is closely associated with temperature in the eastern USA. We did not find evidence that these phenological patterns were associated with vegetation green-up.

In our second example drawn from Kelly et al. (2016), we show that daily data from WSR and eBird cross-validate each other with respect to monitoring of nocturnal songbird migrations. Both methods show expected seasonal trends within locations and across >20° of latitude. We emphasize that the approach we implement here could be substantially improved by exploring different ways that data could be aggregated from both the WSR and eBird networks; the approach we present is proof-of-concept. We think it demonstrates that, when using a conservative approach to data reduction with few assumptions, the signal from nocturnal migration is large enough to be easily detected. Development of better methods could increase the amount of biological information extracted from these data.

Our results regarding correspondence between nocturnal migration phenology and phenology of exogenous environmental cues were consistent with those from the Purple Martin analysis. There were very similar positive trends in maximum migration intensity, maximum species richness, and temperatures across latitudes. There was no positive trend in start-of-spring with latitude and, in fact, the trend had a significantly negative slope in 2013 (Fig. 16.7). This finding from both analyses with respect to SOS was consistent with previously reported patterns for vegetative reflectance across this range of latitudes in the eastern USA (White et al. 2009). Based on these results, we find no support for the idea that there is a green wave across the eastern USA to which Purple Martins or nocturnal migrants could use to adjust their migration timing. A plausible alternative is temperature, which does create a predictable thermal wave in both studies that could be a useful exogenous cue in adjusting the timing migration en route for all trophic levels (Cohen et al. 2012).

The promising results from this approach lead to a couple of options for extending the research. For example, examinations of within- and between-year variation in the phenology of Purple Martins are possible pending the compilation of more data. There are also numerous other species-specific animal aggregations that are relatively easily observed in WSR data for which a similar approach might be employed. Among them are roosts of free-tailed bats (Chilson et al. 2011; Frick et al. 2012), Tree Swallow roosts (Laughlin et al. 2013, 2014), mayfly emergences

(Gauthreaux et al. 2008), and waterfowl flocks (Randall et al. 2011; Buler et al. 2012). Obviously, eBird data would not be available for bats or insects, but in the cases of other birds, it would be valuable to pursue a similar analysis approach.

Our analysis of nocturnal migration also demonstrated the potential to examine the phenology of the mass flow of avian spring migration using both WSR and eBird data (Kelly and Horton 2016). The correspondence between dates with maximum migration across latitudes based on WSR and eBird data encourages us to think about how to extend these analyses to narrower and broader spatial extents (Kelly and Horton 2016). We are interested in examining whether there is variation in the species composition of particular eBird checklists that is more or less associated with variation in radar reflectivity. The potential to leverage the species identity data with the uniform sampling capabilities of the WSR is one of the more promising aspects of this particular analysis method. In the cases we present, the correspondence of the two methods suggests that the information being gathered by both sampling networks is consistent with regard to the phenology of migrant birds, which makes it possible to leverage the strengths of each system to compensate for weaknesses in the other. In particular, the strength of the WSR system is standardized effort, which is the weakness of the eBird system. In contrast, the lack of species-level data from WSRs can be compensated for with eBird data. We suggest this is a promising approach to addressing the future challenges of understanding the scaling of migration phenology from fine spatiotemporal grain to large spatiotemporal extents.

References

Ahas R (1999) Long-term phyto-, ornitho-and ichthyophenological time-series analyses in Estonia. Int J Biometeorol 42:119–123

Alerstam T, Hedenstrom A, Akesson S (2003) Long distance migration: evolution and determinants. Oikos 103:247–260

Bauer S, Gienapp P, Madsen J (2008) The relevance of environmental conditions for departure decision changes en route in migrating geese. Ecology 89:1953–1960

Bellard C, Bertelsmeier C, Leadley P, Thuiller W, Courchamp F (2012) Impacts of climate change on the future of biodiversity. Ecol Lett 15:365–377

Bennett PM, Owens IP (2002) Evolutionary ecology of birds: life histories, mating systems, and extinction. Oxford University Press, Oxford

Bischof R, Loe LE, Meisingset EL, Zimmermann B, Van Moorter B, Mysterud A (2012) A migratory northern ungulate in the pursuit of spring: jumping or surfing the green wave? Am Nat 180:407–424

Both C, Visser ME (2001) Adjustment to climate change is constrained by arrival date in a long-distance migrant bird. Nature 411:296–298

Both C, Bouwhuis S, Lessells C, Visser ME (2006) Climate change and population declines in a long-distance migratory bird. Nature 441:81–83

Bridge E, Kelly J, Bjornen P, Curry C, Crawford P, Paritte J (2010) Effects of nutritional condition on spring migration: do migrants use resource availability to keep pace with a changing world? J Exp Biol 213:2424–2429

Bridge ES, Pletschet SM, Fagin T, Chilson PB, Horton KG, Broadfoot KR, Kelly JF (2016) Persistence and habitat associations of Purple Martin roosts quantified via weather surveillance radar. Landsc Ecol 31:43–53

Bridge ES, Pletschet SM, Fagin T, Chilson PB, Horton KG, Broadfoot KR, Kelly JF (2017) Continental distribution and persistence of purple martin roosts quantified via weather surveillance radar (in review)

Buler J, Dawson D (2014) Radar analysis of fall bird migration stopover sites in the northeastern U.S. Condor 116:357–370

Buler JJ, Moore FR (2011) Migrant – habitat relations during stopover along an ecological barrier: extrinsic constraints and conservation implications. J Ornithol 152(S1):101–112

Buler JJ, Randall L, Fleskes JP, Barrow W Jr, Bogart T, Kluver D (2012) Mapping wintering waterfowl distributions using weather surveillance radar. PLoS One 7(7):e41571

Chilson P, Frick W, Kelly J, Howard K, Larkin R, Diehl R, Westbrook J, Kelly T, Kunz T (2011) Partly cloudy with a chance of migration: weather, radars, and aeroecology. Bull Am Meteorol Soc. https://doi.org/10.1175/BAMS_D11-00099.1

Cohen EB, Moore FR, Fischer RA (2012) Experimental evidence for the interplay of exogenous and endogenous factors on the movement ecology of a migrating songbird. PLoS One 7: e41818

Cohen EB, Németh Z, Zenzal TJ Jr, Paxton KL, Diehl R, Paxton EH, Moore FR (2015) Spring resource phenology and timing of songbird migration across the Gulf of Mexico. In: Wood EM, Kellermann JL (eds) Phenological synchrony and bird migration: changing climate and seasonal resources in North America, vol 47. CRC Press, Boca Raton, p 63

Cotton PA (2003) Avian migration phenology and global climate change. Proc Natl Acad Sci 100:12219–12222

Diehl RH, Larkin RP, Black JE (2003) Radar observations of bird migration over the great lakes. Auk 120:278–290

DiGaudio R, Humple D, Geupel G (2008) Patterns of avian migration in California: an analysis and comparison of results from NEXRAD doppler weather radar and multiple mist-net stations. DTIC Document

Dingle H (2014) Migration: the biology of life on the move. Oxford University Press, Oxford

Dokter AM, Liechti F, Stark H, Delobbe L, Tabary P, Holleman I (2010) Bird migration flight altitudes studied by a network of operational weather radars J R Soc Interface. https://doi.org/10.1098/rsif.2010.0116

Drent R, Ebbinge B, Weijand B (1978) Balancing the energy budgets of arctic-breeding geese throughout the annual cycle: a progress report. Verh Ornithol Ges Bayern 23:239–264

Dunn PO, Winkler DW (1999) Climate change has affected the breeding date of tree swallows throughout North America. Proc R Soc Lond Ser B Biol Sci 266:2487–2490

Dunn PO, Winkler DW, Whittingham LA, Hannon SJ, Robertson RJ (2011) A test of the mismatch hypothesis: How is timing of reproduction related to food abundance in an aerial insectivore? Ecology 92:450–461

Emmenegger T, Hahn S, Bauer S (2014) Individual migration timing of common nightingales is tuned with vegetation and prey phenology at breeding sites. BMC Ecol 14:9

Farnsworth A, Van Doren BM, Hochachka WM, Sheldon D, Winner K, Irvine J, Geevarghese J, Kelling S (2016) A characterization of autumn nocturnal migration detected by weather surveillance radars in the northeastern USA. Ecol Appl 26(3):752–770

Faaborg J, Holmes RT, Anders AD, Bildstein KL, Dugger KM, Gauthreaux SA Jr, Heglund P, Hobson KA, Jahn AE, Johnson DH (2010) Conserving migratory land birds in the New World: do we know enough? Ecol Appl 20:398–418

Fischer RA, Gauthreaux SA, Valente JJ, Guilfoyle MP, Kaller MD (2012a) Comparing transect survey and WSR-88D radar methods for monitoring daily changes in stopover migrant communities. J Field Ornithol 83:61–72

Fischer RA, Guilfoyle MP, Valente J, Gauthreaux SA Jr, Belser CG, Blaricom JW, Donald V, Cohen E, Moore FR (2012b) The identification of military installations as important migratory bird stopover sites and the development of bird migration forecast models: a radar ornithology approach. DTIC Document

Frick WF, Stepanian PM, Kelly JF, Howard KW, Kuster CM, Kunz TH, Chilson PB (2012) Climate and weather impact timing of emergence in bats. PLoS One 7:e42737

Friedl MA, McIver DK, Hodges JCF, Zhang XY, Muchoney D, Strahler AH, Woodcock CE, Gopal S, Schneider A, Cooper A, Baccini A, Gao F, Schaaf C (2002) Global land cover mapping from MODIS: algorithms and early results. Remote Sens Environ 83:287–302

Gauthreaux SA (1970) Weather radar quantification of bird migration. Bioscience 20:17–19

Gauthreaux SA Jr (2006) Bird migration: methodologies and major research trajectories (1945–1995). Condor 98:442–453

Gauthreaux SA Jr, Belser CG (2003) Radar ornithology and biological conservation. Auk 120:266–277

Gauthreaux SA, Livingston JW, Belser CG (2008) Detection and discrimination of fauna in the aerosphere using Doppler weather surveillance radar. Integr Comp Biol 48:12–23

Ge Q, Wang H, Dai J (2015) Phenological response to climate change in China: a meta-analysis. Glob Chang Biol 21:265–274

Horton KG, Shriver WG, Buler JJ (2015) A comparison of traffic estimates of nocturnal flying animals using radar, thermal imaging, and acoustic recording. Ecol Appl 25(2):390–401

Horton KG, Van Doren BM, Stepanian PM, Hochachka WM, Farnsworth A, Kelly JF (2016a) Nocturnally migrating songbirds drift when they can and compensate when they must. Sci Rep 6:21249

Horton KG, Van Doren BM, Stepanian PM, Farnsworth A, Kelly JF (2016b) Seasonal differences in landbird migration strategies. Auk 133(4):761–769

Ibáñez I, Primack RB, Miller-Rushing AJ, Ellwood E, Higuchi H, Lee SD, Kobori H, Silander JA (2010) Forecasting phenology under global warming. Philos Trans R Soc B Sci 365:3247–3260

Jetz W, McPerson JM, Guralnick RP (2012) Integrating biodiversity distribution knowledge: toward a global map of life. Trends Ecol Evol 27:151–159

Jonzen N, Linden A, Ergon T, Knudsen E, Vik JO, Rubolini D, Piacentini D, Brinch C, Spina F, Karlsson L, Stervander M, Andersson A, Waldenstrom J, Lehikoinen A, Edvardsen E, Solvang R, Stenseth NC (2006) Rapid advance of spring arrival dates in long-distance migratory birds. Science 312:1959–1961

Kelly JF, Horton KG (2016) Toward a predictive macrosystems framework for migration ecology. Glob Ecol Biogeogr. https://doi.org/10.1111/geb.12473

Kelly JF, Horton KG, Stepanian PM, Beurs KM, Fagin T, Bridge ES, Chilson PB (2016) Novel measures of continental-scale avian migration phenology related to proximate environmental cues. Ecosphere 7(8):e01434

Kelly JF, Pletschet SM (2017) Accuracy of swallow roost locations assigned using weather surveillance radar. Remote Sens Ecol Conserv

Kelly J, Shipley J, Chilson P, Howard K, Frick W, Kunz T (2012) Quantifying animal phenology in the aerosphere at a continental scale using NEXRAD weather radars. Ecosphere 3(2):16

La Sorte FA, Fink D, Hochachka WM, DeLong JP, Kelling S (2013) Population-level scaling of avian migration speed with body size and migration distance for powered fliers. Ecology 94:1839–1847

La Sorte FA, Fink D, Hochachka WM, DeLong JP, Kelling S (2014) Spring phenology of ecological productivity contributes to the use of looped migration strategies by birds. Proc R Soc B 281(1793):20140984. The Royal Society

Laughlin AJ, Taylor CM, Bradley DW, Leclair D, Clark RC, Dawson RD, Dunn PO, Horn A, Leonard M, Sheldon DR (2013) Integrating information from geolocators, weather radar, and citizen science to uncover a key stopover area of an aerial insectivore. Auk 130:230–239

Laughlin AJ, Sheldon DR, Winkler DW, Taylor CM (2014) Behavioral drivers of communal roosting in a songbird: a combined theoretical and empirical approach. Behav Ecol 25 (4):734–743

Lehikoinen E, Sparks TH (2010) Changes in migration. In: Møller A, Fiedler W, Berthold P (eds) Effects of climate change on birds. Oxford University Press, Oxford, pp 89–112

Lehikoinen E, Sparks TH, Zalakevicius M (2004) Arrival and departure dates. Adv Ecol Res 35:1–31

Liechti F, Bruderer B, Paproth H (1995) Quantification of nocturnal bird migration by moonwatching: comparison with radar and infrared observations (cuantificación de la migración nocturna de aves observando la luna: comparación con observaciones de radar e intrarrojas). J Field Ornithol 66:457–468

Lowery GH, Newman RJ (1966) A continentwide view of bird migration on four nights in october. Auk 83:547–586

Marra PP, Francis CM, Mulvihill RS, Moore FR (2005) The influence of climate on the timing and rate of spring bird migration. Oecologia 142:307–315

Menzel A, Jakobi G, Ahas R, Scheifinger H, Estrella N (2003) Variations of the climatological growing season (1951–2000) in Germany compared with other countries. Int J Climatol 23:793–812

Menzel A, Sparks TH, Estrella N, Koch E, Aasa A, Ahas R, Alm-Kübler K, Bissolli P, Braslavská O, Briede A (2006) European phenological response to climate change matches the warming pattern. Glob Chang Biol 12:1969–1976

Miller-Rushing AJ, Høye TT, Inouye DW, Post E (2010) The effects of phenological mismatches on demography. Philos Trans R Soc B 365:3177–3186

Nisbet I (1963) Measurements with radar of the height of nocturnal migration over Cape Cod, Massachusetts. Bird-Banding 34:57–67

Park HS, Ryzhkov A, Zrnic D, Kim K-E (2009) The hydrometeor classification algorithm for the polarimetric WSR-88D: description and application to an MCS. Weather Forecast 24:730–748

Parmesan C (2006) Ecological and evolutionary responses to recent climate change. Annu Rev Ecol Evol Syst 37:637–669

Parmesan C (2007) Influences of species, latitudes and methodologies on estimates of phenological response to global warming. Glob Chang Biol 13:1860–1872

Paxton KL, Cohen EB, Paxton EH, Németh Z, Moore FR (2014) El nino-southern oscillation is linked to decreased energetic condition in long-distance migrants. PLoS One 9:e95383

Peckford ML, Taylor PD (2008) Within night correlations between radar and ground counts of migrating songbirds. J Field Ornithol 79:207–214

Post E, Forchhammer MC (2008) Climate change reduces reproductive success of an Arctic herbivore through trophic mismatch. Philos Trans R Soc Lond B: Biol Sci 363:2367–2373

Post E, Stenseth NC (1999) Climatic variability, plant phenology, and northern ungulates. Ecology 80:1322–1339

R. Core Team (2013) R: a language and environment for statistical computing. R Foundation for Statistical Computing. Vienna. http://www.R-project.org/

Ramenofsky M, Wingfield JC (2007) Regulation of migration. Bioscience 57:135–143

Randall LA, Diehl RH, Wilson BC, Barrow WC, Jeske CW (2011) Potential use of weather radar to study movements of wintering waterfowl. J Wildl Manag 75:1324–1329

Renfrew RB, Kim D, Perlut N, Smith J, Fox J, Marra PP (2013) Phenological matching across hemispheres in a long-distance migratory bird. Divers Distrib 19:1008–1019

Robinson RA, Crick HQ, Learmonth JA, Maclean I, Thomas CD, Bairlein F, Forchhammer MC, Francis CM, Gill JA, Godley BJ (2009) Travelling through a warming world: climate change and migratory species. Endanger Species Res 7:87–99

Root TI, Hughes L (2005) Present and future phenological changes in wild plants and animals. In: Lovejoy TE, Hannah L (eds) Climate change and biodiversity. Yale University Press, New Haven, pp 61–69

Root TL, Price JT, Hall KR, Schneider SH, Rosenzweig C, Pounds JA (2003) Fingerprints of global warming on wild animals and plants. Nature 421:57–60

Rosenzweig C, Karoly D, Vicarelli M, Neofotis P, Wu QG, Casassa G, Menzel A, Root TL, Estrella N, Seguin B, Tryjanowski P, Liu CZ, Rawlins S, Imeson A (2008) Attributing physical and biological impacts to anthropogenic climate change. Nature 453:353–358

Runge CA, Martin TG, Possingham HP, Willis SG, Fuller RA (2014) Conserving mobile species. Front Ecol Environ 12:395–402

Russell KR, Gauthreaux SA (1998) Use of weather radar to characterize movements of roosting purple martins. Wildl Soc Bull 26:5–16

Russell KR, Gauthreaux SA Jr (1999) Spatial and temporal dynamics of a purple martin pre-migratory roost. Wilson Bull 111:354–362

Russell KR, Mizrahi DS, Gauthreaux SA (1998) Large-scale mapping of purple martin pre-migratory roosts using WSR-88D weather surveillance radar. J Field Ornithol 69:509–509

Saino N, Ambrosini R, Rubolini D, von Hardenberg J, Provenzale A, Hüppop K, Hüppop O, Lehikoinen A, Lehikoinen E, Rainio K (2011) Climate warming, ecological mismatch at arrival and population decline in migratory birds. Proc Biol Sci 278(1707):835–842. https://doi.org/10.1098/rspb.2010.1778

Sauer J, Hines J, Fallon J, Pardieck K, Ziolkowski D Jr, Link W (2012) The North American breeding bird survey, results and analysis 1966–2011. http://www.mbr-pwrc.usgs.gov/bbs/. Accessed 12 Dec 2011

Schmaljohann H, Liechti F, Bachler E, Steuri T, Bruderer B (2008) Quantification of bird migration by radar – a detection probability problem. Ibis 150:342–355

Schwartz MD (1998) Green-wave phenology. Nature 394:839–840

Schwartz MD, Ault TR, Betancourt JL (2013) Spring onset variations and trends in the continental United States: past and regional assessment using temperature-based indices. Int J Climatol 33 (13):2917–2922

Shamoun-Baranes J, Alves J, Bauer S, Dokter A, Koistinen J, Leijnse H, Liechti F, van Gasteren H (2014) Continental-scale radar monitoring of the aerial movements of animals. Mov Ecol 2 (1):9

Shariatinajafabadi M, Wang T, Skidmore AK, Toxopeus AG, Kölzsch A, Nolet BA, Exo K-M, Griffin L, Stahl J, Cabot D (2014) Migratory herbivorous waterfowl track satellite-derived green wave index. PLoS One 9:e108331

Si Y, Xin Q, de Boer WF, Gong P, Ydenberg RC, Prins HH (2015) Do arctic breeding geese track or overtake a green wave during spring migration? Sci Rep 5:8749

Solomon S, Qin D, Manning M, Chen Z, Marquis M, Averyt KB, Tignor M, Miller HL (2007) Contribution of working group I to the fourth assessment report of the intergovernmental panel on climate change, 2007. Cambridge University Press, Cambridge, p 996

Sparks TH, Carey PD (1995) The responses of species to climate over two centuries: an analysis of the marsham phenological record, 1736–1947. J Ecol 83:321–329

Sparks T, Jeffree E, Jeffree C (2000) An examination of the relationship between flowering times and temperature at the national scale using long-term phenological records from the UK. Int J Biometeorol 44:82–87

Sullivan BL, Aycrigg JL, Barry JH, Bonney RE, Bruns N, Cooper CB, Damoulas T, Dhondt AA, Dietterich T, Farnsworth A (2014) The eBird enterprise: an integrated approach to development and application of citizen science. Biol Conserv 169:31–40

Socolar JB, Epanchin PN, Beissinger SR, Tingley MW (2017) Phenological shifts conserve thermal niches in North American birds and reshape expectations for climate-driven range shifts. Proc Natl Acad Sci USA 114(49):12976–12981

Thompson K, Gilbert F (2014) Phenological synchrony between a plant and a specialised herbivore. Basic Appl Ecol 15:353–361

Thorup K, Tottrup AP, Rahbek C (2007) Patterns of phenological changes in migratory birds. Oecologia 151:697–703

Tottrup AP, Thorup K, Rahbek C (2006) Patterns of change in timing of spring migration in North European songbird populations. J Avian Biol 37:84–92

Tottrup AP, Rainio K, Coppack T, Lehikoinen A, Rahbek C, Thorup K (2010) Local temperature fine-tunes the timing of spring migration in birds. Integr Comp Biol 50:293–304

Tøttrup AP, Klaassen RH, Strandberg R, Thorup K, Kristensen MW, Jørgensen PS, Fox J, Afanasyev V, Rahbek C, Alerstam T (2012) The annual cycle of a trans-equatorial Eurasian–African passerine migrant: different spatio-temporal strategies for autumn and spring migration. Proc R Soc B Biol Sci 279(1730):1008–1016

Tucker CJ (1979) Red and photographic infrared linear combinations for monitoring vegetation. Remote Sens Environ 8:127–150

Thorup K, Tøttrup AP, Willemoes M, Klaassen RH, Strandberg R, Vega ML, Dasari HP, Araújo MB, Wikelski M, Rahbek C (2017) Resource tracking within and across continents in longdistance bird migrants. Sci Adv 3(1):e1601360

Van Buskirk J, Mulvihill RS, Leberman RC (2009) Variable shifts in spring and autumn migration phenology in North American songbirds associated with climate change. Glob Chang Biol 15:760–771

Visser ME, Gienapp P, Husby A, Morrisey M, de la Hera I, Pulido F, Both C (2015) Effects of spring temperatures on the strength of selection on timing of reproduction in a long-distance migratory bird. PLoS Biol 13:e1002120

Walther GR, Post E, Convey P, Menzel A, Parmesan C, Beebee TJC, Fromentin JM, Hoegh-Guldberg O, Bairlein F (2002) Ecological responses to recent climate change. Nature 416:389–395

Wang T, Skidmore AK, Zeng Z, Beck PS, Si Y, Song Y, Liu X, Prins HH (2010) Migration patterns of two endangered sympatric species from a remote sensing perspective. Photogramm Eng Remote Sens 76:1343–1352

White MA, Hoffman F, Hargrove WW, Nemani RR (2005) A global framework for monitoring phenological responses to climate change. Geophys Res Lett 32:L04705

White MA, Beurs D, Kirsten M, Didan K, Inouye DW, Richardson AD, Jensen OP, O'Keefe J, Zhang G, Nemani RR (2009) Intercomparison, interpretation, and assessment of spring phenology in North America estimated from remote sensing for 1982–2006. Glob Chang Biol 15:2335–2359

Williams TC, Marsden JE, Lloyd-Evans TL, Krauthamer V (1981) Spring migration studied by mist-netting, ceilometer, and radar. J Field Ornithol 52:177–190

Williams AA, Laird NF (2017) Weather and eared grebe winter migration near the Great Salt Lake, Utah. Int J Biometeorol:1–15

Zhang J, Howard KW, Langston C, Vasiloff SV, Kaney B, Aurthur A, Van Cooten S, Kelleher KE, Kitzmiller DH, Ding F, Seo D-J, Wells E, Dempsey C (2011) National mosaic and multi-sensor QPE (NMQ) system – description, results and future plans. Bull Amer Meteor Soc 92:1321–1338. https://doi.org/10.1175/BAMS-D-11-00047.1

The Aerosphere as a Network Connector of Organisms and Their Diseases

17

Jeremy D. Ross, Eli S. Bridge, Diann J. Prosser, and John Y. Takekawa

Abstract

Aeroecological processes, especially powered flight of animals, can rapidly connect biological communities across the globe. This can have profound consequences for evolutionary diversification, energy and nutrient transfers, and the spread of infectious diseases. The latter is of particular consequence for human populations, since migratory birds are known to host diseases which have a history of transmission into domestic poultry or even jumping to human hosts. In this chapter, we present a scenario under which a highly pathogenic avian influenza (HPAI) strain enters North America from East Asia via post-molting waterfowl migration. We use an agent-based model (ABM) to simulate the movement and disease transmission among 10^6 generalized waterfowl agents originating from ten molting locations in eastern Siberia, with the HPAI seeded in only $\sim 10^2$ agents at one of these locations. Our ABM tracked the disease dynamics across a very large grid of sites as well as individual agents, allowing us to examine the spatiotemporal patterns of change in virulence of the HPAI infection as well as waterfowl host susceptibility to the disease. We concurrently simulated a 12-station disease monitoring network in the northwest USA and Canada in order to assess the potential efficacy of these sites to detect and confirm the arrival of HPAI. Our findings indicated that HPAI spread was initially facilitated but eventually subdued by the migration of host agents. Yet,

J.D. Ross (✉) · E.S. Bridge
Oklahoma Biological Survey, University of Oklahoma, Norman, OK, USA
e-mail: rossjd@ou.edu; ebridge@ou.edu

D.J. Prosser
USGS Patuxent Wildlife Research Center, Beltsville, MD, USA
e-mail: dprosser@usgs.gov

J.Y. Takekawa
Audubon California, Richardson Bay Audubon Center and Sanctuary, Tiburon, CA, USA
e-mail: jtakekawa@audubon.org

© Springer International Publishing AG, part of Springer Nature 2017
P.B. Chilson et al. (eds.), *Aeroecology*,
https://doi.org/10.1007/978-3-319-68576-2_17

during the 90-day simulation, selective pressures appeared to have distilled the HPAI strain to its most virulent form (i.e., through natural selection), which was counterbalanced by the host susceptibility being conversely reduced (i.e., through genetic predisposition and acquired immunity). The monitoring network demonstrated wide variation in the utility of sites; some were clearly better at providing early warnings of HPAI arrival, while sites further from the disease origin exposed the selective dynamics which slowed the spread of the disease albeit with the result of passing highly virulent strains into southern wintering locales (where human impacts are more likely). Though the ABM presented had generalized waterfowl migration and HPAI disease dynamics, this exercise demonstrates the power of such simulations to examine the extremely large and complex processes which comprise aeroecology. We offer insights into how such models could be further parameterized to represent HPAI transmission risks as well as how ABMs could be applied to other aeroecological questions pertaining to individual-based connectivity.

1   Introduction

Movement through the air, once achieved, could be perceived as an entry to all areas of the world touched by the wind. The barriers that affect biological connectivity in terrestrial or aquatic ecosystems are much less pronounced in wide-open skies. Yet, the air column and dynamic processes therein which comprise the "aerosphere" present their own ecological pressures, which can dictate the flow of volant organisms relative to the underlying landscape, other airborne species and materials, and the airmass itself. In the case of vertebrates, sustained movements through the aerosphere are the result of powered flights where the direction, duration, and daily distance traveled are both intrinsically and extrinsically controlled. The balance between a genetically mediated migratory program and phenotypic flexibility produces complex interactions that further mediate the flight behaviors of vertebrates. The ease with which a species can transverse a landscape springs from the sum effect of piecemeal decisions (Taylor et al. 1993). Because of this, characterizing the movements of a species or population likely requires that we examine flight through the aerosphere at the lowest denominator: the individual (Morales et al. 2010).

Advances in tracking individuals using technology such as light-level archival geolocators, satellite transmitters, GPS tags, and tissue stable isotope analyses have begun to reveal much about individual variation and population patterns in vertebrate flight behavior (Cooke et al. 2004; Hobson and Norris 2008; Robinson et al. 2010; Bridge et al. 2011; McKinnon et al. 2013; Kays et al. 2015) or, possibly, in-flight physiology (Gumus et al. 2015). However, scaling such efforts to more widely encompassing levels remains logistically challenging if not prohibitively expensive (Cagnacci et al. 2010; Hebblewhite and Haydon 2010). Because of this, current endeavors to track individual movements may be restricted in their ability to broadly represent movement behavior variation throughout species' regional or range-wide populations. Data-informed models of individual movements within a simulation framework show promise in filling otherwise-unreachable knowledge gaps regarding biological phenomena (DeAngelis and Mooij 2005; Tang and Bennett 2010). Relevant to this volume is how vertebrates move through the aerosphere and, by extension, the ease with which optimized flight strategies can connect populations and biological communities across broad spatial scales (Alerstam 2011).

The implications of better resolving individual movements through the aerosphere as a component of geographic connectivity are likely to enhance various biological disciplines: from evolution, landscape ecology, conservation biology, resource management, and behavior to broadly integrative macrosystem studies. For instance, it is generally thought that capacity for long-distance flight often reduces the impermeability of landscape features that would otherwise be barriers to species with greater movement restrictions (Wiens 1976; Taylor et al. 1993), including non-migratory birds (Harris and Reed 2002). Because of this, flying animals with expansive ranges such as migratory birds are often found to show only modest, if any, phylogeographic variation (Zink 1996; Sutherland et al. 2000; von Rönn et al. 2016). Yet, in many cases migratory constraints within a species appear to have contributed to some degree of geographic diversification (see Chap. 11; Irwin 2002; Pérez-Tris et al. 2004; Delmore and Irwin 2014). By simulating potentially subtle barriers to migratory and breeding dispersal using agent-based models (ABMs; alternatively, "individual-based models"), we may have the opportunity to uncover evolutionarily significant patterns within highly vagile species. This includes integrating the available information about the species' migratory biology and simulating probable real-world scenarios (Bowlin et al. 2010; MacPherson and Gras 2016).

ABM simulations extend our ability to model natural phenomena to include complex interactions among multiple different types of agents. These may include vector agents that are capable of traversing the simulation arena or static agents that occupy the same spatial arena but may dynamically change during the simulation (e.g., a vegetated area which sprouts, blooms, seeds, and senesces over time). Since ABMs can be tailored to any geographic or temporal scale, even slow-developing or locally subtle patterns among agent interactions can be examined, despite the complexity of ecological or evolutionary processes. For example, the seasonal flow of energy and nutrients as a result of animal migration may be difficult to quantify using standard field techniques, since the deposition of waste or carcasses are relatively rarely detected events (outside of huge aggregation sites). Yet, we understand that migratory birds regularly move through the aerosphere to exploit seasonal pulses of resources (Alerstam et al. 2003; Bowlin et al. 2010; Shariatinajafabadi et al. 2014; Si et al. 2015), and so the influx and exodus of billions of individuals, with associated depredations and depositions along the way, certainly must collectively have a nontrivial impact in terms of the redistribution of energy and nutrients (Bauer and Hoye 2014). A properly parameterized ABM could not only outline what this redistribution might entail at stopover sites, it may also predict what transitory impacts this may have where migratory flight paths aggregate over otherwise unsuitable stopover habitat (e.g., over open water of the Great Lakes region of North America).

Perhaps the most pressing cause for studying connectivity among species and the landscape is the potential threat posed by continental-scale transmission of highly pathogenic avian influenza (HPAI) via migratory birds. These long-distance migrants may be the primary conduit by which certain diseases could traverse continents and jump dispersal barriers, such as mountain ranges or oceans. HPAI is of particular concern because recent outbreaks among domestic fowl in East Asia, Europe, and North America were thought to be mediated by wild birds (Okazaki et al. 2000; Hulse-Post et al. 2005; Gilbert et al. 2006; Kilpatrick et al. 2006; Alexander and Brown 2009; Feare 2010). This disease is also naturally prevalent in the environment (Winker and Gibson 2010), and it shows high transmissibility, morbidity, mutability, and potential for jumping to mammalian hosts (Kapan et al. 2006; Olsen et al. 2006; Taubenberger and Kash 2016). A worst-case scenario is that a virulent HPAI strain capable of switching hosts to infect human populations becomes spread across a wide landscape through the flights of wild birds (Tan et al. 2015). Countering disease transmission and host switching is best accomplished through a coordinated detection-and-response network (Wagner et al. 2001; Jebara 2004; Choffnes et al. 2007; Silkavute et al. 2013; Xu et al. 2013). As with any threat to human life, early and accurate detection of a disease is critical if mitigation is to be effective. These actions may include quarantines or inoculations of potentially exposed humans or similar measures among possible animal vectors, with the additional option of population culls of domestic fowl or hazing wild host populations (DeLiberto et al. 2011; Wobeser 2013).

Best management practices dictate that limited pools of funding should be directed to maximize the intended outcome. In the case of HPAI monitoring, this hinges upon one critical need: to detect the disease while there remains an opportunity to halt an outbreak. This "make-or-break" scenario puts tremendous pressure on monitoring agencies to have an expansive focus and to swiftly and accurately diagnose HPAI in the field. Accomplishing this task naively would necessitate a costly outpouring of resources just to capture the earliest signs of disease infiltration or host shifting. Coordinated monitoring networks, such as the United States Geological Service's Wildlife Health Information Sharing Partnership event reporting system (WHISPers; https://www.nwhc.usgs.gov/whispers/), show great promise at early detection of HPAI, although the efficacy of these endeavors is still reliant upon sufficient sampling, accuracy, and timely reporting by field observers.

Fail-safe disease monitoring regimes can not only be expensive and challenging to coordinate, but may be also difficult to grade for accuracy and efficacy. The static models and formulas currently used to determine sampling needs for detecting a disease hinge upon uncertainty parameters which can be broad (e.g., severity of exposure events, distribution and virulence of zoonotic infective source; Yang et al. 2007). In many cases, the assessments of how well monitoring regimes are actually functioning have mostly been limited to preparedness indicator surveys (ECDC 2007; Azziz-Baumgartner et al. 2009) or, when the system fails, a retroactive assessment of shortcomings (Balicer et al. 2007; Scallan 2007). As a low-cost and promising alternative, ABM simulations may be a useful tool for making proactive assessments of HPAI monitoring networks and possibly optimizing their effectiveness relative to cost. With this in mind, we have constructed a realistic ABM which simulates a hypothetical HPAI transmission into North America from Asia and concordantly tests the efficacy of a theoretical monitoring network to detect the disease before it has progressed to an outbreak stage.

The ABM presented in this chapter incorporates underlying biological principles related to both large-scale bird migration and the Susceptible-Exposed-Infected-Recovered (SEIR) model of disease dynamics (Keeling and Rohani 2008). However, instead of using static population-level metrics within the SEIR model, we have leveraged the power of the ABM simulation framework to allow the SEIR factors to dynamically evolve during the course of the simulation. This is likely to more accurately reflect the reality in nature, since viruses such as avian influenza have the capacity to rapidly mutate into more virulent strains just as their hosts have the capacity to alter their individual susceptibility and/or probability of recovery through acquired immunity (Bourouiba et al. 2011; Pybus et al. 2013). In effect, our ABM is designed to parameterize the probability of low-incidence, high-risk disease transmission that would otherwise be difficult to trace, even from extensive sampling regimes. Such a model is not only informed by existing knowledge but can, in turn, guide future empirical studies by exposing knowledge gaps (Harris et al. 2015), aeroecological limiting factors (Lam et al. 2012), or critical nodes within disease monitoring networks (Ferguson et al. 2006; Boyce et al. 2009).

Our entire case-study simulation is couched under the topic of this chapter—the aerosphere as a connector—to illustrate how animal movement through the air may (or may not) facilitate rapid translocations and can effectively reduce the ecological divisiveness of what would otherwise be impenetrable geographic boundaries, resulting in the mixing of biological agents over large geographic extents. Building upon the paradigm of ABM operation described in Chap. 11, we now demonstrate how movement through the aerosphere can drive interspecific dynamics in the form of intercontinental spread of disease.

2   Constructing the Agent-Based Model

The applicability of models of increasing complexity depends upon how well their core components are built to emulate the processes of study. Building an ABM—especially of something as complex as disease transmission, evolution, and monitoring during the intercontinental migration of a very large population of a generalized waterfowl species—required that we first start with basic models of agent movement and then added levels of increasing complexity in stages. In our case, we built the ABM in the following five major stages:

1. Parameterize the autumn migration of a generalized "waterfowl" species according to simple movement and stopover rules

2. Include more complex rules for individual waterfowl stopover and aggregation decisions
3. Add a transmissible waterfowl-borne viral disease (i.e., HPAI) as well as individual variation in disease susceptibility among waterfowl agents
4. Incorporate mutability in HPAI virulence as well as waterfowl susceptibility as a factor of exposure
5. Simulate hypothetical monitoring stations at predetermined locations (i.e., places likely to attract stopover that were relatively close to human settlements) as the means to gauge the efficacy of single versus networked stations to detect and confirm the HPAI outbreak

In the first stage of ABM development, we consulted basic knowledge about general waterfowl migratory biology. We sourced this information from traditional monitoring efforts, such as local field surveys, standardized mark-release-recapture programs, and station-network monitoring (Bellrose and Crompton 1970; Flock 1972; Dau 1992; Moermond and Spindler 1997; Winker and Gibson 2010), as well as individual tracking data gathered using modern technology (Ely et al. 1997; Green et al. 2002; Mosbech et al. 2006; Alerstam et al. 2007; Gaidet et al. 2010; Prosser et al. 2009, 2011; Krementz et al. 2012; Takekawa et al. 2010, 2013; Ely and Franson 2014). From these collective data, we were able to broadly parameterize the timing of population movements, distributions of distances moved, and stopover duration for the generalized waterfowl species being modeled. Since we were not attempting to precisely model one species or replicate an exact situation, our ABM was simply intended to be a starting point from which it can be tailored to specific systems or questions in the future.

The ABM that we constructed was designed from the second stage onward to include more complex daily movement and stopover rules, to introduce individual variation, to model contagious disease dynamics (Brown et al. 2008; Gaidet et al. 2010), and to emulate what was the ultimate focus—evaluating the efficacy of a network of monitoring stations to detect and confirm HPAI arrival by way of aeroecological connections. Because species often have different mechanisms driving large- (i.e., continental) versus small-scale (i.e., local) navigation (see Chap. 6), we allowed for modest spatiotemporal shifts in migratory flight behavior of the generalized waterfowl species throughout the migratory period. We also incorporated simple rules to allow symptomatic birds to be more readily detected at monitoring stations (Brown and Stallknecht 2008) and for both virus and host agents to evolve as the simulation progressed and passed through the stages of the SEIR model. Below we outline the final composition of our ABM which incorporates additions from all five stages of model development.

2.1 Populating the Arena, Agents, and Disease

Our ABM focused on the individual movements of a generalized waterfowl species during a hypothetical 90-day autumn migration period that connected birds between their molting locations in eastern Siberia to wintering grounds in the southern USA or northern Mexico. This reflects a hypothetical situation where HPAI enters the Americas via southward migration of arctic-breeding birds. The arena of possible bird locations included an area extending from eastern Siberia to Iceland, Central America, and Venezuela (Fig. 17.1). This area was projected to a Lambert Azimuthal Equal Area sphere centered on 100°W with latitude of origin at 45°N, which provided a meter-based overlay of the area of interest. At 10 km intervals along both the x- and y-axes, we plotted a grid of possible stopover locations for agents during their migration. We excluded as possible stopover sites those grid points positioned greater than 200 km from land, since the waterfowl agent simulated was intended to prefer land and realistic disease monitoring capacity would be restricted to the near-coast region.

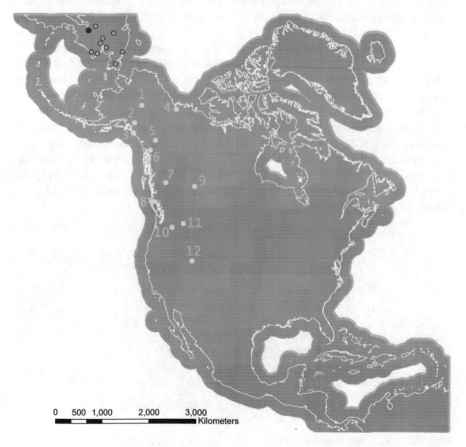

0   500 1,000   2,000   3,000
                                         Kilometers

Fig. 17.1 Map of ABM arena showing 10-km grid of possible stopover points (blue), starting locations of uninfected agents (green), the population of disease origin (red), and 12 monitoring stations (yellow; numbered according to Table 17.2). Rendering the large number of stopover grid points at this scale has created an artificial distortion in the graphic

To reduce processing time, we prepopulated properties for each grid point using data from underlying landscape characteristics maps (Fig. 17.2). These "static variables" included distance and direction to the closest ocean coastline, lakeshore, and river, as well as elevation, anthropogenic biome types, and human disturbance (Table 17.1A). We used these measures during the ABM simulations to allow individual agents to assess the stopover sites to inform avoidance, attraction, and settlement decisions when picking a stopover destination. We allowed certain metrics for each grid point to dynamically change during the simulations, which had implications for waterfowl agent stopover as well as HPAI transmission as the ABM progressed. These metrics included daily values of the total number of occupants, the number of diseased occupants, as well as the mean and standard deviation of disease virulence among infected occupants at each grid point (variables #1 through #4 in Table 17.1B; hereafter referenced in the format "Table 17.1B [1–4]"). Also tracked were individual agent metrics of status (alive and migrating), disease susceptibility, and the virulence and days of infection of any carried HPAI strain (Table 17.1B [5–9]).

At the outset of the ABM (i.e., "Day 0"), the total number of starting agents (10^6) was equally distributed across ten hypothetical molting areas in eastern Siberia (Fig. 17.1). Our first introduction of variation into the simulation came during the disease seeding process when we allowed the ABM to (1) randomly choose the population where the disease originated; (2) randomly choose ~10^2 of the 10^5 occupants (i.e., 0.001 probability for any individual) at that population to be infected; (3) randomly assign each individual's disease susceptibility, regardless of population or infection status (Table 17.1B [7]); (4) select an initial virulence for each diseased individual from a gamma distribution (Table 17.1B [8]); and (5) randomly choose each infected individual's preexisting day of infection from a gamma distribution (Table 17.1B [9]).

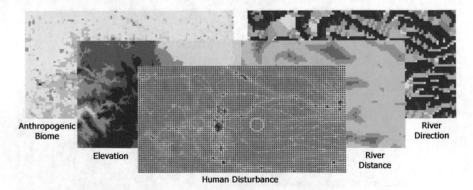

Fig. 17.2 Layer data embedded into each grid point in preparation for the ABM simulation. Shown atop the "Human Disturbance" layer is part of the 10-km grid of possible stopover locations as well as an example neighborhood (outlined in blue) within which an agent would probabilistically select its final stopover location based upon relative weightings of distance to water, human disturbance, and current density of agents

Table 17.1 Summary of how static and dynamic variables used in the ABM were first derived/calculated and how the dynamic variables were updated as the ABM progressed

A. Static variables

	Grid points (stopover locations)					Waterfowl agents
	Identifier	Water boundaries	Elevation	Human disturbance	Anthropogenic biome types	Identifiers
Source	Predefined during ABM construction	Coastlines: World continents (ESRI 2012) Lakeshores and riverbanks: North American Atlas–Hydrography (CEC 2010)	North America elevation–1 km resolution (CEC 2007)	Global Human Footprint Dataset (WCS/CIESIN 2005)	Anthropogenic biomes of the world (Ellis and Ramankutty 2008)	Assigned during intiation of ABM
Metric	[0] PointID	Attraction to: [2] Coasts/ [3] Lakes/[4] Rivers	Avoidance of: [5] Elevation (Ea); [6] Disturbance (Da)		[7] Biome avert (Ba)	[9] BirdID
Calculation	Grid matrix: rows = 844, columns = 1023 Origin(x, y):−5,770,000, −3,930,000 (in meters) $\Delta x, \Delta y = [\text{Gridpoint}(x, y)-\text{Origin}(x, y)]/10$ km PointID = $[\Delta y]*\text{columns}] + [\Delta x-\text{columns}]$	$P_{(\text{attraction})} = 1 - \left\{ \dfrac{1}{1+e^{-[k(\text{Distance}-x_0)]}} \right\}$ where ... k: Coast = 0.05; Lake = 0.10; River = 0.20; x_0: Coast = 125 km; Lake = 60 km; River = 40 km Proportionate attraction values were calculated relative to the total attraction across all features	$P_{(\text{avoidance})} = 1 - \left\{ \dfrac{1}{1+e^{[k(J-x_0)]}} \right\}$ where ... J = elevation at PointID, k = 0.75, x_0 = 3000	where ... J = disturbance at PointID, k = 0.15, x_0 = 30	Residential: 0.9 Cropland: 0.3 Rangeland: 0.5 Forest: 0.15 Barren: 1.0 Ocean: 0.9 [8]$P_{(\text{tAvoid})} = P(Ea \cup Da \cup Ba) - P(Ea \cap Da) - P(Ea \cap Ba) - (Da \cap Ba\,|\overline{Da}) - (Ea \cap Ba|\overline{Da}) - (Da \cap Ba|\overline{Ea})$	Iteratively—assigned integers within sequential starting populations (i.e., 1–100,000 in population #1, 100,001–200,000 in population #2, etc.)

B. Dynamic variables

	Grid points		Waterfowl agents		
	Waterfowl occupants	Population disease virulence	Status change probabilities	Disease-host characteristics	Location
Metric	[1] Total [2] Contagious	[3] Mean [4] Std dev	[5] Death [6] End migration	[7] Susceptibility [8] Virulence [9] Infection day	[0] PointID
Calculation	Running totals increased/decreased iteratively upon each entry/exit of any agent (total) or an agent infected 4+ days (contagious)	Recalculated among contagious occupants of grid point whenever one enters, leaves (by departure or death), or becomes contagious (infected 4 days)	$P_{(change)} = 1 - \left\{ \dfrac{1}{1+e^{-[k(W-x_0)]}} \right\}$ where… $k = 0.9736$; $x_0 = 7$; W = days of consecutive infection [10] P (scpt$'$) = P (scpt) $*$ $[1 - (\text{Vir}'*2)]$; min = 0.0 Probability of mortality/ending migration increases logistically as days infected/southward position increases, respectively	Initial values randomly drawn from function: $P(\text{scpt/vir/infDay}) = 1 - [\text{Gamma}(\alpha, \beta) - 2 \times f]$ where … $f = 1.0$, $\alpha = 6$, $\beta = 15$, Max = 1.0, Min = 0.0 where … $f = 0.1$, $\alpha = 6$, $\beta = 15$, Max = 1.0, Min = 0.1 where … $f = 0$; $\alpha = 6$, $\beta = 1.5$ (as integer) Adjust susceptibility upon clearing disease of virulence P(vir) Assign virulence from local metrics upon infection Update number of successive days alive + infected [11] Vir$'$ = [3] scaled by random draw from normal distribution ($\mu = 1.0$, $\sigma = 0.1$)	Sourced from underlying grid point [12] $W' = W + 1$

Each grid point's dynamic metrics were stored as entries in an assigned ArrayList within a LinkedHashMap

2.2   Daily Disease Transmission, and Mortality

At the outset of each new day, we programmed each uninfected individual to consult its current grid point metrics to determine the probability that it was exposed to a contagious agent (i.e., infection day $\geq$ 4; Brown et al. 2008) given that each individual, regardless of disease status, was arbitrarily set to randomly experience exactly two density-independent interactions that would permit disease transmission with the other occupant(s) of the population (e.g., close feeding behaviors; Brown and Stallknecht 2008). If the agent was indeed exposed to a contagious individual, the probability that it contracted the disease was contingent on the susceptibility of the individual (Table 17.1B [7]) and, since the model did not account for *which* contagious agent was contacted, the mean and standard deviation of HPAI virulence at the population level (Table 17.1B [3,4]). Based upon a normal distribution centered on this population mean virulence and scaled by the accompanying population standard deviation, if a randomly drawn virulence value exceeded the inverse value of the individual agent's susceptibility, then we set the infection day of the agent to 1. At the time of infection, the HPAI virus had the ability to mutate according to a normally distributed multiplier (Table 17.1B [11]). This meant that with every successful transmission, the HPAI strain could become more or less virulent but was most likely to remain unchanged (i.e., multiplier of 1.0). We constructed the ABM to thereafter determine if the agent would move to a new location.

Among waterfowl agents that were already diseased prior to the beginning of each day, we provided the opportunity to clear their HPAI strain or, failing that, we subjected these to an increasing probability of death prior to leaving their current location. Our ABM dictated the probability of clearing the HPAI infection according to three factors: (1) the virulence of the strain, (2) the susceptibility (i.e., inverse resistance) of the individual waterfowl agent to HPAI, and (3) a normally distributed multiplier with mean of 1.0 and standard deviation of 0.1. If the virulence of the HPAI strain scaled by the random multiplier was less than the inverse susceptibility of the individual agent, we allowed the agent to clear the infection. When an HPAI strain was cleared, we adjusted the individual agent's susceptibility by a factor equivalent to the inverse of double the virulence of the strain being cleared (Table 17.1B [10]). In other words, if a strain with a virulence of 0.4 was cleared, the agent's susceptibility would be 20% of what it had been [i.e., $1 - (0.4 \times 2)$] and clearing a strain with virulence of 0.5 or greater imparted complete immunity upon the individual (i.e., new susceptibility $= 0$). This would allow agents to possibly develop immunity to reinfection, especially if they cleared a particularly virulent strain of HPAI.

To simulate mortality in the population as a function of the day of infection, we used a logistic function (Table 17.1B [5]). This calculation meant that the probability of mortality rose with each subsequent day of infection, especially after the HPAI became communicable on Day 4. Yet, this function also meant that even noncontagious individuals had a modestly increased likelihood of death, which allowed us to ensure a background mortality among noninfected agents (i.e., infection day $= 0$)

equivalent to a 40% annual mortality in the absence of any disease (Franklin et al. 2002). If a randomly selected value between 0.0 and 1.0 fell below the infection day-based probability of mortality, we removed the agent from the simulation and, if it was infected, recorded upon its "death" the information about its HPAI strain's virulence (Table 17.1B [8 or 11]) and the number of sequential days it had been infected (Table 17.1B [9 or 12]).

2.3 Daily Movements: Migration

There were two types of possible daily movements that our ABM allowed: long-distance migration or neighborhood searches. The former, which we kept as relatively uncommon events (i.e., mean of 10-day stopover; Takekawa et al. 2010), led to the agent moving hundreds of kilometers in a semi-random fashion. If our ABM selected the agent to migrate, then we drew a random bearing from a normal distribution with mean 170° and standard deviation of 60° (Fig. 17.3). Because of our map projection, the real-world compass direction of movement changed as a function of geographic position; this meant that the agent would generally move in an eastern and then south-southeastern direction from its molting grounds to wintering grounds (see Fig. 17.1). We thereafter selected from a gamma distribution $[\gamma(\alpha, \beta); \alpha = 6, \beta = 0.015]$ a movement distance for the migrating agent. On the rare occasion that the selected value exceeded 1200 km, we folded the

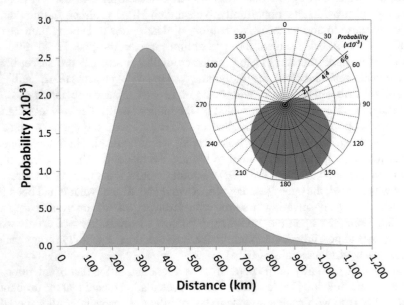

Fig. 17.3 Probability distributions for movement distances (base) and directions (inset) when agents migrated. Distance was calculated from a gamma distribution folded on itself beginning with distances above 1200 km. Direction was drawn from a normal distribution with mean of 170° and standard deviation of 60°

distribution back on itself [i.e., final distance = (1200 km-selected distance) + 1200 km] to prevent unrealistic travel distances (Fig. 17.3).

Prior to finalizing the movement of the agent, we programmed our ABM to check whether the proposed stopping point was valid (i.e., would land the agent within the arena and ≤200 km from land). If the movement was not valid, we set the model to pick a new migratory bearing and distance and recheck the validity of this new proposed movement. If a valid move could not be found after 100 attempts, we denied the agent the ability to migrate and, instead, relegated it to exploring its neighborhood for a more ideal location (see Sect. 2.4).

When an agent completed its migration, we set the ABM to relocate it to the nearest grid point and assess whether this was a sufficiently suitable stopover location. We based this site assessment upon a summation of logistic avoidance functions for elevation and human disturbance (Table 17.1A [5, 6]), plus predetermined avoidance measures for anthropogenic biome type (Table 17.1A [7]). We then calculated the total probability of avoidance (tAvoid; Table 17.1A [8]) as the overlapping probability across all three factors (i.e., probability of avoiding any or all factors). If a random value drawn from 0.0 to 1.0 was less than tAvoid, then we forced the agent to move toward a nearby water feature (coastline, lake, or river) based upon the relative attraction of each from the current point. We determined the probability of displacement to one of the three alternative water feature boundaries using separate logistic functions of attraction for coastlines, lakes, and rivers (Table 17.1A [2–4]). These values were each further adjusted by separate weightings for each water type that we had predefined in the model: in our case, 0.2, 0.8, and 0.5, respectively. We then scaled the relative attraction to each water type against the total draw across all three water types, and the resulting value was used to define successive bins of probable relocation to each water type. Based upon which bin contained a random number drawn between 0.0 and 1.0, we selected the associated nearby water boundary to relocate the agent.

2.4   Daily Movements: Neighborhood Exploration

At either the end of the migratory jump and possible relocation or, if the agent did not migrate, at the original point, we allowed the agent to explore its neighborhood for a "more suitable" location. We defined a neighborhood as the agent's current location plus all surrounding grid points within 25 km ($n = 20$, unless any grid points fell outside the 200 km coastline buffer or the ABM arena; Fig. 17.2). We determined the relative attractiveness of the neighborhood grid points (RelBWwt) based upon a combination of avoidance of human disturbance (Da; Table 17.1A [6]) and the maximum attraction to coast, lake, or river (maxWA; Table 17.1A [2–4]), as well as the total agent occupants (tOcc; Table 17.1B [1]) relative to a predefined optimum (OptD): in our case arbitrarily set at 1500 agents/grid point. In each case, we defined an adjustment factor so that the weighting of each variable could be given more or less importance ("b" for Da, "w" for maxWA, "q" for tOcc). The RelBWwt value was precalculated for each grid point as:

$$\mathrm{RelBWwt} = \frac{[(-\mathrm{Da}^*b) + (\mathrm{maxWA}^*w)]}{b + w + q}$$

Since tOcc was a dynamic value, this required recalculation each day. We also used a slope factor (f) to more strongly weight tOcc as it approached OptD. We did this so that density could vary more in its influence over how agents selected the optimal point in the neighborhood. The resulting formula for density weight (DENSwt) was:

$$\mathrm{DENSwt} = \left\{ 1 \Big/ \left[\frac{|\mathrm{tOcc} - \mathrm{OptD}|}{\mathrm{OptD}^*f^{-1}} + 1 \right] \right\}^* q$$

The sum of DENSwt and RelBWwt provided a total weight for the grid point, which we then scaled against the cumulative weight across all neighborhood grid points to determine a relative bin size. We designated bins according to each one's sequential position in the neighborhood, and using a random draw from a uniform distribution between 0.0 and 1.0, we selected the agent's new point based upon the bin in which this number fell.

Once a grid point was selected, we moved the agent to its new position and accounted for it within the dynamic grid point metrics (Table 17.1B [1–4]). Regardless of disease infection, we added every incoming agent to the tOcc tally. We likewise added individuals diseased at least 4 days to the tally of occupants likely to be contagious (cOcc; Brown et al. 2008) and then calculated a new mean and standard deviation for HPAI virulence across all of the strains infecting contagious individuals. We ensured that these new values for each metric could only be consulted in the subsequent day and retained the day's starting values so that all subsequent disease transmission and migratory movements among the remaining agents would remain unbiased. In this way, movement of a contagious agent into a new population would not result in added probability of disease exposure at this new location until the next day.

At the end of each simulated day, we recorded the total number of HPAI-infected, contagious, and dead individual agents so that we could track the overall disease dynamics throughout the entire simulation arena. We likewise recorded on each day the status of each agent in terms of being alive and/or in a migratory state, its current grid point position (Table 17.1B [0]), its susceptibility to HPAI, as well as metrics of its HPAI infection: day of infection and virulence of the strain (zeros if not infected; Table 17.1B [7–12], where applicable). For each grid point we also logged daily metrics, including number of occupants, number of contagious occupants, and the mean and standard deviation of virulence across all HPAI strains present (Table 17.1B [1–4]. Across the entire 90-day study period, we tracked the distribution of all agents—as well as the subset of contagious agents—relative to the location of the disease origin, in this case: N62.703, W167.117 (red dot in Fig. 17.1). We calculated the median, quartile, and maximum distances among these two groupings at 6-day intervals during the study period and determined the day by which each measure exceeded the minimum distance from the origin of the disease to the contiguous USA (4200 km). These latter calculations were intended to estimate how quickly the agents and the disease, respectively, *could* have possibly reached population centers in the contiguous USA.

2.5   Simulating Monitoring Effort

To evaluate on-the-ground monitoring of disease connectivity, we established a simulated network of monitoring stations at twelve grid points that were located at potentially favorable stopover locations (i.e., relatively low-altitude lakes with little human disturbance within 50 km of human settlements; Fig. 17.1). We allowed the stations to sample daily within a 12.5 km radius of the station (i.e., a 3×3 block of grid points) for signs of communicable HPAI in the agents present. Early-infected agents that were not yet contagious were considered undetectable (i.e., asymptomatic during Days 1–3). To roughly account for density-dependent sampling coverage, we calculated the daily probability of disease detection [$P(dd)$] using the $\log_{10}$ value of tOcc as the base of a root function for the proportion of cOcc in the population:

$$P(\text{dd}) = \sqrt[\log_{10}\text{tOcc}]{\frac{\text{cOcc}}{\text{tOcc}}}$$

This calculation allowed for contagious individuals to experience a higher rate of detection, especially within smaller populations (Brown and Stallknecht 2008). To increase random effects into the probability of disease detection, we then multiplied the resulting value by a normally distributed chance factor with mean of 1.0 and standard deviation of 0.25. If this adjusted detection probability was lower than a random number drawn from a uniform distribution between 0.0 and 1.0, then we determined that HPAI had been detected at this location. We repeated this process 99 more times, each time drawing new chance and random factors. Across all of these repetitions, we then were able to calculate an overall probability of a single detection at each site on each day as the proportion of iterations that produced a positive detection. We likewise calculated the probability of a single detection across any station in the monitoring network. We further calculated the probability of detecting at least one additional contagious agent within a 3-day window (i.e., a "confirming detection"). We calculated this probability separately at each site, as well as collectively across sites, the latter representing a network of stations that would be sharing real-time updates of HPAI detections. This assumed that diagnosis and reporting of the disease could be accomplished in a single day, which may be a departure from reality depending on the testing protocols required.

On the day of migration when the site or network had an 80% or greater probability of detecting HPAI (i.e., a "likely detection"), we determined how far the disease had already spread past the monitoring station, as well as the slope of the 1-day change in the contagious/symptomatic proportion of the surviving population (i.e., the "apparent infection rate"; Van Der Plank 1963). For the former, we calculated this based upon the abundance and relative proportion of still-alive agents that were contagious and located between 215° and 90° from the sampling point (i.e., "escapees"). Among escapees we calculated the mean and median distance from the sampling station, as well as the mean HPAI virulence among contagious agents. In terms of the apparent infection rate occurring at the time of likely detection, we calculated that value based upon the daily change in the proportion of the still-alive population that were contagious, since these were the only diseased individuals that were symptomatic (hence, "apparent"). We then independently assessed the efficacy of each of the 12 sampling locations to detect a disease during the early stages of the disease outbreak and then repeated these same calculations using data based upon 3-day confirming detections. Complete ABM code used within Agent Analyst/RePast available at: https://github.com/rossjd/Migration_monitoring_ABM.

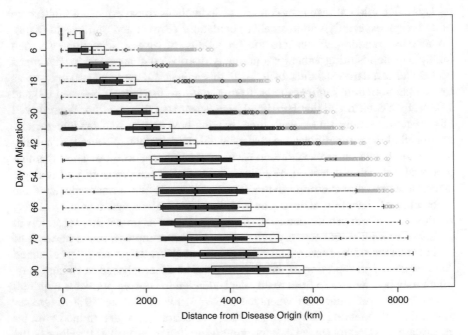

Fig. 17.4 Boxplot distributions of distance from the Siberian population of disease origin among both total occupants (white bars—grey circles) and contagious occupants (red bars and circles) over 6-day intervals of the study period. Note the descending scale on the y-axis

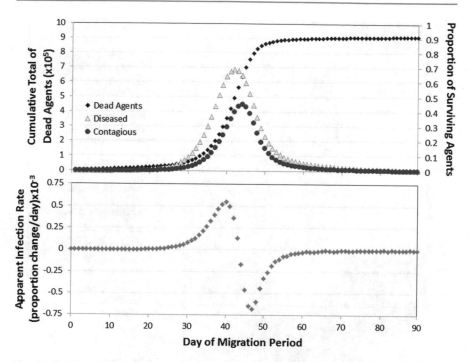

Fig. 17.5 Disease prevalence, infection rates, and mortality metrics across the entire study arena during the ABM-simulated study period. Cumulative numbers of dead agents (black diamonds) are plotted along the upper-left axis, proportions of surviving agents that were diseased (orange triangles) or contagious (red circles) are plotted along the upper-right axis, and day-to-day changes in the proportion of contagious agents (i.e., the *apparent infection rate*) are plotted in the lower panel (blue diamonds). The *x*-axis is consistent for both panels

3   Results

3.1   Breeding-to-Wintering Ground Connectivity

Displacement data, measured as the minimum Great circle route distance over land or ocean from the disease origin, indicated a relatively constant migration and spread of the entire agent population throughout the 90-day simulation period (Fig. 17.4). These patterns were evident as a linear increase of the median distance (black bars within white boxplots in Fig. 17.4) and progressively higher variation within the data (i.e., wider boxplots), respectively. The distribution of contagious individuals, however, appeared to show a different and less-consistent pattern of spread. Specifically, the midpoint of contagious individual locations appeared to spread rapidly outward between Days 42 and 48, which corresponded to the period of rapid collapse in the disease outbreak (Figs. 17.4 and 17.5). The relative distribution of contagious individuals then appeared to largely stagnate, as the

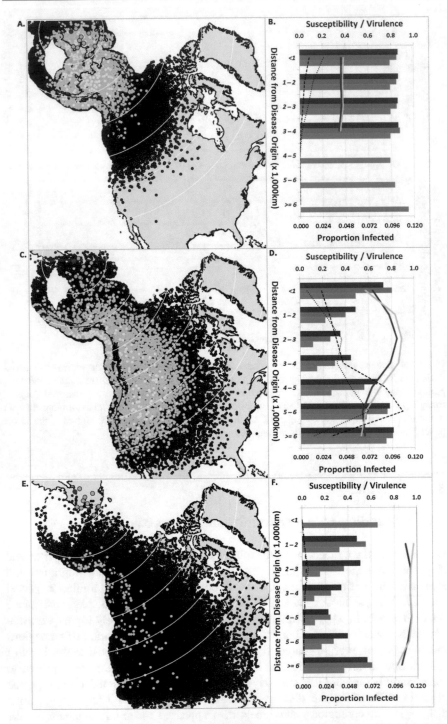

Fig. 17.6 Shifts in spatial distribution of agents (left panels) and disease prevalence/virulence and host susceptibility (right panels) over three select days: (**a**, **b**) Day 23—confirming detection by monitoring network; (**c**, **d**) Day 54—latest day of independent disease confirmation at a

median displacement distance remained between 3300 and 3550 km from Day 48 onward (Fig. 17.3). These distances correspond to approximately northwestern Canada, while the central quartiles during this same time period corresponded to approximately eastern Alaska through western Canada and the Pacific Northwest of the United States (i.e., 2500 and 5000 km from the disease origin; Figs. 17.1 and 17.6).

Cumulative migratory displacements long enough to have reached the contiguous United States (~4200 km from the disease origin) were first observed among a handful of agents by Day 12 of the period; though the second quartile of total survivors did not reach this distance until Day 66 and their median distance did not exceed this distance until after Day 84 (Fig. 17.4). Among contagious agents, the total distance of travel first exceeded 4200 km by Day 18, although this was only one individual. It was not until Day 30 that additional diseased individuals had progressed far enough to possibly reach the contiguous United States (Fig. 17.4). Following the rapid geographic shift in symptomatic agent displacement distances after Day 42, the second quartile of surviving contagious agents overlapped 4200 km by Day 54. Though these agents were still able to normally migrate, the sudden shift in the distribution was more likely attributable to massive die-offs of diseased agents at locations closer to the HPAI outbreak origin than to an uptick in daily migration distances among diseased agents. The median distance of displacement from the HPAI origin site among symptomatic agents never exceed 4200 km; therefore, by Day 90 the bulk of the contagious population could have never reached the contiguous United States under this simulated scenario. However, there were certainly many instances of symptomatic individuals appearing well within that area (Fig. 17.6c), which could have been sufficient to infect resident wild bird, domestic fowl, or even human populations.

3.2 Disease Monitoring

Networked Stations Using a 12-station network of monitoring stations throughout northwestern North America, our model detected with a ≥80% probability the first contagious agent on Day 18 of the migration period. By this point: only 0.16 and 0.56% of the surviving population were contagious or at least infected, respectively.

Fig. 17.6 (continued) monitoring station; and (**e, f**) Day 90—last day of simulation period. Left-hand maps indicate: 1000 km distance intervals (white rings) from disease origin (red dot); locations of uninfected agents (purple points), infected agents (in increasing abundance: blue, yellow, and orange points), and starting populations of the nine uninfected populations (green dots). Right-hand panels summarize within 1000 km distance bins from Siberian disease origin: proportionate abundance and mean disease virulence among contagious (dashed black and red lines, respectively) and early-infected agents (dotted black and orange lines, respectively), and mean susceptibility among contagious (red bars), early-infected (orange bars), and uninfected agents (blue bars)

Table 17.2 Summaries of simulated monitoring station locations and day of likely disease detection (with associated distances escapes had moved past the station and species-wide apparent infection rate) for both single detections (A) and confirming detections (B)

A. Single detection

| | | | | | All escapes | | | Contagious escapes | | | | | |
| Station | | | | | | Distance (km) | | | | Distance (km) | | Mean pop virulence | Species-wide apparent infection rate (%) |
#	Name	Latitude	Longitude	Day of 80% + detection	Total number	Mean	Median	Total Number	Percent of total	Mean	Median		
	Combined network:	n/a	n/a	18	114,986	727	708	80	0.07	730	716	0.35	0.04
1	Minto, AK:	65.209	−149.183	29	316,909	896	830	5728	1.81	800	785	0.39	0.61
2	Tazlina Lake, AK:	61.900	−146.464	22	74,395	607	485	138	0.19	452	417	0.36	0.10
3	Katalla, AK:	60.257	−144.163	26	72,388	669	563	383	0.53	490	396	0.37	0.28
4	Inuvik, YU:	68.278	−134.220	27	104,366	1079	1067	642	0.62	999	1,022	0.38	0.36
5	Laberge, YU:	61.287	−135.176	27	54,721	687	602	284	0.52	537	492	0.38	0.36
6	Chilkoot Inlet, AK:	59.196	−135.042	27	38,141	638	541	180	0.47	470	395	0.37	0.36
7	Rubyrock, BC:	54.704	−125.223	35	23,479	725	639	719	3.06	519	474	0.41	2.69
8	Nimpkish, BC:	50.438	−127.040	37	14,979	731	633	583	3.89	470	388	0.44	3.99
9	Slave Lake, AB:	55.471	−115.456	38	13,623	955	891	519	3.81	751	706	0.43	4.67
10	Columbia, WA:	46.924	−119.325	41	7067	734	657	387	5.48	501	418	0.45	4.98
11	Pend Oreille, ID:	48.101	−116.434	43	8357	812	732	685	8.20	566	522	0.46	1.37
12	Bear River, UT:	41.434	−112.286	50	3074	831	797	228	7.42	550	518	0.48	−3.27

B. 3-day confirmed detection

| Station | | | | Day of 80%+ detection | All escapes | | | Contagious escapes | | | | | Detection-day apparent infection rate (%) |
#	Name	Latitude	Longitude		Total Number	Distance (km) Mean	Median	Total number	Percent of total	Distance (km) Mean	Median	Mean pop virulence	
	Combined network:	n/a	n/a	23	206,283	801	769	607	0.29	729	732	0.36	0.13
1	Minto, AK:	65.209	−149.183	32	367,180	952	870	15,641	4.26	840	813	0.41	1.30
2	Tazlina Lake, AK:	61.900	−146.464	27	132,164	704	570	1071	0.81	531	440	0.38	0.36
3	Katalla, AK:	60.257	−144.163	30	111,415	755	646	2056	1.85	557	487	0.40	0.82
4	Inuvik, YU:	68.278	−134.220	34	188,393	1207	1163	10,328	5.48	1,105	1,104	0.42	2.15
5	Laberge, YU:	61.287	−135.176	30	79,457	733	639	1182	1.49	585	532	0.39	0.82
6	Chilkoot Inlet, AK:	59.196	−135.042	34	87,443	760	650	4093	4.68	559	488	0.42	2.15
7	Rubyrock, BC:	54.704	−125.223	38	33,260	776	676	2494	7.50	561	516	0.43	4.67
8	Nimpkish, BC:	50.438	−127.040	40	21,010	795	700	2128	10.13	531	451	0.44	5.47
9	Slave Lake, AB:	55.471	−115.456	42	20,824	1038	976	2364	11.35	825	776	0.45	3.65
10	Columbia, WA:	46.924	−119.325	44	9588	806	722	1106	11.54	540	471	0.47	−1.63
11	Pend Oreille, ID:	48.101	−116.434	46	10,524	892	813	1469	13.96	607	563	0.47	−6.44
12	Bear River, UT:	41.434	−112.286	54	3966	917	912	252	6.35	669	598	0.53	−1.09

Furthermore, only 80 contagious agents (0.07% of survivors) had escaped beyond any monitoring station by this day, although one symptomatic individual had already reached a point 2320 km ESE past the northwestern-most monitoring station (Minto, AK) and diseased escapees were located an average of 730 km from that station. The mean virulence among strains that had passed at least one monitoring station by Day 18 remained relatively low at 0.35 (Table 17.2A).

If at least one more confirming detection of the disease within a 3-day period was required, then the network of stations was ≥80% likely to confirm the disease by Day 23. By this point, 0.56 and 2.02% of the surviving population were contagious or at least infected, respectively, and 607 contagious individuals (0.29% of survivors) were already present beyond at least one monitoring station. Interestingly, contagious individuals were spread similarly to Day 18, with the furthest present 2042 km SSE from Minto, AK, diseased escapees located an average of 729 km beyond that station, and mean virulence remained relatively low at 0.36 (Table 17.2B). Most of the contagious agents were still contained within or near Siberia or Alaska on Day 23 and the bulk of diseased individuals appeared to be centered on the Bering Strait area (Fig. 17.6a).

Individual Stations Under the scenario of independently operating stations, the rapidity of detecting the disease was understandably lower than if the stations were in constant communication as a network. Certain stations appeared to be more likely to detect the disease than others in the vicinity, particularly in Alaska where the station closest to the Siberian disease origin (Minto, AK) was slower to detect or confirm the outbreak (Days 29 and 32, respectively) than locations further south and east (Table 17.2). This would have meant that this station wouldn't have reported the outbreak until it was well into the acceleration period (i.e., apparent infection rate rising by 0.61–1.30% per day, depending on detection criteria). In contrast, the nearest station to Minto, AK—located only 400 km SSE at Tazlina Lake, AK— rapidly detected the disease outbreak and appeared to do so at an early stage of disease acceleration (i.e., apparent infection rate was rising at only 0.10–0.36% per day, depending on detection criteria; Table 17.2). The primary difference between these stations was that Minto, AK, is located far inland (W of Fairbanks) while Tazlina Lake, AK, is nearer the coastline (NE of Anchorage; Fig. 17.1). Within the structure of the ABM, these characteristics could have favored greater amounts of traffic near the latter station, which would have promoted greater probability of disease transmission within increasingly dense stopover populations.

Across all stations, there was a strong positive correlation between the distance from the disease origin in Siberia (N62.703, W167.117) and the day of earliest disease detection, regardless of detection criteria ($R^2 = 0.92$ for either single or confirming detections; Fig. 17.7). This would be consistent with the disease being propagated relatively evenly across a broad front that progressed, on average, between 135 and 145 km per day (calculated from regression slopes for single and confirming detections, respectively). This rate of disease travel is very unlikely to be simply mediated by individual movements when one considers that (1) the migrations of individual waterfowl agents were set to occur on average every

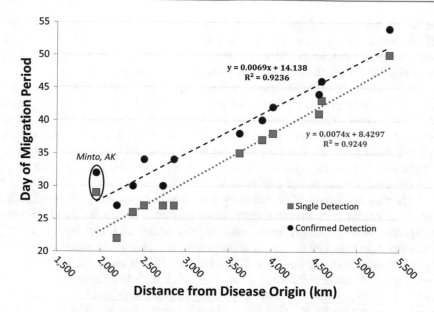

Fig. 17.7 Linear relationship between the distance of monitoring stations from the Siberian disease origin (*x*-axis) and the day of ≥80%-probable disease detection (inverted *y*-axis) for both single detections (grey squares) and confirming detections (black circles). Data from Minto, AK, are circled to illustrate their relative displacement above the regression line, which indicates that disease detections were made at this site later than expected given the proximity of the site to the disease origin

10 days, (2) those movements were selected from a gamma distribution that peaked at ~330 km and was restricted to a maximum of 1200 km (Fig. 17.3), and (3) the probability of mortality quickly rose as a function of the number of days infected. Instead, the most parsimonious explanation is that disease transmission occurred across a series of individuals that were newly infected at successive stopover sites. This aligns with Gaidet et al.'s (2010) similar conclusions which were drawn from actual satellite tracking data among East Asian migratory waterfowl.

3.3 Evolution of Disease Virulence and Host Susceptibility

Much as there was a strong correlation between the distances from the Siberian HPAI origin to each monitoring station versus the days of disease detection, there was also a strong positive correlation between the latter and the mean HPAI virulence among agents that had escaped past each monitoring station. Based upon a linear regression of these data, we found that HPAI virulence appeared to consistently rise over time by a factor of +0.005/day (data not shown). This increase in virulence among surviving HPAI strains effectively reflects the iterative selection pressure imposed on the virus over successive stages of mutation and selection and is likely the result of the following aspects of our ABM: (1) waterfowl hosts can develop resistance through exposure and recovery from less-virulent HPAI strains; (2) clearing the infection is a function of susceptibility and the virulence of the present HPAI strain; and (3) the probability of mortality quickly rises as a function of the number of days an agent has been infected.

The increase in mean virulence was not restricted to only those agents that escaped beyond a given monitoring station; by the time the disease was confirmed at the last station (Day 54 at Bear River NWR, UT; Table 17.2), the mean virulence across the 5686 agents contagious on that day was 0.720 and across 4739 early-infected but asymptomatic agents the mean virulence was 0.799. At the outset of the simulation, 112 of 10^6 agents had been infected with HPAI strains that had a collective mean virulence of only 0.227.

Reflecting the simulated evolutionary arms race incorporated into the ABM, the HPAI susceptibility of surviving agents likewise shifted during the simulation; that is, agents with increasingly lower susceptibility tended to resist the increasingly more virulent strains. At the outset of the simulation (Day 0), the mean susceptibility across all 10^6 agents had been 0.793. By Day 54, the mean susceptibility had fallen to 0.210 among uninfected agents ($n = 101,916$), 0.404 among early-infected agents ($n = 4739$), and 0.536 among contagious agents ($n = 5686$). Among the agents dead by Day 54, 821,658 had died while contagious, 27,232 had died when early infected (Days 1–3), and 38,768 had died without infection (i.e., natural background mortality).

Sticking with the example of Day 54, our examination of locations of diseased agents indicated a dispersed distribution along the Pacific Coast and Rocky Mountain region of North America (Fig. 17.6c). However, when we examined the mean values of HPAI virulence and susceptibility across this area, we found that individuals within a band 2000–3000 km from the disease origin possessed quite virulent strains of the disease, which was counterbalanced by very low susceptibility—especially among uninfected individuals (Fig. 17.6d). At the leading edge of the disease spread (i.e., $\geq$5000 km from HPAI origin), mean susceptibility was high among all agents, regardless of infection status, which indicates that neither natural selection on hosts for high resistance nor a widespread acquisition of immunity had greatly impacted those populations by Day 54. Across all but the smallest 1000 km bins, a consistent pattern was observed for mean susceptibility—that contagious, then early-infected, and then uninfected agents showed stepwise decreases in their levels of susceptibility to HPAI (re)infection. This appeared to fit within the construct of our ABM, wherein the host's susceptibility is roughly proportional to the inverse probability of clearing an infection (hence the higher susceptibility among contagious agents) as well as the probability of being infected after exposure (hence the elevated rate among early-infected agents relative to the uninfected pool).

By the end of our simulation period on Day 90, the suppression of susceptibility remained, though it had shifted further away from the disease origin and peaked at the 4000–5000 km range (Fig. 17.6f). Mean virulence had risen to very high levels across the arena, though the proportion of agents that were diseased had fallen to very low levels. At the end of the simulation period, the original population of one million agents had been reduced to 90,615. This is a very high mortality rate and, as such, may point to the present ABM as a "worst-case scenario" for the emergence and pathology of the simulated HPAI strain. Of the remaining agents, 14,395 had ceased migration at rather broadly spaced wintering locations generally in the mid-continental USA (i.e., beyond 6000 km from the disease origin; Fig. 17.6e). The mean virulence and susceptibility among the few infected wintering agents, either contagious ($n = 19$) or not ($n = 20$), was 0.810 and 0.619, respectively. Mean susceptibility among uninfected wintering agents, although relatively low at 0.362, had not dropped to the levels seen in areas where the disease had been more prevalent (i.e., 3000–5000 km from the origin; Fig. 17.6).

4 Discussion

The exercise we have presented in this chapter was aimed at the general topic of tracking large-scale connectivity among individual animals, and the diseases they carry, as they actively move through the aerosphere to traverse vast landscapes. The ABM used in our theoretical scenario is broadly based upon generalized waterfowl biology, including migration behaviors, conspecific attraction, landscape usage, and HPAI transmission dynamics. Though flight-driven connectivity clearly has the possibility to influence evolutionary and ecological processes related to the species itself, such point-to-point links can drive other ecological processes such as the transfer of energy, nutrients, parasites, or diseases.

A realistically parameterized agent-based model can allow the study of different aspects of natural processes, such as bird migration or disease transmission, perhaps even using the same simulation output in a similar fashion to how scientists of various disciplines could each study different aspects of a natural system. A major advantage of the ABM simulation framework is that we may readily alter input parameters independently, which allows for sensitivity analyses regarding how each factor might disproportionately affect the model output—something that is not easily achieved or even possible under natural or rare circumstances. Even in cases where the input parameters used in ABMs are approximations or generalizations, such as in our case, the scenario evaluations possible through individual-based modeling can still provide valuable biological insights that can inform further data collection, modeling, or management planning (MacPherson and Gras 2016).

4.1 Simulating Migratory Connectivity: Biological Insights

Movement through the aerosphere using powered flight is the fastest way that vertebrate animals can naturally traverse large geographic distances (Alerstam 2003, 2011), allowing for transcontinental transmission of biomass over relatively short periods of time. In the example presented in this chapter, the simulated migration of one million generalized waterfowl from Siberia to the southern USA occurred in some cases within 3–4 weeks. Such rapid movements are not unusual among migratory bird species (Fuller et al. 1998; Kjellén et al. 2001; Alerstam 2003), especially waterfowl (Green et al. 2002; Gaidet et al. 2010; Prosser et al. 2011), though cross-continental, longitudinal movements of individual land birds remain only sparsely studied (e.g., Bairlein et al. 2012).

Long-distance migration facilitated the spread of our simulated highly pathogenic virus. However, migration of the simulated waterfowl host was not in complete lockstep with the dispersal of the HPAI outbreak. In part, this was because of the time required for an HPAI outbreak to achieve critical mass and partly because of apparent selective forces acting on both host and virus. With respect to the former, our model indicated that communicable agents remained rare and the mean virulence of HPAI strains they carried remained relatively low during the first 18 simulated days (Fig. 17.4), by which time nearly half of the total population had moved beyond Siberia (i.e., displacements >1100 km). Even by Day 23, when the disease had spread across Alaska and had been confirmed by the simulated monitoring network, the mean virulence of the disease remained modest (Table 17.2B), although sufficient to readily infect the still quite-susceptible population of agents (Fig. 17.6b).

Subsequent to Day 23, when the HPAI became rapidly more pathogenic and prevalent, the rampant buildup of diseased occupants corresponded to widespread mortality that ultimately resulted in the death of >90% of the entire agent population (Fig. 17.5). A high proportion of surviving agents at distances >4000 km from the disease origin were contagious by Day 54 (7.5%; Fig. 17.6d), which suggests that the virus had disproportionately spread outward during the outbreak phase. The median location of uninfected birds likewise continuously progressed further from the disease origin (Fig. 17.4). However, at no point in our simulation did the median distance of contagious individuals from the disease origin spread exceed 3600 km. In effect, the center point of the disease population appeared to stall after being transmitted only part of the way to wintering grounds. This supports the interesting possibility that migration might initially help but eventually hinder the spread of disease (Altizer et al. 2011), especially when it comes to the geographic extent of HPAI spread (Lam et al. 2012). Such data could only be gleaned from studying the individual interactions between disease and host (Morales et al. 2010) and are an improvement upon well-founded but limited models of disease spread based on very coarse-scale range maps and movements (e.g., Peterson et al. 2009).

The apparent selective interactions in our ABM between host and disease seem to have driven another interesting dynamic—that the surviving populations emerged with greatly elevated levels of resistance (inverse susceptibility) or virulence, respectively. Since we had simulated only a single-species host for HPAI, this meant that equilibrium was eventually reached and the disease returned to relatively low incidence among the surviving hosts (Figs. 17.5 and 17.6). In this respect, the HPAI population had been distilled down to nearly its most potent form (i.e., virulence = 1.0) as a result of selective feedback during stepwise migratory jumps. One may predict that possible alternative hosts for HPAI encountered at points further south could be quite susceptible to such a highly virulent strain of the disease. Such a situation could present a serious risk of rapid infection and widespread mortality in these new hosts, even if there was some degree of native HPAI resistance present. From a competitive standpoint, this could give migratory species an evolutionary advantage on wintering grounds. From an economic standpoint, this could mean that transmission of this HPAI strain into domestic stocks could bring widespread devastation. Such scenarios are possible and warrant further examination using ABMs which incorporate data from targeted quantification studies, such as challenge assays of susceptibility, exposure, infection, and recovery/clearance (Pantin-Jackwood et al. 2007; Brown et al. 2006, 2008).

4.2 ABM-Informed Disease Monitoring and Mitigation

Of significant human health and economic concern is whether confirmed disease detections could be made early enough during an outbreak to prevent further spread and allow effective mitigation efforts within already infected populations (Brown and Stallknecht 2008). Since asymptomatic carriers (e.g., early-infected or resistant individuals) might fly through the aerosphere rapidly and with fewer barriers, the efficacy of a disease monitoring and mitigation program could be drastically altered by delays of just a matter of days (DeLiberto et al. 2011). In our example, the twelve hypothetical monitoring stations distributed from Alaska to Utah varied widely in how quickly they were likely to detect the disease. This pattern was driven almost entirely by the station's distance from the disease origin. However, in the case of the station at Minto, AK, detection and confirmation of the disease lagged far behind the nearest station, Tazlina Lake, AK, despite the former being located closest to the disease origin (Fig. 17.1). The comparative delay of 5 (single detection) to 7 days (confirmed detection; Table 17.2; Fig. 17.7) meant that by the time the disease would have been independently reported at the Minto station, AK, the disease was well into an outbreak, with apparent infection rates 6.2 to 3.6× higher (respectively) than when HPAI was detected at Tazlina Lake (Table 17.2 and Fig. 17.5). Losing this amount of time during the critical early stages of an HPAI outbreak is likely to undermine the efficacy of subsequent mitigation efforts to stem further disease spread (Brown and Stallknecht 2008). Under this scenario, the resources allocated to the Minto monitoring station, AK, would have been much more effective at other sites, such as closer to coastal waterways where the agents appeared to accumulate (Fig. 17.6).

A communicating network of monitoring stations certainly improved the ability to monitor for detectable symptoms of a disease outbreak, though with the caveat that disease detection and information sharing were both instantaneous. Not surprisingly, we found that the early warning efficacy of the network was driven by data generated at only one or two stations in Alaska. Depending on our confidence in the parameters of the model, we could be inclined to say that prioritization of resources should be given to these stations if we wanted to maximize our probability of detecting and confirming a disease during the earliest stages of an outbreak.

However, by reducing the geographic spread of monitoring stations, agencies would have limited capacity to track HPAI dynamics over the course of its invasion of North America. We saw from our simulation that the bulk of the disease outbreak never progressed much beyond Canada. However, our model also indicated that the HPAI achieved a very high level of virulence along the migratory route. Tracing such patterns so that the impending outcome can be more accurately predicted could be critical to avoid an over- or understatement of risk. Such inaccuracies could undermine regulatory agencies' ability to elicit public responses in the face of subsequent disease threats, especially as it pertains to reducing potential avenues of host switching into humans.

4.3   Field Testing and Applying ABMs

Field monitoring is important as a way to continually refine predictive models, as well as to validate the predictions being made in those same models. Such on-the-ground efforts may also be the most effective way to monitor migratory populations for evidence of communicable diseases, in particular those of potential human economic or health impact such as avian influenza (Wobeser 2013). Our model made relatively static assumptions regarding the detectability of contagious individuals relative to asymptomatic or uninfected individuals and that detectability would increasingly rise in smaller populations. However, human observers would likely have much greater intuition about changes in the norm at their sites and could provide the capacity to rapidly adjust monitoring efforts in the face of an emerging disease outbreak. In turn, they could benefit from ABM simulations that focus their search efforts into areas of highest risk for disease appearance and/or escape (e.g., disease may be more likely to slip by at lower-density stopover sites).

For management agencies tasked with preventing potential human impacts from animal-borne diseases, there exists a delicate balance between reducing alpha error (i.e., missing a disease when it is present) at the expense of increased beta error (i.e., false warnings of an impending outbreak). Yet, detecting a disease before it breaks out across large geographic areas, as well as prescribing the appropriate mitigation efforts, could prove impossible without sufficient foresight. Modeling scenarios of biological connectivity from real-world observations could be a powerful, economical, and effective tool to forecast plausible events like the simplified disease transmission and evolution model presented in this chapter. Such information may provide a critical head start in developing and testing potential mitigation strategies. Not only could these models predict what metrics to track during an emerging outbreak or mitigation effort, but they could also define what natural variation may be expected, so that the process can be evaluated in real time to inform where shifts in approach may be needed. It could also pinpoint critical points in the detection and reporting infrastructure, such as a need to rapidly but accurately test for disease within fresh samples at field sites.

Once a disease is detected, possible mitigation strategies could encompass a suite of labor-intensive, socially disruptive, or fiscally expensive management techniques that would be publicly unpopular if they were not objectively supported by a predictive framework. ABMs can repeatedly gauge the probability that, for instance, waterfowl migration would link subsequent stopover nodes within a disease network and what outcomes may be anticipated from various mitigation approaches. These data could provide managers the necessary information to justifiably focus their actions where they could most likely have the intended effect of stopping the disease outbreak. Again, such models could be continually refined and tested as the real-world situation changes and could be accomplished in the field on laptop computers.

4.4   Refining our Migration Connectivity and Disease Monitoring ABM

As with any model, the accuracy of the ABM simulations depends upon the realism of the rules and parameters used in their construction. Our example was, after all, a simulated case study which generalized the migratory characteristics of typical waterfowl species (see Sect. 2). The worst-case scenario presented contrasted prior evaluations of HPAI incidence among migratory birds crossing the Bering Strait, which found little evidence of the disease (Ip et al. 2008; Winker et al. 2007). It may be that avian-borne diseases could enter a continent through a number of different pathways (Peterson et al. 2007), and revisiting our ABM in light of contrasting ground observations is simply another way that this tool can be refined to simulate increasingly realistic scenarios.

Furthermore, our ABM did not account for other possible factors that would likely affect the interplay between migration biology and HPAI disease dynamics. For example, missing were possible mediating factors such as competition between low and highly pathogenic influenza strains (Bourouiba et al. 2011), behavioral avoidance of symptomatic conspecifics (Loehle 1995), stopover duration relative to disease shedding (Gaidet et al. 2010; Feare 2010), impairments among diseased individuals (van Gils et al. 2007; Kuiken 2013), or a full parameterization of density-dependent interactions, especially within small groups (Runge and Marra 2005). On the other hand, the ABM also omitted possible disease-promoting factors such as viral tenacity in the environment (Stallknecht and Brown 2009), cross-species transmission (Kilpatrick et al. 2006; Boyce et al. 2009; Altizer et al. 2011), and in-flight stopover/aggregation decisions (Alerstam 2011). Variables likely to have context-dependent implications for disease spread, such as seasonal or daily variation in stopover suitability (Bauer and Hoye 2014) or host energetic condition (Beldomenico and Begon 2009; Arsnoe et al. 2011), were also not considered in the ABM, and such factors could have drastic effects depending on their timing and spatial patterning along the migratory route. With respect to the monitoring and detection of the disease, here our ABM also disregarded possible extenuating variables, such as interobserver or station variability in search efficiency, local environmental conditions, irregular schedules, and asymmetric detectability of symptomatic individuals. In Box 17.1, we outline these possible additions to our ABM, with brief descriptions of how each might affect the model outcomes.

Box 17.1 Outline of possible extensions of the ABM presented in the chapter that would incorporate increasing realism into the simulation. This list is not intended to be encompassing but a glimpse into how ABM approaches can be as complex as desired (though at the possible expense of resolvability among possible independent variable effects).

Candidate variables	Relevance to host migration or susceptibility	Relevance to disease transmission or detection	Potential data sources (with examples)
Alternate transmission pathways			
Cross-species	Community composition at stopover sites and interspecific interactions may be important	Domestic fowl may be important stores and sources of pathogenic disease (Kilpatrick et al. 2006); passerines are overlooked as potential carriers of avian influenza (Fuller et al. 2010)	Genetic tests of cross-species infections (Lam et al. 2012); disease surveys across multiple species (Winker et al. 2007); species association data
Continental entry points	Multiple pathways by which species might cross, entering different migratory pathways	Greater range of possible disease entry corridors	Circumcontinental monitoring networks (Peterson et al. 2007); integrative individual tracking studies (Bowlin et al. 2010)
Environmental deposition and uptake	Susceptible individuals may not be able to avoid exposure through behavioral avoidance of symptomatic birds	Leading edge of migration may be exposed to latent virus spores at stopover sites	Viral tenacity studies at points across the migratory corridor (Stallknecht and Brown 2009)
Refined monitoring network			
Variable effort and observer effects	Stopover aggregations could be disrupted by observer approach; potentially exposes sick individuals that cannot escape	Increased lag period between disease arrival and detection by site monitors; elevated conspicuousness of symptomatic birds	Existing monitoring station schedules and coverages (Harris et al. 2015); experimental detection probabilities using test drills
East-Asian monitoring sites	Area of flyway overlap and molting grounds; provide	Monitoring effort centered closer to the known locations of	Existing monitoring effort distributions in East Asia (Okazaki

(continued)

Box 17.1 (continued)

	data on initial demography, energetic condition, and community composition	recent HPAI outbreaks (Alexander and Brown 2009; Prosser et al. 2011)	et al. 2000; Xu et al. 2013)
Demographically specific metrics			
Navigation ability	Age, sex, size, and health impacts on the speed and route efficiency of individuals	Certain demographic classes may be more likely to wander further or stopover longer, which could affect exposure and recovery dynamics	Individual tracking studies using satellite transmitters (Gilbert et al. 2010) or geolocators (Bridge et al. 2011)
Disease pathology	Age, sex, size, and health impacts on the disease susceptibility of individuals	Virus transmittal may be aided or muted depending upon the demographic composition of the stopover population	Challenge studies of different age classes, species, or sexes (Pantin-Jackwood et al. 2007)
Density-dependent interactions	Density elevates agonistic interactions and redistribution to suboptimal locations; predation risk tracks population density	Increased disease competition within dense populations as low-pathogenic strains may inoculate hosts (Bourouiba et al. 2011)	Behavioral ecology in migrant aggregations (Runge and Marra 2005); time-allocation budgets (Morales et al. 2010)
Condition-dependent factors			
Disease susceptibility	Migratory birds may be more susceptible to disease during molting periods or under other stressors (Feare 2010; Fuller et al. 1998)	Virus transmittal may be aided or muted depending upon the condition of the exposed individual (Beldomenico and Begon 2009)	Condition-dependent challenge studies (Arsnoe et al. 2011)
Functional impairment	Migratory movements affected by infections, even when asymptomatic (van Gils et al. 2007)	Disease has relatively narrow window to infect hosts and spread to other populations (Gaidet et al. 2010)	Test physiological effects among infected birds (Kuiken 2013); track in-flight physiology (Gumus et al. 2015)
Migration biology factors			
In-flight decisions	Migratory birds are capable of continually assessing underlying landscapes for	Greater potential for disease spread within attractive stopover habitats and among	Real-time individual stopover choices relative to dynamic landscape factors and ground-truthed

(continued)

Box 17.1 (continued)

	conspecific aggregations or suitable habitat	conspecific groupings	population metrics (Kays et al. 2015)
Fat stores and replenishment	Migrants will be constrained to stopover for a duration inversely related to habitat quality	Prolonged stopovers, particularly low-quality habitats, could promote disease exposure	Monitor individual stopover duration relative to underlying habitat (Takekawa et al. 2010)
Spatiotemporal landscape variability			
Seasonal shifts in resources	Climate-mediated resource competition; greater motivation for migration toward suitable wintering habitat (see Chap. 16)	Virus spread must match host's speed toward wintering ground; spatiotemporal patterns in stressors promoting infection	Incorporate seasonal shifts in population distributions and resource availability across the migratory season (Bauer and Hoye 2014)
Daily weather patterns	Locally mediated resource competition; dictates exodus, pathway, and stopover decisions (see Chaps. 8 and 12)	Cycle of host immigration and emigration at stopover sites may ensure a constant supply of potential hosts	Relate archived weather data to existing daily stopover site data for numbers and diversity of birds infected (Winker and Gibson 2010)
Host-disease evolutionary models			
Viral mutability and balancing selection	Greater parameterization of host responses to viral exposure, including immunocompetent plasticity (Beldomenico and Begon 2009)	Underlying genetic variation and spatiotemporal mutation probabilities can drive the pathology of avian influenza disease (Boyce et al. 2009)	Incorporate complex algorithms associated with competition-driven evolution (DeAngelis and Mooij 2005)

4.5   Future Directions for Aeroecological Network Analyses Using ABMs

Biological processes occurring in the aerosphere are diverse and mediated by a plethora of intrinsic and external factors. The application of ABMs to such a large-scale and complex system will necessarily require increasingly greater parameterization based upon real-world measurements. Fortunately, as evinced by the wealth of knowledge presented in the other chapters of this volume, researchers are continuing to increase our knowledge base of ecological processes occurring in the aerosphere, especially as we gain the capability to track how individuals move through this medium and the resulting impacts on population connectivity and landscape permeability.

Aeroecology often revolves around the collection and analysis of so-called big data. While the analysis of such large datasets is daunting, access to increasingly powerful computational hardware is allowing us to fully leverage the power of ABM framework to deconstruct, model, and ultimately understand the complex processes underlying the airspace oddity that is life on the wing. For example, ABMs could readily be applied to broad-scale ecological questions such as the latitudinal redistribution of energy and nutrient resources by migratory animals (Bauer and Hoye 2014). That said, there does exist the potential to overparameterize an ABM at the expense of resolvability among the rapidly expanding realm of variables and interactions. The advantage that the ABM simulation framework allows in isolating specific variables for the purpose of conducting sensitivity analyses could easily be swamped by an overabundance of independent variables. A simple ABM, such as we've presented in this chapter, can provide valuable insights even when it has omitted some variables or has not incorporated all related data during its parameterization (MacPherson and Gras 2016).

The powered flight of animals can rapidly and extensively connect biological communities, including pathogens, across the face of the world. A more thorough understanding of how this connectivity operates can have profound ecological and anthrosocial implications. This is especially true within the modern era, as wild animals are increasingly forced into close association with rapidly growing human population centers and the concentrated animal feeding operations that have become the primary source of domesticated food. How might the future aeroecological patterns affect such human–wildlife conflicts? We believe that ABMs show promise to bind together what we are learning about the global-scale and circannual patterns that comprise aeroecology, as we continue to construct a mosaicked understanding of this yet-emerging scientific discipline.

References

Alerstam T (2003) Bird Migration Speed. In: Berthold P, Gwinner E, Sonnenschein E (eds) Avian Migration. Springer, Berlin. https://doi.org/10.1007/978-3-662-05957-9_17
Alerstam T (2011) Optimal bird migration revisited. J Ornithol 152:5–23. https://doi.org/10.1007/s10336-011-0694-1
Alerstam T, Hedenstrom A, Akesson S (2003) Long-distance migration: evolution and determinants. Oikos 103:247–260
Alerstam T, Baeckman J, Gudmundsson GA et al (2007) A polar system of intercontinental bird migration. Proc R Soc B Biol Sci. https://doi.org/10.1098/rspb.2007.0633
Alexander DJ, Brown IH (2009) History of highly pathogenic avian influenza. Rev Sci Tech 28:19–38

Altizer S, Bartel R, Han BA (2011) Animal migration and infectious disease risk. Science 331:296–302

Arsnoe DM, Ip HS, Owen JC (2011) Influence of body condition on influenza a virus infection in mallard ducks: experimental infection data. PLoS One 6:1–10. https://doi.org/10.1371/journal.pone.0022633

Azziz-Baumgartner E, Smith N, González-Alvarez R et al (2009) National pandemic influenza preparedness planning. Influenza Other Respi Viruses 3:189–196. https://doi.org/10.1111/j.1750-2659.2009.00091.x

Bairlein F, Norris DR, Nagel R et al (2012) Cross-hemisphere migration of a 25 g songbird. Biol Lett 8:505–507. https://doi.org/10.1098/rsbl.2011.1223

Balicer RD, Reznikovich S, Berman E et al (2007) Multifocal avian influenza (H5N1) Outbreak. Emerg Infect Dis 13:1601–1603. https://doi.org/10.3201/eid1310.070558

Bauer S, Hoye BJ (2014) Migratory animals couple biodiversity and ecosystem functioning worldwide. Science 344:1242552. https://doi.org/10.1126/science.1242552

Beldomenico PM, Begon M (2009) Disease spread, susceptibility and infection intensity: vicious circles? Trends Ecol Evol 25:21–27. https://doi.org/10.1016/j.tree.2009.06.015

Bellrose FC, Crompton RD (1970) Migrational behavior of mallards and black ducks as determined from banding. Ill Nat Hist Surv Bull 30:167–234

Bourouiba L, Teslya A, Wu J (2011) Highly pathogenic avian influenza outbreak mitigated by seasonal low pathogenic strains: insights from dynamic modeling. J Theor Biol 271:181–201. https://doi.org/10.1016/j.jtbi.2010.11.013

Bowlin MS, Bisson I-A, Shamoun-Baranes J et al (2010) Grand challenges in migration biology. Integr Comp Biol 50:261–279. https://doi.org/10.1093/icb/icq013

Boyce WM, Sandrock C, Kreuder-Johnson C et al (2009) Avian influenza viruses in wild birds: a moving target. Comp Immunol Microbiol Infect Dis 32:275–286. https://doi.org/10.1016/j.cimid.2008.01.002

Bridge ES, Thorup K, Bowlin MS et al (2011) Technology on the move: recent and forthcoming innovations for tracking migratory birds. Bioscience 61:689–698. https://doi.org/10.1525/bio.2011.61.9.7

Brown JD, Stallknecht DE (2008) Wild bird surveillance for the avian influenza virus. Methods Mol Biol 436:85–97. https://doi.org/10.1007/978-1-59745-279-3_11

Brown JD, Stallknecht DE, Beck JR et al (2006) Susceptibility of North American ducks and gulls to H5N1 highly pathogenic avian influenza viruses. Emerg Infect Dis 12:1663–1670. https://doi.org/10.3201/eid1211.060652

Brown JD, Stallknecht DE, Swayne DE (2008) Experimental infection of swans and geese with highly pathogenic avian influenza virus (H5N1) of Asian lineage. Emerg Infect Dis J 14:136. https://doi.org/10.3201/eid1401.070740

Cagnacci F, Boitani L, Powell RA, Boyce MS (2010) Animal ecology meets GPS-based radiotelemetry: a perfect storm of opportunities and challenges. Philos Trans R Soc Lond B Biol Sci 365:2157–2162

CEC – Commission for Environmental Cooperation (2007) North America elevation 1-kilometer resolution, 3rd edn. http://www.cec.org/Atlas/Files/elevation/Elevation_Layer_Package.zip. Accessed 6 June 2015

CEC – Commission for Environmental Cooperation (2010) North American Atlas – hydrography, 2009 revision. http://www.cec.org/Atlas/Files/Rivers_and_Lakes/Rivers_and_Lakes_Layer_Package.zip. Accessed 23 Sept 2013

Choffnes ER, Sparling PF, Hamburg MA et al (2007) Global infectious disease surveillance and detection: assessing the challenges – finding solutions, Workshop Summary. National Academies Press

Cooke SJ, Hinch SG, Wikelski M et al (2004) Biotelemetry: a mechanistic approach to ecology. Trends Ecol Evol 19:334–343. https://doi.org/10.1016/j.tree.2004.04.003

Dau CP (1992) The fall migration of pacific flyway Brent Branta bernicla in relation to climatic conditions. Wildfowl 43:80–95

DeAngelis DL, Mooij WM (2005) Individual-based modeling of ecological and evolutionary processes. Annu Rev Ecol Evol Syst 36:147–168

DeLiberto TJ, Swafford S, Why K Van (2011) Development of a national early eetection system for highly pathogenic avian influenza in wild birds in the United States of America. USDA National Wildlife Research Center – Staff Publications

Delmore KE, Irwin DE (2014) Hybrid songbirds employ intermediate routes in a migratory divide. Ecol Lett 17:1211–1218. https://doi.org/10.1111/ele.12326

Ellis EC, Ramankutty N (2008) Putting people in the map: anthropogenic biomes of the world. Front Ecol Environ 6. https://doi.org/10.1890/070062. http://www.cec.org/Atlas/Files/Anthro pogenic_Biomes/Anthromes_Layer_Package_GRID.zip. Accessed 20 Sept 2013

Ely CR, Franson JC (2014) Blood lead concentrations in Alaskan tundra swans: linking breeding and wintering areas with satellite telemetry. Ecotoxicology 23:349–356. https://doi.org/10.1007/s10646-014-1192-z

Ely CR, Douglas DC, Fowler AC, Babcock CA, Derksen DV, Takekawa JY (1997) Migration behavior of Tundra Swans from the Yukon-Kuskokwim Delta, Alaska. Wilson Bull 109 (4):679–692

ESRI (2012) World continents. ESRI, Redlands

European Centre for Disease Prevention and Control (ECDC) (2007) Technical Report: Pandemic Influenza Preparedness in the EU. Status Report as of Autumn 2006

Feare CJ (2010) Role of wild birds in the spread of highly pathogenic avian influenza virus H5N1 and implications for global surveillance. Avian Dis 54:201–212

Ferguson NM, Cummings DAT, Fraser C et al (2006) Strategies for mitigating an influenza pandemic. Nature 442:448–452

Flock WL (1972) Radar observations of bird migration at Cape Prince of Wales. Arctic 25:83–98

Fuller MR, Seegar WS, Schueck LS (1998) Routes and travel rates of migrating Peregrine Falcons Falco peregrinus and Swainson's Hawks Buteo swainsoni in the Western Hemisphere. J Avian Biol 29:433–440. https://doi.org/10.2307/3677162

Fuller TL, Saatchi SS, Curd EE, Toffelmier E, Thomassen HA, Buermann W, DeSante DF et al (2010) Mapping the risk of Avian Influenza in wild birds in the US. BMC Infect Dis 10(1):187. https://doi.org/10.1186/1471-2334-10-187. (BioMed Central)

Franklin AB, Anderson DR, Burnham KP (2002) Estimation of long-term trends and variation in avian survival probabilities using random effects models. J Appl Stat 29(1–4):267–287. https://doi.org/10.1080/02664760120108719. (Taylor & Francis)

Gaidet N, Cappelle J, Takekawa JY et al (2010) Potential spread of highly pathogenic avian influenza H5N1 by wildfowl: dispersal ranges and rates determined from large-scale satellite telemetry. J Appl Ecol 47:1147–1157. https://doi.org/10.1111/j.1365-2664.2010.01845.x

Gilbert M, Xiao X, Domenech J et al (2006) Anatidae migration in the western Palearctic and spread of highly pathogenic avian influenza H5N1 virus. Emerg Infect Dis 12:1650–1656. https://doi.org/10.3201/eid1211.060223

Gilbert M, Newman SH, Takekawa JY, Loth L, Biradar C, Prosser DJ, Balachandran S et al (2010) Flying over an infected landscape: distribution of highly pathogenic Avian Influenza H5N1 risk in South Asia and satellite tracking of wild waterfowl. EcoHealth 7(4):448–458. https://doi.org/10.1007/s10393-010-0672-8

Green M, Alerstam T, Clausen P et al (2002) Dark-bellied Brent Geese Branta bernicla bernicla, as recorded by satellite telemetry, do not minimize flight distance during spring migration. Ibis 144:106–121. https://doi.org/10.1046/j.0019-1019.2001.00017.x

Gumus A, Lee S, Ahsan SS et al (2015) Lab-on-a-Bird: biophysical monitoring of flying birds. PLoS One 10:e0123947

Harris RJ, Reed JM (2002) Behavioral barriers to non-migratory movements of birds. Ann Zool Fennici 39:275–290

Harris MC, Miles AK, Pearce JM et al (2015) USGS highly pathogenic avian influenza research strategy: U.S. Geological Survey Fact Sheet 2015–3060, 4 p

Hebblewhite M, Haydon DT (2010) Distinguishing technology from biology: a critical review of the use of GPS telemetry data in ecology. Philos Trans R Soc Lond B Biol Sci 365:2303–2312

Hobson KA, Ryan Norris D (2008) Tracking animal migration with stable isotopes. Terr Ecol 2:1–19. https://doi.org/10.1016/S1936-7961(07)00001-2

Hulse-Post DJ, Sturm-Ramirez KM, Humberd J et al (2005) Role of domestic ducks in the propagation and biological evolution of highly pathogenic H5N1 influenza viruses in Asia. Proc Natl Acad Sci USA 102:10682–10687. https://doi.org/10.1073/pnas.0504662102

Ip HS, Flint PL, Franson JC et al (2008) Prevalence of influenza A viruses in wild migratory birds in Alaska: patterns of variation in detection at a crossroads of intercontinental flyways. Virol J 5:71. https://doi.org/10.1186/1743-422X-5-71

Irwin DE (2002) Phylogeographic breaks without geographic barriers to gene flow. Evolution 56:2383–2394. https://doi.org/10.1111/j.0014-3820.2002.tb00164.x

Jebara KB (2004) Surveillance, detection and response: managing emerging diseases at national and international levels. Rev Sci Tech 23:709–715

Kapan DD, Bennett SN, Ellis BN et al (2006) Avian influenza (H5N1) and the evolutionary and social ecology of infectious disease emergence. Ecohealth 3:187–194. https://doi.org/10.1007/s10393-006-0044-6

Kays R, Crofoot MC, Jetz W, Wikelski M (2015) Terrestrial animal tracking as an eye on life and planet. Science 348:aaa2478

Keeling MJ, Rohani P (2008) Modeling infectious diseases in humans and animals. Princeton University Press, Princeton

Kilpatrick AM, Chmura AA, Gibbons DW et al (2006) Predicting the global spread of H5N1 avian influenza. Proc Natl Acad Sci USA 103:19368–19373. https://doi.org/10.1073/pnas.0609227103

Kjellén N, Hake M, Alerstam T (2001) Timing and speed of migration in male, female and Juvenile Ospreys Pandion haliaetus between Sweden and Africa as revealed by field observations, radar and satellite tracking. J Avian Biol 32:57–67

Krementz DG, Asante K, Naylor LW (2012) Autumn migration of Mississippi Flyway Mallards as determined by satellite telemetry. J Fish Wildl Manag 3:238–251. https://doi.org/10.3996/022012-JFWM-019

Kuiken T (2013) Is low pathogenic avian influenza virus virulent for wild waterbirds? Proc Biol Sci 280:20130990. https://doi.org/10.1098/rspb.2013.0990

Lam TT-Y, Ip HS, Ghedin E et al (2012) Migratory flyway and geographical distance are barriers to the gene flow of influenza virus among North American birds. Ecol Lett 15:24–33. https://doi.org/10.1111/j.1461-0248.2011.01703.x

Loehle C (1995) Social barriers to pathogen transmission in wild animal populations. Ecology 76:326–335. https://doi.org/10.2307/1941192

MacPherson B, Gras R (2016) Individual-based ecological models: adjunctive tools or experimental systems? Ecol Modell 323:106–114. https://doi.org/10.1016/j.ecolmodel.2015.12.013

McKinnon EA, Fraser KC, Stutchbury BJM (2013) New discoveries in landbird migration using geolocators, and a flight plan for the future. Auk 130:211–222. https://doi.org/10.1525/auk.2013.12226

Moermond JE, Spindler MA (1997) Migration route and wintering area of Tundra Swans Cygnus columbianus nesting in the Kobuk-Selawik Lowlands, North-West Alaska. Wildfowl 48:16–25

Morales JM, Moorcroft PR, Matthiopoulos J et al (2010) Building the bridge between animal movement and population dynamics. Philos Trans R Soc Lond B Biol Sci 365:2289–2301

Mosbech A, Gilchrist G, Merkel F et al (2006) Year-round movements of northern common eiders Somateria mollissima breeding in Arctic Canada and West Greenland followed by satellite telemetry. Ardea 94:651–665

Okazaki K, Takada A, Ito T et al (2000) Precursor genes of future pandemic influenza viruses are perpetuated in ducks nesting in Siberia. Arch Virol 145:885–893. https://doi.org/10.1007/s007050050681

Olsen B, Munster VJ, Wallensten A et al (2006) Global patterns of influenza A virus in wild birds. Science 312:384–388

Pantin-Jackwood MJ, Suarez DL, Spackman E, Swayne DE (2007) Age at infection affects the pathogenicity of Asian highly pathogenic avian influenza H5N1 viruses in ducks. Virus Res 130:151–161. https://doi.org/10.1016/j.virusres.2007.06.006

Pérez-Tris J, Bensch S, Carbonell R et al (2004) Historical diversification of migration patterns in a Passerine bird. Evolution 58:1819–1832. https://doi.org/10.1111/j.0014-3820.2004.tb00464.x

Peterson AT, Benz BW, Papeş M (2007) Highly pathogenic H5N1 avian influenza: entry pathways into North America via bird migration. PLoS One 2:e261. https://doi.org/10.1371/journal.pone.0000261

Peterson AT, Andersen MJ, Bodbyl-Roels S et al (2009) A prototype forecasting system for bird-borne disease spread in North America based on migratory bird movements. Epidemics 1:240–249. https://doi.org/10.1016/j.epidem.2009.11.003

Prosser DJ, Takekawa JY, Newman SH et al (2009) Satellite-marked waterfowl reveal migratory connection between H5N1 outbreak areas in China and Mongolia. Ibis (Lond 1859) 151:568–576. https://doi.org/10.1111/j.1474-919X.2009.00932.x

Prosser DJ, Cui P, Takekawa JY et al (2011) Wild bird migration across the Qinghai-Tibetan plateau: a transmission route for highly pathogenic H5N1. PLoS One 6:e17622

Pybus OG, Fraser C, Rambaut A (2013) Evolutionary epidemiology: preparing for an age of genomic plenty. Philos Trans R Soc Lond B Biol Sci 368:540–550

Robinson WD, Bowlin MS, Bisson I et al (2010) Integrating concepts and technologies to advance the study of bird migration. Front Ecol Environ 8:354–361. https://doi.org/10.1890/080179

Runge M, Marra P (2005) Modeling seasonal interactions in the population dynamics of migratory birds. In: Greenberg R, Marra PP (eds) Birds of two worlds: the ecology and evolution of migration. Johns Hopkins University Press, Baltimore, pp 375–390

Scallan E (2007) Activities, achievements, and lessons learned during the first 10 years of the foodborne diseases active surveillance network: 1996–2005. Clin Infect Dis 44:718–725. https://doi.org/10.1086/511648

Shariatinajafabadi M, Wang T, Skidmore AK et al (2014) Migratory herbivorous waterfowl track satellite-derived green wave index. PLoS One 9:e108331

Si Y, Xin Q, de Boer WF et al (2015) Do arctic breeding geese track or overtake a green wave during spring migration? Sci Rep 5:8749

Silkavute P, Tung DX, Jongudomsuk P (2013) Sustaining a regional emerging infectious disease research network: a trust-based approach. Emerg Health Threats J 6. https://doi.org/10.3402/ehtj.v6i0.19957

Stallknecht DE, Brown JD (2009) Tenacity of avian influenza viruses. Rev Sci Tech 28:59–67

Sutherland GD, Harestad AS, Price K, Lertzman KP (2000) Scaling of natal dispersal distances in terrestrial birds and mammals. Conserv Ecol 4:16

Takekawa JY, Newman SH, Xiao X et al (2010) Migration of waterfowl in the East Asian Flyway and spatial relationship to HPAI H5N1 outbreaks. Avian Dis 54:466–476. https://doi.org/10.1637/8914-043009-Reg.1

Takekawa JY, Prosser DJ, Collins BM et al (2013) Movements of wild ruddy shelducks in the Central Asian Flyway and their spatial relationship to outbreaks of highly pathogenic avian influenza H5N1. Viruses 5:2129–2152. https://doi.org/10.3390/v5092129

Tan K-X, Jacob SA, Chan K-G, Lee L-H (2015) An overview of the characteristics of the novel avian influenza A H7N9 virus in humans. Front Microbiol 6:1–11. https://doi.org/10.3389/fmicb.2015.00140

Tang W, Bennett DA (2010) Agent-based modeling of animal movement: a review. Geogr Compass 4:682–700. https://doi.org/10.1111/j.1749-8198.2010.00337.x

Taubenberger JK, Kash JC (2016) Influenza virus evolution, host adaptation, and pandemic formation. Cell Host Microbe 7:440–451. https://doi.org/10.1016/j.chom.2010.05.009

Taylor PD, Fahrig L, Henein K, Merriam G (1993) Connectivity is a vital element of landscape structure. Oikos 68:571–573. https://doi.org/10.2307/3544927

Van Der Plank JE (1963) Plant diseases: epidemics and control. Academic, New York van Gils JA, Munster VJ, Radersma R et al (2007) Hampered foraging and migratory performance in swans infected with low-pathogenic avian influenza A virus. PLoS One 2:e184. https://doi.org/10.1371/journal.pone.0000184

von Rönn JAC, Shafer ABA, Wolf JBW (2016) Disruptive selection without evolution across a migratory divide. Mol Ecol. https://doi.org/10.1111/mec.13521

Wagner MM, Tsui F-C, Espino JU et al (2001) The emerging science of very early detection of disease outbreaks. J Public Heal Manag Pract 7:51–59

WCS/CIESIN – Wildlife Conservation Society and Center for International Earth Science Information Network (2005) Last of the Wild Data Version 2 (LWP-2): Global human footprint dataset (geographic). http://www.cec.org/Atlas/Files/Human_Influence_Terrestrial/HumanInfluenceTerrestrial_Layer_Package_GRID.zip. Accessed 20 Sept 2013

Wiens JA (1976) Population responses to patchy environments. Annu Rev Ecol Syst 7:81–120

Winker K, Gibson DD (2010) The Asia-to-America influx of avian influenza wild bird hosts is large. Avian Dis 54:477–482. https://doi.org/10.1637/9192-874109-DIGEST.1

Winker K, McCracken KG, Gibson DD et al (2007) Movements of birds and avian influenza from Asia into Alaska. Emerg Infect Dis 13:547–552. https://doi.org/10.3201/eid1304.061072

Wobeser GA (2013) Investigation and management of disease in wild animals. Springer Science and Business Media, New York

Xu C, Havers F, Wang L et al (2013) Monitoring avian influenza A(H7N9) virus through national influenza-like illness surveillance, China. Emerg Infect Dis J 19:1289. https://doi.org/10.3201/eid1908.130662

Yang Y, Halloran ME, Sugimoto JD, Longini IM (2007) Detecting human-to-human transmission of avian influenza A (H5N1). Emerg Infect Dis 13:1348–1353. https://doi.org/10.3201/eid1309.070111

Zink RM (1996) Comparative phylogeography in North American birds. Evolution 50:308–317. https://doi.org/10.2307/2410802

Sharing the Aerosphere: Conflicts and Potential Solutions

18

Judy Shamoun-Baranes, Hans van Gasteren, and Viola Ross-Smith

Abstract

As our use of the aerosphere is increasing, so too are the conflicts that arise between our activities and those of aerial wildlife. As a result, numerous stakeholders are interested in monitoring, modelling and forecasting the aerial movements of animals in the context of anthropogenic impacts. Birds can pose a serious threat to aviation, resulting in delays, damage to aircraft, lost flight hours and even the loss of lives. Military and civil aviation use a range of measures to monitor the movements of birds and to try and reduce the risk of wildlife strikes. Increasingly, Unmanned Aerial Vehicles are sharing an already crowded airspace, although just how problematic this may become remains to be seen. The wind energy industry, another important stakeholder, may pose serious threats for aerial wildlife, due to collisions with turbines, or the extra energetic costs and risks entailed with avoiding wind farms. Similarly, other tall structures pose a threat for aerial wildlife. In this chapter, we describe the nature of these different conflicts and provide an overview of the factors that influence the risk associated with aerial movement. We also describe how movement is being studied to provide essential information for these different stakeholders and discuss several

J. Shamoun-Baranes (✉)
Theoretical and Computational Ecology, Institute for Biodiversity and Ecosystem Dynamics, University of Amsterdam, Amsterdam, The Netherlands
e-mail: Shamoun@uva.nl

H. van Gasteren
Theoretical and Computational Ecology, Institute for Biodiversity and Ecosystem Dynamics, University of Amsterdam, Amsterdam, The Netherlands

Royal Netherlands Air Force, Breda, The Netherlands
e-mail: jr.v.gasteren@mindef.nl

V. Ross-Smith
British Trust for Ornithology, The Nunnery, Thetford, Norfolk, UK
e-mail: viola.ross-smith@bto.org

© Springer International Publishing AG, part of Springer Nature 2017
P.B. Chilson et al. (eds.), *Aeroecology*,
https://doi.org/10.1007/978-3-319-68576-2_18

of the solutions that have been implemented to reduce potential conflicts. We conclude by discussing future perspectives for reducing conflicts by integrating different technologies for studying aerial movement, diverse approaches for modelling movement and working across international borders.

1   Human–Wildlife Conflicts in the Aerosphere

Human activities are increasingly encroaching on the aerosphere, the airspace used by wildlife for their daily and seasonal movements, resulting in numerous types of aerial conflicts (Fig. 18.1). Whether it's the large structures we build that extend into this space, such as wind turbines, skyscrapers, power lines, or aircraft flying through the aerosphere, conflicts are bound to occur. This competition for aerial space can be detrimental to wildlife, both directly due to collisions between wildlife and anthropogenic structures or aerial vehicles and indirectly due to disruption of normal movement patterns of wildlife. It can also result in the loss of human lives and heavy financial costs. As a result, there is a diverse array of stakeholders interested in quantifying and predicting potential conflicts and implementing measures to try and reduce them. In this chapter, we briefly review conflicts that arise during animal flight and focus on aviation, wind turbines and static structures (without moving parts, e.g. blades), providing a brief overview of the factors that influence the risk of aerial collisions. We show how ongoing research is helping to provide the information and even solutions needed to reduce human–wildlife conflicts within the aerosphere. Understanding the external and intrinsic factors that influence aerial movement is often the first step in this process.

Fig. 18.1 Schematic representation of the aerial conflicts associated with collisions between aerial organisms and human structures and vehicles that encroach on the aerosphere, such as civil aviation during take-off and landing, military aviation, wind turbines and other tall structures. The most problematic part of the aerosphere is the lowest few hundred metres. However, birds can also regularly fly at altitudes of several thousand metres, especially during migration, bringing them into conflict with military aviation at these altitudes as well

2   Aviation

2.1   Problems for Aviation

Collisions between wildlife and aircraft have been recorded since the earliest days of aviation. In most cases, these collisions are detrimental to wildlife but have little or no impact on the aircraft concerned. However, the small proportion of collisions that do impact aviation can result in flight delays, damage to aircraft, loss of aircraft and on rare occasions the loss of human lives. Tens of thousands of animals die each year as a direct result of collisions with aircraft (McKee et al. 2016) and probably a tenfold more are culled as part of wildlife strike management programmes world-wide (Dolbeer et al. 1993; McKee et al. 2016). Most of the research and development related to wildlife–aircraft collisions has focused on reducing the risk of collisions from the perspective of human safety or amenity rather than species conservation or animal welfare per se, although, ideally, all three issues should be considered simultaneously (Buurma 1994; Froneman 2005; Kumar 2014; Leshem and Froneman 2003). Collisions between wildlife and aircraft resulted in the loss of at least 229 lives and the destruction of 221 aircraft worldwide from 1988 to 2009 (Dolbeer 2011). In the USA, over 82,000 collisions with civil aircraft were reported to the Federal Aviation Authority (FAA) between 1990 and 2007 Anonymous (2015). Birds were involved in 97.5% of these, and bats in 0.3%. The rest involved terrestrial mammals and reptiles on the runways (Dolbeer and Wright 2008).

Economic losses due to such collisions are difficult to estimate as they include direct costs related to damage but also indirect costs related to the disruption of flights, human safety and fatality (Sodhi 2002). It is estimated that worldwide, bird aircraft collisions ("bird strikes") with commercial transport aircraft result in 1.2 billion USD in direct and indirect costs annually due to damage and delays (Allan 2002). Additional costs generally not included in these estimates are costs related to risk assessment and development plans prior to establishing new aerodromes or extending existing airports, money spent on project proposals which are not granted and costs accrued by setting up mitigation measures, all of which should finally result in a net reduction of costs if successful. This section focuses predominantly on birds, the taxa for which the most information is available and related research and development has been conducted. It is important to note that collisions with bats do occur and while they are in general far less common than collisions with birds they can be problematic in certain regions, for example the South Pacific and Paleotropics (Parsons et al. 2009; Biondi et al. 2013). While many of the issues described in this section and examples provided are from bird-related studies, numerous aspects are relevant for bats as well.

The number of bird strikes decreases exponentially with altitude, with most bird strikes occurring below 400 m above the earth surface (Dolbeer 2006; Lovell and Dolbeer 1999). While several species do fly up to altitudes of several thousand meters, especially during migration (Bishop et al. 2015; Leshem and Yom-Tov 1996b; Kemp et al. 2013; Able 1970), flight altitudes during local movements are predominantly within the first few hundred metres of the earth surface (Shamoun-Baranes et al. 2006; Larsen and Guillemette 2007; Avery et al. 2011). Figure 18.2

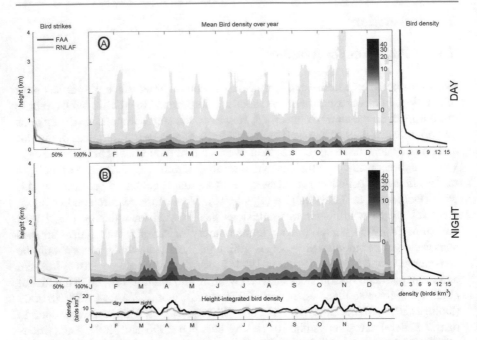

Fig. 18.2 The altitude distribution of bird strikes and aerial bird density (birds/km^3) during the day (**a**) and night (**b**). Left panels: the frequency distribution of bird strikes at different altitudes for the FAA (civil aviation) and the RNLAF (military aviation). Centre panels: 7-day running average of mean daily aerial bird density (birds/km^3) at 200 m altitude bins during the year measured by five weather radar (three in Belgium, two in the Netherlands) from 1 April 2012–1 July 2014; darker colours indicate higher densities. Right panels. Frequency distribution of average bird density with altitude (200 m bins). Bottom panel: the height integrated bird density (birds/km^2) for daytime and night-time. Most bird strikes occur in the lowest 500 m (left) overlapping with altitudes where average bird density is highest (right). Nocturnal bird densities (birds/km^3) are highest during spring (e.g. March, April) and autumn migration (e.g. October)

provides an overview of the altitude distribution of wildlife collisions recorded in the FAA wildlife strike database from 1990 to July 2014 for large aircraft (>27,000 kg) (Dolbeer and Wright 2008) and the Royal Netherlands Air Forces (RNLAF) jet fighter aircraft between 1976 and 2014, showing the concentration of bird strikes at low altitude and the sharp decrease in collisions at higher altitudes. Height distributions of bird movements and mean bird density estimated using weather radar are also shown in Fig. 18.2. To estimate the mean bird density, methods described in (Dokter et al. 2011) were applied to five Doppler weather radar in the Netherlands (two radar) and Belgium (three radar) using data collected from 1 April 2012–1 July 2014. All measurements from five weather radars were averaged for each day (sunrise to sunset) or night (sunset to sunrise). In order to smooth out short-term fluctuations, a 7-day moving average was applied to the data to create the daily density height profiles over the year for bird densities during daytime (Fig. 18.2a) or night-time (Fig. 18.2b). The average density height profile of all five weather radars, averaged over the whole year (Fig 18.2, right), shows a significant overlap with bird strike occurrence (aggregated in 200 m altitude bins) for both daytime and night-time height distributions (day: $R^2_{adj} = 0.994$, $P < 0.0001, n = 20$; night: $R^2_{adj} = 0.847, P < 0.0001, n = 20$). As large commercial aircraft quickly climb above altitudes where birds are present (Fig. 18.2), most bird strikes usually occur in and around airfields, during take-off and landing and at relatively low aircraft flight speeds (Fig. 18.3a). Military aircraft, however, generally operate at low altitudes and at high speeds en route. As a result, there is a bimodal distribution of the aircraft flight speeds at which bird strikes occur (Fig. 18.3b), with bird strikes at relatively low speeds occurring during take-off and landing and bird strikes at high speeds occurring en route. While aircraft flight speeds are relatively low during take-off and landing, ingestion of birds into an

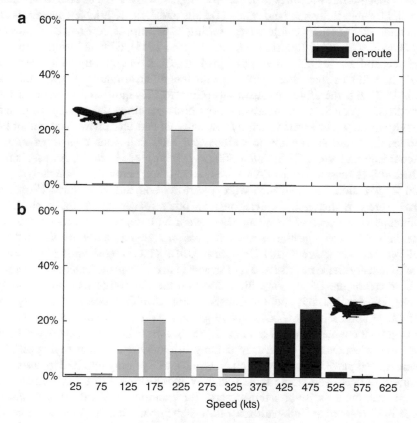

Fig. 18.3 The percentage of bird strikes that have been reported at different aircraft flight speeds. (a) FAA Wildlife strike database 1990–July 2014, large aircraft >27,000 kg ($n = 39785$) and (b) the RNLAF, Jet fighter aircraft, 1976–2014 ($n = 2292$). Note that as civil aviation (a) generally flies above the altitude where birds are found, bird strikes occur locally in and around the airfield, during terminal flight phases (take-off, initial climb, initial approach, final approach and landing) and at relatively low speeds. Military jet fighters experience bird strikes both at relatively low speed during take-off and landing and at higher speeds en route as most of their flights occur within the same airspace shared by birds

engine during take-off will cause more damage than during landing. During take-off, engines are at full thrust, increasing mechanical stress and thus the risk of significant engine damage or engine failure. Thus, the bird strike problem is often divided into local bird strikes, problematic for both civil and military aircraft, and bird strikes that occur en route, relevant predominantly for military aviation.

The severity of a collision depends on several biological and structural factors including the mass of the animal, the number of individuals striking an aircraft, the speed of the aircraft, the point of impact on the aircraft and the type of aircraft (Table 18.1). Direct impact with the aircraft's landing gear, canopy and the engine, especially if there is only a single engine, can all be detrimental. For commercial and military aircraft, airworthiness standards define the impact by birds of different masses that aircraft structures (e.g. canopy, engine, wing) must be able to withstand and still be able to fly and land safely (Dennis and Lyle 2009). New materials and structural designs are being tested to improve impact resistance of aircraft (e.g. Guida et al. 2009; Reglero et al. 2011; Amoo 2013), without losing its specific flight capabilities the aircraft was designed for. Civil and military helicopters tend to fly at altitudes that often coincide with the flight altitudes of birds and bats (Fig. 18.2) and therefore are also susceptible to damaging and sometimes fatal collisions, especially when relatively large birds penetrate the canopy. In general though, military jet aircraft which fly low and fast and often have only one or two engines are more susceptible to serious and fatal collisions than other aircraft (Richardson and West 2005; Allan 2002; Washburn et al. 2013). An assessment of bird strikes reported to the FAA with civil aviation below approximately 152 m (500 ft) revealed that collisions with waterfowl and turkey vultures (*Cathartes aura*) were a higher risk for aviation than other species (DeVault et al. 2011). Collisions with flocks of birds can also have a high negative impact on aviation, especially with the ingestion of several birds in a single engine or birds striking multiple engines (Avrenli and Dempsey 2014). Thus, larger birds and flocking species usually cause more severe bird strikes than other species. One such example is the crash of the US Airways flight 1549 on the Hudson River. On 15 January 2009, the aircraft struck multiple Canada geese (*Branta canadensis*) shortly after take-off at approximately 800 m above ground level and 8 km from the airport, causing both engines to fail (Marra et al. 2009). Fortunately, the experienced crew was able to conduct an emergency landing in the Hudson River saving all crew members and passengers. However, sometimes it takes only one bird to destroy an engine and thus even single birds regularly flying through the same airspace as aircraft can create a flight safety hazard; for example, a Royal Netherlands Air Force F-16 crashed at Volkel airbase on 21 September 2006 due to a collision with one Wood pigeon (*Columba palumbus*, mean mass = 201 g, Dunning 1993).

The flight behaviour of birds is highly relevant when trying to understand and mitigate bird strikes (Table 18.1). The flight altitude, the amount of time spent in the air and how abundant the species is in areas where aircraft are active are all relevant factors that influence the probability of a bird strike occurring. Not all species are equally prone to bird strikes (Both et al. 2010; Allan 2006) and even within a species the susceptibility to bird strikes may change during the course of a bird's

Table 18.1 Factors influencing collision risk during flight posed by human structures and aircraft to wildlife, classified by factors intrinsic to the animals concerned, and those related to their environment or the human structure/activity itself

Factor	Description	Examples
Intrinsic factors		
Morphology	Factors such as body mass, wing loading, aspect ratio and wing span affect flight behaviour and hence collision vulnerability	• Large birds that rely almost entirely on thermal soaring with relatively high wing loading compared to other soaring birds or birds that are less manoeuvrable during flight are more prone to collision with wind turbines and power lines than other species (de Lucas et al. 2012; Janss 2000) • Birds with a higher body mass pose a higher threat for aviation than smaller birds (DeVault et al. 2005, 2011; Dolbeer et al. 2000)
Perception	Limitations of sensory perception and cues attended to may affect species' ability to detect objects with which they might collide	• Visual fields of perception may be limited in frontal vision of birds or leave certain birds blind to structures erected in their direction of travel (Martin et al. 2012; Martin 2011) • Motion smear means birds cannot detect fast-moving objects, e.g. wind turbine blades, at certain speeds and in particular conditions (Hodos 2003) • Bats may be attracted to tall structures which are not readily distinguished from trees (Cryan et al. 2014; Jameson and Willis 2014) • Migrating bats may rely on linear features for navigation and get confused by wind farms (Kunz et al. 2007)
Avoidance	The ability to avoid structures reduces collision risk. Closely linked to morphology and perception.	• Some birds can take last-minute action to avoid wind turbine blades (Cook et al. 2014). • Common terns while often crossing power lines did not collide often, perhaps due to high manoeuvrability during flight and timely adjustment of flight paths despite proximity between the breeding colony and the power lines (Henderson et al. 1996) • Some birds are more strike prone than others, a factor which is incorporated in some bird hazard risk models in aircraft safety (Allan 2006; Both et al. 2010; Soldatini et al. 2010; Shaw and McKee 2008)

(continued)

Table 18.1 (continued)

Factor	Description	Examples
Age	Age and experience may influence flight capacity, collision risk and recognition of danger	• Juvenile common terns (*Sterna hirundo*) fly closer to power lines than adults putting them at higher collision risk (Henderson et al. 1996). • Subadult white-tailed eagles (*Haliaeetus albicilla*) were more active in a wind-power plant than adults (Dahl et al. 2013) • Juveniles appear to be more prone to bird strikes than adults (Sodhi 2002; Kelly et al. 2001)
Behaviour	Flight behaviour and motivation for flight which can change during different stages in the annual routine might affect vulnerability to collision	• The frequency of trips of adult terns increased during the nestling phase compared to courtship coinciding with an increase in the proportion of terns crossing power lines (Henderson et al. 1996) • Male Skylarks engaged in song flights to attract mates are more vulnerable to collision with wind turbines than females (Morinha et al. 2014) • Flocking birds or those engaged in social interaction can be more vulnerable to collisions than solitary individuals (Dahl et al. 2013) • Birds that fly at wind turbine blade height are more likely to collide than those that do not (Drewitt and Langston 2006)
External factors		
Weather	Weather strongly influences animal flight behaviour and hence can influence collision risk. Weather can also influence ability to perceive and avoid risk	• Offshore bird strikes occurred at high levels at night with poor visibility (Huppop et al. 2006) • Species dependent on thermals for lift, e.g. Griffon Vultures, encounter scarcer and weaker updrafts in winter and are more likely to collide with wind turbines as they are flying lower (Barrios and Rodriguez 2004) or on days and in areas where vertical lift was weak (Lucas et al. 2008) • Birds and bats are more likely to collide with wind turbines at particular wind speeds (Cryan et al. 2014; Johnston et al. 2014; Arnett et al. 2011) • Weather strongly influences the intensity and altitude of bird migration and bird migration prediction models

(continued)

Table 18.1 (continued)

Factor	Description	Examples
		are used to warn military pilots against peak migration (high collision risk) (www.flysafe-birdtam.eu) (Kemp 2012; Shamoun-Baranes et al. 2014; Ginati et al. 2010)
Topography	Animals may concentrate in particular corridors, or use topographical features to facilitate flight and navigation increasing the likelihood of collision in these places	• Species of birds that use soaring flight are more likely to collide with wind turbines placed along coastal cliffs, ridges and steep slopes used to generate lift (Barrios and Rodriguez 2004; Kitano and Shiraki 2013; Lucas et al. 2008) • Wind farms constructed in migratory bottlenecks have a high number of avian collisions (Barrios and Rodriguez 2004) • Most military low-level bird strikes occur along the migration flyway from SW Sweden towards SW Europe (Dekker and van Gasteren 2005)
Resource availability	Resource availability such as food, shelter, and breeding sites can strongly influence attraction to particular areas and hence collision risk	• Bats attracted to wind turbines because of high invertebrate concentrations (Horn et al. 2008) • Areas attracting large numbers of birds such as waste treatment sites within the vicinity of airfields greatly increase the probability of bird strikes (Baxter et al. 2003; Leshem and Ronen 1998)
Structural design	Aspects such as height, speed, sound and lighting can influence attraction to and avoidance of structures and aircraft	• Fast-moving objects are more dangerous and harder to avoid than slow-moving ones (DeVault et al. 2015) • Some types of lighting may enhance detection and avoidance of aircraft (Doppler et al. 2015; Larkin et al. 1975) • Birds are attracted to certain lighting regimes, increasing their collision risk, e.g. mass fatalities at lighthouses (Jones and Francis 2003) • Certain structures may attract bats and birds looking for roosting sites, making them vulnerable to collision (Kunz et al. 2007; Osborn et al. 1998) • Specific aircraft design specifications exist to withstand bird impact collisions, and improvements in structural design are ongoing (e.g. Amoo 2013; Dolbeer 2013; Dennis and Lyle 2009; Federal Aviation Regulations 2017)

Many factors described here are interrelated

life (van Gasteren et al. 2014) or due to environmental conditions (Manktelow 2000; Steele 2001; Shaw and McKee 2008). Numerous studies (e.g. Sodhi 2002 and references therein) have reported seasonal patterns in bird strikes with peaks in spring and autumn that correspond to migration in many regions. Peaks in aerial bird densities during spring and autumn migration over the Netherlands and Belgium are also visible in Fig. 18.2 despite data smoothing. Due to the large concentrations of birds in the air during the migration season, the risk of serious bird strikes can be quite high, yet it is predominantly military aviation that specifi-cally addresses the higher collision risk during migration (see Sect. 2.2). The increased risk during the migration seasons also depends on the region, time of day and type of migrants passing through the 3D airspace within which aircraft would be exposed to bird strikes. Passerines are by far the most abundant species groups and travel mainly along broad fronts. While these are very small birds, their sheer numbers during migration can, at times, completely fill up radar screens. As a result, in some regions, such as the Netherlands and Belgium, nocturnal migration is closely monitored. In other areas, especially along leading lines or geographic convergent zones, soaring migrants can pose a major risk to aviation during the day. These are often relatively large birds, flying in flocks (or kettles) through a broad altitude band as they climb through the air to gain altitude and then glide to gain distance. One such area in which military aviation is particularly concerned with soaring migration is Israel, a crossroads for migrants travelling between Eastern Europe and Asia to eastern Africa. On 10 August 1995, for example, two pilots were killed when an F-15 from the Israeli Air Force hit several white storks (*Ciconia ciconia*) during migration. Two storks were ingested into one of the engines and the plane crashed after 2 s (Ovadia 2005).

The risk of collisions with birds and bats during non-migratory flights has mainly been considered in aviation within the boundaries of airfields, where mitigation measures can be applied to reduce the likelihood of damaging encounters during take-off and landing. However, in civil aviation the proportion of damaging collisions just outside aerodrome boundaries has increased, seemingly due to improved mitigation within aerodromes and only limited efforts just outside aerodromes in combination with an increase in populations of hazardous species (Dolbeer 2011). For military aviation and other low-flying aircraft (e.g. rotary wing aircraft), non-migratory movements can also be problematic en route (Fig. 18.3). Again, large aggregations of birds can increase the risk of a bird strike. Species such as common starlings (*Sturnus vulgaris*), which flock during the non-breeding season, and birds like gulls and waterfowl that forage in groups in areas of high resource availability, such as around waste treatment sites or recently ploughed fields, and make regular commuting movements between foraging and roosting sites can pose a high collision risk, especially if these flights intercept airfields (Baxter et al. 2003; Leshem and Ronen 1998). Another example is the aggregation of thousands of Greylag geese (*Anser anser*) in the Netherlands in August foraging on grains left after harvesting. In addition to factors that might result in large aggregations of birds, individual experience and knowledge of the surroundings may also influence how prone a bird is to aerial collisions. Although often not conclusive, studies that summarize bird strike statistics have suggested that migrants passing through an area may be more prone to bird strikes than residents, non-breeding birds more prone than breeding birds and juveniles of the same species may be more prone to bird strikes than adults (Brown et al. 2001; Burger 1983; Sodhi 2002). Similar patterns have been noted in regard to collisions with wind turbines and tall static structures (see e.g. Sect. 3.1.2 and Table 18.1).

In recent years, the use of Unmanned Aerial Vehicles (UAVs) has increased greatly in military and civil applications and UAVs may start replacing many human-based forms of aerial surveillance (Beard et al. 2006; Shahbazi et al. 2014; Marris 2013). The potential for conflict between these relatively small vehicles and wildlife is unclear but likely to increase in the coming years as their use in diverse applications increases, yet the risk of collisions between UAVs and wildlife has still received limited attention. Interestingly, recent engineering studies focusing on on-board systems to detect other aerial vehicles to avoid collisions (Moses et al. 2011) may also help reduce the risk of wildlife UAV collisions. The use of UAVs in ecological research is also expected to increase in the near future and researchers have already begun outlining ethical guidelines on how to minimize the potential impact on animal behaviour when using drones in wildlife research (Vas et al. 2015; Jones et al. 2006).

2.2   Current Solutions in Aviation

There are two general approaches to reducing the probability of collisions between aerial wildlife and aircraft: reducing the number of individuals or types of species present in and around airfields and avoiding flying in areas (location and altitude) or at times when high concentrations of birds and bats are measured or predicted in the air. While commercial aviation usually focuses on a range of solutions that falls into the first class of risk reduction, military aviation often deals with both. Regardless, understanding the ecology of the species in question is often paramount in designing efficient solutions. Methods for managing wildlife populations in and around airfields will not be discussed in this chapter as they have been reviewed elsewhere (Anonymous 2006; Blokpoel 1976; MacKinnon et al. 2004; McKee et al. 2016). Here, we focus on solutions related to avoiding bird strikes by separating the flight paths of aircraft and wildlife in space and time.

2.2.1   Monitoring Aerial Bird Movement

Different types of systems have been applied to remotely monitor aerial movement of wildlife for aviation safety. These systems, as well as visual observations, are used to provide near real-time warnings to pilots via air traffic controllers, as well as advance warnings based on expected aerial densities of wildlife and to collect information on aerial densities in order to develop models that can be used to predict aerial movements. Military aviation has been using medium- and long-range radar for real-time monitoring of bird movements in several countries. For example, the Israeli Air Force has been using air surveillance radar and Doppler weather radar to monitor bird movement during migration in order to warn pilots and either delay or reroute flights where needed. During diurnal migration of large soaring migrants, radar observations are often augmented by visual observations of large flocks either at specific military airfields or along a visual observation network that transects the migration flyway (Alon et al. 2004; Dinevich and Leshem 2010). In the Netherlands, Belgium and Germany, air forces have been using air surveillance radars to monitor bird movement on a daily basis and use this information to produce bird strike warnings during daily flight planning. Due to imminent changes in the military radar monitoring systems, the Netherlands and Belgium air forces have tested the use of operational weather radars for monitoring aerial bird movements en route, as part of an international project (FlySafe) funded by the European Space Agency (Ginati et al. 2010). Currently, both air forces have incorporated the use of operational weather radar into their operational warning system (see http://www.flysafe-birdtam.eu/). One of the major advantages of using existing operational weather radars is that they provide more detailed information on altitude distributions than that given by air surveillance radar previously used. Furthermore, as the weather radars are organized in an international network for the exchange of weather data, there is now a huge potential for an early warning system based on measurements beyond the borders of each country, encompassing the international flyways used by birds (Shamoun-Baranes et al. 2014).

In addition to near-real-time monitoring of large-scale movements, dedicated bird radar systems are increasingly used to monitor local movements in and around airfields. This is mainly to reduce the risk of collisions during take-off and landing. Several different bird radar systems are currently deployed at both military and civil airfields around the world, and this application is quickly growing. These systems are primarily used in three ways: (1) inform ground-based crews who can then take action to reduce the risk through activities on the ground, (2) inform air crews and delay or redirect flights if needed and (3) monitor movements to learn about flight behaviour of wildlife and apply knowledge to develop better measures for risk mitigation (e.g. identify areas of high species abundance, roosts). Currently, there are systems that are being used where individual birds or flocks can be tracked within several kilometres of the radar, and runway crossings can be monitored and identified automatically. These systems include mobile applications enabling ground personnel to monitor aerial behaviour from any location and respond quickly to potential risks (Fig. 18.4).

An additional tool to measure bird movement, which is being used extensively in ecological research but rarely within the context of flight safety, is bio-logging. With GPS tracking technology, the daily movements of some species, depending on their size, can be tracked in 3D and at a very fine scale, providing detailed information on their flight behaviour and aerial space use throughout the annual routine (Bouten et al. 2013; Kays et al. 2015). The time spent in the air, the airspace used by an individual, the number of runway crossings, flight altitudes, response to disturbance and how these aspects may change in relation to external or intrinsic factors can all be measured using GPS and utilized to better inform programmes for flight safety risk reduction. One example is a case study conducted in the Netherlands which monitored the daily movements of adult and juvenile common

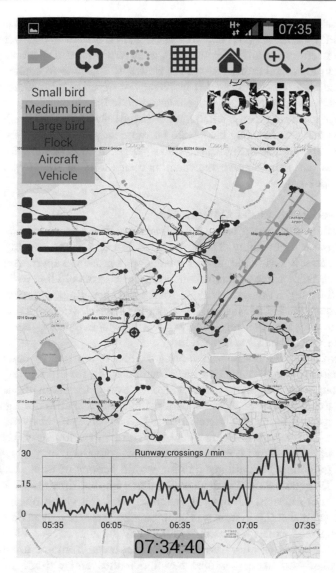

Fig. 18.4 Screenshot (10 March 2015 07:34 UTC) of the bird radar system from Robin Radar systems on Eindhoven airport (The Netherlands) showing bird movements in and around the airport (blue and green linear structures indicate the runway environment at Eindhoven airport). The system has an automated algorithm which distinguishes between the movements of birds from different size classes, aircraft and terrestrial vehicles registered by the radar. The system shows the number of runway crossings per minute as one of the collision probability indicators. On this morning, migration of Chaffinches (*Fringilla coelebs*) was visually observed

buzzards (*Buteo buteo*) monitored with a flexible GPS tracking system in and around military airfields (van Gasteren et al. 2014). Buzzards that were breeding within the Leeuwarden airfield boundaries were territorial and spent about 10% of their time in flight and rarely crossed the runways. This was in contrast to movement measured at another airfield (Eindhoven) in which non-breeding birds were monitored. Birds from Eindhoven spent less time in the air on average but crossed the runways more frequently. Although the sample size was very small, the study did suggest that non-breeding birds were far more likely to cross the runways than breeding adults were. Such findings can have important consequences for relocation programmes and can improve the likelihood of targeting the right individuals within a population.

2.2.2 Predicting Aerial Movements

Understanding how, when and why animals use the aerosphere has important implications for flight safety. This knowledge can be used to develop predictive models to better inform flight planners and adopt a proactive approach to reducing the risk of aerial collisions. Aircraft can reduce the risk of colliding with wildlife by changing their flight times, their flight routes and their flight altitudes, flying for example above the altitudes where the majority of aerial wildlife is found. Predictive models may focus on migration or local daily movements, predicting aerial densities of wildlife in 3D space and time.

Different types of models, with varying degrees of complexity and capacity to capture spatio-temporal dynamics, have been designed for flight safety purposes. When enough data or expert knowledge is available, static models can be generated that reflect mean seasonal patterns in aerial densities during migration. Such an approach has been adopted in Israel where the general flyways, altitude and diurnal and seasonal timing of soaring bird migration have been studied (Leshem and Yom-Tov 1996a, b, 1998) and converted into a static model providing guidelines for military training within periods of time and regions of high risk. This static model is then augmented with real-time and near-real-time warnings from different monitoring systems (radar and visual observations). In the first 10 years that this system was implemented, it resulted in an 88% reduction in damaging bird strikes and estimated financial savings of 30 million USD per year (Leshem 1994). Similar baseline models for broad front nocturnal migration can be established using seasonal aerial densities and altitude distributions (e.g. Fig. 18.2). Information about the distribution of species on the ground has been used to develop Bird Avoidance Models predicting the bird strike hazard for a given region throughout the year (Shamoun-Baranes et al. 2008). In these systems, visual observations from different sources are integrated and used to develop 2D species distribution models at a predetermined temporal resolution (approximately 2 weeks). Expert knowledge was then used to distribute the birds within altitude bands of interest. These models do not account for dynamics in environmental conditions, and while they may be helpful in predicting the densities of birds on the ground, their weakness is extrapolating this information to estimate aerial densities of birds.

Weather conditions influence the migratory behaviour of birds, affecting travel speeds, flight altitudes, flight routes, departure decisions and even flight modes (e.g. Shamoun-Baranes et al. 2010; Vansteelant et al. 2015; Gill et al. 2014). Predictive models can therefore be refined by incorporating environmental dynamics. Forecast models of aerial density of birds have been developed for the Netherlands and Belgian air forces based on the statistical relationship found between local weather conditions at the surface and aloft and measured bird densities (van Belle et al. 2007; Kemp 2012). Once models have been fit to existing data, weather forecasts are used as model input to provide a 3-day migration forecast. These models are site specific and limited in their capacity to estimate the altitudes at which birds are expected to migrate. The use of weather radar has generated new opportunities for modelling flight altitude distributions in response to weather. Predictive models that incorporate the influence of weather on flight altitude distribution of birds during migration have been developed (Kemp 2012; Kemp et al. 2013) and are being tested for operational use by the Netherlands and Belgian air forces. Once models are considered robust enough for operational purposes, they can be used to predict 3D and temporal distribution of birds in the air based on local weather forecasts. This data-driven approach generally requires data on bird movement from several seasons in order to capture enough environmental variability to fit robust models (van Belle et al. 2007). While these models are based on statistical correlations between migration intensity and weather conditions, understanding how birds respond to wind could greatly improve the selection of appropriate derivatives of wind speed and direction to be incorporated in the model fitting procedure (Kemp et al. 2012).

The predictive models described above can generally be considered a Eulerian description of animal movement; they provide information about the flux of migrants through a specific area over time, predicting temporal and altitudinal changes in aerial bird densities at a particular site, but they do not predict the movement of birds per se. Another approach is a Lagrangian description of animal movement in which organisms (or groups of organisms) can be followed in space and time, creating individual trajectories (Turchin 1998). From a modelling perspective, spatially explicit models of animal movement based on behavioural decision rules in dynamic environments could have great potential for flight safety. From a methodological perspective, a rather simple approach would be to interpolate movement parameters in space and time between radar as a type of null model to provide a Lagrangian representation of flow. More complex models that simulate behaviour under varying environmental conditions would be a more advanced approach. Simulation models of nocturnal migration and diurnal soaring migration of birds using information about bird movement derived from radar and other sources have been developed for research purposes and when well calibrated could be applied for flight safety (Erni et al. 2003; McLaren et al. 2012; Shamoun-Baranes and van Gasteren 2011; van Loon et al. 2011). Within the context of flight safety, simulation models could be used for scenario testing, for example, testing the consequence of different environmental conditions en route, to fill in gaps between distant monitoring sensors and if run and updated with near real-time data can be used to predict migration hours or days in advance over a broad scale.

3 Anthropogenic Structures

3.1 Wind Turbines

Humans have been harnessing the wind's energy for thousands of years, but it is only since the 1980s, with the advent of increasingly large and numerous turbines to generate electricity, that the potential of these structures to cause significant effects on the animals that use the aerosphere has been considered and assessed. Commercially available wind turbines now include towers of over 100 m in height, with blades sweeping a diameter of 180 m and moving at up to 320 km/h at their tips (http://www.4coffshore.com/windfarms/turbines.aspx).

As governments around the world are investing in wind power as a "clean" technology that does not emit carbon dioxide and other pollutants associated with fossil fuels, wind turbines are being constructed at unprecedented rates. Global wind generation capacity grew from 1.7 GW in 1990 to 282 GW in 2012 (International Energy Agency 2014). Some of this capacity has been delivered through micro-turbines—small, often single devices, typically found on private land or mounted on buildings—but increasingly policymakers are relying on large arrays of turbines in wind farms, or wind parks, to meet renewable energy needs.

To be effective, wind farms must be built in exposed areas with high average wind speeds. Initially, many wind farms were sited on plains and in coastal and upland areas, including on mountainsides. Recently, largely due to aesthetic objections over wind farms in the countryside, offshore developments have been favoured. These offshore wind farms contain turbines that are greater in number, larger in size and distributed over a wider area than any other developments to date. They are also being installed in ever deeper water and further from the coast, meaning that a wider suite of wildlife in the aerosphere could potentially be exposed to the effects of turbines.

3.1.1 Effects of Wind Turbines on Wildlife

The potential impacts of wind turbines on aerial wildlife are many and complex (Schuster et al. 2015). Birds are the species on which the majority of research to date on the effects of wind turbines has focused, perhaps because of high-profile reports of deaths to charismatic species and those of conservation concern, such as Griffon Vultures (*Gyps fulvus*) in Spain and Golden Eagles (*Aquila chrysaetos*) in the USA, in areas where wind farms have been constructed (Drewitt and Langston 2006; Barrios and Rodriguez 2004). It is also increasingly recognized that bats can be affected by wind turbines in different ways, from fatal collisions to changes in their flight behaviour around wind turbines (Cryan and Brown 2007; Horn et al. 2008; Cryan et al. 2014; Lehnert et al. 2014; Arnett et al. 2016; Arnett and Baerwald 2013).

Birds and bats can be adversely affected by wind turbines in several different ways. The most obvious direct effect is death or injury through collision, which is generally with turbine blades, but can also be with turbine masts and associated structures, such as guy cables, power lines and meteorological masts (Horn et al. 2008; Drewitt and Langston 2006). Mortality due to barotrauma can be a problem, particularly for bats, whereby a reduction in air pressure close to moving blades causes tissue damage and pulmonary haemorrhage (Baerwald et al. 2009). There are also several indirect effects which have been studied in different taxa. Animals can be displaced from an area due to disturbance during construction and/or once wind farms become operational, which effectively amounts to habitat loss (Farfán et al. 2009; Pearce-Higgins et al. 2009). Another form of displacement is the so-called "barrier effect", whereby individuals increase their energy expenditure by flying around a wind farm in their normal flight path (Desholm et al. 2006), be it on migration or during daily journeys between feeding, nesting and roosting sites. Actual habitat loss or change resulting from wind farm development can also occur (Drewitt and Langston 2006; Perrow et al. 2011).

Not all habitat changes associated with wind energy deter wildlife (Table 18.1). Bats and birds may be attracted to wind turbines by these structures' effect on insects, which are drawn in turn by the colours of paint chosen or by the heat generated by turbines (Kunz et al. 2007; Long et al. 2011; Cryan et al. 2014). Wind farms may also attract birds and bats looking for places to roost. The lattice turbine design found in older installations (as opposed to modern monopole designs) was thought to be especially appealing in this respect (Kunz et al. 2007; Osborn et al. 1998), while bats have also been observed investigating turbine blades, both moving and stationary, again possibly in an attempt to find sites for roosting or mating (Horn et al. 2008; Cryan et al. 2014). The presence of wind turbines may actually enhance habitat for some species. For example, offshore wind farm structures not only provide roosting sites, but can act as artificial reefs, thereby attracting fish and their predators, including some species of seabird (Inger et al. 2009; Lindeboom et al. 2011). Wind turbine installations may also act as de facto nature reserves, as other damaging human activities, for example shipping and fishing in offshore areas, can be limited or excluded in the wind farm zone (Inger et al. 2009).

All effects of wind turbines may also be cumulative in time and space (i.e. the effect of multiple installations that an individual encounters over time), such that population level impacts are not seen until several years post-construction (Drewitt and Langston 2006; Hill and Arnold 2012). Effects also vary throughout the year and depend on the season (e.g. Fijn et al. 2015). For example, birds migrating over the sea are more likely to collide with turbines in poor weather, when individuals are more likely to fly at altitudes swept by turbine blades and visibility is reduced (Table 18.1). In these conditions, birds or bats can become disorientated and even attracted to turbines if they are illuminated in particular ways (Hüppop et al. 2006; Cryan and Brown 2007; Ahlén et al. 2007). Similarly, individuals might be more likely to encounter a wind farm site and be vulnerable to negative effects at particular stages of their life cycle, such as during breeding or on migration. For instance, migrating bats are thought to spend a higher proportion of their time flying at blade height and echolocate less frequency than residents, hence increasing their susceptibility to collision [reviewed in (Cryan and Barclay 2009)]. In a more specific example, a Belgian wind farm constructed in the foraging flight path of breeding terns caused significant colony-level mortality as birds were exposed to the risk of collision so frequently (Everaert and Stienen 2007). Individuals may also habituate to the presence of a wind farm (or indeed to mitigation measures designed to discourage wildlife from approaching turbines—see Sect. 3.1.3), such that although initially displacement and disturbance may occur, over time the wind farm area might be used again (May et al. 2015). The benefit of reclaiming this former habitat might be outweighed by the risks associated with collision, although it is possible that residents may learn to avoid turbines, while non-habituated individuals, for example migrants, may suffer higher levels of mortality due to collision (Langston and Pullan 2003). Thorough long-term monitoring is therefore required to fully understand the consequences of wind farm installation (Stewart et al. 2007).

The overall detrimental effects of wind turbines on animals that use the aerosphere have been much debated in the literature, but there is mostly consensus that the levels of mortality are seriously worrisome. The number of bats killed is estimated at approximately 300,000 per year in Germany and 500,000 per year in the USA (Arnett and Baerwald 2013; Lehnert et al. 2014; Voigt et al. 2012; Hayes 2013; Smallwood 2013), while approximately 230,000 birds are thought to be killed annually by onshore monopole wind turbines in the contiguous USA (Loss et al. 2013). Many authors point out that the number of individuals killed or displaced by wind developments is less than that associated with other anthropogenic structures, for instance power lines and buildings (see Sect. 3.2), and also lower than that caused by human activities such as driving and even domestic cat ownership (Erickson et al. 2005; Calvert et al. 2013). It is worth remembering that the number of wind turbines installed is small compared to the number of buildings, cars etc., although this will change if the wind energy sector continues to grow. The negative impacts of wind turbines need to be considered on a species by species basis. For species that are slow to mature and reproduce, any mortality or reduction in fitness brought about directly or indirectly by wind turbines could be critical. Such species include bats, seabirds and raptors, and local population declines have been reported in some studies (e.g. Everaert and Stienen 2007; Marques et al. 2014).

3.1.2 Monitoring the Effects of Wind Turbines

Ecologists assessing the effects of wind turbines often aspire to the "BACI" model (Before-After-Control-Impact), where wildlife surveys are carried out before and after turbines have been installed, in two different sites—the site of the wind farm and a control area with similar characteristics that is far enough away to be unaffected by the development (Pérez Lapeña et al. 2010). This can show changes such as species redistribution or changes in abundance due to wind farms. The number of fatal collisions due to wind turbines can be assessed through counts of carcasses. These counts can be confounded by scavengers removing carcasses, although a correction factor based on experimental removal rates has been calculated in a number of studies (e.g. Krijgsveld et al. 2009). Moreover, carcass counts are difficult or impossible to carry out in some locations, for example offshore (Desholm et al. 2006).

Monitoring of species abundance and behaviour in and around wind farms can be carried out through visual observations, but technology is increasingly being harnessed to greatly improve our understanding of how wildlife responds to installations, especially since effects may occur outside the human visible range (for example, at high altitudes) or in weather conditions during which visual surveys cannot easily be carried out. Such technology includes digital aerial surveys, ranger finders, vertical and horizontal radar, thermal imaging and acoustic detection of animal calls in the vicinity of turbines (e.g. Desholm et al. 2006; Hüppop et al. 2006; Cryan et al. 2014). These methods allow information on a large number of individuals from a broad range of species to be captured over a long time period and (in the case of radar) over a broad spatial scale. However, it can be difficult to separate species, age/sex cohorts and individuals with these techniques (e.g. Desholm et al. 2006).

GPS tracking, whereby small devices are placed on individual animals, overcomes many of these problems. Individuals can be followed, providing fine-scale information on movement patterns (Fig. 18.5), elucidating the connectivity between particular wind farm developments and the population that the focal animal represents, for example an adult seabird breeding at a particular colony.

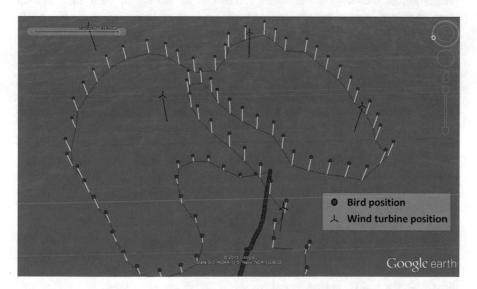

Fig. 18.5 GPS measurements providing location, altitude and ground speed every 10 s of a lesser black-backed gull (*Larus fuscus*) floating first on the sea surface and then flying through a wind farm in the Irish Sea. Vertical line length is proportional to height above sea level. This lesser black-backed gull tagging was funded as part of the UK Department of Energy and Climate Change's Offshore Energy Strategic Environmental Assessment programme

Long-lived GPS tags can also reveal how individuals' interactions with wind turbine installations can vary throughout different stages of their life cycle (Thaxter et al. 2015) and can provide information on whether individuals fly at altitudes that could expose them to collisions with turbine structures (Cleasby et al. 2015; Ross-Smith et al. 2016).

All these technologies can help calculate a species' so-called "collision risk". This is typically determined by models incorporating a number of factors, including the abundance and morphology (including flight altitude) of the species concerned along with properties of the wind farm (Band 2012) (Table 18.1). Collision risk models include a term called "avoidance", which is the rate at which animals take evasive action to successfully eliminate their risk of collision. "Macro-avoidance" occurs when animals avoid an entire wind farm, as has been observed in several species, including migrating Eider (*Somateria mollissima*) (Desholm and Kahlert 2005). "Micro-avoidance" occurs when an individual alters its flight path to prevent collision at the last second. An intermediate stage, "meso-avoidance", allows birds to avoid turbines within wind farms (Cook et al. 2014).

Similar to aviation, collision risk posed by wind turbines varies depending on environmental conditions, aspects of the turbine design and location, behavioural and physiological characteristics of the species concerned, as well as with the time of year, as individuals from the same species can be more exposed to turbines at particular points of their life cycle (Table 18.1). In bats, for example, the majority of deaths appear to occur during migration and migratory bats make up a large proportion of the observed fatalities (Arnett et al. 2008). In birds that rely on thermal convection or slope updrafts to sustain soaring flight, larger soaring birds such as Griffon Vultures that rely almost entirely on soaring flight and rarely on flapping flight are more likely to collide with turbines than more abundant soaring species with comparatively lower wing loading and higher manoeuvrability; Griffon Vultures are also more susceptible to fatal collisions when and where soaring conditions are poorer (De Lucas et al. 2008; Barrios and Rodriguez 2004). Similarly, animals may not perceive the danger of moving wind turbine blades and therefore may not take action to avoid them. Although this is not yet fully understood, it could occur either because a species' sensory abilities are such that individuals cannot detect turbines (which for birds may be compounded at certain blade speeds because of "motion smear") or because individuals are not adapted and attuned to turbines during flight (Martin et al. 2012).

3.1.3 Mitigating the Effects of Wind Turbines

When it comes to mitigation, there is no "one-size-fits-all" approach and several proposed mitigation methods remain under- or untested (Marques et al. 2014, but see Arnett et al. 2011; de Lucas et al. 2012). Although easier said than done, the most effective mitigation is to ensure wind farms are not placed in areas where susceptible wildlife is abundant, especially when the species concerned are of conservation interest. Wind farms can be situated outside protected areas and away from major roosts, breeding colonies, migration bottlenecks and topographic features on which particular species rely for thermals (Drewitt and Langston 2006).

Informed decisions about where to place a wind farm to minimize the impact on wildlife rely on pre-construction monitoring, which is typically carried out as part of an environmental impact assessment process.

Once a wind farm has been built, there are a series of measures that can improve its compatibility with wildlife. Wind turbines need to be "repowered", whereby structures are remodelled and upgraded, and turbines that are more wildlife friendly can be installed during this process. For example, monopole designs and fewer, larger, turbines can generate the same energy as a greater number of small, more densely packed turbines, and may have a lower avian collision rate per megawatt (Barrios and Rodriguez 2004; Johnston et al. 2014; Everaert 2014), although larger turbines may be more dangerous for bats (Barclay et al. 2007). "Micro-siting" can also be effective, whereby particular turbines that cause high levels of mortality are removed (de Lucas et al. 2012; May et al. 2015). Turbines can also be selectively temporarily shut down at times when vulnerable species (both in terms of collision susceptibility and conservation concern) might be present. This method halved Griffon Vulture mortality with only a 0.07% decline in energy production at a wind farm in southern Spain (de Lucas et al. 2012). Similarly, raising the minimum wind speed threshold at which turbines generate power (i.e. reducing turbine operation at low wind speeds) has been found to reduce levels of bat mortality with marginal annual power loss (Arnett et al. 2011).

Measures can also be taken to make the wind farm less attractive to animals, or more conspicuous in the case of enhancing avoidance. Such techniques include altering the paint colour, lighting regime, using lasers, electromagnetic fields and acoustic deterrents (Cook et al. 2011; Nicholls and Racey 2007). However, care must be taken that animals do not habituate to these measures, as has been noted in collision mitigation measures implemented on aerodromes (MacKinnon et al. 2004), and it is difficult to find an effective way to discourage all vulnerable species (May et al. 2015). Areas around the wind farm could also be made more attractive, for instance providing alternative foraging habitat, to encourage wildlife to concentrate there instead of within the wind farm itself (Martin et al. 2012).

In today's world, concerns about climate change mean wind farms play an integral part in reducing humankind's carbon footprint. However, care must clearly be taken to ensure that the impact of this technology on biodiversity is kept to a minimum. To do this, a consistent and holistic approach, incorporating a range of survey and specifically tailored mitigation techniques embedded in a framework of sound aeroecological understanding, is required, ensuring that all negative effects on wildlife are recognized, understood and prevented where possible.

3.2   Other Tall Structures

Tall static structures in the landscape such as buildings, communication towers and their guy wires, large monuments and power lines are also a collision risk for numerous species. Some estimates suggest that millions of birds a year die due to collisions with these structures; for some species, annual mortality has been estimated at several percent of the total population (Bevanger 1998; Erickson et al. 2005; Longcore et al. 2013). Numerous reviews of avian collisions with power lines and other structures are available (Jenkins et al. 2010; Erickson et al. 2005; Drewitt and Langston 2008), and we will not provide a comprehensive overview here. Interestingly, the numbers provided suggest a massive mortality rate, but population effects or adaptive evolutionary consequences are rarely discussed. However, similarities can be found with collisions with wind turbines and even with aircraft, several of which we will briefly highlight.

As with other anthropogenic threats in the air, mortality rates at tall structures are often species specific and not directly related to species abundance (Longcore et al. 2013) with certain species being more susceptible to collisions than others, and numerous external factors influence the probability of collisions as well (Table 18.1). As mentioned for wind farms, animals may even be attracted to such structures. For example, a recent study has shown that several species of tree bats (*Lasiurus sp*) are attracted to tall towers during autumn migration and results support the hypothesis that this attraction is linked to social behaviour (Jameson and Willis 2014). In addition to chronic collisions, collisions with tall and conspicuous structures in landscape can result in episodic mass mortality events, sometimes involving thousands of individual birds over the course of a few days (Erickson et al. 2005). In North America, the highest mortality has been recorded for nocturnal migrating passerines and studies suggest that collisions generally occur on nights with poor weather when birds are attracted to and potentially trapped by lighting of these structures (Longcore et al. 2013; Erickson et al. 2005).

In contrast to collisions with large towers but in common with collisions with wind turbines, the species that appear to be most susceptible to collisions with power lines are large and have low manoeuvrability such as cranes, bustards, flamingos, waterfowl and gamebirds (Jenkins et al. 2010). Familiarity with the surroundings also influences how risk prone birds are, with juvenile and migrants birds being more susceptible than adults and residents (Jenkins et al. 2010). Similarly, the aim of flight might influence the risk of collision, where aerial displays, predator prey interactions or foraging flights to feed nestlings may put a bird at higher risk than when flying for other purposes. Other factors influencing perception are also important, such as ambient light, and flying in flocks when individuals are potentially paying more attention to conspecifics during flight and less to their direct surroundings.

Mitigation measures suggested to reduce the risk of collision with tall structures are related to either improving detection of these structures or reducing potential distraction caused by lighting (Poot et al. 2008). For example, several studies suggest that pulsed lighting on communication towers and guy cables is preferable to constant light, regardless of the colour. For power lines, marking lines to make them more visible is an option although finding a solution that works for a broad range of species and environmental conditions seems more difficult, and field tests have produced different and sometimes conflicting results (Jenkins et al. 2010).

4   Future Perspectives

Aerial conflicts between wildlife and humans are a problem that is likely to persist, especially as human population and economic growth continues, leading to ever further encroachments of human structures and activities in the aerosphere. However, it is also clear that there are numerous types of solutions and that our knowledge and understanding of the internal and external factors that influence animal movement can be put to use in order to reduce these conflicts. While advances in the design of human structures and aerial vehicles may reduce the risk of collisions and other adverse impacts, aeroecologists can contribute a great deal by improving our understanding of aerial movement and finding ways to communicate and apply this knowledge to reduce aerial conflicts.

The different types of sensors and methods used to monitor aerial movement are often complementary in many ways. The integration of these multiple sensors, where possible, to monitor, understand and predict movement, collision risk and the other ways in which wildlife might be affected by human activities/structures is likely to produce better results than single sensors alone. This might become increasingly feasible as individual tracking studies are increasing and biological data from existing sensor networks such as weather radars become more readily accessible (Shamoun-Baranes et al. 2014; Chilson et al. 2012). The development and improvement of algorithms to differentiate between different types of aerial organisms monitored by radar is ongoing (e.g. size classes, species groups, single vs. multiple animals) and will improve the quality of monitoring and real-time warning systems (Gürbüz et al. 2015). This will both reduce the chance of false warnings and unnecessary mitigation measures (e.g. rerouting flights, stopping turbines) and the absence of warnings which are in fact needed, hence exposing wildlife (and humans) to unnecessary risk. Tracking individual birds or bats can often provide the ecological context needed to better understand and model behaviour from a mechanistic perspective, and sensors that scan the aerosphere, such as radar providing information about the densities of animals travelling through the aerospace at a given time, can indicate how representative tracked individuals are of larger scale patterns. Advances in tracking technology will also improve this process, allowing more information to be captured from an ever-wider range of species (Bridge et al. 2011; Robinson et al. 2010).

Tracking data collected for studies not directly related to risk assessment can also be extremely informative, providing information on the general flight behaviour of different species under a range of environmental conditions. By pooling data across studies, we can start filling knowledge gaps about basic flight behaviour per species that would be relevant for a range of stakeholders, such as the amount of time spent in flight, flight altitude distributions and flight speeds. Such studies would be complementary to other comparative studies such as flight speeds measured using tracking radar (Bruderer and Boldt 2001). Tracking studies could also provide information about the timing, routes and altitudes of local commuting flights where risk can be more concentrated. Similarly, ongoing efforts to coordinate the use of weather radar networks for ecological research in Europe and North America can result in large-scale monitoring of aerial movement of wildlife (Shamoun-Baranes et al. 2014; Kelly et al. 2012). Large-scale monitoring efforts should not only focus on migratory movements but also on local commuting flights. Data from different tracking studies and large-scale monitoring programmes can be used to provide baseline information for assessing collision risk at large scales, identifying hotspots of aerial activities, as well as seasonal and diurnal patterns of movement, especially high aerial densities, which are of particular interest from a conservation perspective as well as for flight safety. In time, we can also begin to map the 3D space use of different species for establishing baseline assessments of risk at the species specific level.

Models for predicting aerial movement can be developed with different levels of complexity and capacity to incorporate the diverse factors that influence aerial movement. Gaps between sensors can be filled with a simple model interpolating properties of aerial movement (e.g. density, ground speed) between sensors. Data-driven models can be used to predict temporal variability in aerial movements. When developing models, data-driven and more mechanistic approaches can potentially be combined to provide more realistic and spatially explicit representations of movement through the aerosphere. Modelling need not only focus on what is happening in the air, but also on the ground, the source areas and destinations of flight, which will influence local aggregations and help understand why collision risk may be higher in certain areas and times of the day or year. For example, radar monitoring can also be used to assess habitat use of large aggregates of birds and identify hotspots form which aerial activity will be initiated (Buler and Dawson 2014; Buler et al. 2012). Simulation models that incorporate individual response to environmental conditions can be used to explore changes in collision risk due to specific environmental conditions. Individual-based models have already been developed to address collision avoidance and the impact of group size, social structure and obstacle placement (Croft et al. 2015). Research that improves our understanding of animal perception and how species' life cycles affect their susceptibility to aerial conflicts will also provide information that could fine-tune such models. Animals' sensory abilities must be taken into account when predicting how they will respond to and be affected by human structures and activities in the aerosphere (Martin 2012), while incorporating knowledge of ontogeny and social dynamics will help pinpoint certain times of the year (e.g. the breeding season) when particular species might be especially vulnerable.

We are confident that in the coming years, the combination of improved sensor networks and advances in bio-logging and modelling of animal movement will enormously enhance our fundamental understanding of aerial behaviour. This, combined with the cross-fertilization of ideas and expertise across disciplines and applications, will contribute greatly to establishing a holistic approach to sharing the skies safely for humans and wildlife alike.

Acknowledgements We would like to thank Jeff McKee, Winifred Frick and Phil Chilson for feedback on an earlier version of this manuscript. We acknowledge the support provided by European Cooperation in Science and Technology (COST) through Action no. ES1305, European Network for the Radar Surveillance of Animal Movement (ENRAM), in facilitating this collaboration. The contents of this paper are the authors' responsibility and neither COST nor any person acting on its behalf is responsible for the use which might be made of the information contained in it. JSB's contribution was in part supported by Rijkswaterstaat.

References

Able KP (1970) A radar study of the altitude of nocturnal passerine migration. Bird Band 41(4):282–290. https://doi.org/10.2307/4511688

Ahlén I, Bach L, Baagøe HJ, Pettersson J (2007) Bats and offshore wind turbines studied in southern Scandinavia. Government report published by the Swedish Environmental Protection Agency. ISBN:91-620-5571-2.pdf. Report number 5571

Allan JR (2002) The costs of bird strikes and bird strike prevention. In: Clark L (ed) Human conflicts with wildlife: economic considerations. Proceedings of the Third NWRC Special Symposium. National Wildlife Research Center, Fort Collins, CO, 1–3 Aug 2000, pp 147–153

Allan J (2006) A heuristic risk assessment technique for birdstrike management at airports. Risk Anal 26(3):723–729. https://doi.org/10.1111/j.1539-6924.2006.00776.x

Alon D, Granit B, Shamoun-Baranes J, Leshem Y, Kirwan GM, Shirihai H (2004) Soaring-bird migration over northern Israel in Autumn. Br Birds 97:160–182

Amoo LM (2013) On the design and structural analysis of jet engine fan blade structures. Prog Aerosp Sci 60(0):1–11. https://doi.org/10.1016/j.paerosci.2012.08.002

Anonymous (2006) Recommended practices no. 1 standards for aerodrome bird/wildlife control. International Birdstrike Committee

Anonymous (2015) Embry-riddle aeronautical university wildlife strike download area. Embry-Riddle Aeronautical University. http://wildlife.pr.erau.edu/index.html. Accessed 1 Feb 2015

Arnett EB, Baerwald EF (2013) Impacts of wind energy development on bats: implications for conservation. In: Adams AR, Pedersen CS (eds) Bat evolution, ecology, and conservation. Springer, New York, pp 435–456. https://doi.org/10.1007/978-1-4614-7397-8_21

Arnett EB, Brown WK, Erickson WP, Fiedler JK, Hamilton BL, Henry TH, Jain A, Johnson GD, Kerns J, Koford RR, Nicholson CP, O'Connell TJ, Piorkowski MD, Tankersley RD Jr (2008) Patterns of bat fatalities at wind energy facilities in North America. J Wildl Manag 72(1):61–78. https://doi.org/10.2307/25097504

Arnett EB, Huso MMP, Schirmacher MR, Hayes JP (2011) Altering turbine speed reduces bat mortality at wind-energy facilities. Front Ecol Environ 9(4):209–214. https://doi.org/10.1890/100103

Arnett EB, Baerwald EF, Mathews F, Rodrigues L, Rodríguez-Durán A, Rydell J, Villegas-Patraca R, Voigt CC (2016) Impacts of wind energy development on bats: a global perspective. In: Voigt CC, Kingston T (eds) Bats in the anthropocene: conservation of bats in a changing world. Springer International Publishing, Cham, pp 295–323. https://doi.org/10.1007/978-3-319-25220-9_11

Avery ML, Humphrey JS, Daughtery TS, Fischer JW, Milleson MP, Tillman EA, Bruce WE, Walter WD (2011) Vulture flight behavior and implications for aircraft safety. J Wildl Manag 75(7):1581–1587. https://doi.org/10.1002/jwmg.205

Avrenli KA, Dempsey BJ (2014) Statistical analysis of aircraft-bird strikes resulting in engine failure. Washington, DC. https://doi.org/10.3141/2449-02

Band W (2012) Using a collision risk model to assess bird collisions for offshore windfarms. Commissioned by Strategic Ornithological Support Services (SOSS)

Barclay RMR, Baerwald EF, Gruver JC (2007) Variation in bat and bird fatalities at wind energy facilities: assessing the effects of rotor size and tower height. Can J Zool 85(3):381–387

Baerwald EF, Edworthy J, Holder M, Barclay RMR (2009) A large-scale mitigation experiment to reduce bat fatalities at wind energy facilities. J Wildl Manag 73:1077–1081. https://doi.org/10.2193/2008-233

Barrios L, Rodriguez A (2004) Behavioural and environmental correlates of soaring-bird mortality at on-shore wind turbines. J Appl Ecol 41(1):72–81

Baxter A, St James K, Thompson R, Laycock H (2003) Predicting the birdstrike hazard from gulls at landfill sites. Paper presented at the 26th International Bird Strike Committee Warsaw

Beard RW, McLain TW, Nelson DB, Kingston D, Johanson D (2006) Decentralized cooperative aerial surveillance using fixed-wing miniature UAVs. Proc IEEE 94(7):1306–1324. https://doi.org/10.1109/jproc.2006.876930

Bevanger K (1998) Biological and conservation aspects of bird mortality caused by electricity power lines: a review. Biol Conserv 86(1):67–76. https://doi.org/10.1016/S0006-3207(97)00176-6

Biondi KM, Belant JL, Devault TL, Martin JA, Wang G (2013) Bat Incidents with U.S. civil aircraft. Acta Chiropterol 15(1):185–192. https://doi.org/10.3161/150811013x667984

Bishop CM, Spivey RJ, Hawkes LA, Batbayar N, Chua B, Frappell PB, Milsom WK, Natsagdorj T, Newman SH, Scott GR, Takekawa JY, Wikelski M, Butler PJ (2015) The roller coaster flight strategy of bar-headed geese conserves energy during Himalayan migrations. Science 347(6219):250–254. https://doi.org/10.1126/science.1258732

Blokpoel H (1976) Bird hazards to aircraft: problems and prevention of bird/aircraft collisions. Canadian Wildlife Service, Environment Canada: Pub. Centre, Supply and Services Canada, Clarke Irwin, Ottawa

Both I, van Gasteren H, Dekker A (2010) A quantified species specific bird hazard index. Paper presented at the 29th International Bird Strike Committee, Cairns

Bouten W, Baaij E, Shamoun-Baranes J, Camphuysen KJ (2013) A flexible GPS tracking system for studying bird behaviour at multiple scales. J Ornithol 154(2):571–580. https://doi.org/10.1007/s10336-012-0908-1

Bridge ES, Thorup K, Bowlin MS, Chilson PB, Diehl RH, Fléron RW, Hartl P, Roland K, Kelly JF, Robinson WD, Wikelski M (2011) Technology on the move: recent and forthcoming innovations for tracking migratory birds. Bioscience 61(9):689–698. https://doi.org/10.1525/bio.2011.61.9.7

Brown KM, Erwin RM, Richmond ME, Buckley PA, Tanacredi JT, Avrin D (2001) Managing birds and controlling aircraft in the Kennedy airport's Jamaica bay wildlife refuge complex: the need for hard data and soft opinions. Environ Manag 28(2):207–224. https://doi.org/10.1007/s002670010219

Bruderer B, Boldt A (2001) Flight characteristics of birds: i. radar measurements of speeds. Ibis 143(2):178–204

Buler JJ, Dawson DK (2014) Radar analysis of fall bird migration stopover sites in the northeastern U.S. Condor 116(3):357–370. https://doi.org/10.1650/condor-13-162.1

Buler JJ, Randall LA, Fleskes JP, Barrow WC Jr, Bogart T, Kluver D (2012) Mapping wintering waterfowl distributions using weather surveillance radar. PLoS One 7(7):e41571. https://doi.org/10.1371/journal.pone.0041571

Burger J (1983) Bird control at airports. Environ Conserv 10(02):115–124. https://doi.org/10.1017/S0376892900012200

Buurma LS (1994) Superabundance in birds: trends, wetlands, and aviation. Paper presented at the 22nd International Bird Strike Committee, Vienna,

Calvert AM, Bishop CA, Elliot RD, Krebs EA, Kydd TM, Machtans CS, Robertson GJ (2013) A synthesis of human-related avian mortality in Canada. Avian Conserv Ecol 8(2):11. https://doi.org/10.5751/ace-00581-080211

Chilson PB, Frick WF, Kelly JF, Howard KW, Larkin RP, Diehl RH, Westbrook JK, Kelly TA, Kunz TH (2012) Partly cloudy with a chance of migration: weather, radars, and aeroecology. Bull Am Meteorol Soc 93(5):669–686. https://doi.org/10.1175/bams-d-11-00099.1

Cleasby IR, Wakefield ED, Bearhop S, Bodey TW, Votier SC, Hamer KC (2015) Three-dimensional tracking of a wide-ranging marine predator: flight heights and vulnerability to offshore wind farms. J Appl Ecol 52(6):1474–1482

Cook ASCP, Ross-Smith VH, Roos S, Burton NHK, Beale N, Coleman C, Daniel H, Fitzpatrick S, Rankin E, Norman K, Martin G (2011) Identifying a range of options to prevent or reduce avian collision with offshore wind farms, using a UK-based case study. BTO, Thetford

Cook ASCP, Humphreys EM, Masden EA, Burton NHK (2014) The avoidance rates of collision between birds and offshore turbines. BTO, Thetford

Croft S, Budgey R, Pitchford JW, Wood AJ (2015) Obstacle avoidance in social groups: new insights from asynchronous models. J R Soc Interface 12 (106):20150178

Cryan PM, Barclay RMR (2009) Causes of bat fatalities at wind turbines: hypotheses and predictions. J Mammal 90(6):1330–1340. https://doi.org/10.1644/09-mamm-s-076r1.1

Cryan PM, Brown AC (2007) Migration of bats past a remote island offers clues toward the problem of bat fatalities at wind turbines. Biol Conserv 139(1–2):1–11

Cryan PM, Gorresen PM, Hein CD, Schirmacher MR, Diehl RH, Huso MM, Hayman DTS, Fricker PD, Bonaccorso FJ, Johnson DH, Heist K, Dalton DC (2014) Behavior of bats at wind turbines. Proc Natl Acad Sci 111(42):15126–15131. https://doi.org/10.1073/pnas.1406672111

Dahl EL, May R, Hoel PL, Bevanger K, Pedersen HC, Røskaft E, Stokke BG (2013) White-tailed eagles (Haliaeetus albicilla) at the Smøla wind-power plant, Central Norway, lack behavioral flight responses to wind turbines. Wildl Soc Bull 37(1):66–74. https://doi.org/10.1002/wsb.258

De Lucas M, Janss GFE, Whitfield DP, Ferrer M (2008) Collision fatality of raptors in wind farms does not depend on raptor abundance. J Appl Ecol 45(6):1695–1703. https://doi.org/10.1111/j.1365-2664.2008.01549.x

de Lucas M, Ferrer M, Bechard MJ, Muñoz AR (2012) Griffon vulture mortality at wind farms in southern Spain: distribution of fatalities and active mitigation measures. Biol Conserv 147(1):184–189. https://doi.org/10.1016/j.biocon.2011.12.029

Dekker A, van Gasteren H (2005) Eurbase: military bird strike frequency in Europe. In: 27th International Bird Strike Committee, Athens

Dennis N, Lyle D (2009) Bird strike damage and windshield bird strike final report. European Aviation Safety Agency, Cologne

Desholm M, Kahlert J (2005) Avian collision risk at an offshore wind farm. Biol Lett 1(3):296–298. https://doi.org/10.1098/rsbl.2005.0336

Desholm M, Fox AD, Beasley PDL, Kahlert J (2006) Remote techniques for counting and estimating the number of bird-wind turbine collisions at sea: a review. Ibis 148(s1):76–89. https://doi.org/10.1111/j.1474-919X.2006.00509.x

DeVault TL, Reinhart BD, Brisbin IL Jr, Rhodes OE Jr (2005) Flight behavior of black and turkey vultures: implications for reducing bird-aircraft collisions. J Wildl Manag 69(2):601–608. https://doi.org/10.2307/3803730

DeVault TL, Belant JL, Blackwell BF, Seamans TW (2011) Interspecific variation in wildlife hazards to aircraft: implications for airport wildlife management. Wildl Soc Bull 35(4):394–402. https://doi.org/10.1002/wsb.75

DeVault TL, Blackwell BF, Seamans TW, Lima SL, Fernández-Juricic E (2015) Speed kills: ineffective avian escape responses to oncoming vehicles. Proc R Soc B Biol Sci 282(1801)

Dinevich L, Leshem Y (2010) Radar monitoring of seasonal bird migration over central Israel. Ring 32. https://doi.org/10.2478/v10050-010-0003-z

Dokter AM, Liechti F, Stark H, Delobbe L, Tabary P, Holleman I (2011) Bird migration flight altitudes studied by a network of operational weather radars. J R Soc Interface 8(54):30–43. https://doi.org/10.1098/rsif.2010.0116

Dolbeer RA (2006) Height distribution of birds recorded by collisions with civil aircraft. J Wildl Manag 70(5):1345–1350. https://doi.org/10.2307/4128055

Dolbeer R (2011) Increasing trend of damaging bird strikes with aircraft outside the airport boundary: implications for mitigation measures. Hum Wildl Interact 5(2):235–248

Dolbeer RA (2013) The history of wildlife strikes and management at airports. In: DeVault TL, Blackwell BF, Belant JL (eds) Wildlife in airport environments: preventing animal-aircraft collisions through science-based management. The Johns Hopkins University Press, Baltimore, MD, pp 1–6

Dolbeer RA, Wright SE (2008) Wildlife strikes to civil aircraft in the United States, 1990–2007. Federal Aviation Administration, Washington, DC

Dolbeer RA, Belant JL, Sillings JL (1993) Shooting gulls reduces strikes with aircraft at John F. Kennedy International Airport. Wildl Soc Bull 21(4):442–450. https://doi.org/10.2307/3783417

Dolbeer RA, Wright SE, Cleary EC (2000) Ranking the hazard level of wildlife species to aviation. Wildl Soc Bull 28(2):372–378

Doppler MS, Blackwell BF, DeVault TL, Fernández-Juricic E (2015) Cowbird responses to aircraft with lights tuned to their eyes: implications for bird-aircraft collisions. Condor 117(2):165–177. https://doi.org/10.1650/condor-14-157.1

Drewitt AL, Langston RHW (2006) Assessing the impacts of wind farms on birds. Ibis 148(s1):29–42. https://doi.org/10.1111/j.1474-919X.2006.00516.x

Drewitt AL, Langston RHW (2008) Collision effects of wind-power generators and other obstacles on birds. Ann N Y Acad Sci 1134(1):233–266. https://doi.org/10.1196/annals.1439.015

Dunning JBJ (ed) (1993) CRC handbook of avian body masses. CRC Press, Boca Raton

Erickson WP, Johnson GD, Young DP Jr (2005) A summary and comparison of bird mortality from anthropogenic causes with an emphasis on collisions. Washington, DC

Erni B, Liechti F, Bruderer B (2003) How does a first year passerine migrant find its way? Simulating migration mechanisms and behavioural adaptations. Oikos 103(2):333

Everaert J (2014) Collision risk and micro-avoidance rates of birds with wind turbines in Flanders. Bird Study 61(2):220–230. https://doi.org/10.1080/00063657.2014.894492

Everaert J, Stienen EM (2007) Impact of wind turbines on birds in Zeebrugge (Belgium). Biodivers Conserv 16(12):3345–3359. https://doi.org/10.1007/s10531-006-9082-1

Farfán MA, Vargas JM, Duarte J, Real R (2009) What is the impact of wind farms on birds? A case study in southern Spain. Biodivers Conserv 18:3743–3758

Federal Aviation Regulations (2017) Title 14 aeronautics and space, vol 1. US Government Publishing Office, Washington, DC. www.ecfr.gov

Fijn RC, Krijgsveld KL, Poot MJM, Dirksen S (2015) Bird movements at rotor heights measured continuously with vertical radar at a Dutch offshore wind farm. Ibis 157(3):558–566. https://doi.org/10.1111/ibi.12259

Froneman A (2005) Conservation & industry strategic partnerships - a model approach for the effective implementation of an airport authority bird hazard management program. Paper presented at the 27th International Bird Strike Committee, Athens

Gill RE, Douglas DC, Handel CM, Tibbitts TL, Hufford G, Piersma T (2014) Hemispheric-scale wind selection facilitates bar-tailed godwit circum-migration of the Pacific. Anim Behav 90(0):117–130. https://doi.org/10.1016/j.anbehav.2014.01.020

Ginati A, Coppola D, Garofalo G, Shamoun-Baranes J, Bouten W, van Gasteren H, Dekker A, Sorbi S (2010) FlySafe: an early warning system to reduce risk of bird strikes. European Space Agency Bulletin 144:46–55

Guida M, Marulo F, Polito T, Meo M, Riccio M (2009) Design and testing of a fiber-metal-laminate bird-strike-resistant leading edge. J Aircr 46(6):2121–2129. https://doi.org/10.2514/1.43943

Gürbüz SZ, Reynolds DR, Koistinen J, Liechti F, Leijnse H, Shamoun-Baranes J, Dokter AM, Kelly J, Chapman JW (2015) Exploring the skies: technological challenges in radar aeroecology. In: IEEE International Radar Conference, Arlington, Virginia

Hayes MA (2013) Bats killed in large numbers at United States wind energy facilities. Bioscience 63(12):975–979. https://doi.org/10.1525/bio.2013.63.12.10

Henderson IG, Langston RHW, Clark NA (1996) The response of common terns Sterna hirundo to power lines: an assessment of risk in relation to breeding commitment, age and wind speed. Biol Conserv 77:185–192. https://doi.org/10.1016/0006-3207(95)00144-1

Hill D, Arnold R (2012) Building the evidence base for ecological impact assessment and mitigation. J Appl Ecol 49(1):6–9. https://doi.org/10.1111/j.1365-2664.2011.02095.x

Hodos W (2003) Minimization of motion smear: reducing Avian collisions with wind turbines: period of performance: July 12, 1999–August 31, 2002. Report by University of Maryland, p 35. Document number NREL/SR-500-33249. https://tethys.pnnl.gov/publications/minimization-motion-smear-reducing-avian-collisions-wind-turbines-periodperformance

Horn JW, Arnett EB, Kunz TH (2008) Behavioral responses of bats to operating wind turbines. J Wildl Manag 72(1):123–132. https://doi.org/10.2307/25097510

Hüppop O, Dierschke J, Exo K-M, Fredrich E, Hill R (2006) Bird migration studies and potential collision risk with offshore wind turbines. Ibis 148(s1):90–109. https://doi.org/10.1111/j.1474-919X.2006.00536.x

Inger R, Attrill MJ, Bearhop S, Broderick AC, Grecian WJ, Hodgson DJ, Mills C, Sheehan E, Votier SC, Witt MJ, Godley BJ (2009) Marine renewable energy: potential benefits to biodiversity? An urgent call for research. J Appl Ecol 46(6):1145–1153. https://doi.org/10.2307/25623104

International Energy Agency (2014) Technology Roadmap: How2Guide for Wind Energy Roadmap. International Energy Agency, Paris

Jameson JW, Willis CKR (2014) Activity of tree bats at anthropogenic tall structures: implications for mortality of bats at wind turbines. Anim Behav 97(0):145–152. https://doi.org/10.1016/j.anbehav.2014.09.003

Janss GFE (2000) Avian mortality from power lines: a morphologic approach of a species-specific mortality. Biol Conserv 95(3):353–359

Jenkins AR, Smallie JJ, Diamond M (2010) Avian collisions with power lines: a global review of causes and mitigation with a South African perspective. Bird Conserv Int 20(03):263–278. https://doi.org/10.1017/S0959270910000122

Johnston A, Cook ASCP, Wright LJ, Humphreys EM, Burton NHK (2014) Modelling flight heights of marine birds to more accurately assess collision risk with offshore wind turbines. J Appl Ecol 51(1):31–41. https://doi.org/10.1111/1365-2664.12191

Jones J, Francis CM (2003) The effects of light characteristics on avian mortality at lighthouses. J Avian Biol 34(4):328–333. https://doi.org/10.2307/3677735

Jones GPIV, Pearlstine LG, Percival HF (2006) An assessment of small unmanned aerial vehicles for wildlife research. Wildl Soc Bull 34(3):750–758. https://doi.org/10.2307/3784704

Kays R, Crofoot MC, Jetz W, Wikelski M (2015) Terrestrial animal tracking as an eye on life and planet. Science 348(6240). https://doi.org/10.1126/science.aaa2478

Kelly TC, Bolger R, O'Callaghan MJA, Bourke PD (2001) Seasonality of bird strikes: towards a behavioural explanation. Paper presented at the Bird Strike Committee-USA/Canada, Third Joint Annual Meeting, Calgary

Kelly JF, Shipley JR, Chilson PB, Howard KW, Frick WF, Kunz TH (2012) Quantifying animal phenology in the aerosphere at a continental scale using NEXRAD weather radars. Ecosphere 3(2):art16. https://doi.org/10.1890/es11-00257.1

Kemp MU (2012) How birds weather the weather: avian migration in the mid-latitudes. University of Amsterdam, Amsterdam

Kemp MU, Shamoun-Baranes J, van Loon EE, McLaren JD, Dokter AM, Bouten W (2012) Quantifying flow-assistance and implications for movement research. J Theor Biol 308(0):56–67

Kemp MU, Shamoun-Baranes J, Dokter AM, van Loon E, Bouten W (2013) The influence of weather on the flight altitude of nocturnal migrants in mid-latitudes. Ibis 155(4):734–749. https://doi.org/10.1111/ibi.12064

Kitano M, Shiraki S (2013) Estimation of bird fatalities at wind farms with complex topography and vegetation in Hokkaido, Japan. Wildl Soc Bull 37(1):41–48. https://doi.org/10.1002/wsb.255

Krijgsveld KL, Akershoek K, Schenk F, Dijk F, Dirksen S (2009) Collision risk of birds with modern large wind turbines. Ardea 97(3):357–366. https://doi.org/10.5253/078.097.0311

Kumar S (2014) Nature conservation vs flight safety—the Indian story. Paper presented at the Bird/Wildlife Strike Prevention Conference, Mexico

Kunz TH, Arnett EB, Cooper BM, Erickson WP, Larkin RP, Mabee T, Morrison ML, Strickland MD, Szewczak JM (2007) Assessing impacts of wind-energy development on nocturnally active birds and bats: a guidance document. J Wildl Manag 71(8):2449–2486. https://doi.org/10.2193/2007-270

Langston RHW, Pullan JD (2003) Wind farms and birds: an analysis of the effects of wind farms on birds, and guidance on environmental assessment criteria and site selection issues. By

BirdLife International to the Council of Europe, Bern Convention on the Conservation of European Wildlife and Natural Habitats

Larkin RP, Torre-Bueno JR, Griffin DR, Walcott C (1975) Reactions of migrating birds to lights and aircraft. Proc Natl Acad Sci USA 72(6):1994–1996. https://doi.org/10.2307/64630

Larsen JK, Guillemette M (2007) Effects of wind turbines on flight behaviour of wintering common eiders: implications for habitat use and collision risk. J Appl Ecol 44(3):516–522. https://doi.org/10.1111/j.1365-2664.2007.01303.x

Lehnert LS, Kramer-Schadt S, Schönborn S, Lindecke O, Niermann I, Voigt CC (2014) Wind farm facilities in Germany kill noctule bats from near and far. PLoS One 9(8):e103106. https://doi.org/10.1371/journal.pone.0103106

Leshem Y (1994) Twenty-three years of birdstrike damage in the Israeli Air Force 1972–1994. Paper presented at the 22nd International Bird Strike Committee, Vienna

Leshem Y, Froneman A (2003) Flight safety and nature conservation—the ultimate connection, the Great Rift Valley case study. Paper presented at the 26th International Birdstrike Committee, Warsaw

Leshem Y, Ronen N (1998) Removing Hiriya Garbage Dump, Israel-A Test Case. Paper presented at the 24th International Bird Strike Committee, Stara Lesna, Slovakia

Leshem Y, Yom-Tov Y (1996a) The magnitude and timing of migration by soaring raptors, pelicans and storks over Israel. Ibis 138(2):188–203

Leshem Y, Yom-Tov Y (1996b) The use of thermals by soaring migrants. Ibis 138(4):667–674

Leshem Y, Yom-Tov Y (1998) Routes of migrating soaring birds. Ibis 140(1):41–52

Lindeboom HJ, Kouwenhoven HJ, Bergman MJN, Bouma S, Brasseur S, Daan R, Fijn RC, de Haan D, Dirksen S, Rv H, Lambers RHR, ter Hofstede R, Krijgsveld KL, Leopold M, Scheidat M (2011) Short-term ecological effects of an offshore wind farm in the Dutch coastal zone; a compilation. Environ Res Lett 6(3):035101

Long CV, Flint JA, Lepper PA (2011) Insect attraction to wind turbines: does colour play a role? Eur J Wildl Res 57(2):323–331. https://doi.org/10.1007/s10344-010-0432-7

Longcore T, Rich C, Mineau P, MacDonald B, Bert DG, Sullivan LM, Mutrie E, Gauthreaux SA Jr, Avery ML, Crawford RL, Manville Ii AM, Travis ER, Drake D (2013) Avian mortality at communication towers in the United States and Canada: which species, how many, and where? Biol Conserv 158(0):410–419. https://doi.org/10.1016/j.biocon.2012.09.019

Loss SR, Will T, Marra PP (2013) Estimates of bird collision mortality at wind facilities in the contiguous United States. Biol Conserv 168(0):201–209. https://doi.org/10.1016/j.biocon.2013.10.007

Lovell CD, Dolbeer RA (1999) Validation of the United States Air Force bird avoidance model. Wildl Soc Bull 27(1):167–171

Lucas M, Janss GFE, Whitfield DP, Ferrer M (2008) Collision fatality of raptors in wind farms does not depend on raptor abundance. J Appl Ecol 45(6):1695–1703. https://doi.org/10.2307/20144148

MacKinnon B, Sowden R, Dudley S (2004) Sharing the skies: an aviation guide to the management of wildlife hazards. Transport Canada, Ottawa, Ontario, Canada. Transport Canada, Ottowa

Manktelow S (2000) The effect of local weather conditions on bird-aircraft collisions at British airports. Paper presented at the 25th International Bird Strike Committee, Amsterdam

Marques AT, Batalha H, Rodrigues S, Costa H, Pereira MJR, Fonseca C, Mascarenhas M, Bernardino J (2014) Understanding bird collisions at wind farms: an updated review on the causes and possible mitigation strategies. Biol Conserv 179(0):40–52. https://doi.org/10.1016/j.biocon.2014.08.017

Marra PP, Dove CJ, Dolbeer R, Dahlan NF, Heacker M, Whatton JF, Diggs NE, France C, Henkes GA (2009) Migratory Canada geese cause crash of US Airways Flight 1549. Front Ecol Environ 7(6):297–301. https://doi.org/10.1890/090066

Marris E (2013) Drones in science: fly, and bring me data. Nature 498:156–158. https://doi.org/10.1038/498156a

Martin GR (2011) Understanding bird collisions with man-made objects: a sensory ecology approach. Ibis 153(2):239–254. https://doi.org/10.1111/j.1474-919X.2011.01117.x

Martin G (2012) Through birds' eyes: insights into avian sensory ecology. J Ornithol 153((1):23–48. https://doi.org/10.1007/s10336-011-0771-5

Martin GR, Portugal SJ, Murn CP (2012) Visual fields, foraging and collision vulnerability in Gyps vultures. Ibis 154(3):626–631. https://doi.org/10.1111/j.1474-919X.2012.01227.x

May R, Reitan O, Bevanger K, Lorentsen SH, Nygård T (2015) Mitigating wind-turbine induced avian mortality: sensory, aerodynamic and cognitive constraints and options. Renew Sust Energ Rev 42(0):170–181. https://doi.org/10.1016/j.rser.2014.10.002

McKee J, Shaw P, Dekker A, Patrick K (2016) Approaches to wildlife management in aviation. In: Angelici F (ed) Problematic wildlife: a cross-disciplinary approach. Springer International Publishing, pp 465–488. https://doi.org/10.1007/978-3-319-22246-2

McLaren JD, Shamoun-Baranes J, Bouten W (2012) Wind selectivity and partial compensation for wind drift among nocturnally migrating passerines. Behav Ecol 23(5):1089–1101. https://doi.org/10.1093/beheco/ars078

Morinha F, Travassos P, Seixas F, Martins A, Bastos R, Carvalho D, Magalhães P, Santos M, Bastos E, Cabral JA (2014) Differential mortality of birds killed at wind farms in Northern Portugal. Bird Study 61(2):255–259. https://doi.org/10.1080/00063657.2014.883357

Moses A, Rutherford MJ, Valavanis KP (2011) Radar-based detection and identification for miniature air vehicles. In: Control Applications (CCA), 2011 I.E. International Conference on, 28–30 Sept 2011, pp 933–940. https://doi.org/10.1109/cca.2011.6044363

Nicholls B, Racey PA (2007) Bats avoid radar installations: could electromagnetic fields deter bats from colliding with wind turbines? PLoS One 2(3):e297

Osborn RG, Dieter CD, Higgins KF, Usgaard RE (1998) Bird flight characteristics near wind turbines in Minnesota. Am Midl Nat 139(1):29–38

Ovadia O (2005) Ten years of birdstrikes in the Israeli Air force. Paper presented at the 27th International Bird Strike Committee, Athens

Parsons JG, Blair D, Jon L, Simon KAR (2009) Bat strikes in the Australian aviation industry. J Wildl Manag 73(4):526–529. https://doi.org/10.2307/40208401

Pearce-Higgins JW, Stephen L, Langston RHW, Bainbridge IP, Bullman R (2009) The distribution of breeding birds around upland wind farms. J Appl Ecol 46(6):1323–1331. https://doi.org/10.2307/25623123

Pérez Lapeña B, Wijnberg KM, Hulscher SJMH, Stein A (2010) Environmental impact assessment of offshore wind farms: a simulation-based approach. J Appl Ecol 47(5):1110–1118. https://doi.org/10.1111/j.1365-2664.2010.01850.x

Perrow MR, Gilroy JJ, Skeate ER, Tomlinson ML (2011) Effects of the construction of Scroby Sands offshore wind farm on the prey base of Little tern Sternula albifrons at its most important UK colony. Mar Pollut Bull 62(8):1661–1670. https://doi.org/10.1016/j.marpolbul.2011.06.010

Poot H, Ens BJ, de Vries H, Donners MAH, Wernand MR, Marquenie JM (2008) Green light for nocturnally migrating birds. Ecol Soc 13(2):online

Reglero JA, Rodríguez-Pérez MA, Solórzano E, de Saja JA (2011) Aluminium foams as a filler for leading edges: Improvements in the mechanical behaviour under bird strike impact tests. Mater Des 32(2):907–910. https://doi.org/10.1016/j.matdes.2010.08.035

Richardson WJ, West T (2005) Serious birdstrike accidents to U.K. military aircraft, 1923 to 2004: numbers and circumstances. In: 27th Meeting of the International Bird Strike Committee, Athens, 23–27 May 2005, Athens International Airport S.A.

Robinson WD, Bowlin MS, Bisson I, Shamoun-Baranes J, Thorup K, Diehl RH, Kunz TH, Mabey S, Winkler DW (2010) Integrating concepts and technologies to advance the study of bird migration. Front Ecol Environ 8(7):354–361. https://doi.org/10.1890/080179

Ross-Smith VH, Thaxter CB, Masden EA, Shamoun-Baranes J, Burton NHK, Wright LJ, Rehfisch MM, Johnston A (2016) Modelling flight heights of lesser black-backed gulls and great skuas from GPS: a Bayesian approach. J Appl Ecol 53:1676–1685. https://doi.org/10.1111/1365-2664.12760

Schuster E, Bulling L, Köppel J (2015) Consolidating the state of knowledge: a synoptical review of wind energy's wildlife effects. Environ Manag 56(2):300–331. https://doi.org/10.1007/s00267-015-0501-5

Shahbazi M, Théau J, Ménard P (2014) Recent applications of unmanned aerial imagery in natural resource management. GISci Remote Sens 51(4):339–365. https://doi.org/10.1080/15481603.2014.926650

Shamoun-Baranes J, van Gasteren H (2011) Atmospheric conditions facilitate mass migration events across the North Sea. Anim Behav 81:691–704. https://doi.org/10.1016/j.anbehav.2011.01.003

Shamoun-Baranes J, van Loon E, van Gasteren H, van Belle J, Bouten W, Buurma L (2006) A comparative analysis of the influence of weather on the flight altitudes of birds. Bull Am Meteorl Soc 87(1):47–61

Shamoun-Baranes J, Bouten W, Buurma L, DeFusco R, Dekker A, Sierdsema H, Sluiter F, van Belle J, van Gasteren H, van Loon E (2008) Avian information systems: developing web-based Bird Avoidance Models. Ecol Soc 13(2):38

Shamoun-Baranes J, Bouten W, van Loon EE (2010) Integrating meteorology into research on migration. Integr Comp Biol 50(3):280–292. https://doi.org/10.1093/icb/icq011

Shamoun-Baranes J, Alves J, Bauer S, Dokter A, Huppop O, Koistinen J, Leijnse H, Liechti F, van Gasteren H, Chapman J (2014) Continental-scale radar monitoring of the aerial movements of animals. Mov Ecol 2(1):9. https://doi.org/10.1186/2051-3933-2-9

Shaw P, McKee J (2008) Risk assessment: quantifying aircraft and bird susceptibility to strike. In: Proceedings of the International Bird Strike Committee Meeting, Brasilia

Smallwood KS (2013) Comparing bird and bat fatality-rate estimates among North American wind-energy projects. Wildl Soc Bull 37(1):19–33. https://doi.org/10.1002/wsb.260

Sodhi NS (2002) Competition in the air: birds versus aircraft. Auk 119(3):587–595

Soldatini C, Georgalas V, Torricelli P, Albores-Barajas Y (2010) An ecological approach to birdstrike risk analysis. Eur J Wildl Res 56(4):623–632. https://doi.org/10.1007/s10344-009-0359-z

Steele WK (2001) Factors influencing the incidence of bird-strikes at Melbourne Airport, 1986–2000. Paper presented at the Bird Strike Committee—USA/Canada, Calgary

Stewart GB, Pullin AS, Coles CF (2007) Poor evidence-base for assessment of windfarm impacts on birds. Environ Conserv 34(01):1–11. https://doi.org/10.1017/S0376892907003554

Thaxter CB, Ross-Smith VH, Bouten W, Clark NA, Conway GJ, Rehfisch MM, Burton NHK (2015) Seabird-wind farm interactions during the breeding season vary within and between years: the case study of Lesser Black-backed Gull Larus fuscus in the UK. Biol Conserv 186:347–358. https://doi.org/10.1016/j.biocon.2015.03.027

Turchin P (1998) Quantitative analysis of movement: measuring and modeling population redistribution in animals and plants. Sinauer Associates, Sunderland, MA

van Belle J, Shamoun-Baranes J, van Loon E, Bouten W (2007) An operational model predicting autumn bird migration intensities for flight safety. J Appl Ecol 44(4):864–874. https://doi.org/10.1111/j.1365-2664.2007.01322.x

van Gasteren H, Both I, Shamoun-Baranes J, Laloe JO, Bouten W (2014) GPS-logger onderzoek aan Buizerds helpt vogelaanvaringen op militaire vliegvelden te voorkomen. Limosa 87:107–116

van Loon EE, Shamoun-Baranes J, Bouten W, Davis SL (2011) Understanding soaring bird migration through interactions and decisions at the individual level. J Theor Biol 270(1):112–126

Vansteelant WMG, Bouten W, Klaassen RHG, Koks BJ, Schlaich AE, van Diermen J, van Loon EE, Shamoun-Baranes J (2015) Regional and seasonal flight speeds of soaring migrants and the role of weather conditions at hourly and daily scales. J Avian Biol 46(1):25–39. https://doi.org/10.1111/jav.00457

Vas E, Lescroël A, Duriez O, Boguszewski G, Grémillet D (2015) Approaching birds with drones: first experiments and ethical guideline. Biol Lett 11. https://doi.org/10.1098/rsbl.2014.0754

Voigt CC, Popa-Lisseanu AG, Niermann I, Kramer-Schadt S (2012) The catchment area of wind
    farms for European bats: a plea for international regulations. Biol Conserv 153:80–86. https://
    doi.org/10.1016/j.biocon.2012.04.027
Washburn BE, Cisar PJ, DeVault TL (2013) Wildlife strikes to civil helicopters in the US,
    1990–2011. Transp Res Part D: Transp Environ 24(0):83–88. https://doi.org/10.1016/j.trd.
    2013.06.004

Printed in the United States
By Bookmasters